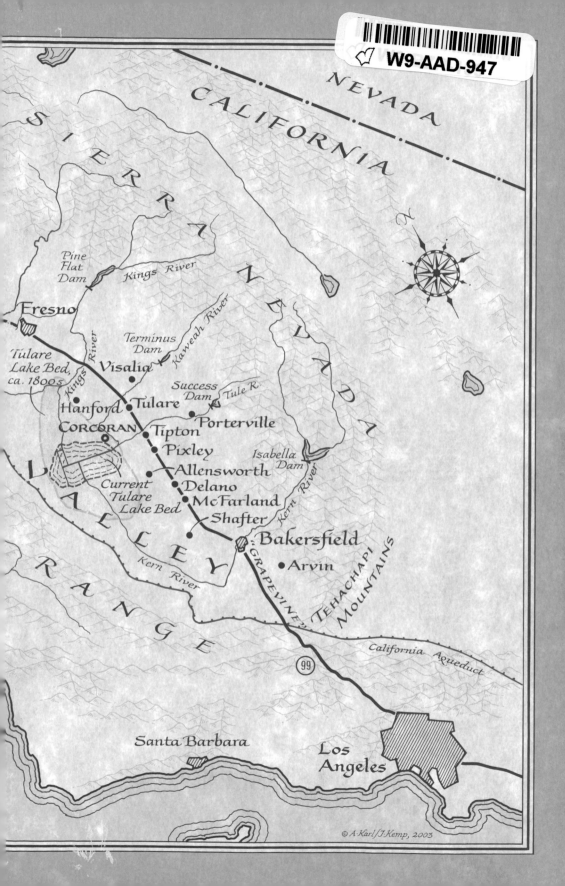

NEVADA

CALIFORNIA

SIERRA

NEVADA

N

Pine Flat Dam

Kings River

Fresno

Tulare Lake Bed, ca. 1800s

Terminus Dam

Kaweah River

Kings River

Visalia

Success Dam

Tule R.

Hanford

Tulare

CORCORAN

Tipton

Porterville

Pixley

Isabella Dam

VALLEY

Allensworth

Current Tulare Lake Bed

Delano

McFarland

Kern River

Shafter

Bakersfield

Kern River

Arvin

RANGE

GRAPEVINE

TEHACHAPI MOUNTAINS

99

California Aqueduct

Santa Barbara

Los Angeles

© A·Karl/J·Kemp, 2003

To J. 11 —

The KING of CALIFORNIA

Mark Arax

Sept, 28, 2004

"on the cotton town"

in the beautiful town

of Helm —

MARK ARAX

&

RICK WARTZMAN

The
KING
of
CALIFORNIA

J. G. BOSWELL
and the Making of a
Secret American Empire

PUBLICAFFAIRS
New York

Published in the United States by PublicAffairs™,
a member of the Perseus Books Group.

Book design by Jane Raese
Set in 10-point Utopia

Library of Congress Cataloging-in-Publication Data
Arax, Mark, 1956–
Wartzman, Rick, 1965–
The king of California: J. G. Boswell and the making of a secret American empire/
Mark Arax and Rick Wartzman.
p. cm.
Includes bibliographical references (p.) and index.
ISBN 1-58648-028-6
1. Boswell, James Griffin. 2. Boswell family. 3. Pioneers—California—San Joaquin
Valley—Biography. 4. Cotton farmers—California—San Joaquin Valley—Biography.
5. Businessmen—California—San Joaquin Valley—Biography. 6. San Joaquin Valley
(Calif.)—History—20th century. 7. Cotton growing—California—San Joaquin Valley—
History—20th century. 8. Agricultural industries—California—San Joaquin Valley—
History—20th century. 9. San Joaquin Valley (Calif.)—Economic conditions—20th
century. 10. San Joaquin Valley (Calif.)—Biography. I. Wartzman, Rick. II. Title.
F868.S173 A73 2003
979.4'805'092—dc22
[B]
2003058495

10 9 8 7 6

For Coby, for being a writer's wife.

 —MA

For Randye, because I love you madly.

 —RW

Contents

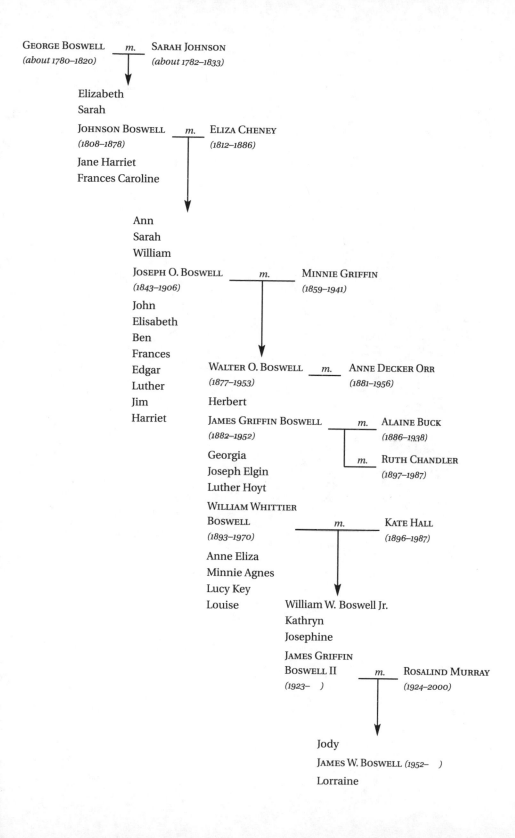

GEORGE BOSWELL   *m.*   SARAH JOHNSON
(about 1780–1820)   *(about 1782–1833)*

Elizabeth
Sarah
JOHNSON BOSWELL   *m.*   ELIZA CHENEY
(1808–1878)   *(1812–1886)*
Jane Harriet
Frances Caroline

Ann
Sarah
William
JOSEPH O. BOSWELL   *m.*   MINNIE GRIFFIN
(1843–1906)   *(1859–1941)*
John
Elisabeth
Ben
Frances
Edgar   WALTER O. BOSWELL   *m.*   ANNE DECKER ORR
Luther   *(1877–1953)*   *(1881–1956)*
Jim
Harriet   Herbert
    JAMES GRIFFIN BOSWELL   *m.*   ALAINE BUCK
    (1882–1952)   *(1886–1938)*
    Georgia   *m.*   RUTH CHANDLER
    Joseph Elgin   *(1897–1987)*
    Luther Hoyt
    WILLIAM WHITTIER
    BOSWELL   *m.*   KATE HALL
    (1893–1970)   *(1896–1987)*
    Anne Eliza
    Minnie Agnes
    Lucy Key
    Louise   William W. Boswell Jr.
        Kathryn
        Josephine
        JAMES GRIFFIN
        BOSWELL II   *m.*   ROSALIND MURRAY
        (1923–)   *(1924–2000)*

        Jody
        JAMES W. BOSWELL *(1952–)*
        Lorraine

The KING of CALIFORNIA

PART ONE
WINTER

Never seen no cotton like this here California cotton.

Long fiber, bes' damn cotton I ever seen.

—**John Steinbeck,** *The Grapes of Wrath*

1

Winter

WE HAD BEEN CHASING JIM BOSWELL for the better part of two years, trying to find the right words that might persuade him to talk. He was a man who had spent half his life safeguarding the family lore. He had once gone to the unusual lengths of hiring a Hollywood screenwriter to gather the stories of his mother and father and kin way back in Georgia. But the stories, covering a century, never saw the light of day. Boswell took the voices of his southern past and locked them away in the basement of his California headquarters where they had been gathering dust for more than twenty years.

Now, at age seventy-six, he had no reason to break the tradition of secrecy that began with his uncle, Lieutenant Colonel J. G. Boswell, the founder of the company. "The Colonel created a culture around the idea that if you ever talked about it, you'd have to give up the game," one Boswell cousin recalled. At family gatherings, they talked about the virtue of stealth this way: "As long as the whale never surfaces, it is never harpooned."

We tried to convince Boswell that his tale was more than a family tale. We appealed to his sense of history and patrimony, to that piece of the quaint South that still seemed to survive in him, even though he had spent a lifetime conquering the West. None of it worked—not our strokes to his vanity, not the sheer persistence that his friends said would win him over, not even the writer's little extortion: "This book is going to be written without you and the voices of your critics will echo only louder." His response was to slam the door shut. We finally decided to appeal to his mortality, a sales pitch he cut short like this: "You don't seem to understand. It won't bother me in the least if I die and this story is never told."

So it came as some surprise when one day we found ourselves sitting right beside him in a beat-up Chevy truck, his hand with the missing

3

fingers guiding the wheel, barreling straight into the forgotten middle of
California. We had met him that morning—"Be here at 5:30 and don't
forget your pencils to take down all my lies"—in the small farm town of
Corcoran, where his land and memories crisscrossed in the clay bot-
toms of an old lake. It sat in the county of Kings on the western flank of
the San Joaquin Valley, a land pinched by mountains and rolled out flat
and never ending. We drove for miles and miles across immaculate
fields where he had hunted Indian arrowheads as a kid, and then he
mercifully stopped to let us take a stretch. We had come to rest on a giant earthen levee, a feat of engineering akin to something in Holland,
and he stood on the dusty dike and gazed out across the empire he had
snatched from a great inland sea.

It was late winter, a few weeks before planting, and everything before
him was open earth. Mammoth machines clawed and leveled the land,
and crop dusters with their jet engines wide open plowed through the
sky. Every square and treeless stretch of it was his, but even as he stood
in the bright sunlight, it was hard reading pride in his face. He had a way
of keeping even longtime friends and his children guessing, something
always held back in the pale blue eyes and thin-lipped smile. The same
Boswell face stared out from the portraits of his Scottish forebears, slave-
holders and gentlemen farmers who had been chased out of Georgia by
the boll weevil in the 1920s and grafted a piece of their cotton plantation
onto this immense land. They had brought the South—its mint juleps,
its chow chow relish, its tarpaper cotton picker shacks—out West.

He wore a Cal Poly Ag cap tucked low, khaki pants, a flannel shirt and
Rockport shoes, not exactly the slops of a farmer about to get dirty with
his land and not exactly the outfit of the absentee corporate landlord
that his critics accused him of being. His voice, like the irrigated fields
that ran past him to the horizon, hardly rose or fell an inch. And yet he
had made it perfectly clear since the first cup of coffee that morning
that he didn't appreciate our line of questions, didn't enjoy this exercise
of show-and-tell. He was here under protest, agreeing to talk only after
we let it drop that the old-timers of Corcoran were portraying his long
dead father as the town drunk, a man who could pound nine straight
beers without going to the bathroom and still had more sense besotted
than a lot of guys do cold sober. Keeping us straight with certain facts
about his father—his drunkenness was explained away by family members as a means of numbing "the misery of hemorrhoids"—was one

thing. Entertaining questions about the size of his farming operation and wealth was quite another.

"What are you, a tax collector?" he said, shooting down our question of how much land he really owned. In a tone of half-lecture, he explained, "I abhor the word 'empire.' It's a word for nations, for civilizations. Why do you guys have to get into this whole damn 'big' thing anyway?"

At some point that afternoon it occurred to us that we had traveled half a day, a distance of some 150 miles, and we had never left his farm. Our route, a slight zigzag across the old lake bottom, took in but half of it. We visited the first cotton patch of his father and uncle and the farm where his brother, Billy, an even harder drinker than his dad, died before his time. We passed the long-gone tent cities of black sharecroppers from Oklahoma and Arkansas, Texas, Louisiana, Georgia and Mississippi who had migrated not north to Chicago or Detroit in the 1940s but followed the cotton trail west to the gut of California. We crossed the levees built by the boom-and-bust pioneers through eighty years of reclamation and flood, each dike a little higher than the next to keep the floodwaters on their neighbor's land. And then we passed the faded camps of Mexican migrants and white Dust Bowlers who journeyed to the San Joaquin Valley for no better reason than they knew how to chop and pick cotton. Each turn of land raised another ghost.

"That was all Rowan, and right behind it was Flynn. Just to the north of it was Chatom. And north of that was Schwartz," he said, rattling off the old farmers and the giant sections of land named for them. He pointed to a Boswell crew raising the levee along the left bank of the Tule River. "Now if you still had all the original owners over on the right side, they'd be panicking today to see us building up that levee."

"Why? Because when the big flood hits, the water's gonna come their way?"

"That's right," Boswell said.

"So why isn't anyone panicking now?"

"We own it all."

We had driven the equivalent of Washington to Philadelphia, though it seemed pointless to measure it that way, and nearly every road, field and irrigation canal belonged to Boswell, and every worker we passed and waved to was his worker and every truck, tractor and leveler for which he politely moved to the side of the road bore the same diamond-B logo.

The goofy grin on a few of their faces made you wonder if they even knew who the old guy waving at them was.

He was the biggest farmer in America and the last land baron of California, and he saw no good in playing it up. His 200,000 acres in the middle of the state, just part of his domain, may have ranked as one of the biggest land grabs in the modern West but, to hear him tell it, the product of guile and vision it was not. A chance encounter here and a little seller's desperation there and, presto, tens of thousands of acres just fell into his lap. The fact that he had built the most highly industrialized cotton operation in the world and had grown more irrigated wheat, safflower and seed alfalfa than any single farmer in the country and was aiming to do the same with onions and tomatoes, well, the fewer people who knew about that, the better. It was not by accident that Boswell had changed a 200-year-old American institution, altered the way cotton was grown, picked, ginned and marketed, and hardly anyone outside Kings County knew his name.

James Griffin Boswell II—"Call me Jim"—had taken a cue from his uncle, the family patriarch, who founded the company from the back of a jitney in 1921 and then married into one of California's most powerful and polished clans, the Chandlers of Los Angeles. Colonel Boswell was a little martinet of a boss who liked to tease his wife, the willful Ruth Chandler, that the men in her family—land developers and publishers of the *Los Angeles Times*—were soft and weak and not half the man he was. The Colonel had no children by Ruth or his first wife, a beautiful and brooding poet who died young, but on long voyages to sell Boswell cotton across the world he groomed his nephew and namesake in the wisdom of lying low.

The Boswell empire was a secret empire that stretched over the years from the middle of California to Arizona to Oregon to Colorado to the outback of Australia, where Jim Boswell had taken his uncle's pioneering spirit and built a world-class cotton industry from scratch. The no-man's-land he inhabited in the San Joaquin Valley, reaching across Kings, Fresno, Tulare, and Kern counties, was a forbidding place—howling desert one day, stinking marsh the next—but it had submitted long ago to his will. The basin had served as a receptacle for three great migrations of workers. The black and white Okies fleeing the sharecropper system were preceded and followed by even bigger and more sustained waves of migrants from Mexico. Each group of workers huddled in the basin's plywood shantytowns, and the growers turned back every union's attempt to organize them. After calming labor for good in the 1950s, in part by treating his workers better than any other grower, Jim Boswell went about creating a marvel of science and automation, hiring top graduates from the best agricultural colleges in the West.

His biotech labs minted new brands of seeds. His precision gins, every bit the industrial wonder of Boeing or U.S. Steel, punched out 400 bales a day of the finest cotton—enough fiber to make 840,000 pairs of boxer shorts. Towels by Fieldcrest, turtlenecks by L.L. Bean and underwear by Jockey began in these fields. He saw to it that nothing went to waste. The oil milked from the leftover seeds went to Wesson and Frito-Lay. Even when he could squeeze no more out of his fields in Arizona, he still turned them into gold, helping build the first big retirement community in the country, Sun City. The extent of his wealth was a matter of mystery, in part because he lived without ornamentation. By some estimates, his water rights alone were worth billions of dollars, a stockpile eyed with greed on the other side of the mountain by a Los Angeles still hell-bent on expansion.

Few of his contemporaries had ever challenged him. On one of the rare times he was sued by fellow farmers and made to answer questions from a hostile lawyer, he seemed to regard the entire proceeding as a joke. He opened his testimony by wisecracking that he was a cowboy—and he had the missing fingers to prove it.

While he liked to play up a rough-grained image, Boswell was a Stanford University graduate who swapped ideas with the erudite at the Aspen Institute, partied with the elite at Bohemian Grove and counted Arnold Palmer as his golf partner. During one round in Australia, Palmer watched a frustrated Boswell hack his way through the first six holes. On the seventh hole, Palmer couldn't resist offering his friend a bit of straightening out.

"Jim, if you move your right hand up a little bit. . . ."

This was Boswell's bad hand. "Palmer," he said, "when I want your advice, I'll ask for it. You play your game. I'll play mine."

When it came time to protect his water rights, he assembled one of the most effective lobbies either the state legislature in Sacramento or the U.S. Congress had ever seen. His boys unseated a congressman who didn't quite see the regulation of big farms his way, and they persuaded the secretary of the interior to tear out the teeth of an eighty-year-old federal reclamation law, all in the name of keeping Boswell territory in one piece.

For fifteen years, Boswell served on the board of General Electric, where Chairman Reg Jones and then Jack Welch came to count on him to provide a view different from that of his more button-down peers. Boswell could dissect a balance sheet with the best of them, but his real value to GE was that he came at problems from the vantage of a man out of the old West. "A shooter, a fisherman, a hiker of trails, a strategist and long-term planner" in the words of Jones. "A very independent, outside-the-mold thinker," Welch said. "Just a maverick sort of guy."

If Jim Boswell had built the quintessential "factory in the fields," it bore no resemblance in his mind to the villainy drawn by writers Frank Norris, Carey McWilliams or John Steinbeck, whose narratives of social injustice and corporate farming all met in the lake bottom. Like his uncle, Boswell had left Corcoran early on and ran things from afar. He chose to live and work in Los Angeles, Arizona, and Idaho, and his style was to give plenty of rope to his top managers and show his face maybe five times a year, shooting in on the company jet. Still, Corcoran was home, the place where his mother and father and brother were buried. The community of 10,000 billed itself as the "Farming Capital of California," though on the outskirts of town one old cotton field had given rise to the deadliest prison in America, a 6,000-bed lockup where guards in high-tech booths shot and killed inmates for fist fighting, the same place that Charles Manson, Sirhan Sirhan, and Juan Corona also called home. During picking time, the cotton clung to the prison's razor wire, chinaberry trees, tumbleweeds, oleanders and telephone wires, and it snagged on railroad tracks and old ladies' hair and floated like gossamer through the thick air and hung like wet socks after a hard rain. Months later, cotton was still blowing, dirty little balls clumped at the side of Highway 43. Cotton here. Cotton everywhere.

That first day with Boswell, we took care to steer clear of subjects that might displease him and put a stop to our conversation right there. The idea of dealing with reporters struck him as so needless and foreign that when you called company headquarters, there was no PR flack to answer even the most basic question. If they were lucky, reporters might be referred to corporate counsel Ed Giermann, a man so dyspeptic that he returned maybe one out of ten calls, and then only to say, with an unmistakable glee, that the company's comment was "No comment."

That morning, looking for ways to break the ice, we mentioned to Boswell how cotton debris from past harvests adorned the drive into town. The bits of white stuff on the side of the road hardly seemed a blight on Corcoran. This was, after all, a place that announced itself to the world with a skyline of grain silos five stories tall and countless metal farm shops, all in shades of industrial gray. Boswell, though, took offense. Leftover cotton, whether it belonged to him or some other ginner, messed with his notion of cleanliness. His crews took care to rake, sweep and pile all errant cotton in the fields and ginning yards, making bales out of detritus. No fiber, he assured us, had blown into town or been lost on the drive to market.

"Now cleanliness is the main thing. Our philosophy is that it's easier to keep something clean than it is to keep it dirty. . . We're not farmers. We're implementers. And nobody can implement better than we can."

The giant levee where we had come to rest that afternoon cut across the confluence of three of the four rivers that met on his land. So complete was nature's bending that the absurdity of where he was standing, smack dab in the bottom of a lake, or rather what was once the largest body of freshwater west of the Mississippi, no longer struck him. He and a handful of farmers before him had sucked Tulare Lake dry and made its rivers run backward. No landscape in America—not the cotton South, not the grain belt of the Midwest, not the cane fields of Florida—had been more altered by the hand of agriculture. It was a landscape every bit as engineered as the Mississippi Valley, and far more intensively cultivated. The scale was so unheard of that they had to invent their own one-of-a-kind machines, monster "moon buggies," to suck water out of canals and hurl it across the fields.

Against the will of two presidents, Franklin Roosevelt and Harry Truman, the Boswells and other Kings County growers had convinced the federal government to make an extraordinary philosophical concession: The four rivers that followed their enduring path from the mountains to the valley weren't replenishing a freshwater lake with origins in the Pleistocene epoch. Rather, powerful congressmen and bureaucrats decided that the rivers were agents of flood and it was productive farmland—the future of the West—they were flooding. Even at the bottom of a lake, floods had to be controlled and controlling them was the domain of the U.S. Army Corps of Engineers. One by one, beginning in the 1940s, they dammed the four rivers in the foothills above the valley, projects underwritten in part by payments from Boswell and other growers. This dam building was eventually topped by an even sweeter prize: The farmers served by the irrigation releases from the dams could grow their crops without any worry about the 960-acre limitation that constrained farmers elsewhere. Here in Kings County, the federal law didn't apply; the irrigated farms of Tulare Lake could grow as big as they liked.

In the event that the dam tenders were forced to release water beyond the needs of irrigation, the farmers had their own elaborate devices downstream to check the rivers. The South Central levee, rising 20 feet high and 120 feet wide and extending more than seven miles, sliced the basin in two and could stop the lake cold. It was the centerpiece, the Big Daddy, of hundreds of miles of dikes and canals that formed a massive crisscrossing network to stave off the proverbial 100-year flood, and certainly anything smaller. But every fifteen years or so, the great phantom lake still managed to stir. The dams of Uncle Sam—and the clay walls and hydraulic plumbing of the farmers—proved no match for the record spring snowmelts that came busting out of the Sierra and down the rivers Kings, Tule, Kaweah and Kern. 1938, 1952, 1969, 1983, 1997, 1998— each year came with a heavy spring thaw that flooded thousands of acres of cotton and grain fields. And so the old Indian lake returned, a land of 10 million geese.

Standing on the levee with Boswell, watching his gaze fix on the irrigated fields, we began to sense that the lake as a concept did not exist for him. He had plenty of respect for the damage that the four rivers could bring, but the construction of the dams had trumped nature, at least a good portion of the time. The rivers weren't so much rivers now as they were precise bands of stationary water, nearly impossible to tell apart from the irrigation canals. And what the four rivers were emptying into was no longer a lake so much as a big sump that every so often got in the way of agriculture.

Like the maps that showed Tulare Lake as a brown or empty spot in the southwest corner of California, the word "lake" had been practically wiped clean from the farmers' nomenclature. Boswell commissioned Hollywood producer David Wolper and screenwriter David Vowell, whose credits included the original *Dragnet* TV series, to make a film praising the death of Tulare Lake. It was the mid-1960s, a time when the phantom lake seemed gone for good. The film, called *The Big Land*, brought to mind a government production with just enough dramatic music to keep awake a high school civics class. It began with an aerial shot of the old lake, the buzz of the plane rousing thousands of shore birds.

"This is Tulare Lake. For centuries it has been a catch basin for waters flowing down from the High Sierra eighty miles away. Once it ranged over 600 square miles, the uncontested master of the valley. Now it's become the lake that was; its waters controlled, its bottom reclaimed. Once master, it's now servant. Once desolate, it's now fertile. The difference is man." It ended thirty-five minutes later with another aerial view, this time of fields gold and green: "Simply by moving into the twentieth century, the American farmer has become a giant. No longer does he serve the land. The land serves him."

The film, to borrow an old farmer's phrase, was a case of premature nut licking. Four more floods, some more epic than others, would follow. The lake, described in company archives as a "geologic freak with no outlet to the sea," had marked Boswell's youth and tested his wits. Mostly, though, it was an object to tame, and he now viewed it from a purely clinical vantage. To conceive of Tulare Lake as its own entity with historical and environmental import—indeed, one of the greatest wetlands in the West—might mean that it wasn't quite the adversary whose stubborn insistence had to be beaten back with all-out assault.

The lake's beauty, the speed with which it found its old self, was a wonder to behold. The spectacle offered no warning. During the back-to-back flood years of 1997 and '98, visitors wanting to glimpse a piece of California's unruly past drove across a flat expanse of land, past vineyards and almond orchards, past dairies and alfalfa fields, until the road suddenly quit at the base of a huge earthen wall. The air filled with the faint smell and sound of ocean. Climbing atop the muddy embankment, gaping at the lake's big belly, you felt lost for a moment and dizzy with vertigo. Was this the heart of cotton country or the New Jersey shore? The wind whipped whitecaps past telephone poles that displayed the high-water stains of past revivals. The lake was brown in parts and pure blue in others, and the sun glinted off flocks of mud hens, pintail and mallard ducks, giant blue and white herons and pelicans scooping up catfish.

For Boswell, such beauty seemed almost beside the point. He found it a lot easier to talk about the floods and the heroic fights to defy gravity and "dewater" the lake. In the flood of 1969, the economic fate of Corcoran and much of Kings County rested on efforts to save the bonanza wheat harvest and keep dry more than 100,000 acres set aside for the spring cotton planting. Boswell said he had never seen a winter like that. The snowpack was a monster, bigger than at any time in recorded history, and more was on the way. By mid-February, the spring thaw, a three-month meltdown, had begun and was already flooding the lowest fields. Boswell and other farmers enlisted shovels, hoes and huge earth-movers to shore up the levees and installed fifty gargantuan pumps along the Tule to keep the river from jumping its banks. In the midst of their efforts, the sky broke open with yet another torrent of rain.

"RAIN HALTS FARMING—FLOOD THREAT GROWS," screamed the headline in the February 20, 1969, *Corcoran Journal*.

"Like the war in Vietnam, our farmers are embattled, alert and doing everything possible to prevent and minimize flood losses sure to be sus-tained in our community this year of the great flood in Tulare Lake." If the writing was a little breathless, biblical even, it was appearing in a weekly newspaper owned by Boswell and a handful of other big grow-ers, all of them facing the same catastrophe.

Boswell's top men knew more about the movement of water—how to make it appear in drought and disappear in flood—than anyone else in the business. But they had never faced a flood of this magnitude, and it wasted no time exposing the folly of their approach. They had been at-tempting to blunt the impact of the waters and spare the crops the old way, by shoving the spring runoff onto junky lowlands and trapping it behind holding cells. All their efforts had been geared to keeping dry 30,000 acres of prime land in District 749, which ran alongside the swollen Tule.

What they didn't figure on was that the water released from the Sierra dams would be so great that it overwhelmed any contrivances they could muster downstream. The river was about to bust through the big levee, and if Boswell lost the Tule, he would lose not only District 749 but also a main office and cotton gins.

In a gesture of near surrender, they decided to tear a gash into the riverbank and flood the best land on purpose. It sounded ass-backward, washing out the very fields they were trying to save, but cutting into the Tule would prevent a series of uncontrolled breaks down river, an even bigger disaster. With the claws of three massive earthmovers, Boswell's men punched a hole into the Tule and watched it surge out. It proved more than a bloodletting. The river was hemorrhaging.

MAJOR FLOOD DISASTER STRIKES TULARE LAKE
WATER COVERS 80 SQUARE MILES
BASIN NOW IN DANGER

The lake had doubled in size, wiped out the fields of District 749 and was threatening to swallow the town. Boswell recalled one desperate Sunday afternoon in the middle of it all when he grabbed a shovel and headed out to a levee that provided the difference between make and break. "I was the only guy out there. The water was rising and it was starting to come over the top of the levee. And I thought if I could just get a shovel here, I might stop it. It might not rise anymore. A friend came by and said, 'What the hell are you doing out here, Jim?' I said, 'If I can hold it and keep it from going over the top tonight, maybe it'll go down tomorrow.'"

The neighbor then turned to Boswell and planted a wild scheme of an idea. All the earth in the world wasn't going to shore up the main levee against the constant lashing of the waves. A black man on the other side of town had forty or fifty junk cars that would make a perfect buffer. "Let's haul 'em out here tonight," the neighbor suggested. The next morning, Boswell put out the word to every wrecking yard in the valley: Fifteen dollars per car. ASAP. Cash on delivery. The line of tow trucks freighted with pancaked jalopies grew so long that it meandered for miles. As the rain poured down, Boswell's foremen, strongboxes in arm, paid every junkman a 10 and a five spot; 5,200 jalopies in all, $78,000. Then, using cranes, they laced eight miles of the big curved levee with Chevys, Cadillacs, El Doradoes, Pontiacs and Thunderbirds, a bumper-to-bumper bulwark. The newspaper took a picture from the air. The junks looked like fine beads on a fancy necklace.

The crazy scheme worked. The lake ended up holding at 130 square miles.

"We saved the El Rico levee that way. You know I've had two 500-year floods in my life and four 100-year floods," Boswell said, without a trace of irony. "You never give up fighting floods in the lake bottom. Don't give up."

THE HEROES OF THE 1969 FLOOD WERE TOTING SHOTGUNS as well as shovels in the middle of the night, farmer against farmer. On one levee, a revered Boswell manager named Les Doan confronted Fred Salyer, the son of a rival farm family. Only days before, Doan had installed a series of pumps to move floodwater off Boswell's best land. Salyer, in the company of an armed guard, wanted the water to stay put. Salyer's land was dry thanks to the big levee that partitioned the lake bottom in two. Salyer wanted the water to remain on the other side of his levee, close enough so that he could dip in his straw when it came time for irrigation. In other words, he wanted his neighbors to bear the entire burden of the flood but share with him the later benefits of its water. Clutching a letter from his attorney, Salyer ordered his tractor driver to upend the Boswell pumps and stop the drainage. They were bobbing like corks in the surging river when Doan showed up and promised Salyer he'd get even. The bad blood from the 1969 flood ended up flowing all the way to the U.S. Supreme Court, where the Boswells and Salyers squared off in a landmark water case.

The Salyer clan was led by a Virginia hillbilly named Clarence, a one-time mule driver who came to Tulare Lake in 1918, a few years before the Boswells. Old Clarence had the misfortune of a mangled eye that wandered so far to the left that all you saw was white. The eye, which earned him the cruel moniker Cockeye, was not the worst of his features. He had a big belly and bulbous nose and not a single pretension when it came to personal style. Whether firing or rehiring one of his employees, which he often did in the same breath, or storming into Governor Pat Brown's office to give him a good chewing out, Salyer dressed the same: slouch hat, chinos, an old brown coat and scuffed cowboy boots. He drove a mud-caked Cadillac in and out of the fields and let his pet collies, Sinner and Saint, make their home on the front seat.

As with Boswell's missing fingers, there were a dozen different stories to explain the disobedience of Clarence's eye, most of them a variation on the staggering amounts of hard liquor he consumed (sixteen dry martinis in one sitting), his savage temper and the sheer merriment he brought to the act of screwing family, friends, employees and his long-time lover. With a fourth-grade education, he had amassed 88,000 acres of prime farmland, second only to Boswell. On his deathbed, Colonel

Boswell rued the day he gave Cockeye his start and the years he loaned him money on each cotton crop. "I'll buy hell if the Boswells put the down payment on it," Clarence boasted.

Money didn't mean a damned thing to him, friends said. Land. Now that's what mattered to Cockeye. He was too busy gobbling up 640-acre squares to worry about debts. Though a millionaire many times over, he papered the town with bad checks, his bank account perpetually depleted from having bought another piece of land. A lifelong Ponzi scheme kept him one step ahead of the bill collector and most of his own workers, too. "Clarence would never miss a payday," one longtime employee recalled. "But sometimes the checks weren't so good." One late afternoon, in the middle of polishing off a quart of bourbon at the Brunswick Club, his foreman, a large man known as Big Swede, grabbed Clarence by the coat and knocked him to the floor with a single stunning blow. Then he reached into the boss's pocket and grabbed some of the back wages owed him. As Clarence dusted himself off, his good eye still silly, he is said to have inquired: "Is there anybody else here I owe money to?"

By some accounts, he could be a lovely host. The few times Ruth Chandler came to Corcoran, she delighted in his company at lunch. During dove hunting season, he made sure that the boys from Los Angeles, Clark Gable and John Wayne, bagged their limit. And before he turned on Pat Brown, sneaking into the state Capitol one night to personally serve the recoiling governor a lawsuit, Cockeye acted as Brown's bagman. His son Fred flew them from dusty airstrip to dusty airstrip, collecting campaign loot from valley farmers. Tired of playing errand boy, Fred told his father that he didn't much care for the grubby-handed Brown and thought even less of his overbearing wife. Cockeye ordered him to keep flying. Years later, after his boys had exiled him to a Delta farm hundreds of miles north and successive generations of Salyers had sued each other until the empire was no more, it was whispered that Cockeye had committed murder and gotten away with it. He shot a man in cold blood during the labor strife of the 1930s, the story went, and Fred helped his father get rid of the gun.

Boswell had heard about the murder from Fred Salyer himself, who confirmed that he took care of the gun in a manner that no cop could ever trace. Boswell had maintained high regard for Fred Salyer through the years, but he couldn't stomach the old man. It was Cockeye who had built the big levee we were now standing on. He had the gall to divert a crop loan from the Boswells in the early 1950s to pay for the cost of construction. Keeping himself dry and his neighbors wet—that's exactly what Salyer pulled off in the flood of 1969. Meanwhile, Boswell and the other growers—the Hansens and Gilkeys and Boyetts—fought like hell to save enough dirt to plant a modest crop.

As it turned out, the price of cotton was high enough that Boswell managed to turn a tiny profit that year. It helped that the federal government cut the farmers a $216,000 check to repair flood damage, and many of them received crop insurance payments as well. Then came the flood's most ironic windfall. Over the next few years, all that corralled water was moved back onto the fields, in measured doses, through the same canal system that had shunted the floodwaters—this time for irrigation.

To Boswell's critics, the whole thing smacked of quite a deal. Suck dry one of America's biggest lakes, carve out a farming empire at the catch basin of four rivers and then, when the flood of the century struck every fifteen years, collect relief from Uncle Sam. Boswell, of course, didn't see it this way. The company, whose stockholders were mostly family and longtime employees, had never finished a year in the red, a remarkable record for an operation where the first intangible was weather.

This success, he assured us, was due not to government subsidies but to guts and teamwork—from the top bosses down to the irrigator making $6 an hour. Still, with our tour of the farm winding down, we pressed him on the notion of welfare for the rich.

"Several years ago, Jim, newspapers were calling you the biggest recipient of federal crop subsidies in the country. And for cotton, a surplus crop."

"That's the most ridiculous, asinine statement. Anytime somebody wants to make an argument, an anti-argument, a non-constructive argument, they use the word 'subsidy' or 'surplus.' They're code words. They're not realistic. Yes, we've been entitled to subsidy checks. But we never did take them."

"You never took the money?"

"No sir, we sent back a check. Came out of Washington."

"How much was that check for?"

"I want to say it was in the millions."

"And you've done that more than once?"

"Yes."

"Let's say the newspapers are half right and you did take a check once or twice. Wouldn't it stand to reason that because of your immense size, it would have to have been an immense check?"

"I hate that word 'immense.'"

"You abhor the word 'empire' and you hate 'immense.' You're gonna take away half our vocabulary before we're done. I mean, you can't get around the fact that you're the biggest farmer in the country. How in the hell are you going to get around that?"

"I've gotten around it damn well without you guys for fifty years."

He farmed the equivalent of more than ten Manhattan Islands with only 300 salaried employees. It was efficiency without compromising excellence, from seed to field to gin to bale yard. Because the picking season was short and no man stayed on the payroll beyond his labors, Boswell made a practice of hiring husbands and wives, parents and their grown children. He placed them in different phases of the operation, some in summer, others in winter, so families could maintain a steady flow of wages throughout the year. In some clans, three generations wore the same Boswell blue diamond on their uniforms.

"You're probably going to think this is a bunch of bullshit, but my biggest concern is the employees," he said. "How are we going to provide work for them and keep them on in good times and flood. We have the best benefits program of any farming operation in the country. Major medical, retirement, profit sharing, stock-option plan. So we keep our people."

He reached down and picked up a clump of hard gray soil and busted it apart with his half hand. The beauty of the heavy clay was that it held the winter rain for a long time. Once the danger of frost had passed, the clumps could be kneaded out of the land and the soil would still maintain enough moisture just below the surface to germinate an entire field. They called it farming the top inch. "That's what we're doing right now. They're knocking down the old borders, and this will all go into cotton."

Below the levee and across the field, a newfangled tractor powered by belts, jeeplike, carved one-inch scallops into the earth. Boswell squinted into a sun that suddenly felt like spring and then lingered for a few moments on a field of wheat bursting to his left and miles of open land ready for cottonseed to his right. "Look at how beautiful that is out there," he said. "You're at the bottom of the lake."

W E HAD SPENT MORE THAN HALF THE DAY with a man who had dodged and charmed us with equal doses of irritation and intellect, and now he began to tire and it was time to head back to town. We hopped into the truck and drove for a while without saying anything. We were gathering up the nerve to ask him the questions that still needed to be asked, even if it meant killing our chances for a second meeting. Boswell seemed braced for anything.

"What you've done here boggles the mind—the level of manipulation, the degree of coordination and expertise. No one can deny that. But what do you say to the person who takes a long look at this thing and concludes, 'So much engineering, so intensive, such a pain in the ass—it really should have never been farmed in the first place?'"

He turned and faced us, his eyes off the rutted road, and thrust his hand with the missing fingers in the air. It took a second to understand his gesture, but he was flipping us the bird—by implication. "It's just somebody that doesn't know their ass from a hot rock. Why should any farmer farm anything? I mean everybody has their vocation. Why the hell are you in journalism?"

"But from a resources standpoint, the inputs are tremendous. Pesticides, water, man versus nature and millions of fish and fowl dead because of toxic runoff from your fields. And for what? More cotton."

"There are four crops out here, son. Four rotated crops. Not just cotton." His pink face grew red, and he began to point. "That over there is wheat. It will be going to a flour mill. Over there is safflower, an edible oil. And over there, we'll rotate barley or alfalfa. Now excuse me."

His voice never rose to anger, but he was bothered enough that he took us halfway down the long list of products that came from the cotton plant alone. A bale of cotton didn't just yield 500 pounds of fluffy fiber. Extracted from the cotton were hundreds of additional pounds of seeds. These kernels fed the better part of the animal kingdom and, in turn, those animals fed us. Cotton made steaks and cotton made milk and cotton made lamb chops, pork chops, fried chicken and shrimp cocktail. The seeds, cooked and pressed, were turned into a refined oil that went into mayonnaise, margarine and salad dressing. The crude oil went into gun powder and wrinkle cream and mole cricket bait. And that wasn't counting what the hairs of the seeds—the linters—were used for. They went into obvious things such as cotton balls and fine writing paper, but also the casing for bologna and sausages, X-ray film, toothpaste, ice cream, solid-rocket propellant and shatterproof glass. Linters went into printing money as well—one bale of cotton pressing out 313,600 $100 bills.

Boswell left unsaid whether this ubiquity proved that cotton was the workhorse of the plant world, or whether there was simply so much of it floating around that man had to put it to use. Was surplus the mother of invention?

For the next twenty minutes, as we drove through the streets of Corcoran, Boswell lingered on the Kate Boswell Senior Citizen Center and the William Boswell Baseball Diamond and the big park and numerous other gifts that he and his parents and the Boswell Company had made to their town. It was an incredible record of philanthropy that ran into the tens of millions of dollars, far beyond the call of duty. But Corcoran as a living, breathing place clearly was not well. The unemployment rate stood at 16 percent, and Main Street looked like a ghost town. Gone were the J.C. Penney and McMahon Furniture and Mercantile, the dress shops, jewelry stores, dry cleaners and three theaters. In their place stood a handful of five-and-dimes with names from Mexico.

"What's happened?"

"The flood. Two years of having half the lake under eight feet of water," he explained. "We still haven't recovered."

"But it must be more than just the flood. What about mechanization? The mechanical cotton picker and your own ingenuity? Surely those are reasons why the town isn't booming like it used to."

"No, no, no. This is one of the most permanently employed towns in the valley without the floods. It's just unique that we've had so many."

"You believe floods are the reason why there are vacant buildings on Main Street."

"Yes, absolutely."

"Where are all the workers? We've been told there was a time when you couldn't walk downtown on weekends because of all the black and Mexican cotton pickers spending their paychecks."

"Blacks?" he said, sounding genuinely perplexed. "We don't really have any blacks in this town. . . Now Harry Watley, he was a big tall guy. That's the only black I can recall. We had a black cook, Laura. And there was Will. He was the prized one. Great huge black fellow who worked for my dad. Came from Georgia. Tap-danced his way across the country. He was my big gorilla one."

He heard the sound of his words rattle around the truck cab and paused. "Hey, I'm trying to give you the truth. I don't care if they're black, red or yellow."

"What do you mean there were no blacks in Corcoran? There were four black churches in this town, black labor camps. Right here next to this field is an old tract where a lot of the blacks first came."

"No, no, no, baloney," he said. "There was nothing out here until the 1970s. . . . You're thinking of Allensworth, twenty miles south of here."

"Wait a second. Blacks may not have been a main work force, but Forrest Riley and others had whole black crews picking cotton here in the 1940s."

"Very small. Now, if you're talking Mexicans, they were the work-force."

He rolled the truck to a stop in front of the Boswell lanai, a gated oasis with thick green grass and a backyard swimming pool under tall mulberry trees. The hedge behind the pool had been pruned in the shape of a big B with a diamond border. It was the only thing about the place that called attention to itself. His wife, Roz, and a Jack Daniels over rocks awaited him inside. We had overstayed our welcome. His face had the pained look of a toothache.

"I hope I'm dead when this book comes out," he said. "I just don't see any good coming out of this."

"This isn't going to be a polemic, Jim."

"I'm the bad guy in agriculture because I'm big. And I'm not going to try and fight it. I can't change an image and say, 'Well, I'm righteous and good and all that.' But I'm telling you, I'm proud of what we've built . . . and I'm not going to apologize for our size."

We had notebooks filled with more questions, but who could blame him if he decided, after one visit, to retreat behind the old family proverb? For almost a century, the secrecy had served him and other Boswells well. He ran the most magnificent farming operation in America and had turned a succession of his top men into millionaires. He paid the highest wages in town to field and gin workers, and he operated Corcoran like a benevolent baron. And yet something had gone terribly wrong. How had a town founded at the turn of the century by the "People's Railroad"—a rail line formed to break up the monopoly of the

Southern Pacific—become the dominion of a handful of growers? How had a place envisioned as the Jeffersonian ideal of small farms and a healthy town square become so sickly? Was it the flood, as he insisted, or was poor economic health the inevitable consequence of large-scale farming, the poisoned fruit of a company town? Was the grip of poverty so strong in the valley that not even Boswell—and all that he had built and given—could prevent such misery?

In the sixty years since Carey McWilliams penned *Factories In The Field*, big agriculture had been portrayed as a single-dimension bad guy. McWilliams and the other writers who followed him, an angry pack of agrarians who believed in the mystical ideal of 160 acres and no more, had succeeded in getting the last word on big agriculture—without ever talking to a single big farmer. The growers were dismissed as fascists and "ribboned Dukes and belted Barons" of the soil whose feudal empires were built on the backs of exploited workers. The rhetoric really hadn't changed since the Great Depression. As for the farmers, they were their own worst enemies, either refusing to talk or whining out of both sides of the mouth about labor shortages and high wages, the perils of good weather and bad weather, too much government interference and not enough government dole. They were forever misunderstood. One of their own had once remarked that they were the worst kind of fools, suffering an excess of suspicion when it wasn't warranted and an absence of suspicion when it was.

Now we had before us the biggest grower in America; and if he was no ribboned duke, he did have a difficult time coming clean on the downsides of his empire. The scale of his farming was truly stunning, but at what cost to the environment and at what benefit and cost to his community? Perhaps Boswell was one of those builders who couldn't afford the impulse to stop and consider the questions. If such an impulse did exist, he might never have built it.

He prided himself on being a rugged individualist who bridled at the hand of the federal government, and yet the headlines said that he had cashed a subsidy check for $20 million in one year alone. The federal water projects that enabled Boswell to tame Tulare Lake raised their own set of questions: Did the dams pave the way for cities and lanes of commerce and navigation, as they had done on the Mississippi? Or did they benefit a handful of gamblers bent on the fancy of growing crops in the bottom of a lake? If cotton was America's best choice for Tulare Lake, did the small profit margins and big equipment costs, not to mention the floods, preclude the little guy? What might Kings County and the Central Valley look like today if Roosevelt and Truman had prevailed, if the biggest farms had been broken up into smaller pieces?

And what about Boswell the man? He said he made a promise early in his life that he'd never read a book on the Civil War or be burdened by the weight of his family's Georgia past. He wanted us to know that he had kept that promise. Yet his occasional "Yesums" and his jarring recollections of Corcoran's black community seemed to reflect something deeper and more disturbing about the South. After twelve years of more or less blissful retirement in Sun Valley, Idaho, he had returned six months ago to once again run the company, pushing out the old board and hiring a new one to his liking. He vowed that his return would be short-lived, just long enough to right the ship after two disastrous flood years. But as 1999 unfolded, his son and successor doubted whether the old guy would ever let go. "My dad's going to die with his boots on."

Jim Boswell stood on the grass in the shade of a mulberry tree. "If you need a bathroom, there's one right over there," he said. He himself didn't bother with the latrine next to the swimming pool. He unzipped his pants and took a leak on the lawn. As to the question of whether we passed the test, he kept us guessing to the end. "I'll be out here again in April," he said finally, bidding us good-bye. "I'll see you boys back here then."

The Cotton Kingdom

WE HADN'T BEEN AROUND JIM BOSWELL VERY LONG before we recognized that to fully understand him, we would have to make a trip back to his Georgia birthplace, back to the Greene County hamlet that had shaped his father and his uncle, back to the original Cotton Kingdom.

The old Boswell farmstead sat near the northern end of the county, on a sweep of greensward just off the Penfield Road. The family had moved into this dwelling, a Queen Anne Victorian with high-hipped roof and gables, after their house on the same site had burned down in the 1890s. More than a century later, long after the Boswells had left Georgia for California, the place retained just enough of its past that you could see backward. Two enterprising sisters from Atlanta had converted it into a bed and breakfast, with wedding receptions spilling into the gardens and deer nibbling at the hydrangea and daffodils.

Just a few Boswell kin, like E. H. Armor, could still be found in these parts as the year 2000 approached. Armor was Jim Boswell's octogenarian cousin and the only one remaining who could resurrect vivid memories of the generation that had gone west to build their cotton empire—the Colonel, J. G.; Jim's dad, Bill; and their brother, Walter. The self-proclaimed docent of Greene County, Armor loved to tell colorful stories about the Boswells—some less doctored up than others.

Armor's older brother, Albert, had headed to California in the late 1920s to work for the Colonel. And in the early 1940s Armor himself had gone off to the flatlands of the San Joaquin Valley. In the lull of the late California afternoon, Armor would evoke the spirit of home by fetching a mint julep for Walter, the Colonel's older brother. "It's five o'clock," Walter would say. "Whip me up one." Or sometimes he would help the Colonel's younger brother, Bill, barbecue a pig, slathering the meat with a tangy Georgia recipe. But Armor yearned for the real Georgia—for the gently sloping hills, for the thick forests of loblolly pine—and after about a year he returned to Greene County. He had been here ever since.

A husky man with a long face, Armor appeared to survive on a diet of Mountain Dew and Brunswick stew—a ground-up hash of onions, green peppers, tomatoes, potatoes, chicken thighs, and Boston butt. Good-humored and gregarious, he back-slapped everyone he met, and as he chitchatted with his neighbors—men and women, white and black—his genuine fondness for them was plain. Yet when Armor spoke about blacks in general, he was often derisive and called them "niggers" or "Ethiopians." The whole display was reminiscent of the Mississippian who told writer Robert Penn Warren: "Some of us, a lot of us, could manage some graciousness to individual Negroes . . . but you know, we couldn't manage it for the race."

Armor lived in Greensboro, the county seat, in the very house where he had been born and raised. He had made one room a shrine to Dixie, and the old man genuflected when he walked toward the shelves that bore the memoirs of Robert E. Lee, a first-edition copy of *Gone with the Wind* and the Civil War diary of his grandfather, a Confederate doctor. Another relative, Armor said, had chased off a band of Union soldiers with pistols so enormous "they looked like stovepipes." But it was when Armor wandered through the Penfield Cemetery that the link between present and past seemed most complete. Dead Boswells were everywhere.

"There's Lawrence Albert Boswell," Armor said, pointing to the dates etched on a headstone: 1882–1964. Armor recalled how Lawrence, a grocer, used to say that he had "a $2,500 thumb" because of the way he manipulated the scale in his store. And then there was a James Boswell, who was known as "Tight" because, said Armor, he embodied a trait that ran through the family. Though they lived well—sometimes extremely well—the Boswells could be "close with a dollar like paint on the wall." Treading between graves, Armor went over to where Frank Adolphus Boswell rested. "He talked like this," Armor said, his drawl suddenly going high and squeaky. "They called him 'Single Nut' Boswell."

Armor became more reverential when he strolled near the tomb holding the Colonel's mother and father—Jim Boswell's grandparents. Six of their twelve children were buried beside them, and several had died as newborns or toddlers—a reflection of how unforgiving nineteenth-century rural Georgia had been, even for a family of means. "Life was hard," Armor said quietly. Close by were the Colonel's grandmother and his grandfather, Johnson Boswell. In later years, the family would trace their lineage all the way back to Scottish royalty of the fifteenth century and to James Boswell, the renowned biographer of lexicographer and essayist Samuel Johnson. However, the story of the modern clan began with George Boswell, the Colonel's great-grandfather.

He had been the first to come to Georgia, enticed by the prospect of gobbling up good farmland. Traveling westward on the Valley Road, he had arrived from North Carolina around 1800. He didn't make it as far as Greene, stopping instead in Wilkes County, about twenty miles to the northeast.

Boswell soon married Sarah Johnson, and they took up farming toward the southern edge of the county along Kettle Creek, where twenty-five years earlier the Americans had routed the Tories in a pivotal Revolutionary War battle. Boswell grew corn in the rich, loamy soil and earned enough income to raise four girls and a boy, and still have plenty left over to procure a regular supply of whiskey and rum. After George and Sarah died, their brood moved to Greene County, a rich farm belt tucked midway between Augusta and Atlanta. It was here that the Boswells' nexus to cotton was forged.

Kɪɴɢ Cᴏᴛᴛᴏɴ. It enriched Alexandria, spurred irrigation in the Sudan, brought fame to Shanghai and became woven into the Indian cultures of Peru, Guatemala and Mexico. It attired the army of Alexander the Great, caught the eye of Pliny the Roman and drew raves from sixteenth-century English botanist John Gerard, who said of its seed oil: "It taketh away many freckles, spots and other blemishes of the skin." It triggered Britain's Industrial Revolution, which pulled people away from their home looms and into the cities with their sweltering factories—the fodder for Karl Marx's *Das Kapital*.

In the United States, cotton inspired a march by Sousa, a painting by Degas, Ivory soap, the Lehman Brothers investment firm and one of the first dozen blue-chip stocks of the Dow Jones industrial average. It even served as an instrument of seduction for playwright Tennessee Williams, whose character, the gin operator Silva Vicarro, uttered to the wife of his greedy rival: "You're soft, fine-fibered and smooth. . . . You make me think of cotton."

And yet it was also cotton that cleaved the country, hurling it into a conflict whose aftershocks would still be felt more than a century later. "Furs, cattle, oil, gold, wheat, corn, railroads—the tale of all these . . . excites the imagination as one perceives with what courage and adventurousness men have bent the resources of nature to their use," one historian has written. "But it is the melancholy distinction of cotton to be the very stuff of high drama and tragedy, of bloody civil war and the unutterable woe of human slavery."

Although the Spanish had grown cotton in America as early as 1556, it didn't have much of a role on the continent until after all thirteen original states ratified the Constitution in 1790. The leaders of the new nation, badly in need of capital, were bent on finding a product that could supplement their scanty exports of turpentine, dried fish and tobacco. The answer sprang from a series of technological breakthroughs in the English textile industry, all of which stimulated a tremendous demand for raw cotton. The way past America's economic uncertainties was suddenly clear: The South would grow the crop, the North would ship it and make it into cloth. Before that could happen on a large scale, however, somebody had to figure out how to speed the practice of extracting cottonseed from fluff—a chore so time-consuming, it took a man a whole day to produce one pound of clean fiber. The solution came in 1793 from a young Yankee who was visiting the South.

Fresh out of Yale University, Eli Whitney was en route to South Carolina to take a job as a private tutor when he met up with Catherine Greene, the widow of the late General Nathaniel Greene, who invited him to her Georgia plantation. An inveterate tinkerer—he had made his own fiddle when he was twelve—Whitney was soon sucked in by the foremost problem confronting area planters: how to efficiently separate out the velvety, green-coated seeds of Upland cotton. The gin (short for "cotton engine") wasn't new. For hundreds of years, Hindus had used a hand-cranked apparatus called the *churka* to wring out cottonseeds, and American planters had been improving the device since 1742. But Whitney, secreted away in a basement room of Greene's Mulberry Grove plantation house, greatly advanced the concept. He was inspired, the legend goes, by watching a plantation cat claw at a chicken carcass and come away with a paw full of clean feathers. His boxy gadget used a spiked cylinder, aided by a series of brushes, to tear away the cotton lint and a slotted iron breastwork to segregate the seeds. Instead of just one pound of seed-free fiber, a slave could now produce fifty or more in a day. By 1800, Whitney's gin and numerous knockoffs had proliferated through the South. The boom was on.

Greene County—named for the owner of the plantation where Whitney fashioned his contraption—was regarded by the 1830s as one of the most lucrative cotton areas in Georgia, having produced two of the state's first three cotton millionaires. Slaves constituted two-thirds of Greene's population and picked 12,000 bales each year.

Johnson Boswell, George and Sarah's only boy and the future grandfather of the Colonel, flourished in the red-clay soil of Greene. He wasn't as well-to-do as some of the huge plantation owners who occupied the land between Richland Creek and the Oconee River—a spot called Prosperity Ridge. But by 1860, when he owned thirty-three slaves and his personal estate was valued at $30,000, Boswell certainly would have been counted as a member of the planter class, not some mere farmer.

In 1829, at the age of twenty, Johnson Boswell had wed Ann Strozier. She died shortly thereafter, leaving him an infant son. He then married Eliza Cheney, who gave him twelve more children over the years. They made their home in Penfield, best known as the cradle of Mercer University. Clearing out a section of wilderness seven miles north of Greensboro, the Georgia Baptist Convention had opened the school in 1833 and named the surrounding village after Josiah Penfield, a Savannah silversmith and church deacon who upon his death had bequeathed $2,500 to the campus.

Johnson Boswell was pious in his own right, a man whose primary pursuit, aside from planting cotton, was reading Scripture. An officer in the anti-alcohol Good Templars Lodge, his only other known pleasure was hunting foxes. He and his family were outsiders of sorts in Penfield—staunch Presbyterians in a Baptist stronghold. Boswell eventually established, just down the road from Mercer Chapel, the Penfield Presbyterian Church—a fine-looking brick and pine structure that mirrored its founder's ascetic style. The two faiths got along perfectly well, and congregants from the churches even made a tradition of attending each other's services, the Presbyterians ambling over to Mercer in the morning, the Baptists visiting Boswell's church in the afternoon.

Greene County thrived through the 1840s and '50s. Penfield developed its own business strip and became a cultural and intellectual hub. It published two newspapers, the *Temperance Banner* and the *Christian Index,* and a literary magazine called *Orion.* At Mercer, the library overflowed with some 10,000 volumes. By 1860, the white community of Greene County enjoyed a per capita wealth of more than $2,200—twice that of the average free Georgian. The key to all this affluence was cotton and, by extension, slavery. More than half the households in Greene owned at least one slave, while 18 percent enslaved twenty or more. In all, slaves accounted for about 60 percent of the county's wealth.

It was hardly surprising, then, to find angst running high in the fall of 1859, when John Brown tried to ignite a mass slave revolt across the South with his raid on Harpers Ferry, Virginia. Although Brown failed in his attempt "to purge this land with blood," President Abraham Lincoln's election the following year heightened concern in Greene County, and the local grand jury asked citizens to watch out for abolitionists trying to stir up trouble. "A number of suspicious white men," the panel cautioned, are "straggling about . . . and visiting Negro quarters and conversing with Negroes and otherwise disturbing the good order, peace and quietude of the country."

Some in Greene County also fretted about the prohibition of slavery in the West and its impact on both the Southern way of life and the market value of those they held in bondage. A decade before, California's prospective admission to the Union as a free state had so inflamed Southern passions that a weak and dying Senator John C. Calhoun of South Carolina roused himself from his sickbed to lash out against what he saw as a power grab by the North. "It can no longer be disguised or denied that the Union is in danger," Calhoun declared. California's fate, he added, framed "the test question" of the North's true intentions—a pronouncement that would make J. G. Boswell's move to the coast just one generation later, to peddle cotton no less, seem rather ironic.

Still, on some level, Southerners felt invincible as they contemplated the disintegration of the United States. And why not? By 1859, the region's cotton production had rocketed to 4.5 million bales, up from a mere 100,000 in 1801, and this fruit of slave labor was the nation's leading export. "No, you dare not make war on cotton," Senator James Henry Hammond of South Carolina roared to his colleagues. "No power on earth dares make war on cotton. Cotton is King!"

IN LATE 1860, about a month before Georgia seceded from the Union, an infantry unit called the Dawson Grays was mustered in Penfield. The following April, the Grays thronged on Mercer's front lawn and received a battle flag that had been stitched together by a coterie of townswomen. From there, they marched toward the trains that would take them to Virginia. Among them was Johnson Boswell's son Joseph, the father of the Colonel. Joseph was just seventeen years old.

As the Grays prepared to embark, a certain self-righteousness over-came the people of Greene County. "Heard today that 'Old Abe' had called for 75,000 troops to subjugate the Confederate States," a Mercer astronomy professor who had drilled the Grays jotted into his journal. "Our people do not seem to be much disturbed by the news. We are firmly persuaded of the justice of our cause, and trusting in the God of Battles, it will require more than 75,000 of Abe's Myrmidons to conquer us."

Actually, in Joseph Boswell's case, it required far less.

The Grays fought at Malvern Hill and Manassas, Sharpsburg and Fredricksburg. For Boswell, the pivotal moment of the war came in 1863, after he had switched to another unit and was promoted to second lieu-tenant. Dug in at Cumberland Gap, Tennessee, in September of that year, Boswell and his 2,500 Confederate comrades were up against 1,700 Union soldiers who had just run out of bread and were down to their last thirty rounds of ammunition. The Confederates, however, didn't have an inkling how overmatched their foes were. And what the com-mander of the Federal troops, Colonel John Fitzroy de Courcy, lacked in firepower, he more than made up in cunning.

Knowing that Confederate spies would infiltrate his camp, de Courcy rigged the regiment numbers on his men's caps so that the four units on hand appeared to be sixteen. Then, when Confederate field glasses were trained in his direction, de Courcy marched his troops down the moun-tainside, and when they reached the bottom, he had them double back under tree cover and march down again. They carried out this ruse four times, fooling the Confederate lookouts. A little later, de Courcy ordered three half-manned cavalry regiments to gallop along a dry road, and they kicked up so much dust, it added to the illusion of strength.

On the morning of September 8, de Courcy sent the Confederate brigadier general, John W. Frazer, a demand to lay down arms. De Courcy professed to be reluctant to attack because he didn't want to bring about a "cruel loss of life." The following day, in what the Confederate brass immediately characterized as one the most ignominious episodes of the entire war, Frazer gave up without a single shot having been fired. "Where are the rest of the men?" some of the Confederates asked, as they lined up to surrender before the small Federal force.

Boswell spent the remainder of the war being shuttled from one Union prison to another.

Things hadn't gone much better for the whole of the Confederacy. When conflict first broke out, many were confident that the world's dependence on cotton would soon engender aid for the Southern cause. It was said that Confederate President Jefferson Davis even welcomed the Union's naval blockade as a "blessing in disguise" because it would bring France and Great Britain to his side.

It didn't turn out that way. For one thing, English mills had amassed an ample reserve of the crop. For another, the textile workers of England—despite the hardships it could bring them—openly sided with President Lincoln and his "providential mission" to emancipate the slaves. That made it tough for any British politician to support the Confederacy. Others simply resented the South's ploy. The Confederates "thought they could extort our cooperation by the agency of King Cotton," sniffed the *Times* of London.

By the middle of 1864, the North's supremacy on the battlefield was evident. Early the next year, transportation problems and successful blockades began to cause severe shortages of food and supplies throughout the South. By the spring of 1865, it was all over: Jefferson Davis had been captured and all remaining Confederate troops had been defeated.

In June, Joseph Boswell was released from his prison cell at Fort Delaware after taking an oath of allegiance to the United States—a ritual known as "swallowing the yellow dog." Returning to Penfield, he found his world turned upside down. Sherman's army had mostly skirted Greene County, sparing it the torching that occurred to the west and south. Nevertheless, the end of slavery had wiped out most of the county's wealth overnight. Emancipated blacks so outnumbered whites that they could now decide the outcome of elections.

Black power was short-lived, however. Over the next decade, whites in Greene County waged a crusade of terror against their former slaves and through a combination of legal-system abuses, political deceit, election fraud and Ku Klux Klan violence, they seized control again. "Governor, the Cu Clux is found riding in our county every night," a leader of Greene's black community wrote to the governor in a desperate yet unanswered plea for help. Meanwhile, the South remained economically dependent on the North. And, most significantly for the Boswells of Georgia, cotton kept pushing its way west.

It had been pointing in that direction since the advent of Whitney's gin. As the soil tilled by the Tidewater planters wore out, they found that they could migrate west with their slaves and their cottonseed and start anew. By 1810, cotton had moved into Tennessee, and by 1830 it had sparked a golden age of Mississippi River trade that would make New Orleans the nation's premier cotton market. By 1850, the Cotton Kingdom—as Frederick Law Olmsted called it—stretched for more than 1,000 miles, from South Carolina to near San Antonio. After the Civil War, so many Georgians headed for the kingdom's far reaches, they tacked signs on their doors saying, "G.T.T."—Gone to Texas.

Back home in Greene County, retailers were rapidly replacing planters as the most prominent men of society—a transition the Boswell family would come to play a big part in. By the early 1900s, two department stores in Greensboro jostled for customers: Boswell & Mc-Commons and Williams & Boswell. Joseph did fine for himself, as well. In 1877, he had married Minnie Griffin, the daughter of a physician and Greene County school commissioner. Among their dozen children were the three who would move out to California: J. G. (the Colonel), W. W. (Bill) and Walter.

Joseph farmed 1,500 acres and chaired the local division of the Southern Cotton Growers Association. He also invested beyond the land, buying real estate and helping start up a cotton mill in Greensboro. In 1900, he served a term in the state legislature. Joseph died in 1906, at age sixty-three, a leading citizen of Greene County. His obituary noted that he had "accumulated quite a fortune" through his business acumen. But in time, the riches of his son J. G. would dwarf his own.

The Little Sahara

EVERY TIME HE MADE THIS TRIP, rattling atop the planks that rolled out across the California–Arizona desert, Lieutenant Colonel J. G. Boswell couldn't help but marvel at how strange the whole thing was. Only sixteen years before, in 1906, as a young army sergeant, he had been asked to name the best cotton lands in the United States as part of his examination to become an officer. Raised as he was on a cotton farm, Boswell had ticked off the answer to this question with utter confidence: Texas, Louisiana, Mississippi, Alabama, Arkansas, Oklahoma, Indian Territory and, of course, his native Georgia. California and Arizona were nowhere on the list. Yet here he was, crossing between these two Far Western states on a road made of wood, with the sun sinking low and the sand blowing all around him, a cotton merchant traversing America's new cotton frontier.

It was October 1922, the start of the ginning season, and Boswell's trek took him along what by now had become a familiar route. He slipped out of bed at his home in Pasadena, just north of Los Angeles, shortly before dawn, taking care not to awaken his wife, Alaine, as he dressed and ducked into the early morning darkness. He had recently opened an office in downtown L.A., renting space in the Santa Fe Building, a well-appointed, nine-story structure that sat amid the hubbub of South Main Street. But Boswell was hardly there. He spent so much time on the road that he was fond of telling people, "I carry my office in my hat." In fact, not long before, he had no real office at all and stored his ledgers and other important papers behind the counter of his favorite diner. The friendly fry cook who looked after the cache, R. L. Curtis, would become Boswell's first employee.

At age forty, James Griffin Boswell was a handsome man, with blue eyes, rosy cheeks on fair skin and a head of hair fast changing from brown to white. He stood only five feet seven and walked a bit unsteadily, still hobbled by the bad back that had led to his reluctant retirement from the military. Sometimes, when the pain got too much, he was forced to walk with a cane, but no one would mistake him for being infirm. The Colonel had a commanding presence, heightened by the gold braid and gold-leaf cluster that adorned his pea-green officer's uniform, the one he continued to wear two years after returning to civilian life.

Stepping stiffly out of his big ranch-style house, set along the grand watershed known as the Arroyo Seco, Boswell steeled himself against the break-of-day chill. He clambered into his Buick sedan—the kind of luxury that Alaine's considerable wealth allowed for, even as his own career as a cotton-market middleman was just getting under way.

He turned east on Foothill Boulevard and, before he knew it, reached San Bernardino and then Redlands, where lush green citrus groves quilted the countryside. All in all, it was shaping up to be a rather good season for a cotton man like Boswell. A few years earlier, the crop had plummeted from thirty-five cents a pound to as low as fourteen cents as the government released a torrent of World War I stockpiles, but prices were now on the rebound. Boswell liked to boil down his own business plans to a single maxim: "Never try to buy anything the other fellow doesn't want to sell; never try to sell anything the other fellow doesn't want to buy." Truth be known, there was much more to it than that, and as he rode along, his car putt-putting at thirty miles an hour, his mind juggled the intricacies of freight rates and futures contracts.

By the time he maneuvered through the San Gorgonio Pass, the desert sun was blazing in a cloudless sky. A strong wind tossed sand onto the road, and Boswell bore down and tried not to steer onto the soft, treacherous shoulder. Mesquite and creosote bushes, which broke up an otherwise pallid landscape, whipped side to side. As he motored beneath the craggy summit of Mount San Jacinto and pulled into Palm Springs—a hundred miles and nearly four hours from home—his back ached terribly. It was a pain that might have caused some men to let fly an expletive or maybe sip from a flask of whiskey, but not Boswell. He prided himself on never uttering a swearword and never indulging in a drink. For the Colonel, discipline mattered above all else, and he wore his discipline right down to his manicured fingernails.

He stopped for a rest by the Palm Springs Hotel, where visitors took dips in the famous curative waters of an old Indian spa, but he was too impatient to loiter. He navigated south into the heart of the Coachella Valley, where date trees hung heavy with brown-sugared fruit and snow-white cotton fields rose incongruously from the desert floor. This 100-plus-mile stretch of the West that lay before him, from Coachella to Arizona's Yuma Valley, had become his main turf. Boswell's job was to scour the region, farm by farm, and find cotton that matched the requirements set forth by his customers. His biggest customer was Goodyear Tire & Rubber Company, which spun cotton into tire cord at a sprawling factory in Los Angeles. Even though Boswell had been in business only for a couple of years, Goodyear trusted his keen eye for hunting down cotton with the perfect strength and length.

Cotton had been grown in the Coachella Valley since 1906, when the wife of the railroad stationmaster happened upon some cottonseed left in a boxcar and stuck it in the ground. The crop grew so well that the Southern Pacific Railroad featured it in an exhibit in London. Since then, this patch of desert had been transformed into a prime source of cottonseed production, and Boswell would often stop in the town of Thermal, where a gin had been constructed, to gather intelligence on the market. Today, however, he was making a beeline for Yuma, eager to close out a deal.

About halfway there, Boswell began to glimpse one of the great landmarks of his desert journey: the Salton Sea, an opalescent oasis shimmering in the distance. He had passed this body of water countless times since he began working his territory. Still, its breadth—forty-five miles long, seventeen miles across—dazzled him every time.

Through Westmoreland and Brawley and Imperial and, finally, on to El Centro Boswell drove. In the two decades since the California Development Company had first dug the canal that directed water to the Imperial Valley, this unwelcoming expanse of caliche had blossomed into an agricultural wonderland—America's version of the Nile Delta. Trains hauled away load after load of cantaloupes, watermelons, grapes, peas, spinach, lettuce, grapefruits, milo, barley and alfalfa. This time of year, though, most of the activity revolved around one crop: cotton.

Boswell steered down Dogwood Road and came to Main Street. Cotton fluff, the flotsam of wagons going to the gin, hovered in the late afternoon air. Ten miles to the south was the Mexican border. To the west lay a borough called, appropriately enough, Dixieland. To the east was Holtville and, about sixty miles beyond that, the Arizona line and Yuma.

Yet in between loomed the most arduous part of the trip: the Algodones dunes. The Mexicans had dubbed this stretch *La Palma de la Mano de Dios*, "the Hollow of God's Hand." Anglos preferred "the Little Sahara."

The dunes—some of them six miles wide and towering as high as 300 feet—were a traveler's nightmare, and the only way across them was on a splintering wooden thoroughfare called the Plank Road. The road was born of San Diego's attempt to make itself a transportation hub after the city lost out to Los Angeles in the race to become a terminus for the transcontinental railway. Without a means for people to get over the dunes, San Diego had little chance to draw commerce or tourism from Arizona and other locations to the east. So, a San Diego businessman and booster (and later state senator) named Ed Fletcher raised the money to buy 13,000 planks of Oregon pine—a first step, he hoped, in establishing a highway that would one day run clear from the Pacific to the Atlantic. In early 1915, owing to the unflagging efforts of Fletcher and a supervisor from Imperial County, workers assembled parallel wheel paths—each twenty-five inches wide—and placed them on top of six and a half miles of shifting sand.

The next year, state engineers made slight improvements to the Plank Road, putting in new wooden cross-ties and installing turnouts every thousand feet. So crazy was the notion of crossing the desert by wood that people from the Imperial and Yuma valleys would venture out along the planks for fun, packing their cars with food and drink for "desert parties." As legend had it, one motorist spied a hat resting in the dunes after a particularly fierce windstorm. He went over to snatch it and was startled to find a man's head sticking up underneath. "I'll get a shovel and dig you out," the motorist said. "Better go back to town and get a tractor," the buried man replied. "I'm still at the wheel of my Model T."

As Boswell made his way along this same road, the wind howled and the sand danced over the boards. A line of cars bobbed up and down in front of him. The pace was maddeningly slow, and every so often he had to pull over to let the westbound traffic through. Off to the side of one turnout, a man tried to calm a team of braying donkeys, which had just helped rescue a hapless driver from the undulating terrain.

At sundown, Yuma was still nowhere in sight. The last fifty miles of the trip would take Boswell more than four hours to complete. Bouncing along the boards he couldn't keep the grit out of his Buick. He was exhausted and his back throbbed, and as he stared out at the ocean of sand, it struck him the way it always did: How could this be the middle of cotton country?

AS EARLY AS THE 1850S, California farmers had planted cotton fields around the Central Valley and near Los Angeles. And by the 1870s, some were hopeful that California would be the next great cotton state. Even "labor is abundant," reported James W. Strong, who had come from Mississippi to raise cotton along the Merced River. "White men can be hired for one dollar per day, with board. Chinamen in any quantity at $25 per month, they boarding themselves."

Strong regarded the Chinese as even "more efficient . . . than the Negro labor of the South. It is only employed when actually needed and is therefore less expensive. It is controlled with less difficulty, and is universally conceded to be industrious and painstaking." That said, some thought it best to duplicate the conditions of the South as closely as possible. A San Francisco tycoon named James Ben Ali Haggin, for one, hatched a plan to recruit 1,100 blacks to pick cotton near Bakersfield. Nowhere near that many ever made the trip, and most of those who were wooed west from South Carolina and Tennessee quickly forsook cotton picking for jobs in the cities. By 1885, barely a year into it, the whole effort went bust.

Even with the enthusiasm of men like Strong, cotton all but died out in California in the late 1880s, and it wouldn't be resuscitated until the early 1900s. That's when scientists from the U.S. Department of Agriculture, tugged west by the completion of the Colorado River irrigation canal, began to experiment with sowing long-staple Egyptian cotton in the Imperial and Yuma valleys.

In 1917—six years after Imperial Valley cotton had claimed a silver cup at an exposition in New York—the horticultural commissioner in El Centro counted thirty gins in the area. And in 1920, the year Lieutenant Colonel J. G. Boswell was honorably discharged from the army, more than 100,000 valley acres were planted to cotton.

Some 300 miles north, past Los Angeles and into the San Joaquin Valley, cotton would soon have an even more profound impact, thanks to a South Carolinian by the name of Wofford B. Camp. Bill Camp, as everyone called him, was a newly minted Clemson College graduate in 1917 when the Agriculture Department sent him to the valley to see if long-staple cotton could be grown there. It was an urgent undertaking. German U-boats were sinking ships carrying cotton—the material used to make the fabric that covered the wings of American warplanes.

Camp encountered his share of skeptics, who told him that cotton couldn't grow in the San Joaquin and that, even if it could, the last thing the valley needed was the inevitable influx of blacks to pick it. Cotton will "ruin California economically and socially," argued B. H. Crocheron, the director of the University of California's agricultural extension service. "Nobody picks cotton but Negroes. . . . Orientals can't pick cotton because their fingers are too short, too delicate."

But Camp persisted. "I had more nerve in those days than I had sense," he'd later say. Like Johnny Appleseed, he spread the crop from Kern County to Tulare, Fresno, Madera, Merced and Kings—the fecund lake bottom that, in short order, J. G. Boswell would make his own.

THE LONG-DRAWN-OUT TOOT of a passenger train pulling into downtown Yuma sent J. G. Boswell scurrying from his bed on the second floor of the Southern Pacific Hotel. He was still drained from the previous day's drive across the desert, but there was no time to waste. Boswell scrubbed his face at the bathroom sink, quickly donned his army uniform and set off for the lobby to pick up a copy of the *Yuma Morning Sun.*

Boswell's eyes didn't focus on the headline blaring across the front page—"Allies, Turks Meet Today." As was his habit, he zoomed in on the little box to the right of the date, Tuesday, October 3, 1922. That's where the spot cotton price from New York, the benchmark of his business, was faithfully recorded each day.

The Colonel digested the information—"Cotton closed easy at net advance of 7 to 20 points"—and climbed into his Buick. He was intent on getting to Gadsden, about twenty miles southwest of Yuma, where a cotton gin owner named J. B. Long had set aside several 500-pound bales of Pima cotton for Goodyear Tire. Boswell wanted to be sure that he had a peek at the stash before any of his competitors did.

He knew firsthand just how cutthroat the industry could be. A year or so before, Boswell had wrestled away the Goodyear account in Los

Angeles from the two giants of the business: Anderson, Clayton & Company and George H. McFadden & Brothers. McFadden had been around since 1872, Anderson Clayton since 1904. Both were truly global enterprises, with webs of agents and offices selling cotton to the spinning mills of Bremen, Liverpool and Le Havre.

The two firms had been lured to California by the irrigated cotton farms spreading through the state and by the Los Angeles tire plant and cotton mill that Goodyear had erected in 1919. Toiling away under the slogan "Los Angeles Made for Western Trade," Goodyear workers operated 33,000 spindles as part of a manufacturing process in which Sumatran rubber converged with Imperial Valley and Arizona cotton.

Boswell had won the account by talking his way inside the fortress–like facility when, luck would have it, Goodyear wasn't getting all the cotton it needed through its regular channels. "It's just not available," the Goodyear manager had told Boswell. But the Colonel was not one to back away, and he wouldn't leave until the manager handed over an example of the kind of cotton the company was seeking. For the next couple of weeks, Boswell hustled all over the Imperial and Yuma valleys and ultimately tracked down enough of the cotton to make a triumphant return to the Goodyear factory. It was the break his fledgling business needed.

As Boswell pulled away from the hotel that morning and swung onto Madison Avenue, Yuma was bustling. Cotton had put its unmistakable imprint on the town, and the flavor of the South mingled with that of the West. Yuma farmers had their tractors repaired at the Cotton Belt Garage, and when they got hungry, they crowded into Ham Elliott's Old Plantation Lunch Room. The Yuma Title, Abstract & Trust Company hawked cotton insurance, and the Yuma Hardware Company stocked jute bags for picking. At least eight gins around Yuma, including J. B. Long's in Gadsden, were now open for business.

The Colonel headed up Eighth Street and out of Yuma, dropping down off the mesa and onto a carpet of cotton fields. Crews of Mexican pickers were hunched over the rows, their fingers a blur grasping at the lint. Twenty miles later, past Somerton, he reached Gadsden.

Entering town, Boswell cruised by a drugstore, meat market, lumber company and mercantile, and then slowly approached J. B. Long's gin. A transplanted Texan, Long had helped teach Boswell the ins and outs of the industry when the Colonel first came to the Yuma Valley. Though he grew up around cotton, Boswell had a lot to learn, and Long knew all the angles: where to find the highest-quality crop, whom to ship it to and how to milk the system to maximize profits.

Long greeted the Colonel at the big open door, and the two men immediately went to tackle the task at hand. Over the din of sharp-tooth saws separating soft, white cotton fiber from seed and leaf and twig, Long showed the Colonel what he had for him. Boswell unsheathed his knife, cut into the side of a bale and removed a large tuft by hand. He held up the cotton and, squinting at the filaments floating in a shaft of light, checked for "neps"— pesky knots in the fiber that could trip up Goodyear's spinning machines.

He then inspected a bale destined for export, scrutinizing its color, luster and trash content. Eventually this cotton would be labeled "middling fair" or "strict good middling" by an official government classer. Inspecting cotton was considered a job so sensitive that, as one merchant expressed it, a classer's hands "should be as soft as a debutante's and as supple as a violinist's." But Boswell didn't need any official stamp to discern what was before him. Long's cotton was gorgeous.

J. G. then started to haggle. He loved this part of the business, the hammering out of the deal, and he wouldn't let up until he felt he had completely worn down the other side. After a while, the Colonel and Long came to terms, and then the gin man proffered a bit of friendly advice. Send the bales going abroad right to the Port of Los Angeles, he suggested, and they will soak up so much moisture sitting out on the docks, the inflated weight will cover the cost of shipping.

The Colonel's lips curled into a tight smile at the thought, and then he launched into a well-rehearsed riff—the one he always used when something sounded too good to be true. "You know," Boswell began, "there was this time when my friend Henry was riding around the back roads of Georgia, and all of a sudden he saw ol' Rastus out in the field. He stopped the car and walked up to Rastus and said, 'Now, Rastus, you make about twenty cents an hour.'

"'That's 'bout right.'

"'And how'd you like to make fifty cents?'

"'Oh, I would.'

"'Now, Rastus, I got a plantation over here. How'd you like to come over and live in one of my houses?'

"'Oh, boss, that'd be great.'

"'And, Rastus, maybe I can arrange to let you have the buckboard and go to town on Saturdays.'

"'Oh boss,' he said, 'that'd be wonderful. But just one thing, boss: You sure is peeing on my leg.'"

THAT COLONEL BOSWELL chose to wear his army uniform for his meeting with J. B. Long on this day said more about him than he might have wanted known. Boswell had taken up the cotton trade more by default than anything else; the military remained his first love. His years in the service, in many respects, were the happiest he had known. Until he had injured his back, he figured on making it a career, following the lead of his brother Walter.

J. G. stood a bit in awe of Walter, though he emulated only half of him—the sober half. Indeed, Walter Osgood Boswell was polished enough to get into West Point and wild enough to get kicked out. The dismissal, in 1900, might have hurt his climb through the military, except for the fact that he married well. His wife, Anne Decker Orr, was the best friend of the wife of General John J. "Blackjack" Pershing.

Walter Boswell became Pershing's aide-de-camp in 1909 and sailed for the Philippines to battle Moro tribesmen. Six years later, he stood at Pershing's side again, this time on the Mexican border to size up the revolutionary Poncho Villa. The general and his aide became so close that Walter named one of his three boys John Pershing Boswell. And when a fire ripped through the Presidio in San Francisco in 1915—killing Pershing's wife and three daughters—the general dispatched Boswell to see who was to blame. No villain, save for ill fortune, ever surfaced; a stray coal had evidently rolled out of the fireplace. "General," Walter wrote, "my heart bleeds for you." Later, during World War I, Walter gallantly crossed the Rhine. And now, four years later, he was poised to ascend to the upper ranks of the War Department in Washington.

J. G.'s own stint in the army hadn't been so heroic. As Walter marched through Germany, J. G. lay in a hospital bed in New Jersey, mending his broken back. He had dreamed of flying combat aircraft for the Signal Corps, but he hurt himself before that could happen. The injury occurred not in the skies over Europe or on a daredevil training mission, but in a freak accident in which he bounded off a Washington streetcar carrying too much luggage. "It was damn embarrassing," his stepson, the son of Ruth Chandler, would recount years later. "When people saw him with a brace on his back, they assumed he had been shot down in France, and he had to own up to the fact that he hadn't seen a minute of action."

Still, J. G. Boswell cherished the military life, and nearly two decades in the army had made him what he was—resolute, hard-edged, a tad cocky. He had signed up in 1903 after two years at the University of Georgia, where, in the words of the chancellor, his academic performance was just "fair." Boswell was detailed as a private to the 118th Coast Artillery in Virginia, but all along, he had designs on becoming an officer.

In 1905, Boswell was elevated to second lieutenant, just like his older brother, and he and Walter were soon reunited in the Philippines. J. G. was an able officer, although he earned higher marks for his "zeal" and "general bearing and military appearance" than for his "intelligence," which was rated "average." And though he didn't distinguish himself on the battlefield—before he hurt his back, his most serious injury came from being knocked on the head with a bowling ball at Fort Logan, Colorado—he showed a toughness that would become a hallmark of his business career.

By 1920, his back in sorry shape, Boswell faced a no-win proposition. He could remain in the army as part of the Quartermaster Corps—where, as he saw it, he'd be relegated to pushing paper—or he could retire and be bumped up one rank, from major to lieutenant colonel. Boswell chose the latter, although, apparently feeling awkward about this last promotion, he continued to refer to himself as "the Major" for more than a decade.

Out of the military for the first time in seventeen years, Boswell didn't have a lot of choice about what to do or where to go. Besides the army, he knew only one thing: cotton. And yet there was no hope of starting a business back in Georgia. The boll weevil had seen to that.

A quarter-inch, hard-shelled, humpbacked insect with the regal-sounding scientific name *Anthonomus grandis*, the weevil had stormed across the cotton fields of the South, puncturing flower buds and chewing up young bolls. The pest made its first appearance in Mexico in 1862 and by the 1890s had spread into Texas. Over the next thirty years, the weevil worked its way east, destroying so much cotton, it was said to be as devastating as the Civil War. "Mr. Weevil . . . is almost a counterpart of the elephant," a frustrated Senator Ellison DuRant Smith of South Carolina, better known as "Cotton Ed," once remarked. "He looks like one and his proboscis is his mouth. He sucks his food. You can use all the poison in the world—we have spent millions and millions on calcium arsenate—and he seems to fatten on it." Carl Sandburg viewed the weevil as having emerged straight out of "America's traditions of tragedy," while Southern blacks sang of its capacity to render even the white man powerless:

> Boll weevil said to the farmer,
> "Better sell yo' old machine;
> When I get thru with you
> You can't buy no gasoline

In Greene County, farmers first spotted the weevil in 1917. Within two years, more than one hundred farms were infested, and growers turned under their cotton fields to try to get rid of the problem. Nothing helped. "The boll weevils and politicians have played hell with the country," the *Greensboro Herald Journal* opined. A group of Greene County businessmen offered $50 in gold to the child who could catch the most weevils over a six-week period. The winner, Mary Lou Ashley, bagged 6,896 of the critters. In the early 1920s, Greene's normal annual cotton crop of between 18,000 and 25,000 bales plunged to about 300, and the county witnessed the start of a mass exodus. Within a decade, a third of the population would pack up and go, in many instances abandoning their farms and leaving their houses to rot.

With the promise of nothing but failure back in Georgia, J. G. Boswell looked to the only other place he had ties: California. His mother-in-law had moved to Pasadena, and in 1921 the Colonel and his wife, Alaine Buck, settled in the fast-growing city. They had met in 1909 while Boswell was posted at Fort Logan in Denver. The next year, J. G. left for another tour of the Philippines, and he and Alaine kept in touch. When her first husband died of pneumonia, Boswell swooped in and proposed. They married shortly before he exited the army. A graceful and slender beauty with a taste for fine clothes and diamonds, Alaine treasured music and theater and wrote poetry and plays. She had little interest in J. G.'s business dealings, but proved critical to his new cotton venture for one simple reason: She possessed the cash to get it going.

Except for a modest military retirement check, J. G. didn't have much money of his own. He had never held a job outside the army, and his father, Joseph, had left him and his siblings only $2,500 each when he died. Alaine, though, was in a whole different position. Three years younger than J. G., she had grown up as the adopted daughter of a wealthy Denver insurance man, and then married into an even richer line. Alaine's first husband, Raymond McPhee, was a highly respected lawyer and the son of one of the biggest lumber barons in the West. Alaine inherited $188,000 from him—the equivalent of almost $2 million today.

The money from Alaine was all the start that J. G. needed. He proved a quick study and soon devised a strategy to thwart the gamblers playing havoc with the cotton market. These were farmers who would pledge to sell to merchants like Boswell but then fail to deliver, holding on to their cotton in the hope that prices would spiral higher. Boswell's scheme was to lock in a guaranteed supply, and to do that he brought out his younger brother, Bill, to farm some land around Somerton.

Bill Boswell was nothing like the Colonel. He had acquired his brother Walter's roguishness—and then some. Their mother, Miss Minnie, had sent Bill off to a Presbyterian college in North Carolina with the idea that he'd become a minister. More interested in baseball than the Bible—he'd later brag that he had once made an unassisted triple play—he lasted there only two years. Back in Greene, the chain-smoking Bill fostered the image of a rascal. "When Bill walked home from school with me one day and carried my skates, my mother saw us, and when I got inside, she said, 'I don't want you to see that boy anymore,'" remembered Kate Hall. "The first time he called for a date, she wouldn't let me go out with him. She said, 'That boy has no good intentions.'" In spite of those early warnings, Kate married Bill in 1915 at an altar festooned with palms, ferns and Easter lilies. She had just turned nineteen; he was twenty-one. They would have four children: Bill Jr., Josephine, Kathryn and Jim, the boy who would one day take over for the Colonel and become the nation's Cotton King.

Bill Boswell agreed to come to Arizona to help his brother, but after one season of wilting in the desert heat, he retreated to Greene County. The Colonel tried to convince him that there was money to be made in the West—especially now that the boll weevil had so ravaged the South—but money didn't motivate Bill the way it did J. G. And besides, Bill's life in Georgia was pretty sweet. He and Kate lived in a three-bedroom farmhouse between Penfield and Greensboro in a part of the county that came to be known as Boswell's Crossing. They raised cotton and vegetables and butchered hogs, and every night, at Kate's insistence, they'd take turns reading from the Bible. In the morning, the peal of a bell summoned their black servants and sharecroppers to eat. "How those Nigras ever lived on what we gave them, I'll never know," Kate would say many years afterward. "They had what we called blackstrap molasses, and would make these great big pans of biscuits—the great big thick biscuits—and what they'd call streak 'o lean. It was a streak of fat, and then they'd put that grease or that gravy—I mean, they'd throw it together and they'd sop. It must have been very nourishing 'cause that's what they got. That was breakfast.

"And then anything in the way of vegetables or things that we had at the plantation for them to eat, that was all right. But you know, it's remarkable. Have you ever seen a Nigra that didn't have beautiful teeth? Stop and think about it and watch it sometimes. . . . They all have beautiful teeth."

In the summer of 1919, Bill sold the farm for a hefty $60,000, and he and Kate and their three children moved to Greensboro, where he started a sawmill and was elected chairman of the County Commission. But for all his accomplishments, Bill clearly wasn't a go-getter like the Colonel, and he'd just as soon fritter away the afternoon fishing or pitching horseshoes or kicking back with Ol' John Barleycorn. "He was so different from J. G.," said Kate, "as different as night and day."

On a Sunday late in the summer of 1923, J. G. and Alaine Boswell checked into the Sunnyside cottage at the Hotel Miramar, a beloved resort that hugged a wide swath of beach near Santa Barbara. It was early September, just a few weeks before the Colonel was to set out once again on his wearying annual pilgrimage in search of the perfect cotton bale.

The Miramar was a first-class hotel with an array of amenities, a beautiful spot from which he and Alaine could watch the Pacific lap the shore or meander the grounds, strolling hand in hand among the marguerites. The Colonel was not an affectionate man, but he was lovestruck by Alaine.

At night, the Miramar's guests poured into the dining room, where fresh fish was the specialty, and when the plates were cleared, the women retreated to the Blue Room to socialize and the men gathered around the fireplace in the library. It was in that library, with its red-wood-paneled ceiling and walls turned dirty yellow by ceaseless puffs of pipe and cigar smoke, where Colonel Boswell met Harold S. Doulton, the son of the owner of the Miramar.

As it happened, Doulton was into cotton himself. A dapper fellow who sported a wide-brimmed felt hat and a carnation in his buttonhole, Doulton worked on the family's Los Posos Ranch near the San Joaquin Valley town of Corcoran. Deemed one of the top locales in California for cotton, the ranch's white fields had even been shown to theatergoers around the country a few years earlier on a newsreel. Doulton listened to Boswell's story, and it became clear that the Colonel could benefit from a trip to Corcoran—and Corcoran might benefit from a cotton man of Boswell's caliber.

Rather than go down to the Yuma Valley again, Doulton asked the Colonel, why not check out a new place, a place called Tulare Lake?

The Lake of the Tules

So MUCH OF EARLY CALIFORNIA WAS PURE SUGAR that it sold itself on first bite. Tulare Lake, with its mosquito swamps and salt grass desert, took imagination. When it came to reshaping man and land, no one had more imagination than the padres of imperial Spain, boomers of immense vision but empty pockets. In the early 1800s, the padres were looking to extend providence beyond a chain of missions they had built along the California coast. Each of the twenty-one missions stood a day's journey from the other on *El Camino Real*, the King's Road, which ran from San Diego to San Francisco. On the other side of the mountains, just out of God's reach, were the Yokut Indians of California's interior. Here, amid a labyrinth of lagoons and tule reeds, were 30,000 fresh souls ripe for that mix of slavery and salvation that would take root behind great adobe walls.

First, though, the padres needed to persuade the military men who oversaw Spain's interest in California of the wisdom of building a new chain of missions in the San Joaquin Valley. They argued that there was no better spot for these sanctuaries than the land of four rivers that fed the *Laguna de los Tulares*, the Lake of the Tules. A rift soon grew between the friars and the commanders over the basin's worthiness for divinity. It may well have been one of the first debates between California pragmatist and California dreamer.

With the exception of Mount Whitney, the tallest peak in the continental United States, Tulare Lake was the most dominant feature on the California map, an immense sheet of water that extended out over the desert some 800 square miles. The villages of four distinct tribes of Yokuts, who had called the basin home for nearly 7,000 years, were concealed among the tules that grew ten feet tall and surrounded the lake in mile-wide bands, as impenetrable as a bamboo forest.

Some considered it a cursed land. One gung-ho padre making the case for its salvation called it a "republic of hell and diabolical union of apostates." The reeds fell and rotted in the marshes and the water belched a foul gas, and it mixed with the tule fog that hovered all winter long over the lake. The water, thick and brown, was full of "mystery and malaria." The tules were California's version of the boondocks, and they hid all manner of fugitives, bandits, Spanish and Mexican Army deserters and Indians who had rejected catechism and fled the coastal missions—neophytes turned backsliders. Even the horses stolen from the missions had returned to their feral state, indistinguishable from the millions of wild mustangs that ran in magnificent herds across the valley. The lake basin in summer was a "huge broiler where the sun rose and fixed its hot stare" and never once blinked the entire day. Such was the popular superstition—that California's interior was the province of beastly condors and giant dragon flies and Satan himself—that not a single Spanish land grant was ever sought or given inside the valley.

The commanders, a practical if feckless lot, knew that an inland mission would demand a full military presence, and they seized on all sorts of reasons to avoid civilizing the tules. For the Franciscan friars in their earnest gray robes, however, the entire purpose of being in California was to save the souls of its 150,000 aborigines. The more savage the tribe and the more wicked the landscape, the higher the calling.

Father Francisco Garces carried a quadrant and compass and was the first man of any race to actually record his observations about the San Joaquin Valley for the outside world. He had a rather innovative way of testing the suitability of individual Yokuts for conversion to Christian belief and Spanish allegiance. Everywhere he went he carried a piece of canvas that was two feet wide and three feet long. On one side was a painting of the Virgin Mary. On the other, the devil was pictured reposing in hell. No sooner would Garces arrive in a village than he would unfurl the banner. So vivid were the renderings that usually no other cue was needed. Those ready to be converted shouted their equivalent of "good" at the raising of the Virgin Mary and "bad" at the specter of Satan.

Garces and his brother friars were remarkably gifted men who served as teachers, musicians, weavers, carpenters, masons, architects and "physicians of the soul and body." Needing water, the padres at Mission San Buenaventura on the southern coast toiled alongside their Indian neophytes to construct the first aqueduct in California, a five-mile-long canal of dirt and intricate brickwork. They weren't above putting their hands to the plow, and fields of oranges, lemons, figs, dates, olives, wheat and even cotton flourished in the shadows of the missions. They milled grain, fermented grapes, pressed oil, flailed hemp and transformed cows into candles. It wasn't so easy turning certain natives—inveterate gamblers who would risk their last possession and wives, too, in games of strip poker—into good Christians. In a practice known as clerical mortification of the flesh, the friars found it necessary to become master flagellators, virtuosos of the "*disciplina* whip."

While Garces and the other friars awaited official word on a new mission near Tulare Lake, they used the valley as restocking grounds. They recruited Yokuts to replace the Chumash and other mission tribes who were fleeing in droves or dying by the hundreds of smallpox and syphilis, the scourge of the European. The friars bribed their way inside Indian culture by waving the prosaic and dangling the temporal: ornaments, beads, colorful garments and granaries brimming over with acorns. Some Indians took a fancy to a particular item, say red sashes or handkerchiefs, and adorned their arms, legs and loins with nothing else.

Indulged in this way, the Yokuts were sometimes enticed to give up their children to the padres and lieutenants passing through the lake basin, though they occasionally had second thoughts once they heard of the cruelties visited on their kin by mission soldiers. The Yokut raids against the missions often ended in the killing of neophytes, padres, soldiers and Indians. Almost always the savage Tulare tribes were blamed, whether guilty or not.

In 1804 Father Juan Martin and two soldiers set out on horseback from a mission called San Miguel Arcangel and headed inland to Tulare Lake. They traversed ground so punctured by the catacombs of gophers and kangaroo rats that their horses frequently sank to the knees. Three days later, reaching the southeast shore of the lake, they stumbled on a peaceful Yokut village made entirely out of willow poles and dried woven tules. Each beehive-shaped hut was built next to another with a single flat roof extending overhead—the Yokut version of row houses.

The lake itself was home to an astonishing bounty of fish and fowl. The Yokuts had built tule rafts buoyant enough to carry an entire family for days at a time on the lake and haul hundreds of pounds of rainbow trout, perch, catfish, pike, carp, salmon and sturgeon—caught with bare hands or speared through a hole in the bottom of the craft. The oyster-shaped inland sea was so shallow, two or three feet deep in many parts and never more than forty feet at its deepest, that a fierce northwest wind would whistle through the reeds and blow the waters another mile or two across the savanna. From the shore, the women would wade in, fifteen and twenty abreast, treading for a mile and more, all the while feeling with their toes as they scoured the lake bottom for clams, mussels and terrapin. It took no time to fill the conical baskets slung across their backs. So abundant was the lake's fish life that a century later, Chinese fishermen from San Francisco were still plying its waters. One crew using crude nets scooped up 3,000 pounds in a single haul, and the fanciest restaurants in the city served a prized turtle soup featuring Tulare Lake terrapin.

To describe their world, the Yokuts found language in the throats of swans and the hooves of antelope. The billy owl gave out a tiny squeak when it bobbed its head, and the human imitation of this sound, *peek-ook*, became the word for billy owl. The word for ducks was the gabbling noise they made while feeding, *wats-wats*. Tulare Lake was the *Pah-ah-see*, the pulse of its ebb and flow. It took something different, though, to capture the sound of the blue sky as it turned dark and deafening from the wings and cries of millions of native and migratory birds—Canadian geese, mallards, swans, pelicans, cranes, teal and curlews. How to mimic the sudden flight of flocks so immense they extinguished the sun? One of the first white men to camp along the lake could think of only one noise, the roar of a freight train, that compared with the takeoff of the birds. But the Indians had no way of drawing on the railroad for inspiration. By the time the Southern Pacific arrived in the San Joaquin Valley, the land no longer belonged to the Yokuts and their language had stopped breathing new words. So their word to describe the great honking sky of geese was no sound at all, but a number. *Tow-so, tow-so*. A thousand thousands.

Upon his return to San Miguel, Father Martin made an appeal to his superiors. He had found the perfect spot for the first inland mission, right there at the Yokut village called Bubal. The only shame was that the Indian chief would not allow him a head start by taking a few callow souls back over the mountains. "Their disposition is wonderful and they have shown good will toward the soldiers," Martin wrote. "What I am sorry for is that many are dying from the continuous wars and also from many diseases. Now if a mission is not given to them soon, they will not be held in subjection."

Despite the plea, Joaquin de Arrillaga, the tall, red-headed son of Spanish aristocracy who governed California between 1800 and 1814, couldn't decide on the best method to conquer the natives: by military might or missionary zeal. Into the tules the governor sent teams of so-called civilizing squads led by one lieutenant and one padre who clashed over the best way to reform the marauding Indians. In search of both thieving neophytes and an ideal spot for a first mission, the civiliz-ing squads visited twenty-four villages populated with 5,300 Yokuts, but precious few children. The Tulare Indians, ravaged by syphilis, were fast becoming an infertile race. The padres often returned to their missions without a single Christian convert, young or old. With precious little to show for their journeys into California's heartland, the friars began blaming the military officers who accompanied them. "I do not know that any conversion has been effected when the ministers of the gospel have taken troops along," one missionary wrote Governor Arrillaga. "In-dians begin to look upon them with dread, for the first sight of troops makes them think that . . . loss of liberty will follow. The Indians have no fear or dread when the missionaries come alone."

Still, Governor Arrillaga remained unconvinced. Frustrated with the padres and their weak-willed policy of reconverting the sinners and for-giving all crimes, he dispatched a band of soldiers led by Gabriel Mor-aga on a "punitive expedition" in the winter of 1805. Their orders were clear: comb the tules for neophytes and military deserters suspected of stealing horses, cattle and guns from the missions and presidios—and execute anybody who resists arrest.

Moraga was an explorer and poet at heart, and while he summarily put to death a handful of fugitives along the way, he was much more taken with forging a trail that reached the bank of a broad river never before crossed by any white man. It drained more than a million acres of granite mountain and came roaring down one of the world's deepest glacial canyons, a great forest of pine and giant redwood that towered 8,000 feet above the riverbed. From mountain to lake, the river mean-dered a distance of some eighty miles through the heart of the Yokut na-tion before it flared out in a delta fan and came to a dead halt at a ridge formed from its own scourings. Moraga made a camp along the river and when he awoke the next morning, it was January 6, the Feast of the

Epiphany that commemorates the visit of the Three Wise Men to infant Jesus. In honor of the day, he christened the river *El Rio de los Santos Reyes*, the River of the Holy Kings.

Through the 1820s, the Spanish who traversed the basin were joined by a succession of American explorers, naturalists and fur trappers whose names would resound through the generations: Kit Carson, Grizzly Adams, Ewing Young, Jedediah Smith, John C. Fremont, John W. Audubon and Peg-leg Smith. The Yokuts invited the first of these white men inside their huts and served them smoked eel and a mud hen that tasted like fine duck. Peg-leg Smith had amputated his own limb after a Platte River Indian shot and nearly killed him. When whiskey made him nasty, Smith unhinged his wooden replacement and wielded it like a club. He befriended the Yokuts and taught them horse stealing on a scale far grander than they had ever known. His course ended with the biggest heist in the history of the California missions. Why Peg-leg needed to steal horses from the coastal missions when the Tulare Plains abounded with the fastest and most beautiful of their breed was a matter of expedience. In the time it would take him to break a few wild colts, Peg-leg had sold 3,000 tamed mission mustangs to cowboys on the Santa Fe Trail.

The case for an inland mission continued to be made by padres who journeyed to Tulare Lake. One father even succeeded in baptizing two dozen natives, though they were all old and dying. The friars' reports were sometimes pessimistic about a landscape infested with tarantulas and horned toads, but they invariably found a great need among the Yokuts for godliness. In the end, however, not a single Spanish mission, much less a chain of them, would get built in the land of the tules. The Spanish themselves would be gone in a few years, replaced by the Mexicans and then a motley crew of Pikes, Confederates and forty-niners. Yet the selling of Tulare Lake as a place of evident destiny never stopped and what the padres had begun, the wiping clean of a native culture, the Americans would all but finish.

OF ALL THE HUCKSTERS AND HALF POETS who preceded Colonel J. G. Boswell to Tulare Lake—railroad boosters, cattle barons, wheat kings, San Francisco capitalists, Hollywood developers, canal builders and canal saboteurs—no man had more fire to his pitch than James Henry Carson. That he had no gold or sweat invested in the basin, that he would proclaim its gifts even as he lay dying of a disease borne in its soil, made his vision all the more emphatic.

He was barely eighteen years old when he waved a final good-bye to his native Virginia and set sail around Cape Horn in the summer of 1846, one of 117 rowdy recruits aboard the *USS Lexington*. They were destined for Alta California to snuff out whatever remained of Mexico's resistance. The crew included a few black-haired Irish boys who hailed from the Shenandoah Valley such as Carson and a few old hands who had braved the savages and swamps of the Florida Indian wars. Most of the rest came from the coal mountains of Pennsylvania, young bucks who called themselves the Frosty Sons of Thunder.

They were all there for the same purpose: to ensure the inevitability of California's succession to the United States. Leading them were two men who would go on to become military legends, albeit in another war, Lieutenant Edward O. C. Ord and Lieutenant William Tecumseh Sherman. Their journey around the tip of South America to the Bay of San Francisco took six months, and by the time they arrived in Monterey and pitched their tents on a hill in the driving rain, Mexico had all but surrendered.

As the first army unit ordered to serve in California, F Company did its best to keep busy. Garrison mop-up was hardly the sort of duty envisioned by recruits who signed on thinking they would see action against Mexican forces in southern Texas and, if lucky, revel in the Halls of Montezuma. When gold was discovered in the Sacramento Valley not long after their arrival, nothing short of all-out war could have kept them at their posts. By the dozens they deserted, pick axes and jerked beef in hand, and headed on donkey back for the hills of El Dorado. The most admirable thing one could say about the soldiering of Sergeant James Carson, who came from a proud and distinguished family of Virginia militiamen, was that he had the decency to wait for a general furlough to let the gold lust seize him.

It came one day in May 1848, courtesy of a codger named Billy. The old guy had been one of the first to throw in his lot with the mines and now he stood before Carson in a flannel shirt riddled with holes and buckskins that reached only to his knees. He looked like the devil, Carson later recounted, all beard and hair hanging from his hat. Out of his great bag of temptation the gold tumbled and tumbled, not in mere dust or pitiful scales but in clean nugget sizes ranging from pea to hen's egg. It was just five weeks' worth of dry diggings, all picked with the edge of a simple knife.

"There was before me proof positive that I had held too long to the wrong side of the question," Carson said. "I looked on for a moment; a frenzy seized my soul; unbidden, my legs performed some entirely new movements of Polka steps. . . . I was soon in the street in search of the necessary outfits; piles of gold rose up before me at every step; castles of marble dazzling the eye with their rich appliances; thousands of slaves bowing to my beck and call; myriads of fair virgins contending with each other for my love. These were among the fancies of my fevered imagination. The Rothschilds, Girards and Astors appeared to me but poor people."

He was seen an hour later with washbasin, shovel, blanket and rifle, heading at "high-pressure mule speed" for the diggings. Carson took the usual route to Mormon Island but found the ground there picked over, and so he trekked south to Calaveras County, where he struck a rich vein on what became known as Carson Creek.

He spent a year hunting the lucre, encountering the best and the worst in man. There was Dutch John, the Yiddish-cussing barkeep who brought his own twist to the custom of dipping thumb and forefinger in a miner's pouch and extracting just enough dust to cover a $1 shot of jackass whiskey. Rather than reach straight into the pouch, John would first slide his hand into his mouth. And what a pinch he'd then bring up—$4 to $8 worth of gold flakes sticking to his saliva. There was the Strapping Son of Oregon who swung his pickax night and day. Each time the earth disgorged another nugget, he would cry for joy over the material comfort he had just secured for his parents back home. This nugget's for Dad's winter coat. This one's for Mom's new stove. "Few men with a heart like his have ever come to California," Carson said. And there was the drunkard known as "the parson," once a powerful preacher in the East who had turned altogether carnal digging for gold. He consented to officiate at the funeral of a miner named George on the south fork of the American River, and things were going reasonably well until it was necessary to sing a psalm. Midway through the second verse the parson stopped all at once and muttered, "The Good Lord has obliterated my memory." He motioned the mourners to kneel beside the freshly dug grave, and they had been praying for a good ten minutes when they discovered that the soil was "lousy with gold." The discovery created quite a stir but the parson, eyes closed in prayer, thought it was simply the spirit of Jesus that had swooped down and was infecting the boys. In the presence of such giddiness, the parson "warmed up," his supplications for the dead man's soul now echoing across river, valley, hill and mountain. Then he suddenly stopped preaching, opened one bloodshot eye and caught a glimpse of the real reason for all the hub-bub. "Boys, what's that?" he shouted. He very well knew the answer. "Gold by God! The richest kind of diggings—the very dirt we've been looking for." It goes without saying that poor George wasn't buried there. He was taken from his rich hole and a grave made for him "high up in the mountain's side."

On his own deathbed fourteen years later, voice mute and lungs shot through with a fungal infection known as Valley Fever, the thirty-two-year-old Carson penned one of the first memoirs of California. In his febrile state, he had written a masterpiece of a miner's life and an ardent narrative of his travels through the desert and swamps of the Tulare Plains—and what he regarded as its vast potential for vineyards, or-chards, wheat and cotton fields. It was a vision at least seventy-five years ahead of its time, right down to the dam that would have to rise in the magnificent Kings River canyon for it all to work.

Carson's memoirs were published first as a newspaper series and then as a book, gaining wide readership in Stockton, San Francisco and several East Coast cities. He dared to heap praise on one particular tribe of Yokuts living at the junction of the Kings River and Tulare Lake, an en-tire village filled with "intelligent, hospitable" natives who chronicled their history through story and song and tracked time by cutting notches in a stick. "Great friends of the white man," he called them. There's no telling how many broken-down miners and starry-eyed set-tlers came to the Tulare flatlands holding on dearly to Carson's vision of friendly Indians and gold and green fields.

Given what others had written about the place, these newcomers would have been taking a giant leap into the quixotic.

Lieutenant George Horatio Derby was an army topographer not known for his writer's flair, but when it came time to report on his 1850 survey of the valley, he saw a different land than did Carson. He could scarcely hide his contempt. "The most miserable country that I ever beheld. The soil was not only of the most wretched description, dry and powdery and decomposed, but was everywhere burrowed by gophers and a small animal resembling a common house rat. . . . With the exception of a strip of fertile land upon the rivers emptying into the lake, it is little better than a desert."

That wasn't the worst of it. Thomas Jefferson Farnham, in an 1851 travelogue titled *Life, Adventures and Travels in California*, likewise found a valley that bore no resemblance to Carson's, even though they had come in the same year. "The climate of this valley is its greatest misfortune. This intense heat, poured down so many months upon the submerged prairies, evaporates the water . . . and converts the lakes into stagnant pools . . . which send out most pestilential exhalations, converting this immense valley into a field of death." As for the wild horses, elk, antelope, deer, grizzly bears, coyotes, beavers, mink, squirrels, opossums and kit foxes, the critics dismissed the multitudes as either vermin or curiosity.

It was easy to misgauge the valley. Many surveyors had come during drought or at the tail end of summer, when the rivers ran low and the soil had baked dry. Behold that same ground in spring, a mountain-to-mountain meadow of every color of wildflower, and it held all the virtues of loam. John Muir, the naturalist who immigrated to California from his native Wisconsin to "study the inventions of God," stood frozen in stupefaction the first time he laid eyes on the valley, a sweet bee garden in the flutter of spring. It was "one smooth, continuous bed of honey-bloom, so marvelously rich that, in walking from one end of it to the other, a distance of more than 400 miles, your foot would press about a hundred flowers at every step."

For Carson, turning these fleeting fields of purple, gold and blue into another Italy or France was a simple matter of alchemy, requiring little more than the bending of water. The spoils, as Carson saw it, went to those who controlled the movement of rivers. Whether India, Egypt, China or ancient Mesopotamia, it was water that turned dust into civilization. Stand in the middle of the valley and gaze east and see the towering peaks and bottomless canyons of the Sierra Nevada. These mountains weren't blue for nothing. They happened to hold the greatest water fields in the West.

Perhaps it took a son of the Shenandoah to understand it. Carson had grown up in a valley of small wheat, corn, oat and barley farms fringed by apple orchards and small herds of sheep and cattle. The limestone belt ran 200 miles long and 30 miles wide, crossed by the horseshoe bends of the Shenandoah River and walled off by the Blue Ridge Mountains on one side and the Alleghenies on the other. In Carson's eyes, it was no more of an Eden than this valley on the opposite end of the country, if one factored in the wonders of irrigation and reclamation. Canals would dole out the snowmelt from river to "wretched earth." Levees would hold back floods, and in summer turn the lake from "pestilent pond" to the "garden of California."

"I saw in 1850 a crop of barley raised on the Tulare Plains that was equal to any I ever saw in the country," Carson wrote. "Cotton and sugarcane could be brought to high perfection anyplace within the plain. . . . And the whole of this valley could be made one vast vineyard and orchard."

The hands of Chinese laborers and the rails of the Iron Horse, one following the other, would attract the German capitalists of San Francisco. Carson was sure of it. He saw only one possible snaffle in what he called his "fancy pencillings" of neat cottages and church spires peeping out from "fields of flowing grain." Carson feared that the Indian Affairs Commission in Washington would give away the best of the valley to buy peace with bands of natives who were nothing like the Yokuts he had encountered along Tulare Lake but were "wild beasts of the field in human shape."

On his last trip to the headwaters of the lake, Carson stood on Venice Hill surrounded by oak trees and blackberries and the roar of the Kaweah River, a panorama that made clear the limitless potential of the plains below. The fires of a camp where the Indian Commission and the natives had met to negotiate a settlement—a treaty too magnanimous in Carson's eyes—were still smoldering. Nearby were twelve hillocks of fresh earth, the graves of John Wood and other white settlers who founded the first county seat in a cabin beside the river.

"Can these treaties stand? Will the settlers in California submit to it? No!" Carson declared. "Look among the graves there! One looks greener than the rest! It is poor old Wood's grave! He was my old companion. We, together, explored the plains around, where the feet of white men had never trod before. He now sleeps there, murdered by the Indians, who instead of being punished have been pampered, fed and enriched by the Christian hands of the Indian Commissioners. Now the demon of

Revenge has seized my soul and the blood runs boiling through my veins."

Carson never got his chance at avenging John Wood's death. A short while later, after the miners had elected him to the California legislature, Carson lost the ability to speak and he succumbed to Valley Fever on December 12, 1853. There was a family graveyard at Pleasant Green in Frederick County, Virginia, the 380-acre farm where an impressive line of politicians and brigadier generals would be buried. James Carson, who preceded even his grandfather in death, chose instead the sod of the Tulare Plains. So fleeting was honor and fame in this new California that only three people saw him to his grave.

"I am one of those who have pitched their tents in California to remain in it forever," Carson wrote. "Her interests are mine, and to thee, California:

My voice, though but broken, was raised for thy light;
And this heart, though outworn, has a throb still for thee."

THE MURDER OF JOHN WOOD, an act that forever altered relations between natives and newcomers, was committed in the winter of 1850 when a group of fifteen settlers began clearing land and building five log cabins along a stretch of the Kaweah River. The group had completed one cabin when an agitated Chief Francisco, a mission dropout, showed up on horseback flanked by several of his men. He told the settlers that the Four Creeks region still belonged to the Yokut nation and they had ten days to pack up and leave, an ultimatum to which the settlers apparently agreed.

Perhaps they figured there was no harm in waiting until the last minute. Whatever the reason, the settlers were still in the process of moving when the Yokuts came back as promised on December 13—the tenth day. If Chief Francisco was inclined to give Wood and his men a one-day reprieve, he must have changed his mind after glimpsing the settlers burying farm implements and other supplies and surmising that they intended to return soon to their spot beside the river.

Wood was working near the log cabin when the massacre began in the grassy field just beyond a stand of oak trees. The death of his ten compatriots was accomplished with great stealth because Wood and another settler had no inkling of any ill will as the Indians approached the cabin. They even indulged the natives in a game of target practice, the settler holding up a mark for one Indian's bow and arrow. The Indian then pulled out a gun, took aim and shot the settler to death. Wood ran inside the cabin and fastened the door. He had a single rifle and a short supply of ammunition—enough for the Missouri native to put up a gallant fight. Firing through crevices in the logs, he killed at least seven Indians before his ammunition ran out.

A group of attackers rammed the front door until it gave way and there stood a helpless Wood. The Indians tied him down and began to skin him alive. Partway into the job they seemed to lose interest or decided to give chase to two survivors fleeing on horseback. A large patch of skin from Wood's back was later found nailed to an oak tree next to the river. His body was discovered at the edge of the creek, where he had apparently crawled while dying.

More than 100 Indians had taken part in the attack, according to some accounts, and they fled into the Sierra backcountry with a herd of 600 stolen cattle. The massacre roused several other foothill and mountain tribes still clinging to their land and shuffling between plunder and treaty. Hunkered down in the foothills, they awaited a reprisal.

California was now a land caught between, no longer Indian or Spanish and still a good ways from being white. The headiest days of the Gold Rush were over, and many miners were looking to settle down and find a homestead to plant wheat or raise cattle. For those seeking direction, at least one Bay Area newspaper pointed south—to the large body of water called Tulare Lake. In a view that was right in line with Carson's, the *San Francisco Picayune* described how the lake was skirted on one side by an "earthly paradise" of luxuriant grasses and riparian forests that, once cleared, could be forged into the finest agricultural district in the new state. "The abundance of fish of all kinds is absolutely astonishing," the paper said in November 1851. "The quantity of game . . . is immense. If game abounds, so also do rattlesnakes. . . . The Indians kill them by means of a tule and a forked stick."

The Yokuts—those who hadn't died of measles or consumption or venereal disease—were living in perpetual exile. Herded away from the lake, they were driven across the prairie from reservation to rancheria. They had survived the fur trapper and mountain man, picaresque vagabonds who more or less respected their ways. These new settlers, mouths full of tobacco juice and profanities, were different.

They were known as Pikes, whether they hailed from Pike County, Missouri, or not—mostly poor white trash who had landed in the valley with a few meager head of cattle after a lifetime of wanderlust. They had come to their last chance at the end of the American plain, with an eye to stay and a sense of entitlement. Many were out-and-out squatters who showed little regard for the Yokuts. They didn't bother to distinguish the valley tribes inclined toward peace from the foothill tribes bent on war. To the Pikes, all the Indians were the same diggers, naked savages and every act of theft or vandalism visited on their cattle or little box houses was reason enough to round up a posse and hunt some down. By gun or rope, they meted out the only brand of justice they knew, sometimes even getting the real culprit.

On January 24, 1851, in the wake of the Wood massacre, California Governor John McDougal issued a call to form a special brigade to smother the fires of Indian rebellion and broker a series of peace agreements. The Mariposa Battalion's first order of business was to select a leader both feared and respected by valley Indians. Two of its members, Walter Harvey and Jim Savage, stood out from the rest. Harvey was the judge of the local court and helped run the fledging county from a building that now carried great symbolic weight: Wood's bloodstained cabin. Judge Harvey was a fine horseman who didn't back down from a challenge, and he would have made an able leader for the Mariposa Battalion. But he found himself dwarfed by the legend of Savage, a man regarded by the Indians as a wizard and known far and wide as *El Rey Tulares*, King of the Tules.

Savage's wife and baby girl had perished on the trip west from Illinois in 1846, and he befriended several Indian scouts in California who taught him the language and culture of the Tulare tribes. He could speak five Indian tongues in addition to German, French, Spanish, and English—the perfect polyglot to cash in on a Gold Rush that lured dreamers from every nation. Along the Los Angeles to Stockton miner's trail through the valley, Savage opened three trading posts and set out to build friendly ties with each tribe in a 100-mile range.

He was strange enough, a mountain man with blond hair spilling past his wide shoulders who spoke their language and heeded their ways. But the Indians, suckers for magic, swore that Savage was a sorcerer who had ridden in on a moonbeam. With their own eyes, they said, they had seen him catch bullets fired at pointblank range with his bare hands. Several tribes bewitched by his powers took the remarkable step of naming the twenty-five-year-old Savage their chief, an honor that bestowed, among other gifts, a harem of young *mokees*. He took no fewer than five Indian brides and put their fathers and brothers to work in the gold mines. He then happily traded for the gold they found, raking in tens of thousands of dollars of precious ore in exchange for cheap trinkets.

Savage's role as fierce Indian protector wasn't purely selfish, however. He furnished plows, seeds and implements to tribes thrown off their hunting grounds and taught them the ways of farming. He took up their cause, sometimes violently, when they had staked out new ground only to be driven off by land-mad whites. But being both exploiter and savior proved to be an impossible juggling act, even for a man as protean as the blond king. Just a few days after the Wood massacre, to the surprise of no one but Savage, the suddenly cheeky Indians staged a coup.

One of the foothill tribes not in his clutches led an attack on Savage's trading post in Fresno, killing three clerks and stealing every shelf bare. They ran the cattle, horses and mules into the mountains. They stripped all of the clerks of their clothing and then crushed them with rocks and shot them full of arrows. "It was a horrid scene of savage cruelty," one eyewitness wrote. "One of the men had yet 20 perfect arrows sticking in him." Savage's wives had been warning him for weeks that just such an attack lay around the corner, but Savage had ignored them. The tribes loyal to him had no choice now but to flee the camps near his trading posts, head into the hills and join the uprising.

During the next several months, through blinding snow and over trails known to no white man but him, Major Savage led the seventy-five men of the Mariposa Battalion in pursuit of the recalcitrant tribes. To track down Chief Tenaya, Savage and his boys became the first known whites to set foot among the sheer granite domes and towering falls of the mystical place that they would name after Tenaya's tribe—the Yosemites. The battalion of soldiers marveled at Savage's preternatural skills. He seemed to come from another world, half man and half wolf. Some believed he could smell the smoke of an Indian fire miles away, discern from the pulse of a song or dance the likelihood of peace or battle. He required not a moment of sleep and took a catnap only as a nod to convention. "No dog can follow a trail like he can," one soldier remarked. "No horse can endure half so much."

In the name of peace, Major Savage and his men did not hesitate to set fire to Indian camps and villages and kill scores of natives in skirmishes along the way. When it was over, in the spring of 1851, they had met with 120 bands of Indians and brought home nineteen signed treaties. There was only one small problem. What looked so good to the tribes on paper—they were to receive 7 percent of state lands and a whopping 900,000 acres along the Kings River and in the Tulare Lake basin—turned out to be fool's gold. Because so much of this land sat in the direct path of any settler trying to make his way from southern to northern California, the reservations conflicted with destiny. They were breached before the ink was ever dry.

Judge Harvey, who had never forgiven Savage for beating him out as leader of the Mariposa Battalion, organized a posse of men to raid the Kings River rancheria. The ostensible purpose was to arrest Chief Francisco and others involved in the Wood massacre. The Yokuts, armed with bows and arrows, ran and took shelter in thick weeds. Rather than negotiate, Harvey and his men sent a volley of rifle shots into the distance, killing nine Yokuts and wounding scores of others.

Major Savage denounced the raid as an act of butchery, a criticism echoed in newspapers whose usual stance was to rabble-rouse for tough measures against the Indians. Savage had been in a state of despair ever since it became clear that the treaties he had brokered amounted to little more than an intolerable last surrender for *his* tribes. In an act more of atonement than revenge, he rode to a trading post along the Kings River and confronted Judge Harvey face-to-face. The judge had been badmouthing Savage in public, telling people that the major had no room to talk when it came to the treatment of Indians.

"There is a good horse, bridle, spurs and leggings which belong to me. I fetched them for the purpose of letting you have them to leave this country with," Savage told Judge Harvey.

"I have a fine mule, and I will leave the country on my own animal when I want to leave it," Harvey replied.

Savage tied up his hair and rolled up his sleeves. He took out his six-shooter and stuck it under the waistband of his pantaloons. He then asked Harvey to take back his words and pronounce him a gentleman in the affairs of Indians.

"I think you're a damned scoundrel," Harvey sneered.

Savage struck Harvey on the side of the head, and the judge fell down on a sack of flour. He tried to shake off the blow, but Savage swarmed over him, punching and stamping. As a peacemaker tried to separate the two, Harvey kicked Savage and his gun fell out of his pants. Savage shook off the peacemaker and charged straight into Harvey, knocking the judge to the ground again. This time, Harvey rose to his feet, drew his pistol and shot the unarmed Savage through the heart. The major fell without uttering a word. To ensure that the great sorcerer was dead, Harvey stood over him and fired two more bullets into his chest.

News of Savage's death spread quickly throughout the state. The Indians had to see his body to believe it. "Our Indian shoot, bang! Major Savage he grab in air with his hand. He don't fall down. Our people look each other. They say he big medicine man," recalled one old Yokut. "White man shoot Major Savage like our man shoot 'em. Major Savage he die. Ha! Ha! Why he no catch 'em bullet?"

They buried him at the rancheria, and all over the valley that night the Yokuts let loose with their death chant and spun like dervishes in sorrow. Wrote one observer: "I have never seen such profound manifestations of grief. The very blood within one curdled at the scene."

Judge Harvey's acquittal at the hands of a colleague and friend may have been a travesty of justice, but it cleared him in enough eyes that he went on to marry the daughter of Governor John Downey and win a lifetime appointment as superintendent of immigration for the Port of San Francisco. For the Yokuts and other valley Indians, the death of Savage seemed to kill the last bit of hope left in them. Whom could they trust to negotiate a new set of treaties? What white man would give them anything but the most desolate of lands?

THE ELK AND ANTELOPE WERE GONE and the grieving Yokuts watched in disbelief as the settlers began chopping down the sprawling oaks that furnished the only staple they knew—the acorn—for the purpose of building fences and fires. In midsummer, the Yokuts and Miwoks used to gather in long, noisy processions to feast on the blackberries and grapes that grew wild on the banks of the Four Creeks. Now they were rounded up and forced to live beside the mountain tribes that they had scarcely tolerated. Forbidden from roaming, they were no longer free to collect even pine nuts or trap quail. Tulare Lake, their fishing ground, had been turned into a pigsty. On one of the lake islands where a Yokut village had stood, Visalia Judge A. J. Atwell began raising razorbacks and purebred Poland china hogs. Breeders came from all around to buy the high-grade studs that made such sweet, lusty meat.

This was the California of the 1850s that confronted Yoimut, a child of the Chunut tribe of Yokuts who inhabited the eastern shore of the lake. She would live past eighty-five, the last full-blooded survivor of her people, and speak eight Yokut dialects, along with Spanish and English. The photograph snapped late in her life shows a small, round woman with copper-colored skin and hair so thick and low over the brow that it looks like a bonnet. She has the huge hands and flattened nose of a prizefighter, and her face is quiet and full of pride, right down to the scars between her eyes from the cuts made by the medicine man.

A few years before Yoimut was born, soldiers on horseback had rounded up the Chunuts and other lake Yokuts and driven them eighty miles north to a reservation in the making along the Fresno River. Yoimut had heard the story of the long walk a hundred times, always told the same way by her mother. "None of the Indians wanted to go. But the soldiers beat the Indians with whips and hit them with their swords and ran their horses over them. My mother said she saw 12 Indians killed. It took about 10 days to get all the Indians together. Then it took four days to drive them to the Fresno River reservation. . . . About 10 Indians more died along the way. Three or four babies were born."

The acorns stockpiled at the reservation were half eaten by worms. The soldiers took pity and killed a few cattle, but the meat could only sustain the Yokuts for two months. They were rounded up again and taken on a 150-mile march down the valley and past the lake to a new reservation in the foothills. "None of our people liked it. They all wanted to go back to Tulare Lake."

As darkness fell one evening, a group of Yokuts sneaked away and followed the old trail toward home until they reached an abandoned Indian village near the Kaweah River. They managed to stay there unmolested for seven years, until Yoimut was born and a measles epidemic killed the chief and half of the already decimated tribe. The survivors scattered and scratched out lives as farmhands and housemaids for the whites who continued to make their way to the valley.

In a last stab at repatriation, a handful of Yokuts eventually returned to a few huts still standing beside Tulare Lake. As they settled in and got to thinking they were home again, a Pike showed up with a gun. Kicked from one side of the lake to the other, they decided finally to leave for good. They built one last tule boat and took a final trip across the water still teeming with trout, geese, swans, herons and ducks. They fished what they could and said good-bye.

In the years that followed, Yoimut returned to the lake once or twice to watch the big boats—the *Alta*, the *Water Witch*, the *Mose Andross*, the *Alcatraz*—plow through the choppy waters with their great masts or engines of steam and snare by the thousands what her father had speared by the one. The turtles that her mother had found with her bare feet now sold for $4 a dozen. The eggs of ducks that she and her brother once hunted had become items of delicacy in fine hotels up and down the coast. She watched the steamers and schooners ferry hogs from one island to the next across the lake. The pigs had become so fond of clams—they would eat nothing else—that they dove like ducks in the shallows until all Yoimut could see was their pink tails wiggling skyward. The pig farmers on Judge Atwell's island were forced to disband when picky diners began complaining that their once sweet pork now tasted of fish. The farmers freed the razorbacks and Yoimut watched them became wild again, guerrilla bands of tall, gaunt and ferocious swine attacking hunters from hiding places deep in the tules.

Like the Yokut fishermen of Yoimut's childhood, the sailors now crossing Tulare Lake were sometimes caught by surprise in the sudden winds howling down the Coast Range. For most boats, though, it wasn't the wind but the thicket of tules that played havoc. Just once, during the high waters of 1868, did the channels run swift enough through the reedy marsh for a vessel to go all the way from the lake to the Pacific Ocean, a 170-mile journey across a land that the year before was desert. A sixteen-foot scow carrying a cargo of valley honey worked its way up the north fork of the Kings River into the San Joaquin River, up through the Stockton delta and out to the San Francisco Bay. No other craft ever made the return journey without getting stuck in the tules. The sailors hired vaqueros who toiled day and night to dislodge the vessels, heaving their anchors sideways and forward only to see the channel waters evaporate into mud.

For years, the rotting hulk of the little steamer *Alta* sat beside the highway not far from Yoimut's last home. She would live her final days with her daughter and Latino son-in-law near the shore of the old lake. The big farmers had made it go dry, and in its place stretched a sea of white fields that seemed to have no end. She wondered if it would ever be blue anymore. "You ask me, 'Will Tulare Lake ever fill up again?' I got only one thing to say. Yes. It will fill up full, and everybody living down there will have to go away. I'd like to see that time myself. I am the last full-blood Chunut left. My children are part Spanish. I am the old one who knows the whole language. When I am gone no one will have it.

"All my life I want back our good old home on Tulare Lake. But I guess I can never have it. I guess I can never see the old days again. Now my daughter and her Mexican husband work in the cotton fields. Cotton, cotton, cotton. That is all that is left. Chunuts cannot live on cotton. They cannot sing their old songs and tell their old stories where there is nothing but cotton."

The Little Kingdom of Kings

THE SETTLERS STOOD IN TWO LONG FLANKS across a treeless stretch of the basin and waited in the hot sun for the bugle call. When it sounded, they began to sweep the prairie in a precise pincers movement, a steady rumble that roused thousands of hopping-mad jackrabbits from slumber. As the rumble grew and the flanks of the settlers squeezed tighter, the rabbits, struck with terror, were shooed into a giant pen built the night before. Inside the corral, the settlers had little need for military precision. Overcome with frenzy, hundreds of men, women and children proceeded to club and stomp to death the thunderous mass, a tangle of leaping and flailing and writhing. There was no call to end. It ceased only when the movement of rabbit limbs ceased. The settlers, dead tired, sank waist deep into a sea of fur, urine, blood and feces. This is what passed for entertainment in the valley of the 1870s and '80s, the frontier West's version of the fox hunt. Twice a year, the settlers gathered in the middle of the desert for the slaughter. When they finished, they posed for a snapshot beside the carnage, 20,000 dead jackrabbits stacked high or hanging from the rails of their makeshift corral. Then they broke out their lunch baskets and had a picnic. The hides were sold to fur makers in San Francisco, their coats marketed to society matrons as fine English hare.

In between rabbit hunts, the settlers took aim at each other: Pikes pitted against Sandlappers in a bitter feud over their starkly different visions for the valley. The Pikes laid claim to the riverbeds where they fed and watered their cattle. The Sandlappers took the dry ground where they planted small plots of grain. Because they relied on the valley's meager rain for irrigation, the Sandlappers got their nickname from the local bird that scratched in the sand for crumbs. The two groups knocked heads over a law that allowed cattle to forage anywhere a Pike pleased, even if his herd was caught mowing down a Sandlapper's grain field. Only if the Sandlapper had erected a fence around his crop—an expensive proposition—could he claim a loss and file suit against the Pike. Otherwise, his appeal to the cattleman for crop damages typically met with the same dull response. "Surry 'bout yur wheat, but if ya'll want a side of beef to ease yur pain, help yurself."

Many Pikes had congregated in Visalia, a bustling town on the east side of the valley where dueling newspapers, one staunchly Confederate and one resolutely Union, chronicled the cultural turn from livestock to bonanza wheat to fruit farms. Long before the boll weevil drove Colonel Boswell and other southern cotton growers to the Tulare basin, county voter rolls showed a decidedly Dixie bent: saloon keepers from Kentucky, miners by way of Tennessee, farmers from North Carolina and Virginia and stockmen from Arkansas, Texas and Missouri. So virulent were Visalia's southern loyalties during the Civil War that the federal government had to build a fort on the outskirts of town and man it with pro-Union soldiers. Even so, a guerrilla band of southern sympathizers, operating from a tule hideout along the lake, stole hundreds of horses from local ranchers and delivered them to the Confederate Army.

Among the ranks of Sandlappers looking for better ways to water their crops were men of impressive roots and first-rate education, though it wasn't always easy picking them out, hidden as they were under layers of grime. Only at night, after they had finished another long day digging holes and connecting those holes in a manner that would slowly reveal they were building an irrigation canal one foot at a time across the prairie, did they give themselves away. In a remarkable account of ditch diggers trying to grab their fair share of the Kings River in 1875, one crew member wrote of the collective restlessness that teased their evenings as they tried to find sleep in a converted barn. "Tired, aching limbs would not allow their possessors sleep without first an hour of relaxation, and that hour of wakeful rest was devoted to song," William Sanders wrote. "Favorite airs from Beethoven, Mozart, Franz Abt would ring out on the evening air from our improvised choir. Our Sundays . . . were spent reading about and discussing events passing in the world around."

Sanders was a botanist and schoolteacher who had fought in the Indian wars in Arizona before coming to California to prospect for gold. He was six feet tall and handsome, a facile writer and tooth-and-nail fighter who had traveled the globe gathering facts on raisin drying, fig tree propagation and grafting. From Europe he brought grapevine cuttings to the Tulare basin, arguing that the sandy loam of the valley was much too fertile to be squandered on wheat, barley and alfalfa. In time his 320-acre farm would feature 100,000 vines and fruit trees of dizzying variety. Given the otherwise headlong rush toward cattle and wheat spreads, the Sanders farm would come to represent a kind of demonstration project for the heretical. A good rain or two in February and March was plenty to see the wheat farmer through to summer and fall harvest. But for Sanders's vision of permanent fruit fields to endure, he had to tap into the Kings River and import water through jackrabbit and horned toad country.

To obtain his water, Sanders struck a deal with the valley's most formidable force in river diversion, Moses J. Church. A one-time Napa blacksmith, Church had come to Fresno in 1868 while searching for pasture land for his 2,000 sheep. He took one look at the Kings River and abandoned his newfound profession of sheepherder. Church became the first man to erect a dam—albeit a brush and cobblestone one—across a section of the Kings River. Once he had accomplished his goal of water storage, he built a canal that shipped river water to the fledging town of Fresno twenty-five miles away. The grain fields sprouting from Church's canal system caught the attention of the Big Four: Leland Stanford, Charles Crocker, Collis Huntington and Mark Hopkins. To transport wheat to market, the railroad tycoons decided to run the tracks of their Central Pacific down the heart of the valley and rename it the Southern Pacific.

The deal between Church and Sanders called for the botanist to double the width of Church's canal through high ground. In return, Church would grant Sanders and his twenty-two Mozart-loving ditch diggers enough water to fill their own canal and irrigate hundreds of acres of crops. The brutal job of deepening and enlarging Church's canal began in the summer of 1875. "Weeks passed away, months passed away, each but a repetition of what had preceded," Sanders recounted. "Under the constant wear and strain, horses finally became poor; men became haggard, sunburned, morose and irritable. Still, the daily toil went on, on, on . . . the ditch gradually becoming lower and the hills of earth on either side becoming higher. We now saw nothing, realized nothing and cared for nothing, only that long ditch of loam and rock. This was our prison, our whole earth and heaven; we knew nothing beyond it."

When they finished the canal six months later, Sanders and his crew took a short break and began the process all over again, this time building a second set of ditches that would carry Church's water over twenty miles of dry grass and salt brush to their own farmsteads. They worked with shovels and crude metal earth scrapers drawn by teams of mules and horses. There was no mortar or brush to adhere to the earth, just earth that they packed and packed only to watch a solid mound run off like molasses when the first rain found a weak spot. When they finally had it up and running, the water rushed too fast, and the canal had to be graded again and checked with wooden weirs that governed the flow. Even then, amid the canal's midnight roar, the water didn't quite reach their farms but instead tumbled down a sinkhole made massive by the burrows of mammals and reptiles, lairs generations old. At last, the disappearing water gurgled and hissed, and the dirt sank like sugar until the air came out of the earth and the honeycombs were no more. That's when the water bubbled to the surface and their fields filled with the spoils of irrigation.

It took Sanders and his men a year to finish. Their channel was soon joined by other canals and branch ditches that grabbed the water that once reached Tulare Lake and spread it out over a plain transformed. "The wildest, maddest, most enthusiastic enthusiast never dreamed of such a change as has taken place in this region," the *Selma Irrigator* observed in 1886.

Once unleashed, the compulsion to dam and canal the rivers of the San Joaquin Valley would continue for the better part of 100 years, culminating in a system so elaborate that it would rank as one of the most advanced hydraulic societies in the world. The Kings River alone would come to irrigate over 1 million acres, more than any other river in the world except the Nile and the Indus. It hardly mattered that Sanders, Church and all the other early appropriators were breaking with the rich tradition of river rights, which went all the way back to England. California had adopted in 1850 the English common law doctrine of riparian rights, which held that the flow of water belonged to the owners of land along a river. Unfortunately, the law did a terrible job of distinguishing between wise and unwise use. It mattered not whether the holder of river rights was a farmer living along the banks and raising crops for the larger community or a cattleman who roamed far and wide and used his section of riverbed as a sewer.

The wheat growers and fruit farmers whose crops grew beyond the river's edge argued that riparian law made sense in the East, where rainfall spread its abundance. But its application could not be defended in a West defined by aridity. Besides, the greatest windfall of the Gold Rush had come from miners laying claim to river water and then moving the flow beyond the river. The wheat growers and fruit farmers, by attempting to tap into the Kings River, were only continuing the forty-niners' tradition of appropriation, they argued.

With no government regulators to give teeth to the pro-riparian law, the farmers and other appropriators simply built their upstream canals and took the water anyway. Few of the cattlemen and other riparian landowners downstream seemed to mind these diversions in years of heavy spring melt. There was plenty of water to go around. Only when drought hit and the downstream rivers ran dry did the riparian lawyers and saboteurs go to work. Thousands of cattle had died of thirst in the droughts of the 1860s and 1870s, their famished frames plunging into the muddy waters of Tulare Lake. Too weak to extricate themselves from the clay bottom, they drank their fill of hot alkali water and lay down by the hundreds to die. For years after, their bleached bones marked the steady retreat of the lake.

At some point in the 1880s, as the courts filled with Kings River lawsuits and counter-lawsuits, it was no longer merely Pike against Sandlapper. The entire basin found itself in the clutches of a water war: farmer against farmer, neighbor against neighbor, appropriator against appropriator.

On Independence Day in 1883, the canal company founded by Moses Church, whose diversions to distant Fresno still rankled local farmers, made a sudden grab for more Kings River water. Church decided to raise the level of his dam and shut off the river's flow to the irrigation company just down from him, the C&K Canal.

The superintendent of the C&K, William Shafer, was not a man to be taken lightly even in good times, and it so happened that 1883 opened as a bad drought year. Burly and all business, Shafer was a history buff and self-taught civil engineer who would later shoot a man dead in self-defense. He naturally read contempt in Church's action, coming as it did in the depths of summer when his own farmers downstream needed irrigation water the most. He made a direct appeal to Church to lower his dam, but Church refused. Shafer wanted to pursue the matter in court but his attorney told him it would take years, a delay that would turn dozens of farms in his twenty-mile service area into dust. Then the attorney suggested an alternative: dynamite.

The explosion that followed was a milestone in the long history of rebellion along the Kings River, a harbinger of the levee feuds between the Boswells and Salyers seventy-five years later. While other incidents of sabotage may have proven more damaging and incendiary, the destruction of Church's dam has managed to live on in part because of the mysterious involvement of a student from the University of California at Berkeley, a young man named Samuel Moffett. How the twenty-two-year-old Moffett landed in Kingsburg and why he was enlisted in an act of dynamiting may be a question only his uncle and mentor, Samuel Clemens, better known as Mark Twain, could have answered with clarity.

Twain's only son had died as an infant and he regarded Moffett, his namesake and the son of his older sister, Pamela, as his male heir. Moffett grew up in St. Louis and dazzled his uncle by displaying an encyclopedic knowledge of historical facts. In his autobiography, Twain lovingly recalled his nephew's prodigious gifts of instant recall, a "large and varied treasure of knowledge" that the boy showcased while visiting the author and his wife in Buffalo in the summer of 1870.

At age nineteen, Moffett had already decided on a career as a newspaperman, and Twain encouraged him to try his luck in San Francisco, though he made a point of refusing to write any letters introducing his nephew to friendly editors. Moffett went on to become a great newspaperman under the surly watch of William Randolph Hearst and later a writer of serious tomes such as *The Tariff*, works that could not have departed more from his uncle's crackling style.

But about his movements leading up to the canal bombing on August 2, 1883, only this much is known: Moffett was spotted in Visalia at a hardware store called the Sol Sweet Company. He purchased a twenty-five-pound box of dynamite, wrapped it in layers of cotton and shoved it under his buckboard. Then he rode along the foothills to a prearranged spot on the Kings River near Centerville. There he was met by Shafer, the chief of the C&K canal, and told to step aside. Shafer would alone plant the dynamite and light the fuse to blow up Moses Church's dam.

Moffett protested, insisting that he was plenty brave to do the job himself.

"I do not doubt your courage at all," Shafer replied. "But there may be shooting when I get to the dam, and I cannot take the risk of having you with me."

Shafer was setting the last sticks of dynamite in place when Church's superintendent confronted him. "What the hell's going on here?"

"We're going to dynamite your dam."

"You're headed for a pack of trouble."

"I know that, but this is war and no war was ever pleasant," Shafer said. "We are entitled to get the water we are entitled to, trouble or no trouble."

Shafer then lit the fuse and blew up Church's dam, sending rock and brush sky high. The river surged downstream, and the C&K headgate was opened. For two weeks, under the constant watch of armed guards, the water flowed to the fields of distant farms and saved the crop. Shafer freely admitted his role in the bombing and was arrested and hauled before the Fresno justice of the peace. Even if the judge had wanted to side against the civil engineer and squeeze him for the names of Moffett and any other accomplices, he dared not in the contentious world of California water rights circa 1880. To do so might have opened a Pandora's box.

Unimpeded by the courts, the diversions of the Kings River flowed on. The town of Sanger sprouted peach fields. Parlier became the home of the Thompson grape. And Selma used its allotment of funneled snowmelt to build a raisin dynasty.

HAD THE TRANSFORMATION of the San Joaquin Valley been allowed to progress on its own gradual terms, the Indians and whites, not to mention the riparian cattlemen and ditch-bank farmers, might have reconciled in a manner that would have honored treaties and curtailed violence. They may even have found a way to slow, if not prevent, the devastation of Tulare Lake. But what happened in the years before the ascendancy of the southern cotton culture in the West was one of the biggest land grabs in U.S. history, a fire sale that concentrated the best and most fertile parts of California in the hands of a grubby few. The easy villain was the Southern Pacific railroad, which owned, by government fiat, alternating sections of land along its proposed routes. But this great land grab began at least a decade before any tracks were laid and was jump-started by a series of federal giveaways in the 1850s and 1860s.

On the eve of America's takeover of California, the Mexican government had rushed to reward its close friends, both Mexican and American, by placing more than 8 million acres of land in the hands of 800 people. Unlike what happened in the Spanish days, thirty such grants had been awarded in the San Joaquin Valley, including a 48,000-acre ranch along the main branch of the Kings River. Many grantees hadn't bothered with surveys or filing paperwork; some couldn't muster a single document backing up their claims. Into this mess stepped the U.S. Land Commission for California, a final arbiter to determine which Mexican grants were bogus and which ones would carry the new government's imprimatur.

The land commission ultimately became an agent for monopolization, approving twenty-four of the thirty grants in the valley and setting the stage for more wild speculation through a series of government programs. Each of the land distribution mechanisms—the Military Bounty Act, the Swamp and Overflowed Lands Act, the Morrill Act—spoke to the notion of thwarting concentrated holdings. Each program had the potential to be an instrument for small farmers to acquire 160-acre parcels and build the villages envisioned by James Carson. But from the start, speculators perverted their intent, and federal and state officials refused to intervene to curb the abuse.

The selling of $1.25 government scrip to help war veterans buy land or to fund state agricultural colleges became riddled with fraud and deceit. The scrip were sold in large blocks for as little as fifty cents apiece and hoarded by San Francisco capitalists such as William Chapman and Isaac Freidlander. Chapman and Freidlander teamed up with Moses Church and amassed 170,000 acres in the Tulare Lake basin and another 80,000 acres around Fresno. It would take years for some of these holdings to be subdivided into twenty-, forty- and eighty-acre parcels and sold to small- and medium-sized farmers for a ten-fold profit. Often the land was never broken up but simply passed from one behemoth to another. By 1871, after twenty years of statehood, California found itself a more stubborn oligarchy than at any time during Mexican rule. Nine million acres of its best land were held by 516 men.

No one man shaped early California—its land and its water—quite like Heinrich Alfred Keiser, a.k.a. Henry Miller. To call him the "Cattle King," the title of a biography that scarcely did justice to his incredible rise and dominance in the West, was to ignore that Miller governed more land and more riparian water rights than any other citizen in America. At the height of his holdings, he controlled two rivers and owned 1.3 million acres and more than 100,000 head of cattle, a principality that extended across three states. Miller's empire was so huge that people used to say he could ride his horse from Canada to Mexico and sleep every night on a ranch that was his own.

He left his family in Germany at age fourteen and arrived in New York City in 1847. He had learned how to cut meat from his father and within two years he owned a flourishing Manhattan butcher shop that he sold on a whim to come west to search for gold. He nearly died on the long voyage over, depleting his nest egg to get well. He arrived in San Francisco with a new name stolen from an acquaintance and $6 in his pocket.

His idea, hardly novel, was to mine the miners, and within three years he had peddled enough pork sausage to become the second largest meat dealer in San Francisco. This naturally attracted the attention of the city's largest meat packer, a German immigrant named Charles Lux, and the two men became partners and set about building a barony of beef. They employed 1,200 workers, one of the largest labor forces in the country, and organized them in assembly-line fashion from corporate headquarters to butcher shop to slaughterhouse to grazing land and watering hole.

The Miller & Lux partnership owned 900,000 acres in the San Joaquin Valley and another 200,000 in the Santa Clara Valley, a good chunk of it coming with Miller's purchase—at dirt cheap prices—of old Mexican land grants. Miller understood that the true value of these tracts wasn't the acres they encompassed so much as the rivers they contained or skirted. Controlling the rivers meant controlling the two variables that made the valley a risky investment: drought and flood. He grabbed a 120-mile stretch along both banks of the San Joaquin River and a fifty-mile run along the Kern. Not satisfied with riparian rights alone, he seized giant tracts of government-subsidized swampland between the two rivers. Much of this land, due to drought and upstream diversion, was no longer swamp. Even so, Miller had to prove he had traversed the land in a rowboat to qualify for the government subsidy. Legend has it that Miller employed a boat all right, but the boat happened to be perched atop a wagon powered by horses. Whether he employed this scam or another, Miller managed to grab vast tracts of dry land for the swampland price of $1 an acre.

In his travels up and down California, *Harper's* magazine writer Charles Nordhoff went looking for the mercurial Miller, a dusty buccaneer riding horseback across the valley's sprawling west side. Miller was known to get up before the sun, eat breakfast with his vaqueros and oversee every minute detail of his empire, right down to the chow house where he insisted that no tramps or winos be turned away hungry. Of course, they had to be willing to eat after the employees and off the same dirty plates. This way, Miller pointed out, the dishes had to be washed but once. To the hobos, the footpath to Miller's ranches became known as the Dirty Plate Trail.

Nordhoff, one-time editor of the *New York Evening Post* and a barker for the Southern Pacific, trailed Miller for weeks before catching up with him at the west side outpost of Firebaugh. Here was where California, as the writer knew it, suddenly swerved into Texas. Miller's ranch hands were in the process of erecting hundreds of miles of fence to rein in his cattle. The decision to enclose his kingdom had the effect of altering the balance of power between cattlemen and farmers. Recognizing the growing clout of wheat growers, Miller sensed that the state legislature was about to change the No Fence law and place the onus of erecting fences on cattlemen and not farmers. Sure enough, a new law was passed, giving farmers the green light to sue for crop damages from runaway steers, even if the farmer hadn't fenced in his fields.

In the raw, open countryside of Firebaugh, Nordhoff came upon a storefront filled with eighteen or twenty vaqueros seated about an open fire, smoking and chewing tobacco, spitting and swearing, with "great gravity and decorum." They were Miller's cowboys, straight from Missouri and Texas and Mexico, outfitted in jingling spurs and riding trousers. "I began to think I should have to take some care not to be spit on," Nordhoff wrote, "but the accuracy, neatness and precision of their aim presently re-assured me. One fellow, lounging on the counter behind me, spat over my hat; a vigorous cross-fire was kept up by two others across the toes of my shoes; a scattered but un-intermitting rain fell upon the centre of the floor and occasionally the fire received a douche."

In the middle of it all sat a short, neatly dressed man with a long, pointed nose and a low forehead whose face registered nearly every expression as he smoked and quietly instructed the men in the next day's cattle drive. He got up from his chair and wandered about the room, dodging with some finesse the salvo of tobacco juice. As a vaquero took his chair, the little man squatted down in a wood box next to the fire and rested for a few minutes, and then he went off to bed. It was Henry Miller himself, caught at the crossroads, trying to figure out how to grow his cattle kingdom in a time of falling beef prices and a prairie now sprouting in every direction with "wheat, wheat, wheat and nothing but wheat."

From the shores of Tulare Lake—which were fast receding thanks to all the ditches and canals—the largest grain farm in America burst forth. As fast as the water drew back, the farmers rushed to plant seed using horses and mules shod in wooden shoes to keep them from getting stuck in the mire. Throughout the basin and beyond, the wide-open flatland and dry summer climes conspired to make conditions perfect for farms of king-size proportions. The acres devoted to grain kept increasing through the 1870s and '80s until California led the nation in wheat production—57,420,188 bushels by 1884.

Unlike the South, where the summer rains and modest farms negated industrialization, this was a land that gave itself over to the giant scythes of the combined harvester. A single machine, drawn by as many as thirty-six jackasses and driven by a wily mule skinner, could duplicate the labors of more than a hundred men, harvesting up to thirty acres in a summer's day. Like a military advance, the big growers staggered eight to ten of these monsters across a horizon shimmering of new California gold. The mass of reaping, threshing, blowing and bagging—one continuous rattle and clang—shook the earth. Three or four men straddled each machine and read the vibrations for something amiss as the big bull wheel clawed into the land and gave power to the forty-foot sickle. The mule skinner could move his entire team right, left or forward with a simple "gee dock" or "haw dock" and the slightest jerk of his line that laced each bridle. If his team hit a bad spot, he'd work his way down a long list of curse words in search of the perfect expletive to lift them out.

Even the Tule River Reservation got into the act. Indians bitten with grain fever harvested 600,000 pounds of wheat, 50,000 pounds of barley and 10,000 pounds of rye in one season. Rising from the tracks of the Southern Pacific, the town of Traver shot up overnight, a rollicking hamlet with sixteen saloon keepers, eleven blacksmiths, one preacher, one teacher, two physicians, one undertaker, two whorehouses, one Chinese gambling den, two Chinese laundrymen and a Mexican tamale maker named Jesus. Traver held the world record for the greatest amount of grain shipped from a producing point during a single season. The line of wagons unloading wheat in summer stretched for more than a mile outside the town's three warehouses.

Moving that golden stream to tidewater took the rails of the Southern Pacific, and the railroad held farmers in an iron vise, squeezing for every last penny. What goodwill the Big Four had earned by going deep into debt to bring more than 10,000 laborers from China and shiploads of iron and gunpowder and heavy machinery around Cape Horn, all in the name of building a railroad through 800 miles of granite mountain and frontier desert, was now being squandered by greed. The Southern Pacific grew fat on wheat, exploiting its monopoly to collect ever higher rates. "All that the traffic will bear" became Southern Pacific's mantra. The large growers produced enough crops to pay the higher tribute and still make a killing. As a favor to land barons such as Henry Miller, the Big Four took pains to ask him what he considered to be a reasonable fee. It was the small farmers in little communities along Mussel Slough who had trouble covering the price of shipping.

Mussel Slough farmers, many of them southern Confederates, had acquired some of the sweetest loam in the Kings River delta through the well-worn practice of squatting. The land actually belonged to the Southern Pacific, a gift from the federal government to induce the railroad to lay its tracks through the state's heartland. The Big Four were looking to maximize the millions of free acres, and what better way than to populate the land with farmers whose wheat needed to be hauled to San Francisco. The farmers of Mussel Slough argued that they had been lured to the lake basin after reading the breathless circulars of the Southern Pacific, which offered to sell the land for $2-$5 an acre and implied that they could settle now and pay later. They believed that their illegal grab had become less illegal by virtue of their unbroken tenancy and improvements to the land.

When it came time to buy the land, however, Southern Pacific wanted to charge the settlers $35 an acre and essentially make them pay for the houses and irrigation canals they had built and dug with their own hands. The settlers refused and on the forenoon of May 11, 1880, the U.S. marshal and a railroad man, armed with a court edict, rode into town to take back the land.

What happened on Brewer's homestead that day—who tried to keep the peace and who fired the first shot—engendered years of speculation and controversy across the nation. It became an overwrought symbol of the evil of industrial American monopolies and the righteousness of the small farmer. The bloody gun battle in the wheat fields of the Tulare Lake basin took the lives of seven men, five of them settlers lined up against the Southern Pacific and two of them settlers working in concert with the railroad to seize land for their own taking. The Mussel Slough tragedy would live on in the pages of one of America's greatest novels of social protest, Frank Norris's *The Octopus*.

Joining the fight to break up Southern Pacific's monopoly was the king of California sugar, Claus Spreckels, and his son John. They approached none other than Miller to help build an independent rail line, the "People's Railroad," through the valley. Miller granted passage through his land, a route that eventually became the Santa Fe and gave rise to the town of Corcoran. Then Miller, ever the conniver, turned around and handed the Southern Pacific an even choicer right of way.

For Miller & Lux, however, the real future lay not in playing one railroad off the other but in capturing a bigger share of California's snowmelt and using that windfall to grow more grain. Toward that end, Miller and a handful of San Francisco land speculators lent their names and pocketbooks to a colossal irrigation and navigation project rising along the valley's west side. The plan called for a canal running from Tulare Lake to the San Joaquin-Sacramento delta, a 150-mile artery through the state's midsection. It would tap into California's three biggest rivers, carry grain to compete with the railroads and shunt water to 3 million acres of land.

Nothing like it had ever been tried before in the West, a vision lifted from India where the British had built 6,000 miles of irrigation canals to claim 10 million acres of desert. Indeed, the British engineer who oversaw India's hydraulic miracle, Robert Brereton, had come west and was now working for Miller and his group at a salary of $1,000 a month in gold. Like Carson before him, Brereton took one look at the expanse of salt grass and marsh and saw the potential for a garden unparalleled, the richest and most productive farm region in America.

A forty-mile stretch of the canal had already been built by the labor of 1,000 Chinese workers, a project that Brereton regarded as shoddy and unsuitable for either irrigation or navigation. The new and expanded version would be the "people's canal," built by the toil of white men under the eye of Henry Miller.

Over the next several months, an immense dust cloud on the horizon marked the advance of the San Joaquin & Kings River Canal and Irrigation Company, its teams of men and mules pulling an army of Fresno scrapers across the barren west side. The scraper, a five-foot long hunk of sheet iron invented in the valley, would revolutionize the movement of dirt throughout the world in the years before the gas-powered bulldozer. In its natural state, the valley resembled a rolling savanna not unlike the Serengeti. The scraper reconfigured the land field by field, leveling knolls and hog wallows and filling in gulches, a huge continuous flattening beyond anything nature ever conceived.

From the seat of his horse, Miller watched over the canal construction, an industrial cowboy directing the Italian and Irish immigrants and tramps recruited for the job. By year's end, a fifty-eight-mile rut had been carved into the earth and a 350-foot-long dam erected across the San Joaquin River, diverting the riparian flow onto Miller's wheat and alfalfa fields. It hardly mattered to the boss that the great canal itself was about to die of its own ambition. Brereton's ego had gotten the best of him as he tried to sell his ever grander vision—and the need for government funding—to the state legislature. Miller became irritated with the engineer's gassy talk of a canal system running down both sides of the valley, carrying barges and irrigation water. California newspapers, having cut their teeth on the railroad monopoly, railed against such a subsidy for the "land and water sharks." Wrote the *Sacramento Bee:* "They have monopolized the land, and now they want to monopolize that other great element of life—water. Having the water of the San Joaquin Valley in their control, they would rule it forever."

The canal's cause wasn't helped by the fact that Tulare Lake, the plan's centerpiece, had begun to turn dry from upstream diversions. In the end, the canal investors, some of the wealthiest men in California, went belly up. Only Miller won. He acquired at one-third cost the only stretch of completed canal and could now irrigate a good chunk of his land.

Brereton died embittered, wondering what his life would have been like had he remained in India to engineer a new railroad and never come to the San Joaquin Valley.

His vision of a grand canal, though, lived on. A century later, it would be realized in the massive Central Valley Project that wheeled water from north to south in a path nearly identical to Brereton's. His failed canal also succeeded in bringing to the valley the first engineers of the U.S. Army Corps, who would all but ordain the hand of the federal government in making the valley bloom. In an 1873 report, the Army Corps paved the way for the building of Pine Flat and other dams seventy and eighty years later, ushering in a program of water subsidies that helped a handful of growers create cotton and wheat fiefdoms in the lake bottom.

Miller died at the ripe age of eighty-nine but not before taking on one last challenger to his throne: San Francisco mogul James Ben Ali Haggin. Their feud (over water, of course) would reach the state Supreme Court and help reconcile long-standing conflicts between riparian and appropriative rights. Haggin, the so-called Grand Khan of the Kern, boasted a portfolio that included mines, utility companies and 413,000 acres south of Tulare Lake—most of it acquired through fraud and shameless manipulation of federal law. When the Kentucky-bred Haggin wasn't adding to his stable of the finest thoroughbreds in America, he was busy building another canal to draw more water off the Kern River.

For Miller & Lux, these water diversions under the doctrine of appropriation presented a challenge that couldn't go unmet. It did little good for Miller to own all that riverfront land if Haggin simply planted his spigot upstream and diverted the flow onto faraway farms. Miller & Lux sued under the 1850 statute that seemed to give priority to riparian rights, and the state's highest court eventually ruled in its favor, though the court also confirmed the rights of some appropriators who got to the river first. Because the tradition of appropriation exercised by Haggin and countless others was so embedded in California, the busy work of diverting rivers and digging canals carried on.

Over the next four decades, through legislative act and constitutional amendment, the state found a way to blend both riparian and appropriative rights under a formula of "beneficial and reasonable" uses. What this meant on the ground was that the water generally went to those with the most creative lawyers and engineers. While lawsuits on the Kings River continued for years, the war over the Kern River more or less ended when Miller and Haggin, big men who aimed to get bigger, shook hands and became partners in an upstream reservoir that put even more control in the grasp of a few.

Whether riparian or appropriative, it no longer mattered. It all flowed into the same deep pockets.

THE SANTA FE PASSENGER CAR pulled up to the Corcoran station on a spring day in 1905 and out stepped the rather large and irrepressible frame of Hobart J. Whitley, the father of Hollywood and a hundred Podunk towns across the Great Plains. Beneath the valley sun, Whitley stood for a moment in his fine suit and broad hat and surveyed the land that he and his Los Angeles partners, some of the most powerful men in the West, had come to civilize. Truth be known, the farmers along the four rivers and throughout the lake bottom had already done much of the heavy work. For the first time since the white man stepped foot in this valley, the middle of Tulare Lake had nearly evaporated in the April sun. The center of Yokut civilization had been subdued by a web of upstream canals and wooden dams. What little water now reached the lakebed in all but the wettest years sat in shallow pools trapped behind levees. In fact, because this catchment fed the wheat, barley and alfalfa fields and dried up quickly in the 110-degree summer heat, the problem for the farmer was no longer flood but drought.

It had been four years since the last heavy snowmelt, and the stagnant lake was choking on salt and its own foul belchings. Ducks were dying from worms in their eyes, and the leftover waters had become putrefied with millions of rotting fish, their stench riding on the breeze to farmhouses miles away. "Tulare lake is gone," the San Francisco Chronicle lamented. "So dry that a mosquito could not find in its bed enough water wherewith to moisten its parched bill. Once the largest body of fresh water west of the Mississippi is now a grain field."

Whitley stood at the edge of that grain field amid a thousand longhorn steers and tried to fathom the 60,000 acres of California white velvet wheat shivering from mountain to mountain. The harvest was a month away and in the middle of the lake bottom, a team of 60 men and 300 horses and mules were pulling dozens of Fresno scrapers across the lumpy gray soil. They were building a master levee four feet high and twenty miles long, a rampart that would provide "complete control" if and when the floodwaters struck again. When finished, the big levee would enclose thirty-six square miles of the country's finest grain fields.

The forty-five-year-old Whitley had started out as pitchman for the Great Northern Railroad as it crossed the plains from Dakota to Oklahoma to Texas. Along the tracks that pried open the prairie, he platted town after town with names such as Chickasha, El Reno, Steele and Ellendale. He had come to Los Angeles on doctor's orders to mend his sickly lungs and sold jewelry to stake his first Southern California development, a project that turned 400 acres of truck farms into the first grand neighborhoods of Hollywood. Whitley Heights in the Hollywood Hills would become the domain of celebrities, a cluster of Mediterranean-style villas where Valentino practiced his swoons and Hearst first bedded Marion Davies, just down the block from Chaplin and Fields, Faulkner and De Mille. Years later, after his powerful Los Angeles syndicate had finished subdividing one of the biggest tracts of land in American history—47,500 acres at the south end of the San Fernando Valley—Whitley would wind up near broke and one of the villains of the legendary Chinatown water grab from Owens Valley. His embittered widow, in an ode to the man she called the Great Developer, would pronounce her husband a genius whose only failing was greed. "'Just one more million,' he kept saying."

If Whitley's vision for Corcoran—a Jeffersonian farm town mixing small- and medium-sized spreads with a thriving downtown—proved a little naive, he and his partners had come amid another supplanting of cultures in the San Joaquin Valley. Gone was the cattle baron. Except for herds owned by Miller and a handful of others, the steer had been relegated to the foothills on both sides of the valley. Gone was the bonanza wheat farm. Outside of the lake bottom, wheat growers had fallen victim to drought and their own failure to fertilize and rotate crops. As fast as it had sprouted from the plains, the busy little wheat town of Traver had vanished in a cloud of alkali dust, its once fertile topsoil poisoned by salts bubbling to the surface.

In their earnestness to sell the new promise of grape and orchard farming, a group of bankers and agricultural tradesmen paid a visit to an orchard outside Visalia to see for themselves whether the claims of an old man named Briggs were true. Farmers told their own fish stories, and Briggs seemed to be telling a whopper about a French plum tree that each year yielded 1,000 pounds of prunes. After surveying the field and corroborating the tree's curious fecundity, the bankers and tradesmen swore out an affidavit for any doubters to read. "The product of said French prune tree was gathered and weighed, and know of my own knowledge that all the statements made in said affidavit are true."

Eyeing a future of diverse crops, Whitley was counting on the levee construction to stop the four rivers from ever flowing into the lake again. Wheat farming had been a fine start, but its habit of concentrating money in the hands of few was no way to build a real town. The new steam-powered grain harvesters, marvels of industrialization, knocked down the harvest in record time but replaced even more workers. As long as the new levees could be made to hold, Whitley reasoned, farmers such as J. W. Guiberson, George Smith, Nis Hansen and Henry Cousins would have every reason to plant peaches, plums, apricots and other permanent crops. This kind of high-value agriculture would lure a more stable workforce to Kings County, and the men and women needed to prune and pick the crops would spread the farmer's payroll downtown—to Corcoran's merchants and home builders.

Whitley had plucked his directors from the very top of the L.A. business world and sold them on the idea of turning middle California into a fount of new villages surrounded by five- and ten-acre orchards and, farther out, forty- and eighty-acre grape, wheat and sugar beet fields. Whitley's Security Land and Loan Company board included the president of Homes Savings of Los Angeles, O. J. Wigdal, and the most feared and despised man in Southern California, General Harrison Gray Otis, owner of the *Los Angeles Times*.

Otis was a fourth-rate publisher and first-rate bully who used the columns of his disgraceful newspaper to spill bile and venom at organized labor and an infinite list of enemies, real and imagined. To his credit, Otis wasn't a poser hiding behind the moniker "General" or "Colonel" like so many transplanted Midwesterners and Southerners reinventing themselves in California. Otis had fought with distinction first in the Civil War and then in the Spanish-American War, and he wore his battle wounds with justifiable pride. Problem was, the general's war never ended. The front had simply moved from the Philippines to southern California. In his fight to keep Los Angeles free from the "corpse defacers" who ran the unions, he united all the city's merchants and business leaders under one anti-labor umbrella and took to the streets in a car mounted with a cannon. He referred to his elegant house as the bivouac and his *Los Angeles Times* headquarters as the fortress. The newspaper staff became the phalanx. Otis braced for battle, and it came when union leaders bombed the *Times* printing plant, an explosion that rocked Los Angeles and killed twenty employees. The general, it so happened, was off checking his land holdings in Mexico and rushed home eager to get even. "O you anarchic scum, you cowardly murderers, you leeches upon honest labor, you midnight assassins," he editorialized.

In Corcoran, Otis's "indefatigable and fruitful labors" so impressed Whitley that he resolved to change the town's official name to honor the general. At a July 1905 meeting, the Security Land directors unanimously agreed that "Corcoran"—originally named for the local railroad superintendent—tied the tongue and struck the ear as harsh. From now on, the town would be known as "Otis." Whether the general thundered his disapproval or the idea died for another reason, the change never took place. Only a main entrance into town, a boulevard lined with tall palm trees, would carry the Otis name into Corcoran's future. Otis would be long dead when his granddaughter, Ruth Chandler, would marry the man who would change Corcoran again, Lieutenant Colonel J. G. Boswell.

Whitley's land company proceeded to purchase 30,000 acres along both sides of the Santa Fe and lay out everything in precise detail. The plots would sell for $110 to $130 an acre and the company would offer small rebates for planting stone fruit, grapes and alfalfa. The houses would be constructed of adobe or concrete with thick walls and wide verandas to make the brutal sun more tolerable. Everything was proceeding as planned when the flood of 1906 struck. The levees melted and the lake in all its wonderful plenitude returned, and the plans of Whitley had to be put on hold, at least until the work of dredging and erecting a better levee system could be completed.

That job was enthusiastically taken up by D. W. "Daddy" Lewis, a pistol-packing nurseryman and farmer who had turned 640 acres of lake bottom into an asparagus plantation and was aiming for more. Lewis may have been a half-pint dwarfed by his ample wife, but he commanded fear and respect as a visionary. People said he could see three decades into the future but cared not a lick for what stood right in front of him. Fence posts, 400-pound sows, fine cars, calves—he ran into all of them as he drove from machine shop to lake bottom. He was a man in a hurry trying to keep up with his monster floating dredges that were moving tons of water and earth to make levees. The beauty of the floating dredge was that Lewis didn't have to wait until the lake bottom dried out to begin his levee work. His operators would swing the long-armed boom to one side and open the wide mouth of the clamshell bucket and claw up enough muddy earth to form the base and sides of the earth wall.

Short of building a dam, erecting a levee was engineering at its grandest. Nothing—not the hundreds of skulls of an Indian burial ground or the iron mast of the steamer *Alcatraz* or the gigantic bone of an ancient mastodon—could stop the digging and trenching by the iron and wood leviathans. It went on for ten years, from 1907 to 1917, a massive torching of the tules followed by the building of nearly 100 miles of levees that stood seventeen feet tall and wide enough at the crown to permit the passing of vehicles.

No longer content with simply siphoning water from the Kings, Tule and Kaweah, farmers began altering the very bend of the rivers, turning meanders into lines as perfect as a draftsman's, straitjackets of banked earth. So faultless was this remaking that years later, *Ripley's Believe It or Not* would feature Tulare Lake as the "Square Lake" in its syndicated cartoon, right above the drawing of Enrique Cuda, the man who ate twenty-six hot dogs and drank forty-six bottles of pop at a picnic in Buenos Aires.

There was no small hubris in the choice of word to describe the lake's retooling: "reclamation." It literally put man ahead of nature in the chain of possession. Nature wasn't being "claimed" for the first time by man but "reclaimed." Man's right was a first right, and if nature had the audacity to take over for a time, man was now back, staking what had been his all along.

"They're digging long, wide canals and building huge dikes to hold back the surplus waters of the Kings River and at the same time permanently reclaim the lake bottom," the *Corcoran Journal* wrote in January 1915. "Within three years, possibly a shorter space of time, much of Tulare Lake will be no more, and the land will be added to the fertile soil of Kings County."

The next year, the floodwaters returned anew and cooled the notion that permanent orchards and Jeffersonian farms would have a role in

Corcoran's future. For now, it remained a place of big growers and even bigger gamblers who could ride out three years of disaster for one year of riches. Wildcat farmers, oilmen at heart. On the north side of Corcoran in 1914 stood the biggest haystack in the world, twenty-eight tons of dairy feed piled tall and broad, proof that whatever else was taking place in the rest of the valley, Tulare Lake remained the emperor of grain. One night the stack mysteriously caught fire and the townsfolk of Corcoran, absent a fire department, formed a bucket brigade that worked in vain to put out the flames. The fire burned for weeks, a huge cloud of smoke on the horizon, and when it finally exhausted itself, the ashes had been blown by the wind all over the surrounding land. The cause of the fire was never determined, and it marked the most important year of Corcoran's young life, the same year it was incorporated with a population of 400 hardy souls.

Back-to-back dry years once again got the farmers thinking that floods, if properly spaced out, were a good thing. Floods washed down the deadly salts in the soil and allowed the growers to bank water for future irrigation. And Corcoran still prospered in spite of the occasional washing out, perhaps not in the ideal of Whitley but enough that by 1922 county boosters published a fancy promotional booklet featuring a cover shot of an apricot orchard in full February bloom. "The Little Kingdom of Kings," as they called it, offered every opportunity for "success, happiness, contentment and health." Corcoran now boasted a modern cheese and butter factory, and the county's list of top agricultural commodities was led by dairy products, grain, dried fruits and alfalfa. Cotton hadn't even made the list.

That, however, was about to change.

PART TWO
SPRING

I see mom and dad with shoulders low

Both of 'em picking on a double row

They do it for a living because they must

That's life like it is in the Tulare dust

—Merle Haggard, "Tulare Dust"

Spring

"WHAT WOULD IT TAKE TO GET RID OF YOU GUYS? How much cash are we talking about?"

It was early April, time for a return trip to the lake bottom, and Jim Boswell wanted to know if we might change our minds about pursuing his story. On the phone or in person, he seemed to begin every conversation with a bribe of one kind or another. For many reasons—the tone of his voice, the wink of his eye, the jabbing way he presented it—we never took the offers to be serious, although he would later swear that they were. This offer, like the ones made before and after it, sounded more like a grumble than a solicitation. Two men, neither one on his payroll, were rooting around in his past, and he needed to hear himself object. After so many gentle protests, it made you wonder if it was a protest at all.

So we headed back to Corcoran, this time to meet with him and his son, Jim Jr., the company president, and to witness what people were calling a miracle of modern agriculture: the spring planting of 90,000 acres of cotton—140 square miles—in nine days.

We were coming back, as well, to piece together the narratives of the men and women who had once picked and chopped the cotton, including a group of black field hands who didn't exist in Boswell's memory. We were searching for artifacts, it seemed, for the migrants who had streamed into town every harvest, a ribbon of brown, white and black curled out along Highway 43 and Avenue 10. They were a half century gone, their labor replaced by mechanical pickers that swept the earth like giant anteaters, each snout doing the work of 320 hands. The harvest that once took four months now finished in a mere six weeks.

Driving up and over the mountain on a clear day, the well-worn traveler, even one who had crossed every corner of America thrice over, found his breath taken away at the scale of what unfurled beneath him.

An hour north of Los Angeles on the backside of Tejon Pass, the mountain suddenly opened and revealed a big-mouthed flatland of gold and green checkerboard 2,000 feet below. Here was one of the more stunning demarcations in the nation's landscape, a last leg of ridge that separated the sprawl of Southern California from the perfect farm fields of the Great Central Valley. Behind you beamed the lights of Hollywood; before you stretched the state's big middle, 400 miles long and 60 miles wide, 7 million acres of lush vineyards and orchards and stinking dairies from Bakersfield to Redding. This was the West's Mason-Dixon Line, veiled as it was by some of the worst smog in the country.

It was not by accident that this valley ranked as the most productive farm belt in history. No other place on earth brought together its length and flatness, a climate that presented almost no danger from frost, and a rain that fell only thirty days a year, usually arriving in such timid drops that it played no havoc with harvest or spade work. The 105-degree sun could be a fierce force but it bore down only when the peaches and grapes needed ripening, so that it took just three weeks of late summer to blister a Thompson grape into a Sun Maid raisin. Only then did it begin to rain.

The valley qualified as desert but it had to be the damnedest desert in America. It sat at the foot of an immense watershed, the Sierra Nevada range, and the mountains gave rise to not only the four rivers that emptied into Tulare Lake but a half dozen others that flowed straight across the desert to the sea. If the sky above the valley was a miser, the sky over the Sierra just an hour's drive away sometimes gave too much. Each spring brought a torrent of snowmelt—misplaced rain, the boomers called it. The trick was to hem in all that water, to keep it up on the mountain when the farmer didn't need it and summon it down in careful portions when he did. Back in the 1930s and '40s, Colonel Boswell and the other pioneer growers from the Confederate South leaned on the government they despised to do the right thing, to blunt the forces of drought and flood. They wanted something in the perfect middle. It took four decades for the federal and state agencies to erect an interlocking system of dams, reservoirs and pumping stations that seized every river flowing out of the Sierra. As one hardhat who worked on the historic project put it, "We moved the rain." The desert between Los Angeles and San Francisco was no more.

The valley that bloomed from those efforts could hardly be considered a thing of beauty, though. It bore only a faint physical resemblance to the rolling farms and red barns of the Midwest. A whole host of west side farmers lived not on their land in the agrarian ideal but forty-five minutes away in an exclusive suburb of northwest Fresno—zip code 93711—that received more federal crop subsidy checks than any other zip code in America. The valley did have pockets of bucolic splendor, especially on the east side where the vineyards and farmhouses sat in forty- and eighty-acre chunks at the foot of the Sierra. For the most part, however, the scale was king-size and industrial, and Highway 99 ran straight through its heart. Alongside the four-lane zipper of road rose a thin long strip of unclaimed junk land. It was a peek into California before the miracle of irrigation: giant tumbleweeds, black-eyed Susan, salt cedar brush and a dry hot loam that kicked up whenever the Southern Pacific blew down the tracks.

For two decades now, the valley had ranked as one of the fastest growing regions in the country, and more than a few demographers were predicting a doubling of the population—to 12 million—by the year 2040. Already, the land between Bakersfield and Sacramento was the king of fast food and big-box discount houses. When corporate McDonald's wanted to introduce a new flourish to its Egg McMuffin or test a new high-tech way for customers to pay, it did it here. The retail wars in Fresno left more K-Mart blood on the floor than in any other city, three of its stores shutting down in the same month. Civic leaders had talked for years about luring a high-end department store like Nordstrom, but the developers couldn't deliver. So many middling franchises were competing for the same space that in one mall parking lot there were actually two McDonald's: a McDonald's that stood on the corner and a McDonald's that hung its golden arches on the Wal-Mart next door. One had a drive-through and the other didn't. Otherwise, their fries came out the same.

Still, such was the sweep of the fields that all the strip malls and housing tracts hardly seemed to make a dent, if you took the wide view. County by county, just beyond the fake-tiled roofs, the farm still reigned. Town after town vied for bragging rights. Selma was the "Raisin Capital of the World," Patterson the "Apricot Capital of the World," Pixley the "Home of the Black-eyed Pea," Mendota the "Cantaloupe Heart" and Yuba City the "King of the Prune." No one said different. Each November, townsfolk in Yuba City celebrated the sticky dried fruit, pitted and otherwise, with festival and folly and a dance through the orchards by the Prune Princess in her purple satin tutu, singing this tune:

Prunes, when you're feelin' blue
Prunes, that's the thing to do
They give you lots of Vitamin C,
Making us happy, healthy so we eat
Prunes, when the day is through
Prunes, you just need a few
And if you want a prune without pits,
We have that, too!

The valley grew everything—250 crops in all, specialty to breadbasket—and in such staggering amounts that there was no such thing as a single harvest season but rather a constant pick and pack and shipment of fruits, vegetables and grains. Valley farmers pulled more milk from their cows than any other region in the country. The same with grapes, tomatoes, peaches, nectarines, plums, figs, almonds, walnuts, pistachios, pomegranates and olives. Farm sales in the valley weighed in at $14 billion a year—more than Texas, Iowa, Nebraska or Kansas.

California's farming miracle, lost as it was between Tinseltown and Silicon Valley, didn't always get its due. From the vantage of L.A. and San Francisco, the big middle belonged to hayseeds and clodhoppers, a truck driver's Shangri-La of chicken-fried steaks, tri-tip sandwiches, biscuits and gravy, drag racing, Merle Haggard and Buck Owens, Mormon temples, Jehovah's Witness halls and dust bowler holy roller churches. Who knew that high-tech agriculture employed one out of nine Californians or that the valley drew an ethnic mix so stunningly varied—more than eighty nationalities in all—that no labor union could ever hope to organize its fields.

When one group of peasants stopped picking and began growing their way into the landed class, another showed up to grab their lowly wage. The Armenians and their raisins, the Slavs and their table grapes, the Basques and their sheep, the Azores Portuguese and their milk cows. Swedes in Kingsburg, Filipinos in Delano, Punjabis in Caruthers and Assyrians in Turlock. The Dutch Reformed, German Mennonite, White

Russians, German Russians, Hmong. "Mexicans, Chinese, Japanese, dirty knees." They had congregated in such numbers that the older immigrants could go the rest of their lives in this farm belt and speak nothing but the mother tongue. No one culture, though, quite dominated like the culture that emerged from the wind and dust of Oklahoma more than six decades before. Every kid grew up speaking a little Okie. No matter what foreign nation had deposited them, it took but a generation for the valley's twang to set down on the tongue.

A FEW MONTHS BEFORE that second encounter with Boswell, as a kind of preparation and challenge, we had made our way to Gardner Street and Pickerel and an eighty-acre patch called Bootville that sat on the far side of Corcoran, all alone. This was where the black community, numbering 200 people at most, lived and worshiped in an ebb and flow that no longer had anything to do with the farm seasons. Planted in the alkali were shacks of tin and wood, battered trailers, stucco houses and three churches, two of them Baptist and one a storefront Pentecostal. It had the feel of rural Mississippi with decades of junk piled high in yards and fields, an intricate gathering of abandoned cars and buses, rusted washing machines, refrigerators, stoves, camper shells and piles of wood and tires. There were chicken coops with too few chickens and hog pens with no more hogs.

The old black farm hands, refugees from Oklahoma and Arkansas, Texas and Louisiana, had come west in the 1940s looking for a new South, a place where, in the words of one migrant, the cotton grew a little taller and the white folks were raised up a little nicer. From Corcoran to Pixley, across the thirty-mile-wide belly of Tulare Lake, their tiny hamlets sat at the edge of towns that long ago locked them out. The city cops didn't come here, and neither did the city sewer lines. There were no stoplights, no schools, no businesses except for a soda machine. Every third house had been lost to fire from the blowing embers of old wood stoves.

"Don't feel sorry for me," said Martha Williams, the eighty-six-year-old widow of an Arkansas sharecropper. "Yes, this is a shack but it's my shack. God gave it to me. I ain't got nobody coming to me saying, 'You owe me rent.' I sleep as long as I want to and get up when I'm ready, and when the beautiful wind gets to blowing, I can flap my wings when I want to flap them. I sleep easy at night, right here in my little rundown shack by the highway. It may not be your dream but it's mine. Now you can just turn around and leave us alone again."

It was a Sunday in February, the morning sky breaking clear after a night of hard rain. The Sierra, capped in snow and lit by a brilliant sun, looked to be only a field away. The air smelled sweet even past the dairies of Hanford. There were no crop dusters in the sky, no pesticide tankers blowing their way down the rows, no big rigs swishing their fannies up the highway. Without the buzz of agriculture, Corcoran seemed asleep, or maybe dead. Whitley Avenue, where thousands of workers used to converge each Saturday night, hardly uttered a sound.

The only bustle—and it was a small bustle—stirred outside the small stucco church where 11:00 A.M. services were about to begin. We had driven by the church any number of times and once or twice knocked on the door, but no one ever answered and we weren't sure if the building, like so many others in Corcoran, had been left abandoned. This time the door swung open and we stepped inside, gingerly. It was a one-room rectangle that smelled stale, and the plaster walls were crumbling and the windows were so dingy that the sun had a hard time coming through. We took seats in the back. The parishioners, like the apostles, numbered twelve, most of them old folks. Some stood and others sat and there was a general clearing of the throat. This was a signal for the choir of the First Baptist Church of Corcoran to sing an opening hymn. Six parishioners left their seats and took a place up front and began singing to the six who remained. The preacher, Reverend Raymond Scott, arrived twenty minutes late from Fresno with no collar and no belt and no apology. The choir was a little off-key, and the only voice to speak of was the booming voice of Jeremiah King, an old hulking cotton picker from Louisiana who fished crappies out of the scant remains of the lake. The choir was accompanied by a white man with longish hair and a pink shirt, who did his best to keep up on the gospel piano. His black wife, the program director, got up to announce the activities for February and she apologized for having none in this month, Black History month. We're just too small and too old, she explained, as if they didn't already understand the two things that defined them. Then she called up her mother, Hattie White, the oldest parishioner and one of Corcoran's longest living residents, a tiny, beautiful woman dressed stylishly in black and gold.

Hattie White *was* Black History month. She removed from her purse a piece of paper on which her fifteen-year-old granddaughter had written a poem, an ode to the black woman, and leaned in on her cane. Ain't nothing about the black *man*? Reverend Scott teased. He got a laugh, but not out of Mrs. White, and she began to read with gusto. The poem was strong and completely without artifice and it brought tears to the eyes, and that's how the whole hour went, a dozen people moving about the small room without show or glory, parishioners one moment and deacons the next, filling two and three roles with humor and humility. They had little news to announce, and most of it was about the sick and dying. Sister Patrick was doing about the same, it was reported. Sister Johnson was still in the hospital and so was Eddie Hyman. You might want to go see them if you get a chance. Next Sunday is Pastor's Day. We want everyone to come and invite a friend. Because there's so very few of us, we'll have to double up on what we bring. They passed around the collection plate and it came back with twelve one-dollar bills. Eventually it came time for Reverend Scott to preach, and you wondered how he would manage. He was coming off a long night working the graveyard shift at a supermarket in Fresno. When he wasn't wiping sleep from his eyes, he was downing another glass of water. He didn't seem the least bit prepared. Where did we stop off last week? Let's try Mark. How about Verses 1 through 12? Splendid day, isn't it? Weather is nice, good. The whole parish was coughing. His sermon was about the rigidity that infects some true believers. They know by heart the list of rights and wrongs that will get them to heaven or take them to hell, he said. They follow the rules but miss the whole point, the love of religion. Those of you who picked cotton, you know what it's like to have your hands turn to callus. I used to work the watermelon fields with a man who wore no shoes. He didn't feel anything because of the dead skin. Can you imagine someone whose heart is surrounded by so much dead stuff that they can't be moved by a homeless person or a person addicted to crack? Have we become so dead and so religious, so impenetrable in our rightness, that we would not welcome that addict if he took a seat in our church? Don't let dead skin build over your heart. His sermon was quiet and filled with humor and the one time he raised his voice, it wasn't about fire and brimstone, but about irony. The gospel is about feeling, about love and laughter, he said. If I couldn't laugh and have a good time, I wouldn't be a Christian. Don't forsake the laughter for rules. They sang a last hymn and prayed for the afflicted, and then they filed out with handshakes and pats on the back.

You walked away thinking it was the most uplifting church service you had ever witnessed, and yet wondering why they did it, why they continued to carry on in a church so far past its prime. The roof and ceiling were falling apart, and they were going to have to dig deep in their pockets to fix them. Why not just close the doors and join the St. Paul's Baptist Church a few minutes away in Bootville, where other longtime black families, once members of this church, had gone? Jeremiah King was asked what brought him here each week. The gospel, he said. God and glory and old friends. You can find it here as well as anywhere.

Reverend Scott stood outside under a big chinaberry tree and marveled at the stubbornness of the black tenant farmers who came west in the 1940s and 1950s to keep alive their rural souls. They were the great exception to America's great migration. Unlike millions of other blacks fleeing the South, liberation for them wasn't Detroit or Chicago or New York City. Liberation wasn't even Los Angeles or San Francisco, though some had worked in the dockyards and shipyards for a time. Liberation was the cotton fields of old Tulare Lake.

It didn't take long for the big machines of California cotton to idle them. They drove trucks, picked grapes, bucked hay, pitched watermelons and swamped onions. They had been mostly overlooked by the War on Poverty and the civil rights movement—just like the irrigation water flowing past their land to the big farms. Instead of protesting this neglect, the reverend said, they had accepted it with a sharecropper's shrug of the shoulders. Any anger they felt was turned inward and this led to a split that became part of the landscape, two Baptist churches standing on opposite sides of the tracks.

The bad blood started with Reverend Huff, a ladies' man who took over the First Baptist Church in its heyday. The good mothers of the parish, half of them smitten, voted him a salary of $800 a month. It was an ungodly amount. All these years later, Reverend Scott was making only $200 a month.

"There were two black communities even back then. The men who ran this community were the Stigers, the Hatleys and the Patricks," Reverend Scott said. "They were the kings and studs. They made the most money, they made the most babies and they were the most violent. They didn't fancy to preachers getting that kind of attention or money. They came in from the fields one day and confronted Reverend Huff. 'We're not going to give you anything but the boot.' They pulled guns on each other, had a showdown right here in the church. Huff wouldn't budge. He padlocked the door and said, 'I'm going to be pastor here until I'm paid everything I'm owed.' Huff was every bit as tough as those guys.

"The softer men, the Toneys and the Whitfields, wanted a church of their own so they didn't have to come to this side of town. Because right here was a bunch of juke joints and clubs. Willie Montgomery, he was one of the godfathers. Prostitution and illegal gambling. And Sheriff Ely would take his cut. If he didn't get his cut, you got shut down."

In the 1960s, a bar on the white end of town refused to serve blacks. It wasn't an issue, Reverend Scott said, until the owner turned away Curtis Stiger. "He and his brother-in-law and cousins just went out and shot up the place. . . . They had a different way of doing business, a different way than other black folks. So the split in the church was inevitable."

The mechanical cotton picker was initially cheered by many blacks because it meant no more backbreaking labor. But nothing much came along to replace the fieldwork. Most of the children had to leave home to find jobs in Fresno, Los Angeles and the Bay Area. Working for the Salyers was never an option for black folks, not unless they were looking to be maids, Scott said. The Boswells were only a slight bit better.

When he first came to preach here in the 1980s, Scott was surprised to find the plantation South still alive. The same attitudes right down to the hymns sung in church, old slavery songs. "I tried to change that," he said. "I wanted it all gone. But it wasn't going to happen. I lost the battle in that regard. I lost some people. Some went to the other Baptist church. Some went home. You're going to find that this is a strange place. We haven't done what we needed to do as a black community. The whites? It's the same story. They don't see us. As far as the big growers and other whites are concerned, we're the ghosts of Tulare Lake."

IT WAS BARELY LIGHT OUTSIDE, but Jim Boswell was already groaning about the "enviros" in Washington who cared more about Delta smelts and kangaroo rats than about food and fiber. The old cowboy, whose favorite magazine was *Range*, a periodical devoted to

Western cattle grazing, was still chuckling over a bumper sticker glimpsed at the Tulare County Farm Show. It read "Earth First . . . We'll graze the rest of the planets later."

He had spent the night in Corcoran in the Boswell lanai, which was something more than a veranda but not quite a house. Across the kitchen table sat his only son, James Walter Boswell, and his right-hand man, Ross Hall. Years ago, Hall met Boswell's oldest daughter, Jody, on the ski slopes and fell in love. Their marriage ended after seven years, a breakup that found Boswell siding with his son-in-law. In his eyes, the irreconcilable differences came down to one: cigarettes. His daughter smoked and Hall didn't. Before he would consider having children, Hall insisted that she quit. "Hell no, I didn't hold it against him," Boswell would later say. "Marriages come apart over little tiny things just like that."

Instead of firing Hall, Boswell kept promoting him. The kid who grew up on a cattle ranch in the foothills and used to wander the Boswell feedlot with his father was now a company director who oversaw all processing, from cotton gin to oil mill. Hall returned the loyalty and then some. Whenever a conversation veered from, say, the tons of safflower oil crushed at the mill to something touching on the personal about Boswell, Hall would clam up. "You're not going to get me to say a negative thing about J. G. . . . I would die for the guy."

Jody, now in her fifties, was still something of a beautiful, if defiant, flower child. She lived in Lake Tahoe and didn't see her father much. He found himself drawn closer to his younger daughter, Lorraine, a cowgirl who resided in Idaho near her parents. She rode and boarded horses and seemed to live to please her father, still running to him with girlish glee to show off a badge she had won in a local skeet-shooting contest.

It was Boswell's relationship with his only son—whose silence and slouch that morning hardly fit the picture of a CEO—that confounded folks. For starters, nobody quite knew which one of them was running the show. Boswell was telling friends that he had come out of retirement only to help his son manage a few headaches that followed back-to-back flood years. But rumors persisted that Jim Sr. had shoved aside Jim Jr. at a bloody board meeting the previous summer, and things were still bad between them. The board itself, which included junior's buddies, resigned en masse. Boswell wasted no time filling the positions with Hall and other top employees, faithful soldiers who wore jeans and work boots and knew their way down every levee and across every ditch.

Anyone meeting the two Jim Boswells for the first time would strain to find the father in the son. Jim Jr. was taller but he had none of his dad's commanding presence. He had a big gut and a gap between his teeth, and he gave the impression of having been dragged to Corcoran that day. Only later did we learn that he had wanted the family story kept sealed in the boxes and file cabinets that gathered dust in the basement of their Pasadena headquarters. He couldn't understand how his father could ignore his own axiom about privacy. Boswell and his men had always shared a phrase to describe newspaper reporters whose friendly manner surely masked sinister intent. In a company memo twenty years earlier, Boswell water engineer Stanley Barnes had detailed for the boss a rare encounter with two "very personable" reporters from the *Los Angeles Times*. Barnes branded the reporters "practiced good guys" who inevitably showed their true hand when their meeting on irrigation broke up and they brazenly asked for Boswell's annual report. This was a document that the private company guarded closely, and Barnes felt burned. "I will be exceedingly surprised," he wrote, "if the product of their effort is anything more than an elaborate 'credible' castigation of the 'giant' J. G. Boswell Company."

Now the old man was offering bagels to two more practiced good guys and heading out the door to lead them on a tour of the last pristine part of the lake basin, and then into the cotton fields where Boswell was attempting to do something that no other grower had ever done.

"I'll tell you a little funny story," he said, his pickup digging into the dust. "I said to my dad many, many years ago: 'Now look, Dad. Run this by me again. You left Georgia in the 1920s. You came across the country and you passed up every beautiful river valley there was. You passed up the Mississippi, you passed up the Colorado, you passed up Imperial, you passed up the rest of the San Joaquin. And you ended up in a god-forsaken salt lake in a place called Corcoran? Now tell me that story again, Dad.' 'Shut up, you little bastard,' he'd say."

He wore the same Rockport shoes, plaid workshirt and beat-up chinos (he owned three pairs in varying states of fray), but this Boswell seemed different than the one we had encountered during the winter: more animated and surefooted and funny. He was full of old stories about the men who carved fields out of the lake, and when cussing wasn't enough punctuation, he pounded his hand with the chopped-off fingers on the steering wheel.

As we crossed the big levee, the one that Cockeye Salyer built in the 1950s with money the Boswell Company had loaned him, money intended for that year's cotton crop, Jim Boswell marveled at its insolence. Salyer had built the divide right where the Tule River met the Kings. Here, in the lowest spot in the lake bottom, where the greatest pressure concentrated, Salyer had erected a levee wider and longer than any other to displace all that pressure onto his neighbors.

"Now everybody knew Clarence for what he would do. He'd run his combines right over in your field and take a big fat cut of your wheat. So we heard Clarence was taking dirt and building a levee. I drove out and, lo and behold, it was big and it was 80 percent done.

"I said, 'You son of a bitch. You're not putting the money into the crop. You're building a levee to flood the rest of us.' He said, 'Well, I got her just about done.' He kind of cocked one eye at you and you never knew what he was going to do. I said, 'Clarence, I'm through with you. You've just fucked us long enough. We'll see you through this crop and then you can go someplace else for your loan.' And that was the end."

Only it wasn't the end, Boswell said, smiling. A few months later, after the flood of 1952 had come and gone, a man rowed six miles in a small boat to a spot near the base of the Salyer levee. Whoever it was carried a few sticks of dynamite and some short fuses. Cockeye had posted two guards on the levee and when they broke for lunch, the man in the boat made his move. "He rowed in close and was about to blow the levee to pieces when one of the guards came back," Boswell said. "He had forgotten his lunch bucket. When the guard saw what was happening, he sounded his alarm and the man in the boat rowed like crazy the other way. He got away, but the scheme was foiled."

"That mysterious man wouldn't happen to be you?"

Boswell ran the fingers of his good hand through the wisps of his gray hair. He had dodged the question for years and wasn't about to be pinned down now. "Cockeye later came up to me and said, 'You little prick. It was you in that boat.'" Boswell said he didn't hesitate. "I don't know what you're talking about, Clarence." More recently, Fred Salyer, Cockeye's son, approached him and asked the same question. "Before I

die, Jim, I just wanna know one thing. Was that you in that boat?" His reply didn't waver. "I don't know what you're talking about, Fred."

He and Fred Salyer had been pals growing up. They were seven or
eight when they nearly drowned while swimming in a water tank atop
the Salyer ranch house. Cockeye happened to hear their screams and
pulled them out like two opossums from a ditch. Years later, whenever
his son or Boswell caused him grief, Cockeye would mutter that if he
knew then what he knew now he'd never have climbed into that water
tank and saved them.

"Down through the years there's been nothing but cooperation between our two families," Boswell said. "I can't think of one instance
other than the levee building where we opposed the Salyers."

We waited for his punch line, but it never came. His face stayed
straight as he made a beeline toward the Kern River and the sand ridge.
Finally, we spoke up, reminding him that the fight over water and power
between the Boswells and Salyers had consumed a good part of a century, and one case had made it all the way to the U.S. Supreme Court.

"Aaah, that was just a pissing match between lawyers," he said. "You
have to understand something. When it floods, the best of friends protect themselves. So don't get too caught up in the water thing."

We had talked to the old Salyer attorney in Virginia, a bow tie and sus-
penders patrician named T. Keister Greer, and he remembered it differently. We tossed his words right in Boswell's face.

"Greer says the bad blood was considerable. And whenever there was
a crisis in the lake bottom, be it flood or another fight, you weren't there.
You were gutless, he says. Unlike the Salyers, you let your men lead the
charge."

"I could kick the shit out of him," Boswell snapped. "In fact, I went
around the desk at him one day and he wanted no part of me."

Boswell wasn't a big man, but his physicality—the way he hiked trails
and chopped wood and skied everyone's rear ends into the mountain—
was a big part of who he was. And yet so many of those feats had taken
place in the distant past. He was nearing eighty, and his eyes were
bloodshot and rheumy and talk of kicking ass, directed at a Virginia barrister who liked to brag about winning the 1938 state debating championship, seemed a little sad.

"Look, I've never pretended to be anything other than what I am,"

Boswell said. "I'm a great believer in giving people pride in what the hell they do. Let them feel that this is their baby.

"But I've never backed away from anything in my life. If there was a flood or a labor strike, I'm the first guy out there." Then he trotted out the old story about the '69 flood, just as he had done the last time: "I was standing on the levee with a shovel, the only guy left, and I wasn't gonna quit because I'd been taught never to quit on a flood."

Now Boswell steered the old Chevy toward the leftover waters from a new deluge, the floods of 1997–98. All that remained of the year-old snowmelt was a small blue lake whipped white by the wind. It sat bottled up on the backside of the basin, awaiting the call of irrigation. This is where the Kern River, at the farthest reaches of its run, deposits its silt and sand and loses its steam in all but the wettest years. "In the wet years," Boswell said, "the Kern is indiscriminate coming in down here." Before the dams, this stretch of the valley was a series of depressions dug out by the four rivers: Kern Lake, Buena Vista Lake, Goose Lake, and Tulare Lake. The diary of one old river rat tells of a spring snowmelt so vast that he was able to travel half the distance of California by navigating a boat from river to lake. Whether fish story or not, Boswell liked to tell people that he, too, had taken a canoe one summer with his high school buddies and skipped from one lake to the next, all the way to San Francisco. This spring, though, offered no such adventure. Unlike the previous two years, the snowmelt had been easy to contain. The Kern hadn't made it into the basin. Quiet, too, were the Tule and Kaweah and Kings.

Boswell took his foot off the pedal and gazed at the ducks, geese and gray-blue herons nesting at water's edge. "Look," he said in a tone of exoneration, "there's your birds, your ducks, your herons, your mud hens. Around this water and in every ditch, birds are nesting."

His earliest memories of his father, he said, weren't playing catch on the front lawn but hunting ducks in the lake bottom and fishing for trout in the Sierra rivers. He was a card-carrying member of Ducks Unlimited and a one-time director of the state board of the Nature Conservancy. Twenty years earlier, he had stunned the conservancy with an out-of-the-blue donation to purchase land along a famous fly-fishing creek in Idaho. The conservancy, by coincidence, had been eyeing Boswell's 3,300-acre Creighton Ranch on the other side of the basin. It was a rare piece of bygone San Joaquin Valley, a melding of grasslands, valley oak, desert scrub and wetlands. The conservancy needed only to let Boswell know of its interest and he turned over the whole ranch as a preserve. All told in the 1980s, he had given $1.5 million to the environmental group, making him one of its biggest contributors in the West.

In the next breath—and this was the paradox he would not concede—his farmland had become a killing field for the migratory birds and fish and the other wildlife he was so lovingly pointing out. A story in *Audubon* magazine had described it this way: "If there is a Hell for people who love wild things and wild places, it is California's Tulare Lake—at any time of year, but especially in late October, when the first ducks filter in from the north, stirring visions and even memories of how things used to be here." The irrigation runoff that Boswell and other farmers collected in giant evaporation ponds, a brew of selenium and other toxic salts, was killing thousands of black-necked stilts and American avocets still in the egg. From the vantage of migratory fowl looking for water, the killer blue of the evaporation ponds looked no less inviting than the pure blue of captured snowmelt.

"I don't deny that we get a lot of dead ducks, mud hens and so forth. But it isn't the selenium in the water," he said. "They're ingesting a seed from the wheat and safflower that hasn't germinated yet. It swells up inside them until they die."

This sounded like a version of what sly big brothers tell naive little brothers the first time they eat watermelon: don't swallow the black seeds or a plant will pop up from your belly. But a decade of environmental study by state and federal scientists blamed the deformities—twisted beaks, missing wings, missing eyes, missing heads—not on the seeds of grain but on the farmers' evaporation ponds, full of irrigation runoff.

It wasn't clear whether Boswell knew this and chose to reject the science or truly had been cushioned by the layers of agronomists and pest control advisers and farm managers who stood between him and what his land had become—the perfect deniability. This was a man who not only pleaded dumb to the tons of fish found dead in his canals, but lived in a world so much of his own making that he had somehow missed the news of the Columbine High School massacre two days before. This was a builder who saw no contradiction in killing a lake in Kings County while fashioning a fake lake, stocked with real fish, in his masterplanned community in San Diego County.

Now, each time we tried to press him on the environmental toll of his farming operation, he replied by pointing his finger out the window—at the wildlife before us. "Look at those pelicans. And those are mallards right there. There's another pair and another pair. It's mating season, you know."

We had driven the breadth of the most intensively farmed basin in the country, if not the world, hopped across a roadway and landed on some shore, some crazy shore, with salt cedar and sheets of moss hanging from cottonwood trees and purple tumbleweeds blowing as in the movies. We were bouncing on top of the sand ridge that for centuries separated the Kern River from Tulare Lake, a piece of beachfront that was too hostile for even Boswell to touch. He called it a "wasteland," but it was more out of respect than contempt. Maybe it wasn't California at its earliest or most pristine, maybe the bunches of foxtail weren't native and the wildflowers weren't the wildflowers that John Muir stepped on, but to glimpse what nature had designed and Boswell had engineered— side-by-side—was to understand, in the deepest way, the beautiful incongruity of the West.

On this side of the line, the land belonged to the jackrabbit and coyote and horned toad. The only farmer was the farmer behind the wheel of an old Chevy truck fast losing traction.

"That's sand, boys," Boswell said. "Pure sand. I've got it wide open."

Pedal to the floor, he tried to work up a head of speed but we were stuck at fifteen miles an hour, moving side to side as much as straight forward. With the engine revving and wind howling, it was hard to make out his words. Beneath his hat, he had this funny look that we took for genuine concern. Only later, after hearing all the stories of his physical triumphs over chunks of nature far more forbidding than a five-mile spit of sand dune, did we wonder if getting stuck was another of his practical jokes. Like his offers of cash or making us wake up at four in the morning to meet him at sunrise, he was having fun playing the city boys. Meanwhile, the beach beneath our spinning tires was growing deeper.

"What's the problem, Jim? Trucks cross the beach all the time, right?"

"Yeah, but this isn't the beach you guys know. We're not driving on wet sand. This is dry sand."

If need be, he said, we'd let air out of the tires to get better traction. As for driving us to the other end, the full tour, he was changing plans. "We're gonna have to find a spot to turn around and head back."

"Come on. The next thing you're gonna tell us is we've got a thirty-mile hike back to civilization."

"Lookit. I don't want to talk for five minutes. Let me drive and you guys stay dumb. Ignorance is bliss, isn't it?"

He sounded perturbed and one of us tried to crack a joke. "This is where he pulls out the shotgun and takes care of both of us and says, 'I don't know what happened to those two journalists.'"

"That's right," he said.

"They were supposed to be here at six in the morning but they never made it."

"I've done that to some Japs, but I probably wouldn't do it to you guys. More likely, we'd die of thirst walking back. A helluva story just the same."

"Yeah, 'J. G. Boswell, the biggest farmer in the world, missing for two days, was found dead on the last bit of natural habitat left in a lake bottom he spent a lifetime transforming.' Ten 'graphs down, it might mention two reporters found shriveled up alongside him. A sand dune that led to a pond that led to selenium poisoning."

He was too busy looking for a dirt shortcut he took years ago to crack a smile. "Aha, look there. That's the road. Hold onto your hats and I'll move us out of here."

Then, just as seamlessly, we crossed over from pure desert to the lake he had created out of cheap state water and last year's snowmelt. To go from dry sand to the wet rim of a faux lake in the bottom of what was once a real lake now dry, it was all pretty simple to Boswell. He had no patience for our talk of wonder.

"Well men, we claimed an old lake bed," he said matter-of-factly. "Now you've seen the Boswell flood control area. Now you can say that you had an environmental tour with the old fart."

On the road back to town, we drove through the fields where fifty Boswell workers had entered Day 9, the last day, in their race to sow 90,000 acres of cotton—surely a record if anyone was counting. A one-of-a-kind armada of eighteen-row planters rolled over the combed land depositing the hard, pointy seeds that, if all was right with moisture and heat, would elbow their way through the gray clay in five to seven days.

All those elbows turning their two leaves to the sun, that's what a cotton farmer called his stand.

A stand wasn't a stand if it had bald spots, a sign of sloppiness that brought a tsk-tsk from the neighbors. If a stand went thin, it usually happened at field's fringe where the soil ran too dry or too wet or the mechanical planter skipped a beat. That didn't happen in a Boswell field, or if it did they brought back the planter and did it all over again—one seed every inch, thirty inches between rows, 65,000 plants to the acre.

"They always say that you plant to the first knuckle," he said. "Deep enough to capture the moisture and heat zone and no deeper."

No plant was more unforgiving than the cotton plant. A farmer had to wait until the soil temperature was perfect, 56 degrees or a little higher, to sow the seed. A few days too soon or too late and the plant would never catch its potential. A small mistake in the first eight hours would rear its head 150 days later in a lower yield. One farmer compared it to raising kids. "If you beat 'em up when they're young, they won't amount to much when they're older." There was a formula to calculate this neglect. Each successive heat unit that a farmer failed to capture in the soil brought an ever diminishing return at the gin. Throwing away thermals, throwing away cotton.

Yet for all its temperamental qualities, the cotton plant was a survivor, having overcome the arid wilds of Africa and Arabia for thousands of years. The genus *Gossypium* had churned out so many different species that it took seven pages of taxonomy tables to list them all. There was an Old World gene pool and a New World gene pool and a lot of debate among botanists over where the twain had met. There were cottons native to Mexico, Peru, the Galapagos, and the Hawaiian Islands, the deserts of eastern Somalia, the outback of Australia and the Indus Valley of Pakistan. There were cottons that preferred wetter soils and cottons partial to drought and one cotton that deposited an extra layer of fat on its seeds so that ants might find it more alluring and transport it far and wide. What the biotech genies were doing now in their labs, the cotton plant had done quite on its own way back when. Its leaves actually expressed a poison known as gossypol, a first line of defense against the pink bollworm and other pests.

Killing a cotton plant was nearly impossible. It tolerated salt and thirst and grew into a tree if you let it. The white fiber, long before it fueled world commerce and civil war, was little more than armor for the seed. If nothing else, it guaranteed a soft landing. Then man got hold of it. No plant, it was said, had been more fussed over. Agricultural advances had transformed the tree into a stout bush, and the small hard seeds bearing a few pathetic fibers had been turned into fat bolls sprouting a thick fur coat.

Together, science and nature had teased more than 7,000 varieties out of those original seeds, and they covered 101 countries across a third of the globe. It hardly mattered that cotton wasn't native to the continental United States. What seeds the Hopi Indians of New Mexico hadn't snatched up from the Europeans had crossed the border in a pirate's pouch. One of the more famous imports of Acala cotton had made its way from Mexico City to the port of Mississippi in 1806, disguised as stuffing inside dolls. Finding varieties that thrived in the middle of California wasn't difficult.

Boswell pulled over into one field and hopped out, waving us to follow him down a freshly planted row. He bent over and started scratching at the bed until his finger unearthed a seed painted blue. What we thought was the ultimate expression of control—Boswell seed coated Boswell blue—turned out to be the hue of an anti-fungal dip. Eventually the seedlings would produce their own tannins that acted as an antibiotic, but it took a few weeks for their immune systems to kick in. In the meantime, the fungicidal bath would keep the seed out of pathogen's path. Already, Boswell said, the plant's DNA had found an answer to the first and most important question: Where am I? The answer was almost always the same in a Boswell field: You're in a good place. The seed coat had begun to absorb the moisture still trapped in the soil from a pre-irrigation weeks before; the signal to send down the taproot, which would grow up to four feet long, had already been made. By week's end, the brown earth would turn new green.

"Our whole premise, and it goes back to the Colonel, is the man, the land, the water. This isn't a place for small guys. It's a land for expertise."

We wondered what separated Boswell from other farmers, what decisions in the field distinguished his cotton from all the other cotton, so much so that he got a two- to three-cent premium per pound on the market. Boswell, though, begged off.

"I'll play any game you want to play, but you know, I'm not a farmer."

"Come on. In your heart and soul? You don't feel that spine tingle when the water gurgles down the field at night?"

"No. I got no interest in farming. I never had the time. I was too busy building, growing, perfecting, getting the right people in the right spots. Now if you want to talk farming, you need to talk to the Wild Ass of the Desert."

"Who?"

"The Wild Ass of the Desert. He's the guy who pulled this miracle off." Boswell took a look at his Seiko watch. He had a dinner date with an old friend who lived among the cattle on Boswell's Yokohl ranch on the east side of the valley. "It's getting close to my cocktail time. I'm going to drink a little Jack Daniels and eat some calf nuts with my cowboy. I'm done with you guys."

MARK GREWAL, a.k.a. The Wild Ass of the Desert, wasn't an easy man to contain. Six feet two, a black belt in karate, he had a gift of gab that made you wonder and a smile that made you want to believe. Everything he did was full-tilt. "I'm a wild man," he boasted. "I work hard and I play hard." His energy had so captivated Jim Boswell that a year earlier he fired his top field man and handed over the entire Corcoran operation to Grewal, where the swing between a good season and a bad season was $30 million.

Now the forty-three-year-old farm manager was talking about changing the culture of the lake bottom, moving away from the thick, puffy white Acala cotton that had dominated for seventy-five years and shifting to a longer, stronger and creamier variety called Pima. So few acres of Egyptian Pima were grown—it made up only 2 percent of U.S. cotton production—that you couldn't place a Pima bet on the futures market. What's more, Grewal was playing heretic, planting tomatoes, onions, garbanzo beans and lettuce on land that had known only fiber and grain.

"People say I'm crazy. 'Vegetables in the lake bottom?' But J. G. has given me the room to experiment. Wherever I've gone in this company, I've made money. In 1989, they sent me to the Homeland district, our toughest ground, and I said, 'Welcome to hell.' I not only made money, but I turned it into the highest-yielding ranch in the whole company. Three bales an acre. I slammed it. We call it 'God's country' today. I've never lost money in my twenty-one years with Boswell. Last year, with half this ranch under water, I still didn't lose money."

If he well deserved the "wild ass" part of the nickname that Boswell had given him, the "desert" part didn't exactly fit. Boswell loved poking fun at Grewal's camel-and-sand Bedouin roots. Trouble was, Grewal had no Arab in him. His father had come from one of the world's oldest cotton belts, a land fed by five rivers in the Indian state of Punjab. Tejinder Grewal was one of the first Sikhs to enroll at Fresno State, a foreign-exchange student who arrived in the summer of 1951 wearing a saffron turban wrapped around his uncut hair and an ornate sheath holding a two-foot long sword, the *kirpan,* which symbolized his religion's 500-year fight against injustice.

No landlord in Fresno was willing to rent him an apartment. He picked tomatoes and peaches to pay his tuition and got a job washing dishes at Mar's Drive-In. The elder Grewal would go on to become a half owner of the drive-in and earn an engineering degree from Fresno State and two master's degrees—one from UCLA and one from USC. He would help build nuclear power plants and go to work for the L.A. Department of Water and Power.

The younger Grewal, who grew up in Southern California but made the journey back to Fresno to study agriculture, had almost no ties to his father's past. Around his wrist he wore a *kara,* the bracelet signifying a Sikh's bondage to the truth but his freedom from every other entanglement. The bracelet and his love of rice, he said, were all that he retained of the Punjabi culture. He dressed each day in the same get-up: blue jeans, Wolverine boots, a golf shirt and a Callaway golf hat. Gone were his father's turban and sword and the special cotton undergarments that symbolized a Sikh's commitment to sexual purity and fidelity to family. Grewal had married and divorced and married again, and this second wife understood what his first wife had not—that she and the children came second.

"I flat-out told my wife when we married that my first priority is the J. G. Boswell Company. Tommy Lasorda used to say he bleeds Dodger blue. Well, I'm a Boswell man. I bleed 'Diamond B' blue. I leave home at six in the morning and I get home at six at night. I've already driven 224 miles today, and I've still got two hours left. This is my passion. This is a way of life, and it's my life and my love. I've been farming ever since I was twelve. It's like breathing. I can't remember ever wanting to do any-thing else."

The choice of what to keep and what to leave behind, Sikh or Ameri-can, had been made easy for him. Long ago, his father had shaved his beard and given himself a butch haircut and married a blond girl named Adams from the little farm town of Exeter. His father's break with tradition could not have been more bold. Even today, a half century later, only one out of ten Sikhs in the valley married outside the tribe.

It was Grewal's maternal grandfather, a peach and orange farmer, who taught him that irrigation was art. A farmer could only do so much with too little water. But too much water, that's when a farmer got stu-pid and greedy. A good farmer was stingy when it came to irrigation. A good farmer knew a plant required a little stress to push out more crop. The art on the front end was in trying to make a plant think that it was dying of thirst. You actually fooled it into believing that it needed to hurry up and churn out its little babies before it was too late. The art on the back end was in not being too cruel and knowing when the flowers had set into fruit—before ever seeing the little buds. That's when it was time to come back in with the water.

"I grew up in Sylmar but I spent every summer in the valley with my grandfather. He had only twenty acres, but he was a great farmer. That's where I learned that stress could be magic. In cotton, there's a fine line between good stress and bad stress. How much stress is needed to get the buds? How much is too much? I've grown every crop and none com-pares with cotton. Cotton is a mother. If you don't watch out, it'll break you."

Grewal sat in his Chevy Tahoe on the levee bank and waited for a crew to pass. Like Boswell, he yielded to the workers in the field. It was one thing for the hourly guy to take an order from the boss and another thing to eat the boss's dust. He flashed the thumb's up to let them know he appreciated their good work in a week that began with the planting of 7,000 acres on Easter Sunday.

Covering 90,000 acres in nine days wasn't the usual drill. The arrival of spring had been delayed by a brutal March. The wet and cold had put Grewal two weeks behind the normal planting schedule. He knew the longer he waited, the more he shoved the harvest into late fall—into the rainy season. So as the heat units began to build in early April, Grewal decided to make up for lost time. Because his men couldn't plant in the

dark, he devised a plan to utilize every moment of light. They would begin at 6:00 A.M. and end at 7:00 P.M., dividing up the crews so that there
was always a rover to cover for lunch or breaks. If a tractor had to be
serviced, it would be serviced at night, when the repairmen were no
longer busy with planting. It helped to have $20 million in spare equip-
ment sitting in a yard in the fields.

The success of the planting was nothing more than the execution of a
simple equation, Grewal said. He had reduced the whole of farming, if
not life, to the formula of E + R = O. The E—*environment*—had put him
two weeks behind. There wasn't a thing he could do about that. But he
knew the *outcome*—the O—he wanted. He wanted to take advantage of
the heat units the moment they built up. But because this wasn't a 200-
acre cotton farm, he knew he had to capture the average optimum temperature for all 90,000 acres. The R was the way he chose to *react* to the
challenge that the environment had placed in front of him. The R was
the one thing he could control—the one thing that could turn bad luck
into good—and it's what he spent weeks designing.

"To me, it was like a battle plan. Like our own little war. Everyone
knew his mission. All my foremen knew they had to put in sixteen-hour
days. The rovers had to be staggered. They literally ran from truck to
truck, tractor to tractor, relieving this guy and then that guy. No matter
what came at us, everyone was determined to get the same outcome.
We slammed it."

Manning the five channels of his two-way radio, guiding the movements of his five district managers, Grewal was more choreographer
than dirt farmer. The month of April was one fat dance. In the field to his
right, a tractor pulling eighteen rows of airplane tires put a final seal on
the cotton beds. Why airplane tires? Because they were plump and
round, and kissed the ground with just the right force. Too much pack
might choke off germination. Too little pack might let in air and dry out
the moisture. In the fields to his left, workers were irrigating the hard red
wheat they had planted the previous fall, right after the cotton harvest.
No land ever went idle and no time seemed to pass between crops. As a
Boswell worker plowed under the stubble of one crop, another Boswell
worker ripped, harrowed and smoothed out the earth for the next crop.
In the field behind Grewal, almost in the same movement that his men
planted the cotton, they were sowing the seeds of yellow and red onions,
lettuce and tomatoes. As soon as a row was planted, another driver

came through with a side dressing of fertilizer and herbicide. Some of the planters were calibrated to drop a tomato seed every fifteen inches and others were calibrated at a two-inch space. More plants squeezed tighter meant fewer tomatoes per plant, Grewal said. But how many fewer? This was one of the experiments they were conducting. If you quadrupled the number of plants and only suffered a one half drop in output, it still meant twice as many tomatoes rolling off the field, twice as many tomatoes going to the paste plant and coming out sauce for Domino's Pizza. This was the laboratory where Mark Grewal and his five district managers, one pitted against the other, played for keeps.

"I've got five studs working under me, five guys capable of kicking my ass. Each one of them farms 25,000 acres and each one of them, every year, sets a goal to beat the other guy. Who can get the most yield for the least cost? That's our game, and nobody in the world plays it better."

White Gold, Black Faces

THE FOLKS INSIDE THE BIG TENT on the outskirts of Corcoran were hooting it up when another minstrel stepped forward from the semicircle of performers and doffed his top hat. "A Georgia Negro spent several years as a servant in California," he began, his deep voice washing over the dusty showgrounds. "One day, he returned to his home and attempted to instruct the members of his family in correct usage, especially in their language. Sitting at the table, his brother said to him: 'Gimme some 'lasses, Sam?'

"'You mustn't say 'lasses,' corrected Sam. 'You must say molasses.'

"'What is you talking about?' grunted his brother. 'How's I gwine to say mo' 'lasses when I ain't had none yit?'"

The audience burst out laughing again. Before they could settle down, one of the end men—those positioned at the far edges of the arc of entertainers—started up with a new round of patter.

> Bones: "I knew a fellow who had over 50,000 men under him."
> The interlocutor: "He must have been a great general."
> Bones: "No. He was in a balloon."

More laughter. With that, Bones lifted two sets of ivory sticks and struck them together, creating a clackety-clack-clack like castanets. Tambo, the other end man, shook his tambourine, and the twang of a banjo pierced the air.

> Interlocutor: "Who were the first gamblers?"
> Bones: "Adam and Eve."
> Interlocutor: "How so?"
> Bones: "Didn't they shake a paradise?"

On and on it went through this feel-good night, October 2, 1929, as the all-black troupe of Georgia Minstrels—"40 fun makers, band and orchestra"—played to a rollicking crowd sprawled over 2,000 seats on the western edge of town. For many of those in attendance, mostly whites with some blacks scattered around the back of the tent, the show's lambent material evoked a nostalgia for the South, albeit a fairy-tale South where life on the plantation was romanticized as a happy time for all.

Even those who had no link to that part of America could surely see the logic of bringing a minstrel show to Corcoran, so thick had the town become with the culture of cotton. Earlier in the year, the Dixie Jubilee Quartet had presented the "cotton field favorites of the Sunny South" at Corcoran's Harvester Theatre. As one promotion put it, "What would you rather hear than a good colored male quartet harmonizing on the Southern plantation melodies?" And a few weeks from this night under the canvas tent, the Corcoran Thursday Club, a social group, would put on a play at the town high school: *A Southern Cinderella.*

Minstrel acts had, in fact, been popular in Kings County since the early part of the 1920s, though these productions typically featured white casts with their faces blackened by burnt cork. And their depictions of African Americans were more cruel caricature than playful parody. In the spring of 1922, a year and a half before Colonel Boswell had come up to Tulare Lake, the Corcoran fire department staged a minstrel show of its own, the ensemble straddling the old South and the new West. In the middle of the program, the Darktown Quartet sang a series of "Southern Melodies," while the whole company closed with a rendition of "My Heart's in California."

The night the Georgia Minstrels frolicked, there was much to be merry about in Corcoran—and far beyond. The era of "Coolidge prosperity," stoked by a frenzy of consumer installment buying, had placed all sorts of new luxuries—fancy duds, appliances, cars—into American hands. But there were signs suggesting that bleaker days lay just around the bend. In the summer of 1928, the high-flying stock market suddenly swooned, giving weight to those who warned that rampant speculation on Wall Street presaged total economic failure. Yet by late in the year, the market had rallied again, and, at least for the moment, the nation continued to find itself drunk with "unparalleled plenty," as journalist Frederick Lewis Allen termed it. "The prosperity bandwagon rolled down Main Street."

In Corcoran, that main street was Whitley Avenue, where clothier Walter H. Wright sold $70 tailored suits and Cross Hardware vended the latest Frigidaire ("The QUIET Automatic Refrigerator"). Over at Ahern Motor Company, smooth-talkers pushed the new Model A Ford. And in a few weeks, a brand new J.C. Penney would open as well, store no. 1,274 for the chain.

At the Hotel Corcoran, packed with businessmen from as far away as New York, Seattle and Santa Fe, proprietor William Cromlie sneered at Prohibition and surreptitiously served up a stiff "nip on the hip" for thirty-five cents. For those so inclined, vices beyond bootleg liquor were also readily available: Charlie Quong ran blackjack and poker games out of the back of the Eastside Pool Hall, and Mrs. Richmond's whorehouse turned away no man. Those with more genteel tastes, meanwhile, could head to the Harvester to see the "talkies" that had just come to town. Or they could pad along newly paved sidewalks to the baseball park, where a dozen lights—thousand-watt lamps attached to thirty-foot poles—illuminated the night sky of Kings County, luring hundreds of spectators to after-dinner games. During football season, fans of the local high school team were treated to the play of quarterback Ernie Caddell, who later starred for Stanford and the Detroit Lions, and Byron Gentry, who went on to fame with Southern Cal and the Pittsburgh Steelers. "Our line," Gentry would later recall, "averaged 130 pounds; our rooting section, 190, including the girls." A surge in home building and commercial construction was also under way. Uneasy about all the bustle, the *Corcoran Journal* editorialized about the need for smart growth: "It cannot pay any community in the long run to over boom its resources."

Outside the San Joaquin Valley, most of rural America was beset by a very different problem. Incomes on the nation's farms—about a third of which grew cotton—had skidded during the early part of the decade, and by the late 1920s growers in many parts of the country faced weakening export markets and stiffer domestic and international competition. The great bull market on Wall Street "concealed the fact that farmers were losing their farms by hundreds of thousands," wrote Henry A. Wallace, the Iowan who would become Franklin Roosevelt's agriculture secretary and later his vice president. For growers in the main of the Cotton Belt, forced to brave a drying up of European demand and the tireless assault of the boll weevil, a glum feeling had set in.

Around Corcoran, the mood was the opposite, the atmosphere go-go-go. About the only challenge—and it was no small challenge—was drought. Tulare Lake had been dry through most of the 1920s, and the canals feeding the farms had become choked with weeds and willows. "We are vitally interested and in hearty approval of any means for increasing the available water supply for the district," Colonel Boswell told the residents of Corcoran in the summer of 1928. The dry conditions were so severe at one point that local farmers scraped together $4,000 and hired "Hatfield the rainmaker." Charles Mallory Hatfield, who had first experimented with rainmaking on his father's San Diego farm around the turn of the century, was hired all over the state in the 1920s to raise the level of municipal reservoirs. Hatfield was humble about his handiwork, explaining that he didn't really "make" rain. "I merely assist nature," he said. "I only persuade the moisture to come down." In Kings County, he seeded the clouds with secret chemicals from high atop a platform that he had erected in the parched lake bottom, and some would later claim that a subsequent downpour was the result of Hatfield's intervention. But it was more conventional remedies—drilling ever deeper wells to tap the groundwater table and importing and impounding water—that really made the difference.

Through hard work and resourcefulness, the growers of Kings County continued to irrigate through the dry years, fueling the unremitting development of farmland. In 1919 Kings County had 187,868 acres under irrigation; by 1929, that had jumped to 267,350 acres. Besides, while the drought was difficult to cope with, at least Tulare Lake wasn't in danger of flooding. One valley resident, remembering what the area looked like when it had been submerged in 1909, couldn't believe the progress twenty years later. "Had anyone been foolhardy enough to have suggested this," he said, "he would have been sat on by the insanity board."

Others had caught oil fever. In October 1928, a gusher blew across Tulare Lake, and for twenty days a yellowish-green fog of gas vapor enveloped the area. What amounted to pure gasoline spewed from the earth. One news account predicted, a bit hyperbolically it turned out, that several towns near Corcoran would see oil riches as fabulous as anything in Oklahoma or Texas. This "will make a second Tulsa of Hanford," the report said.

Yet it was cotton, above all, that had Corcoran agog. "King Cotton Now Holds Stage Center," the *Journal* crowed in August 1928. By late November, with five gins having already turned out 17,500 bales, the paper announced: "Revise the Corcoran cotton yield upward again!" Such a steady boost in output can often be a prescription for oversupply and, as a consequence, falling prices. But this time, there wasn't the usual comeuppance. With unprecedented numbers of boll weevils appearing in the South and a drought wilting the Texas cotton fields, irrigated and weevil-free California was sitting pretty.

Not everyone shared equally in the good times. The men and women of the cotton fields endured an often merciless existence. They did backbreaking work for little pay, and sometimes they died in the most violent way. Jasper Palacio was found murdered at a Boswell ranch, his drink laced with strychnine. An unidentified black man, his skull crushed by an iron bar, was discovered floating in an irrigation ditch, "the fourth victim of death in a lewd manner," according to the *Journal*.

As the big wheels of Corcoran watched the Georgia Minstrels sing and dance and jest, they couldn't help feeling jubilant: The current cotton crop looked even better than the previous year's, and two more gins had opened to handle the harvest. But it wasn't only quantity that had Corcoran farmers so upbeat. They found that their high-quality bales commanded a premium of 3–4 percent over the rest of the market. Guilty of less exaggeration than usual, officials at the Corcoran Chamber of Commerce started referring to the town as "the center of the cotton industry of the San Joaquin Valley."

It was an environment well suited to big dreamers. Clarence Salyer, the Virginia hillbilly whose empire would one day rival the Boswell's, was by 1929 building up his acreage and pocketing a little extra cash on the side. "He'd haul cottonseed into Los Angeles," said Don Keith, who ran a Corcoran trucking company, "and he'd dig down to the bottom and put about 10, 15, 20 gallons of moonshine in there. . . . Now they can all say what they want to, but Clarence was real smart."

In a couple of years, a trailer loaded with cottonseed would get the best of Salyer. He was attempting to unhook the trailer from the back of a stalled truck, when someone smashed into him from behind, leaving his face a bloody wreck, his left eye barely hanging on. This was the eye that would forever wander, the iris lost inside the socket. Myth making—much of it from Clarence's own foul mouth—ensued. Salyer insisted to one friend that the battered eye was the result of being trampled by horses when he was a child. Another big grower swore that he had seen the eye poked out during one of Salyer's barroom brawls. The story that Cockeye told his daughter-in-law was closer to the truth. She had come to Corcoran for the first time, fresh off the train from Chicago, when Clarence got in her face. "I guess you want to know what happened to my eye?" he asked, a bit menacingly. A polite woman, she said that she hadn't noticed anything amiss. "Well," said Clarence, "let me tell you how it happened. I ran into a tit!" It took her years to learn that he meant a "teat," as in a metal protuberance.

Outdoing Salyer's eccentricities wasn't easy, but Elmer von Glahn sure tried. He was born in San Joaquin County, 150 miles north of Tulare Lake, and came to Corcoran in 1915, a young man intent on becoming a mega-farmer. At the time of the minstrel show, though, he was down on his luck. He had lost a fortune during a grain market collapse in the early 1920s and tried to rebound by launching a toy airplane business. The venture nose-dived and now he was working for Boswell as a machinist. Before long, however, von Glahn would set out on his own and make it big—55,000 acres in all. In later years, as *Newsweek* magazine reported on von Glahn's "harvest from the lake," he would give his ranch foremen Cadillacs to drive; townies called them "von Glahn pickups." He also became the first in the lake bottom to inspect his land by airplane, winning the sobriquet "The Flying Farmer."

A jolly sort, von Glahn would carry an accordion into his wheat fields and play. On Sundays, right in the middle of church services, he'd take a seat at the organ and break into a rousing rendition of "Roll out the Barrel," his sausage-fat fingers thumping on the keys. Pressed into the duties of Sunday school teacher, he'd mesmerize his students—little Jimmy Boswell among them—not with stories of Jesus but with tales of striking gold in Mexico's mother lode. In 1940, von Glahn built a new warehouse and office, a mahogany bar as its centerpiece, and he and his friends would go there and get loaded. Sixty years later, folks would still talk about the party he threw for more than 3,000 people from across the valley, who came to hear Jimmy Grier and his orchestra do their thing.

The halcyon times would come crashing down for von Glahn. In the 1950s, a couple of con artists convinced him to sink his money into a dry oil well, and he never recovered. He died penniless in 1972.

It was hard to say if rascals such as Sandy Crocket were simply drawn to boomtowns like Corcoran or if the rush of a boomtown brought out the devil in any man. Crocket came from Canada in the mid-1920s with all of $26 to his name, took one look at the boundless stretch of level land that Tulare Lake offered and muttered to himself: "Crocket, you're going to make a lot of money here. You roll up your sleeves and work hard for 10 years, and you'll make enough so you can retire." He wasn't far off. A few weeks before the minstrel show, Crocket bought one of his first sections of land, and within twenty years, he and a partner would control more than 80,000 acres—then the largest farming operation in the valley.

Like many of Kings County's giant growers, Crocket was a strange mix. A strapping man, he could knock you on your ass one day and then play fancy-fancy with the high-society crowd the next. In his unpublished memoirs, written in his nineties, Crocket would devote a good number of pages to the punishment his fists gave out and, at the same time, relate how he had advised President Eisenhower on agricultural policy, golfed with Bing Crosby and toured Europe in a Rolls-Royce.

And then there was Lieutenant Colonel J. G. Boswell. By the fall of '29—and long thereafter—no name in the lake bottom was bigger.

NOBODY TODAY KNOWS exactly what hotelier Harold Doulton told J. G. Boswell during their fireside chat at the Miramar back in the summer of 1923. It's safe to say, though, that the Colonel wasn't disappointed once he traveled to Tulare Lake and saw the 1,000 or so acres that had been sown to cotton that season. Turning up in Corcoran, the Colonel found the autumn fields thick—better than a bale to the acre, even before they'd been picked over a second time—with some of the best-looking lint he had ever laid eyes on.

In October, he purchased two train-car loads of cotton from the Doultons' Los Posos Ranch and shipped them both to France. By the end of November, having bought up most of the cotton Corcoran had to offer, an epiphany seems to have struck him: Trooping back and forth to Yuma may have been fine for a tyro. But by anchoring himself here in Kings County—in what he instantly concluded was "the richest land next to the Nile Valley in the world"—he had the chance to become a titan. To get there, the Colonel decided he needed to go beyond the business of merchandising cotton. In the near future, he would also gin the crop and mill its seeds. And he'd grow cotton, too.

The Colonel's timing was impeccable. The price of cotton in the Imperial and Yuma areas was about to fall into a tailspin; an invasion of insects would soon send the industry along the Mexican border practically into oblivion. In Gadsden, where the Colonel had transacted business with J. B. Long, the town's cotton gins would lie deserted in just a few short years, their tin cavities turned into homes by destitute Cocopah Indian families.

Cotton had blasted onto the scene in Corcoran in 1919 when the local farm bureau began to encourage the planting of it. By the following year, a number of farmers—some of them reportedly from "the best cotton states of the South"—were piling into town to take part. Though some skeptics viewed cotton cultivation in the lake bottom as a fool's mission and others worried about the ill effects of migratory labor, a growers association was formed, and by late 1920 a gin was up and running. Over the next couple of years, ever more cotton was planted in the area. Now, the arrival of Boswell sealed it: Cotton was destined to be as much a part of the landscape as wheat or floodwater.

The only thing missing was the Colonel's brother Bill. The Colonel himself knew almost nothing about farming, and so he again called on his younger sibling to leave Georgia and come to California to help him. "You'll make a killing," the Colonel assured him. After his experience in the Arizona desert, Bill was a little leery, but he agreed to give the West another try. That very first year in Corcoran, Bill invested in 800 acres just south of town, and he and the Colonel turned a tidy $36,000 profit on the land. The brothers, said the *Corcoran Journal*, "know the 'game' from every angle."

Bill had come out to California alone, leaving his wife, Kate, and their four small children in Greensboro. The success of that first harvest in Corcoran meant that Bill wasn't going back home, and his family would have to be uprooted. The South that had forged the destiny of five generations of Boswells would now give way to the West. But Kate was reluctant to move, and she did her best to forestall the transition. Back in Georgia, she remained a fixture on the Greene County society page, throwing parties and lunching on finger food with the ladies of the Entre Nous Club. To her, faraway California seemed like "the jumping-off place of the world." She had been born of the South, and a daughter of the South she would always be.

"I have blue blood," Kate explained years later in a remarkable account of a life that bridged the memories of her slave-owning grandmother and the California cotton culture built by her son Jim one hundred years later. "I'm very proud of my ancestry, my tradition, my heritage." She recalled sitting on the floor at the feet of her grandmother, listening to her talk about the old days, how the slaves cried when they were freed and they "didn't want to leave their white folks—Master So-and-So."

Kate's father, George Hall, had died when she was young. Of her scattershot memories of him, the one that stood out was how "all the old Nigras would come to his bedroom window and show him samples of cotton to grade for them." His passing left it to Kate's mom, Eudora, to raise her right, and in the South that meant teaching her how to sew a fine seam. The lesson ended, as it always did in rural Georgia, with the making of a whole quilt. Every girl, as a rite of passage, had to sew one and then stash it away in her hope chest. "My mother gave me up as hopeless," Kate said. "I'm afraid I can't sew." What she could do was play basketball and she loved to jump center.

Back then, Greene County did its best to keep a safe distance between the boys and girls. "We had picnics, particularly church picnics. But properly supervised at all times. We never went anyplace without a chaperone."

For his part, Bill Boswell wasn't the kind of boy who had much use for chaperones. Before Kate, he began seeing a girl in Greensboro, a wild girl named Marie Wright. She came from a rich family, and she had it in mind to marry him. "She had her own little buggy and horse and snappy little outfit," Kate recalled. "But that didn't work. In those days, if a boy kinda strayed away from home and didn't go right down the straight-and-narrow path, they would all say, 'We're going to send him to Texas.' . . . They thought that maybe Bill would have to go to Texas."

So Bill's mom and Kate's mom conspired to get their kids together. "It was a little undercover work going on, I think, between the two mothers," Kate said, "'cause they were raised in the church together and the Lady's Aid and the Missionary Society."

One day, when Kate was a senior in high school, Bill finally came courting in his horse-drawn buggy. He picked her up and took her to his house to meet his family. On the way there, though, one of the ashes from the cigarettes that he chain-smoked landed on Kate's skirt. "I didn't know it was burning 'til I felt it," she remembered, "and I had to go there in front of all the Boswells—poor little thing, scared half to death anyway—with that big hole in my skirt." Bill's mom, Miss Minnie, saw the hole, fetched a safety pin and pinned it over. "It's fine," she said. "Nobody's going to see it." That was Kate Hall's first experience with the Boswells.

Bill reminded Kate that she was no Marie Wright, at least in the looks department. She had gotten downright plump from all the boys sending her boxes of candy. "Bill said the first time he saw me he couldn't tell whether I was going or coming, I was so fat. And I said, 'Well, why'd you ever fall in love with me if I looked so funny?' He said, 'I can't tell you, but I did.'"

They caught the train to Atlanta and honeymooned at the old Ainsley Hotel for a few days. Then they returned to Greene County, settling in at Boswell's Crossing, in the bungalow that Bill had just built. That first year of marriage, Kate didn't know a thing about keeping house. "I was just a scared little brat," she said. "I didn't know what was going on. But I settled down. Didn't take long. I had those wonderful servants, and they were so sweet and good to me. . . . I can remember being terribly frightened. I wanted to do the right thing. And everything was 'Miss Kate' this and 'Miss Kate' that. But I got by all right 'cause they were so wonderful to me."

If Kate was a little unsure of herself, Bill's mom was more than happy to barge in and assume control. Soon after their wedding in February, Miss Minnie came calling. "It's about time to start kindly making blackberry jelly," she said. Kate didn't know where to begin. "Well, I never made jelly in my life," she said. "And Miss Minnie came out, and every little Negro on the plantation had to go pick blackberries. . . . She wanted to boss the Negroes. She just had a field day. She loved it."

Kate got pregnant—"'fore I could bat an eye." After Billy Jr. was born, she was pretty much tied down to the house and making things pleasant for Bill. "I ate by lamplight every morning with him. The Negroes fed the other Negro workers from the kitchen, and then Bill would go to the field with them. Come back for lunch. We called it dinner. We had breakfast, dinner and supper."

When Billy Jr. was ready to start school, Kate and Bill decided to move from Boswell's Crossing into Greensboro. There they had their second child, Kathryn, and then a third, Josephine, whom everybody called Jody. And then horror struck: Kate found herself pregnant again with her fourth baby. It was the last thing in the world she had figured on.

Kate's sister-in-law in Washington, D.C., had sent a scolding letter after the third child was born. She told Kate that "you are too young and too attractive to be having babies so fast. You're going to ruin your health." She sent along what she called a "never-failing remedy," and Kate thought, "Oh boy. I've got it made." Kate took the concoction faithfully, but now, here it was just a year or so later, and she was heavy with child once again. "I thought the sky was falling in on me. I thought everything was wrong. . . . I didn't want that baby."

At one point, Kate found herself at the family physician, asking—begging, pleading—if there was anything he could do to make the child inside her go away. "Well, I'll give you some good advice," the doctor replied. "When you go to bed tonight, put a glass of water on your bedside table; and during the night you drink that and nothing else." Kate was livid. "I could have killed him," she said. "I think if I'd had a gun, I'd shot him right there!"

That fourth child, who was named after her husband's older brother, James Griffin Boswell II, would go on to take the reins from the Colonel and build America's biggest farming operation. From the day he was born, March 10, 1923, Kate couldn't get over the guilt about not wanting little Jimmy. She lived in fear that God would one day exact revenge.

Decades later, when Jimmy was sent off to war, Kate was convinced that fate's cruelty had waited all that time to strike. She got a call from one of Jim's buddies at Fort Ord: They were being shipped out in two days. Kate was determined to bid him a final good-bye in person. She got in her Buick and tried to persuade her husband to join her on the 170-mile drive up the California coast. "I'm not going," Bill said. "That's a wild goose chase." So Kate raced off on her own.

"I got in that old Buick of mine, picked up every little solider boy hitchhiking—you know, the road was just full of 'em. And I got over there and stayed in one of those barracks, shared a room with a little pregnant gal who just wept her heart out all night. And I was crying right with her. Took Jimmy out to dinner; and we sat and talked as long as we possibly could. And he had to go back. He said, 'You won't see me anymore.' So I got in my car and came home the next day, and I just thought, 'This is the way the good Lord is gonna punish me now, for being so ugly before he was born.'"

Jim Boswell survived the war, leaving Kate to conclude that the good Lord had decided to punish her in another way: by making Jimmy so good to her. This compounded by a thousand times the shame she felt for having wanted to get rid of him. "Jimmy's been the sweetest child I've had," she said. "I love 'em all alike. Don't misunderstand me; it's like my mother used to say: 'They're like the five fingers on my hand.' I need 'em all. But I mean he was more thoughtful. . . . He was the last one. . . . And we have always been so close—a very beautiful relationship. Always, always."

AS THE MONTHS PASSED with Bill in California, Kate Boswell found it tougher and tougher to be left behind in Greensboro. The situation was made all the more difficult because Jimmy was sickly as an infant. He had asthma, and Kate despaired as he struggled and wheezed. "It sounds to high heaven," Kate recalled. "You didn't know where his next breath was coming from."

At last, in January 1924, Kate agreed to make the trip to Corcoran. She had been expecting, as she put it, "all sunshine and roses," but the California that greeted her was shrouded in a thick tule fog, and a light mist fell on the valley floor. "Oh my God," she thought, staring up at the melancholy sky. "This is California?" Bill picked them up at the train station, and as they drove through downtown Corcoran, Kate peered out at the largely vacant stretch that was Whitley Avenue. "There was nothing," she later recollected. "Nothing."

As time went on, Kate would become the grande dame of Corcoran, graciously entertaining her friends and dispensing her rules on proper conduct—"Mom Kate's no-nos"—to the women of town: No lady should wear white shoes after Labor Day. No lady should wear a sleeveless dress past age fifty. No lady should wear a short-sleeve dress past sixty. But on that first day, when they pulled up to the house that Bill had rented, Kate couldn't see any kind of life for herself, and she began to sob. "My cook in Greensboro has a better home to live in," she told him. "I'm not going in there."

Instead, they moved into a suite in the Hotel Corcoran, where Kate fell into an even deeper funk. "I got to my room on the second floor, and looked down that row of palm trees in front of the old Santa Fe depot and just wept my heart out. I didn't think it was possible for California to look like that. It was just awful. It was one of those dark gloomy weeks, week after week, you don't see the sun. I couldn't have landed at a worse time." She never forgot how those palm trees mocked her, and she detested palm trees ever after.

Kate tried to adjust to life without her usual retinue of Georgia servants, but when the family's clothes came back from the laundry and needed pressing, she was bewildered. "I've got an ironing board and an iron down in the basement," Miss Drew, the hotel manager, told her. "Just go down there."

"But Miss Drew," Kate replied, on the verge of tears again, "I don't know how to iron."

Her introduction to Corcoran society, such as it was, came with an invitation to the Hansens', a pioneer family that had arrived in the lake bottom so early they remembered how the 1906 San Francisco earthquake made the water in Corcoran's irrigation ditches flow in reverse. Getting ready that afternoon, Kate grabbed a badly wrinkled white shirt of Bill's, got out the iron and thought, "Well, I'll do the best I can." She smoothed out the collar and the cuffs, and then something struck her: "Now, this is just wasted energy. All this part doesn't show under his coat." So she left the rest of the shirt a rumpled mess, told Bill to put it on and off to the dinner party they went. Bill, though, was eager to get the last laugh. "I saw him kind of whisper to the men. And the first thing you know, one man said, 'I'm hot. Let's all take off our coats.' My poor Bill is going around with the rough dried shirt on. . . . If he wasn't the worst looking thing you ever saw in your life."

They lived for nearly a year at the hotel before moving into a two-story home where Kate started to host her own parties—bridge games, mostly, with little Jimmy Boswell making his entrance at the end of the evening, his wagon full of prizes for the high scorer. Bill tried to make things easy on Kate by hiring a maid to do the housework, but when the employment agency in Fresno sent the woman along, Kate was in for another shock: She was white.

"My God," Kate said, "I never had white help in my life. . . . So then I said something about the kitchen floor needing to be polished and waxed, and she said, 'My contract doesn't call for that.' And this just went from bad to worse. She ate at the table with us. Well, I guess that was all right, but I don't know, usually domestics don't eat at the table with the family."

Finally Kate couldn't take it any more. "I'd rather get down on my hands and knees and scrub," she told Bill. "Get that woman out of my sight."

Bill obliged, bringing in a black woman named Laura to help around the house. She was as wide as she was tall, leading Jimmy and his sister Jody to dub her Five-by-Five. Bill also sent back to Georgia for a "dear old Negro couple," Maggie and Will—the man whom, many years later, Jim Boswell would reminisce about as the "prized one" who had "tap-danced" his way to California.

In reality, things didn't turn out so well for Will Weaver and his wife. Their relationship with Kate and Bill began on a good note. But, as Kate recalled, it soured when Will became sick. "The doctor said you've got to put him in the hospital. So here he goes to a white hospital with white nurses taking care of him." It was then that Will and Maggie "realized, for once in their life, that they had been distreated. . . . They were not mistreated. But maybe they weren't getting what they should.

"And I just couldn't go to the hospital to see him with those little white nurses taking care of him. To see that black face lying up there in that bed, I just couldn't take it."

Will got better, but Maggie lost her head, doing things that never would have been countenanced in Georgia. For instance, Kate said, Maggie invited "her Negro friends" to come through the front door of the Boswells' house. "I'd say, 'Go around to the back.' Well, I didn't know any better. That's what we did back home. . . . All of a sudden, they just developed into two of the worst characters. They got very impudent. And I can never forget when Bill went out one morning, and Will said something to him. I don't know what it was; I didn't hear it at all. But Bill said, 'Nigger, you get yours and Maggie's things packed. There's a train that leaves here at four o'clock this afternoon. I've got your ticket, and you be on that train or else I'll have a dead nigger on my hands.'

"Bill took them down there, put them on the train and sent them back" to Georgia.

BY THE HARVEST OF 1924—as an entourage of minstrels de-scended once again on Corcoran, this time to celebrate the town's first-ever cotton festival—the Colonel was firmly enmeshed at the top of the industry in the West. He leased the Corcoran gin, though he'd laugh that the town's original cotton men were so inexperienced that they had cobbled the thing together wrong and it was running backward. He quickly built another gin, this one bigger and faster than the first. And by 1925, with Bill overseeing work in the fields, the Colonel had erected yet another gin and an oil mill staffed by two twelve-hour shifts of work-ers. The Boswells, as one newspaper pointed out, now boasted the "largest industrial unit" for cotton in the valley.

The opposing styles of the two brothers made for quite a team. J. G., the Colonel, "was very nervous, very excitable, moved fast, talked fast—jerk, jerk, jerk," Kate remarked. "Bill goes sauntering along with that cig-arette . . . in the side of his mouth. Takes his time. . . . You know what Walter said: 'If you could get half of J. G. and half of Bill and put them to-gether, you'd have a perfect man.'"

Within two years, their partnership dissolved, changing the family dynamic forever. The Colonel wanted to incorporate as the Boswell Brothers. Bill, though, had debts in Georgia to pay off, and he wasn't one to embrace the responsibility of running a big company. So, standing on a ditch bank, the two of them cut a deal: The Colonel would own the new corporation, and Bill would take some cash up front, go back to Greensboro and settle his affairs. His debts cleared, he'd then return to California and work as the Colonel's employee. There was only one wrinkle: As part of Bill's compensation, the Colonel would provide the best education he could for his brother's children.

The Colonel financed the company through two sources: His wife, Alaine, chipped in more than $100,000 (equal to almost $1 million to-day), and a Los Angeles bank president named Dabney Day loaned an additional $190,000. Day's Citizens National Bank was so gung-ho on the prospects for Western cotton that it displayed a bale right in its downtown lobby, giving the suit-and-tie set a glimpse at an industry that Day believed would "take its place as one of the greatest factors, if not the greatest factor, in the economic life of California."

Despite such a sanguine assessment, the business was far from trou-ble-free. Some California gins had taken to blending together different varieties of cotton—Acala, Durango, Mebane, Meade, Pima—producing bales that the textile mills found tough to spin. Bill Camp, the U.S. Department of Agriculture employee who had shepherded cotton into the San Joaquin Valley during World War I, attacked the problem with a religious fervor: "I made it part of my job every hour . . . everywhere, with everybody," he said, to preach the virtues of standardizing the industry. The result was that in 1925, the legislature in Sacramento passed one of the most significant farm bills ever: the One-Variety Cotton Law, which mandated that only Acala, which Camp judged to be best suited to California, could be planted in the state. Everything else was strictly verboten, and county agriculture officials stood ready to pounce on and quarantine contraband cottonseed.

Wild fluctuations in the market were another problem. In 1923 California cotton fetched an average of more than thirty-one cents a pound, but this slipped to less than twenty cents two years later and dipped under fourteen cents in 1926, before climbing back to the twenty-cent mark the following year. The instability forced some companies out of business. But Boswell was able to thrive even during the tough times because he never took on debt and he never speculated hog wild in the futures market. His was a steady hand.

The Colonel also had a great eye for talent. Fred Sherrill, who was among Boswell's first employees, would become one of the most influential cotton men in America. A pure intellectual, he had a penchant for quoting Burke, Emerson and Lowell, and he did so in a voice that was almost regal. Listening to a tape recording of him many years after he had died, you could picture his jaw stiffening as he waxed lyrical about the lessons that modern-day politicians could learn from Richard the Lionhearted. "If a course of study," he once wrote, "can be made available in these three great sources of wisdom—Shakespeare, Milton and the Bible—we might avoid the charge that there has been no improvement" in mankind over the years. He turned out long letters to everyone from Walter Lippman and Will Rogers to a host of public officials, readily sharing his ideas for solving the world's problems: taxes, trade, banking, Prohibition, you name it. But Sherrill wasn't all earnestness. A West Point man, he had been the last commanding officer at Fort Apache, and he loved to tell war stories about his days there. "I was everything. Doctor, lawyer, merchant and Chief—that was what all the Indians called me." He liked the ponies and hand-rolled his own Bull Durham cigarettes. With his slick-backed brown hair and sharp features, he cut a dashing figure.

Sherrill's official title at Boswell by the early 1930s was "treasurer," but he really functioned as the Colonel's political operative. In the years ahead, he'd play a huge part in getting the government to capture the flow of the Kings River. Late in life, he'd advise Barry Goldwater and Richard Nixon. Sherrill had joined the firm at the urging of Walter Boswell, who had gotten to know the young lieutenant at the War Department in Washington. Walter informed Sherrill about the promising new cotton company the Colonel had recently started up in California, and suggested that he could use a bright guy to help run the Los Angeles headquarters. Intrigued, Sherrill took the train west and found a "wonderful opportunity" awaiting him. A Long Island native, Sherrill didn't know much about cotton when he started. But "this brother of yours," he told Walter, "is a great one" and the business "seems to be growing by leaps and bounds."

Most of the others the Colonel brought on board were from the South. R. L. Curtis, the fry cook turned cotton marketing man, was out of tiny Paragould, Arkansas. (The R. L. stood for Robert Lee.) Georgia, naturally, sent a number of its sons to hitch up with the Colonel and Bill. "I was used to growing about a half a bale to the acre . . . and they called me up and said, 'You can grow two bales out here.' I said, 'Here I come!'" recalled Louis Robinson, one of at least a half dozen men, including his brother Reg, who departed Greene County to become a top manager for Boswell. Once they signed on with the company, many remained fiercely loyal. The Hammond brothers—Gordon, Tommy, Julius and Joe—were typical. They left Summerville, Georgia, in the mid-1920s to work at the Corcoran gins, and by the time they retired, they had given a combined 146 years to Boswell.

While Corcoran remained the heart of Boswell's operation, the Colonel spread his reach to other towns in the San Joaquin Valley, including Tipton, Porterville, McFarland, Mendota, Guernsey and Tulare. He also stretched into the Salt River Valley of Arizona, where during World War I Goodyear Tire had begun farming "cotton that grew taller than a man's head and was comparable in quality to the best Egyptian grades," as Chairman Paul Litchfield remembered it. In 1929 Goodyear invited Boswell to build a $125,000 oil mill there, and in due time the

Colonel also farmed and set up as many as ten gins in Arizona, mainly around Phoenix.

Along with a handful of other cotton processors, Anderson Clayton foremost among them, Boswell began lending money to growers. Because the banks wouldn't loan directly to the farmers, financing from the gins became the lifeblood of California's cotton communities. Many small growers complained about the power the ginners possessed, and some felt they were being squeezed by terms that verged on usury. That opinion wasn't universal, however, and some relished the ready capital. Financing from the Boswells in the late 1920s was "the biggest bouquet I ever had handed me," said Jesse Anderson, who grew cotton in the lake bottom. "I never knew better people." The Colonel directed his managers to base their loans on three criteria: an individual's character and creditworthiness, the potential output of his acreage, and his access to irrigation—or, more simply, "man, land and water."

By doing it all—from making farm loans to growing to ginning to marketing cotton across the world—Boswell set itself apart from the rest of the industry. This all-embracing approach—what business gurus call "vertical integration"—also helped shape an odd sort of corporate culture, one defined as much by wheeling and dealing as by tilling and plowing. Bill Boswell regularly wore a suit and garters to work, not bib overalls like other ranchers in town. The redheaded Louis Robinson, who shaded his fair skin with a straw hat whenever he happened outside, was so office bound that his family used to joke that he "farmed with a sharp pencil."

Every few weeks, the Colonel would drive the Ridge Route from his home in Pasadena to Corcoran—in due time, his chauffeur would take him—but his orbit remained Los Angeles. As many as fifteen cotton dealers, including the Colonel, occupied offices in downtown L.A., most of them at the Cotton Exchange Building on West Third Street, where telegraphs clicked away with the latest quotes from New York and New Orleans. At home, the Colonel lived in increasing comfort. He and Alaine bought up more and more land around their house, which she turned into a showcase of gardens. With a flair for the dramatic—she liked to appear suddenly from the balcony in their living room, her head wrapped in a black tulle—Alaine tended the flowers and dreamed up short stories and poems. She liked a good brandy now and then, even as the Colonel

adhered to teetotalism. While the Colonel fixated mostly on the worka-
day issues of his business, Alaine's verse spoke to other passions:

> Hips protruding and nether parts fearfully rounded,
> Teeth bared in a grin and a smell of rum,
> And cocktails imperfectly digested.
> A usually kind and familiar face distorted and veiled with liquor;
> Thickened tongues and the glitter of sequins;
> Marcelled hair and massive ankles.
> Quick! Quick! There is a young nymph,
> Lithe and clean-limbed and exquisite!
> How did she come here?
> Ah, she too with the grape already in her veins.
> A glimpse of garter—one turns away surfeited with flesh.

The Colonel paid himself $25,000 a year (about $235,000 today) and
hobnobbed with other leading business executives over meals at the
California Club and rounds of golf at the Midwick Country Club. The
Colonel's ambit was such that in the summer of 1928, he entered into a
partnership with Cecil B. DeMille, telling the great filmmaker that farming out West was virtually immune to the normal vicissitudes of nature.
"A crop failure in this territory," he said, "would be equivalent to a
bumper crop down South." Attracted by the Colonel's reputation for
"honesty and integrity," the director who had made *The King of Kings*
put up more than $30,000 to grow cotton with the king of Kings County.

The Colonel came to expect the royal treatment. When he went to
Arizona, the train would often make an unscheduled stop just for him,
so he could get off exactly where he wanted. And in the cotton towns
where Boswell operated, the children made sure they had on their Sunday best when the Colonel called. "We had shiny faces and nice clean
clothes because J. G. was coming," said Mandy Durand, whose father
ran the general store in the town of Marinette, just northwest of
Phoenix. "Let me tell you, I remember the first time I saw him. He was
dressed in a beautiful black suit and beautiful, shiny black leather
shoes. He would come in and sit down on the counter at the mercantile
and wait to be picked up. That's when I could see his shoes. A child no-
tices things like that. He had on a little felt hat. It was dark gray. And you

never saw anything so beautiful in your life. Wow, to see a man like that get off in cotton-pickin' country."

For the men who worked under the Colonel, the sight of him could stir fear, and there was always a lot of commotion when he showed up in Corcoran. "They'd have us shine up everything," said Frank Medina, a longtime employee. "They wanted everybody on their toes and everybody working. He didn't want to see anybody standing around." Once Medina watched as the Colonel came upon a loitering foreman at the Ryan Ranch. "Fire that man," the Colonel barked to the foreman's supervisor, without pausing to hear who he was or whether he was merely on a break.

The Colonel's most senior men weren't immune from his tirades, either. "Louis Robinson used to get frustrated and say, 'I'm going to put my hat on and go home, and I'm not coming back tomorrow.' It was just too hard," said Alta Ober, whose husband, Rice, worked as a Boswell executive starting in the 1930s. "But then he'd wake up, and look at his three little girls and put his hat back on and go to work." Knowing that Robinson had a standing golf game at noon on Thursdays, the Colonel always phoned a few minutes before 12:00, just to ruffle him.

Only Alaine's soft touch seemed to soothe the Colonel. One night, on the way to a movie, he discovered that the car was low on gas. He cursed the chauffeur for not filling the tank and waited impatiently while the gas station attendant filled up another car that was ahead in line. Then he started laying on the horn. "Uncle Jim didn't want to wait," recalled Bill's daughter Jody, who was sitting in the backseat. "All of a sudden, Aunt Alaine said, 'Jimmy, I can just hear it. When you die and go to heaven, you're going to be up there and you're going to be rapping on those gates saying, 'Saint Peter, I'm J. G. Boswell. Let me in.' . . . And you know, he just calmed right down."

The Colonel's haughtiness played most poorly with his relatives. In 1930 Walter Boswell retired from the War Department and went to work for the company, eventually taking charge of the Arizona ranches after a brief stop at the L.A. office. His wife, though, didn't get along with the Colonel, and tensions mounted. The Colonel "would tell my mother how to run her children," Walter's oldest son, James O. Boswell, said. "She didn't like that worth a God damn." Even during social occasions, when the Colonel seemed to be relaxing, he'd still come across as a petty tyrant. He would "have a dinner and everybody would enjoy themselves," said

James O., "and the next morning, out would come the orders: 'Well, Walter, you got to go to Phoenix.' 'Bill, you go do this and that and the other.' And you just felt that instead of enjoying the dinner, he'd been thinking all night what orders he was going to give out in the morning."

The Colonel was "the little king," said Jody Boswell. "He liked that part of it, whereas daddy didn't want that part of it at all."

Certainly, Bill Boswell led a much different life than did his brother, the Colonel. He could be hard-nosed and he had his conceits, but Bill also exhibited a certain sweetness absent in the Colonel. His lack of outward ambition, when set beside the Colonel's hard-charging style, could be charming. As the Colonel traveled the globe by luxury ocean liner, meeting with Boswell agents in Europe and Asia, Bill was content right where he was. Go abroad? Heck, he'd say, referring to a town just across the Kings County line, "I haven't really seen Alpaugh yet."

In the early 1930s, Bill started up another important piece of the Boswell empire: a feedlot that fattened thousands of head of cattle with cottonseed cake. Unlike his brother, he relied on classic country cunning to get the job done. "Don't try to outsmart people," he once counseled his son Jim. "Just outdumb 'em." The Colonel asked Bill at one stage to move to the Los Angeles office, but he declined. To her dying day, Kate was bitter about this, feeling that she had been cheated out of all the grandeur that the big city could offer. Yet Bill saw the virtues of raising his kids in a small town. He could go out with hound dogs and run rabbits and hunt 'coons. Or perhaps it was the booze that bound Bill to Corcoran, a fear that he couldn't hide his drinking from the Colonel if they lived in the same city, much less worked in the same office. Kate, for one, would implore the operator at the Corcoran telephone exchange to keep Bill's drunken calls to his brother from going through. "I have shielded him from J. G. so many times, you have no idea," said Kate, "because he couldn't talk and I didn't want J. G. to hear it."

As the Georgia Minstrels strutted about the stage in October 1929, Bill Boswell was conspicuously absent from the show. Just a few months before, he had experienced what the Porterville paper called a "miraculous escape from death," when his car overturned. Now he was home in a full body cast, recuperating.

He had been driving on a Saturday night up to the family cabin at Doyle Springs in the Sierra, 4,000 feet above the valley, a place for the Boswells and their friends to escape the summer heat. As Bill's Ford made its climb, it veered off the road and fell a hundred feet to the riverbed below. The only other passenger—a pig, set to be barbecued— tumbled out. Two men on horseback rescued Bill, who lay unconscious for more than three hours. Once revived, his first words to Kate were, "I guess I was driving too fast." He had broken his neck, and the Colonel sent a well-known orthopedic surgeon from L.A. to patch him up.

Bill swore he hadn't been drinking that night, but nobody quite believed him.

Whatever the truth, Bill never took another drop of whiskey again— though he immediately filled the bathtub in his hospital room with beer and over the years consumed astounding quantities of the stuff, frequently finishing off an entire case of Coors in a single afternoon.

ON OCTOBER 29, just twenty-seven days after the minstrel show in Corcoran, the stock market crashed in what became known as Black Tuesday. Even the California cotton industry, which had floated so blithely above the crisis that befell most of agricultural America through the 1920s, couldn't escape. "Agriculture is sick," said C. O. Moser, vice president of the American Cotton Cooperative Association. "This sickness is not temporary nor seasonal, nor merely in one locality."

For the Boswell Company, the nadir came in 1931, when the average price of cotton sank to about six cents a pound. "It seemed as if the very bottom would drop out of everything," Fred Sherrill remembered. In the end, the only choice for Boswell was to liquidate its futures contracts at an enormous loss.

The company fell into the red for the first time, ringing up a deficit of nearly $114,000, and gossip swirled that Boswell might not survive. The Colonel cut his own salary to about $15,000, and he relinquished his customary $3,000-plus bonus for the year. Desperate for cash, he also turned to his friends, some of the more prominent and wealthy citizens of Southern California. Six men, most of them the Colonel's golfing buddies, came to Boswell's rescue by buying up $128,000 worth of company stock.

Still worried about a shortage of capital, the Colonel tried to end his partnership with Cecil DeMille in early 1932, but the movie mogul insisted that Boswell keep farming their land in Arizona. During a tense meeting at Metro-Goldwyn-Mayer studios, the Colonel said that he couldn't meet DeMille's demands. The Colonel also told DeMille that he hoped the matter wouldn't go to court because, frankly, he didn't have the money for a lawsuit. "But I do," DeMille curtly replied. The two men eventually ended their relationship without litigation, but more pain would come.

That same year, Boswell closed down his Corcoran operation and furloughed nearly all of the workers there. Only a handful of men, such as gin supervisor Gordon Hammond, stayed on, knowing that they couldn't be paid for a while. Scores of others, including Gordon's brothers, were left to wonder about their fate. "There just wasn't any work for anybody," recalled Tommy Hammond's wife, Ruth. "Tom didn't know whether he'd get hired back or not."

To the Colonel, it was clear that the blame for all this misery resided in Washington, with its misguided policies of collecting European war debts and insisting on a high protective tariff. Together, these actions had clogged overseas markets for cotton farmers.

The core of President Hoover's farm program was to promote agricultural products through marketing cooperatives and, as the Depression persisted, to stabilize prices by buying and holding cotton and other commodities. None of these steps, though, was enough to keep cotton prices from falling further. Then, in August 1931, Hoover's Farm Board called for "immediate and drastic action": Growers were urged to rein in production by plowing under every third row of cotton. The farmers' response to the plan was overwhelmingly negative; one Southern editor suggested "plowing under every third member of the Farm Board" instead.

It wasn't until Franklin Roosevelt assumed office that the effort to pare cotton plantings finally worked. Under the Agricultural Adjustment Act of 1933, farmers were persuaded to plow up about a quarter of the nation's cotton crop. Though the cutback was voluntary, like Hoover's, it had one critical advantage that the former president's plan didn't: cash. The AAA issued benefit payments to the growers who participated. The Colonel supported the thrust of the legislation, calling Roosevelt an "honest, intelligent and capable directing physician for this sick country of ours."

Not everybody fared equally under the program. Throughout the South and Southwest, the AAA had the effect of pushing sharecroppers and tenants off their land, but the lot of many farmers improved, especially bigger operations like Boswell. Not only did the Colonel gladly take tens of thousands of dollars from the AAA for limiting his acres, but the loans that Boswell made to other farmers who took part in the program were effectively guaranteed by the government.

With the balm applied by Washington, Boswell could now take advantage of other farmers' weaknesses. As the 1930s dragged on, many smaller farmers and ginners in California went out of business, and Boswell grabbed up their assets at bargain prices. Over a five-year stretch, starting in 1933, Boswell bought thousands of acres around the lake bottom; by 1938, it operated thirteen different ranches. And just four companies, including Boswell, now ginned two-thirds of the cotton in California. Boswell executives also acquired key water rights at this time, in part by wheedling Portuguese farmers along the Kings who were starved for cash.

"I think the Depression had some advantages to the Boswell Company," Louis Robinson acknowledged. "We probably were able to do things very advantageous to us with very little or no competition."

Although they fared well under Roosevelt, Boswell and his top men ultimately lost their fondness for the president, vilifying his fiscal policy and attempts to pack the Supreme Court. "This New Deal is too much for me," Walter said in 1937, "and an old Georgia Democrat is looking for somewhere to hang his hat." The Colonel, who as a diehard Southerner remained a registered Democrat all his life but usually voted Republican, also regretted having cast his ballot for Roosevelt in '32. "Every man is entitled to one mistake," he'd say, accusing the president "and his crowd" of steering the government toward "socialism, communism or some other type of ism."

Of course, some of that socialism, in the form of subsidies, had helped save the Boswell Company—an irony that inspired a little doggerel from a crony of the Colonel's. Before an annual golf outing in 1934, his pal penned a poem to remind the group that, even during the Depression, "Our friends are still around us and the fairway nice and green. And things are sometimes not so bad as they are apt to seem." The Colonel then added this couplet:

> And there is Jimmy Boswell, who is letting out his belt,
> Anticipating all he'll get from Franklin Roosevelt.

La Mordida

THE FIRST TIME HE GLIMPSED the long line of cars pouring into the basin in late September, Rudy Castro thought the most important man in the world had died and they were coming to Corcoran to bury him. There must have been 2,000 people packed on that last stretch of country road leading to town, and as they got closer and closer in the wavy heat, he could see the washing machines and bed frames and wood chests hitched to the battered Fords and Chevys, the provisions of a ragtag army of Mexican men, women and children chasing the crops of the valley.

Now, a decade later, on the eve of the 1933 cotton harvest, many of the migrant faces rushing into the camps were familiar to Rudy. Mexico's rapid industrial expansion had uprooted its rural peasants and sent them across the border. At least 150,000 Mexican nationals, by far the largest agricultural workforce in the West, slogged away on California farms. They were the most recent in a procession of exiles from around the globe who had come and gone in these fields for seventy-five years. The constant flow of fresh hands kept wages low, about sixteen cents an hour for picking cotton, and the workers hustling. J. W. Guiberson, who had come to Corcoran as part of Hobart Whitley's development team and was one of the first men to plant cotton in the lake bottom, saw it as one big parade of nomads. "The class of labor we want is the kind we can send home when we get through with them," he said.

The families who filled the Boswell camp run by Rudy's father had just finished picking thousands of acres of Thompson grapes and laying them down on wood trays to dry in the vineyards. Visalia to Fresno smelled like one giant bakery cooking raisins in the late summer sun. The end of the grape harvest and the beginning of the cotton pick had been marked, as always, by a one-day fiesta to celebrate Mexico's independence. Now it was time to return to the fields waist high in white gold.

Twelve-year-old Rudy was no stranger to this rhythm. He, too, had spent the last six weeks picking grapes sunup to sundown near Selma with his older brother, Roberto, and sister-in-law, Aurora, and he thought it would never end. It was filthy work, parting the curtain of vines only to be hit in the face with sulfur dust and gnats, and then dodging the black widows and lizards. Summer evenings in the valley brought the most wonderful breeze, cool and light, compensation for those 100-degree days, and Rudy hated going inside the barn where the three of them ate and slept beside the farmer's horses.

"In the eyes of the growers we were just about equal to his mules," Rudy said. "I mean here are these animals defecating and right next door, separated by a rope and nothing more, we were cooking our meals and sleeping. It was horrible, but somehow we thought ourselves lucky. Whether it was just beans and potatoes and tortillas, we never knew hunger."

Even as an old man, Rudy never forgot the freedom of those Sunday afternoons during the raisin harvest, taking his lunch outside under the shade of a big cottonwood tree, sitting beside the Southern Pacific tracks and getting lost in the endless rumble, wondering where all those families riding the rails were headed. North and south, north and south, they all looked the same: scared and desperate.

Back home at the camp in Corcoran, Rudy watched his father and brother set up the outhouses and clean the makeshift school. The camp was a rectangle of forty tents and cabins, each one measuring twelve by fourteen feet. The hog pens that Boswell had built alongside the camp were now gone, but the stench would stay with Rudy through the years as he made a proud career with the Boswell Company. The pens were replaced by another field of cotton so that the mothers and children had only to walk across the dirt road after school to pick their nightly allotment. As Rudy gazed out to the dusty caravan with license plates from Texas, Arizona, Colorado and New Mexico, and greeted his old buddies from Los Angeles and Corona, Riverside and Oxnard, he had no inkling that the 1933 cotton harvest would change everything.

THE CASTROS HAD ARRIVED A DECADE EARLIER and were one of the first Mexican families to settle in Corcoran. On their heels came other Latino clans who would reshape the town: the Medinas from the railroad camps of the Southwest, the Maganas from the plantations of Guanajuato, the Banales family from the citrus groves of Anaheim, and the Dominguez family from a tiny cotton patch in Texas. There, the earth was so rocky that sparks flew whenever their sharecropper father struck the ground hard with his hoe.

Even as the baby of the family, Rudy Castro understood that his father, Mateo, and his mother, Maria, enjoyed a security not known to the *campesinos*. Mateo had journeyed up from Michoacan in 1910, an accomplished horseman who had learned enough English as a fruit tramp to act as a conduit between the growers and workers. He had been plucked out of nearby Hanford by Bill Boswell himself and brought to Corcoran to oversee the camp of cotton pickers. Roberto would become his dad's assistant. In the early years, when the migrants lived in tents with dirt floors, the Castros were afforded the luxury of a wood cabin with a lean-to for a kitchen. When the pickers' tents were later upgraded with plank floors and then replaced altogether by tiny wood shacks, the Castros were given three "deluxe" cabins side-by-side. They used one for cooking and bathing and the other two for sleeping, enough room that the girls and boys had their own separate quarters.

Rudy slept next to Roberto, and there were enough years between them that Rudy came to regard him more as a second father than a brother. The day Roberto came home with his bride, Aurora, he made a bed for Rudy out of a wood trunk and one of their mother's handmade quilts. Rudy slept beside the newlyweds on their wedding night and remained there for weeks after, until they secured a cabin of their own. "They didn't have any money to go on a honeymoon or anything like that. Their wedding night was there in the bed. I was just a kid and I didn't know anything about sex. Whatever happened, I never knew anything about it. I loved Roberto. He was the only brother I had. Very gentle, always reading, very intelligent, never caused any problems, no drinking, just a great guy, very obedient to my mom and dad."

For the workers—whether crew boss, irrigator or picker—the difference between food on the table and hunger in winter was cotton. No other valley crop employed whole families through the slack months of November, December and January. The average cotton picker netted 400 pounds a day, a haul that would bring him $1.60 worth of tickets that he'd cash in each Saturday—the money to be quickly swallowed up by the merchants on Whitley Avenue. It was all a matter of stretching the 100-pound sacks of pinto beans, flour, rice and cans of lard and getting to February, when the warmth of early spring heralded a new round of work in the plum orchards along the coast. Experienced pickers earned more in cotton than in raisins, and as demanding as the stooped labor could be, it was nowhere near as backbreaking as working for the railroads, the other industry that welcomed Mexicans.

The patriarch of the Medina family, Fidel, had a job with the Santa Fe that had taken him from Kansas to New Mexico to Colorado and then to California. He arrived in Corcoran in 1932 with his wife and seven children and worked as a gandy dancer, cutting track by hand with a hacksaw and filling up railway warning lights with kerosene. Each night, he returned home to the railroad's barracks-style housing on Otis Avenue and collapsed, his blue work shirt caked salt white and his knees swollen with water. He teased his wife about the necessity of even washing the shirt when the next day would bring only the same deposit of sweat, and then he moaned himself to sleep. "At night he would groan like somebody had beaten him up so much he just couldn't hardly stand it," Frank Medina, his middle son, recalled. "One day he had enough and told my mother, 'Let's go back to Mexico' and she said she would rather die here with her kids than take us back. She said, 'Let's call it quits and move to the camps.'" So as a compromise, Fidel Medina left the Santa Fe and began picking cotton for Boswell. His six sons would go on to log a collective 160 years of employment for the company.

For the migrants accustomed to the fields of Texas or even Mississippi, the cotton rising out of Tulare Lake was a curiosity to behold. The only variety grown here, Acala, had been lifted from Mexico itself. Each plant stood tall and loaded with bolls, and each boll spilled a fiber puffy and pure white. There was no word in English, let alone Spanish, to describe the mile upon mile of cotton. The word "field" just didn't do. Gus Dominguez and his family arrived in Corcoran in midwinter when thick fog obscured the entire basin. That first day out, as the fog began to lift and the horizon broke into thirds—the bottom third cotton, the middle third fog and the top third sky—he couldn't believe his eyes.

"I saw what I was in the middle of, and it was nothing but cotton. Cotton to the east, cotton to the west, cotton in front and in back of us. We grew up on a cotton patch in Texas so small you could see from one end to the other. These fields were huge. The cotton was waist high and there were a whole lot more bolls. Back in Texas, the cotton was scrawny. We took one field, row by row, and that was it the first day. My dad picked a few days after and then he went to the owner and began driving a tractor. I had to pick over 400 pounds a day just to make a living. Some nights I'd go to bed and still be dreaming about walking down the rows picking. You couldn't stop it in your sleep. After a while, my brother and I both went to tractors."

Working as a tractor driver or an irrigator wasn't an option for the women and children. And if the children got a break each day inside a classroom where the teacher spoke no Spanish and the sixth-grade library amounted to a single copy of Jack London's *The Call of the Wild*, the work never ended for the mothers. The matriarch of the Marin family, Felicitas, grew up on a small cotton farm in Rosebud, Texas, and was barely sixteen years old when she met Manuel Marin at one of the September 16 celebrations. He was ten years older and proposed to her on the dance floor, and the next thing she knew she was on the road headed to the camps of Corcoran, a baby on the way. For the next forty years, her husband would drive a tractor and irrigate cotton for one of Boswell's neighbors, the Gilkeys, never making more than the minimum wage.

"I would get up first in the morning, 4:00 A.M., and do my own tortillas by hand," she said. "I'd start the dough and plop it on the wood stove one tortilla at a time. After the first batch, I'd cook a big pot of pinto beans and then fix my husband's lunch for the fields. Burritos or tacos. The children were still asleep, and I'd wake them up for school. In the beginning, we had only four or five children and breakfast was no big deal. After we had thirteen kids, I used to fry ten pounds of potatoes every morning and cook two dozen eggs. They'd go off to school and I'd do chores. When they got home, we'd walk or drive to the fields to pick cotton. It was cold and my hands turned numb. Red like a tomato. I carried a ten-foot sack. Some of the men, the best pickers, carried twelve- and sixteen-footers. I wasn't that fast. We'd get our 200 pounds and go home and then I'd start dinner right away so they could go to bed and go to school the next day."

It took three separate picks over four long months to haul in all the cotton. A lavish first pick early in the season was followed by a second round in which the wages per pound increased a few cents to compensate for the slimmer pickings. If a grower wanted to grab any leftover fiber, he turned the workers loose on a third run. Picking for "bollies," they called it, the half-opened bolls that were more green waste than white cotton.

Each picker developed his or her own technique, a calculation based on size and strength and how best the hands and mind worked together. Some walked down the middle, splitting two rows of cotton, and their arms were long enough that each hand could pick a different side. Others found the juggle difficult and concentrated on one row and one plant at a time, attacking the bolls with all ten fingers. Some pickers yanked and some pulled. A yank might ensure that no "feathers" were left behind, making the growers happy. But if pickers yanked too hard, the sharp edges of the bolls bit their fingers, and the nicks pestered them all season long. Gloves weren't an option because you needed to feel the cotton. Mothers and grandmothers came to the rescue, slathering hands with corn oil and other lotions and potions. By the 9:00 A.M. break, you were soaking wet from sweat or fog's drizzle, and your back either gave out or you found a new way to bend like a pretzel. The ache began at the kidneys and worked down to the thighs and shot up the spine.

How a cotton picker carried his or her sack, where best to attach the canvas to the body, became a matter of much debate. Some tied the sacks around their necks or shoulders and placed the open end to one side. Others stooped extra low, their sacks strung around their backs, and mowed their way down the field. The open end rested between their legs and the closed end trailed eight feet behind like a bridal train. There was two-sack Chavira and two-sack Gonzalez, men known for their ability to wield a pair of sacks at the same time and pick 900 pounds of cotton a day. The champion pickers knew how to shake their sacks so the cotton packed itself. The less skilled ones had to use a leg to stomp it down, wasting time and energy. Some pickers developed a feel for 100 pounds—no more and no less in their sack. Those who picked light paid for it at day's end. It pained a worker to drag his or her sack all the way from field to weigh station, climb a ladder to the top of the trailer and walk across a plank to dump the cotton, knowing it was ten or fifteen pounds light.

Ray Magana was introduced to the cat-and-mouse game between picker and farmer the first day his parents dragged him to the fields. The crew boss and labor contractor had turned the screw on the tripod so that the scale, unknown to the pickers, began at five pounds in the negative. The math was simple. Fifteen hundred sacks rigged to register five pounds less per sack equaled 7,500 pounds of hidden cotton at week's end. Pure cream for the boss man. Magana said the workers would compare notes and respond in kind, sticking dirt clods and wild melons into their sacks, adding back five or ten pounds to the scale. The clever ones took pains to wrap each clod and melon inside a clump of cotton so the supervisor standing atop the trailer would see only white pouring out.

"They'd deduct for mud on the sack or fog because it made the cotton wet," Magana said. "They'd turn the scale so we couldn't see the weight. That's when we started sending two pickers to the station, one to deliver the cotton and the other to stand behind the crew boss and check the scale."

The growers liked to keep the workers guessing in other ways. There were thirty or so camps scattered throughout the basin, and the farmers provided every cabin rent-free. They made sure that the firewood for heating baths and cooking meals never ran out. As the outside voices pushing for improvements grew louder, Boswell and other farmers began furnishing enough electricity to heat shower houses and string up a bank of lights. At Thanksgiving and Christmas, they would hand out turkeys or roasts, and old man Gilkey himself liked to drive through camp, his big trunk spilling chocolates. But when it came time to draw water, the workers were forced to drive several miles into town and find a communal spigot to fill their metal drums and milk pails. In the fields, the growers didn't provide outhouses, and the workers had to squat between the rows. On the second pick, you'd find dry droppings everywhere. The growers filled the big field tank but once a season, and by the time January rolled around, the workers were drawing more polliwogs than drinking water. Likewise, the farmers didn't furnish cotton sacks, and the pickers tried all sorts of ways to shore up their store-bought brands. They sewed on patches and coated the canvas with tar or paint. Even so, the sacks scarcely lasted two weeks.

Sometimes the migrants' frustration came out in folk songs, ranchera music. Suddenly the fog would lift and one worker and then another would unleash their *gritos*—iiiiiiiyayayayyaaaaaaaaa—across the field. The gringo may have taken this trill as a joyful shout, but he'd have been mistaken. It was more a tear, a cry that got stuck in the throat. The pickers had no where else to turn to lodge their complaints. The only ear belonged to the labor contractor—the same man often shaking them down, demanding bribes in return for job security. They called it *mordida*, the bite, and if a picker desired a new season of work as an irrigator or a tractor driver, he had to deliver adequate boodle: new cowboy boots from Mexico, or a side of beef, or a case of tequila, or 20 percent of his paycheck, or sex with his wife. "It's life," said Chon Rivas, a picker who landed in Corcoran in 1927 and worked for Elmer von Glahn and then Boswell. "That's the way it went."

IT HAD RAINED ALL NIGHT, and the pickers were huddled around wood stoves in the Boswell camp, waiting for the fields to dry. Rudy saw a column of smoke and then fire and heard a woman imploring his father to hurry. "*Mi nino esta adentro!*" she shrieked. "*Mi nino esta adentro!*" Her baby was inside the burning tent, and Rudy watched in horror as his father covered his head with his fleece jacket and disappeared into the flames. Rudy stood next to the woman and her husband and counted the seconds.

"My father didn't ask any questions. He just ran right into the heart of that fire. He found the baby on the dirt floor. It was no more than three or four months old. He put his left hand beneath the neck and with his right hand he tried to pick him up. But the baby was cooking. The skin was already melting off. I remember my father coming out of those flames, trying to keep his coat over his head, holding the baby and crawling real low. He handed the baby to the mother and it died in her arms. I want to say this family stayed the rest of the season and picked cotton till the harvest was over."

By the time the 1933 season arrived, the Castros had every reason to believe that their camp would stand impervious to the seductions of a union that was now making a great deal of noise in the fields. After all, if other labor contractors and crew bosses demanded a bite from their workers, Mateo Castro refused to be part of the game. He and his wife had always regarded the migrants living under their watch as extended family, and whenever the single men would drink up their wages, a condition that seemed to befall seven or eight field hands at a time, Maria Castro would find a way to feed them until the next payday. The workers had grown to trust the Castros, and that trust extended to the entire camp. No one bothered to hide his belongings because no one bothered to take anything that wasn't his. The only locks to speak of were pieces of knotted rope that tied together the flimsy flaps of the canvas tents.

On behalf of the workers, Castro dispatched his youngest son, deft in both Spanish and English, to function as a translator and liaison. From the tender age of seven on, Rudy Castro acted as a bridge between worlds. If his father was busy or sleeping, the workers knew to knock on the boy's door. From jailhouse to hospital, he stepped in as their voice. When a young mother couldn't stop her vaginal bleeding, an embarrassed Rudy had to describe the symptoms to Dr. van Voris. When one outlaw picker faced charges of rape, Rudy found himself in the courtroom of Judge Nonhof, old "Ninety days, none off." The judge had a bad leg and walked by taking a step and a half, and his sentence for winos was ninety days suspended if they helped pick the cotton. Rudy translated a proceeding or two before the accused rapist was set free and run out of town. When an old woman died in the fields, Rudy accompanied her husband to the funeral home in Hanford. He watched the shame register on Jesus Garcia's face as he was made to wait in the parking lot to say good-bye to his wife. It took so long for the mortuary men to open the door and lead Garcia to the casket that Rudy picked up a telephone in the parking lot and for the first time in his life lectured an adult on the notion of decency.

Even before the '33 harvest, as more acres were planted to cotton and more Mexicans were called on to pick the crop, Rudy began to sense an odd resentment taking shape among the whites in Kings County. They understood, of course, that their own prosperity relied on the smooth flow of Mexican labor. Likewise, the economic well-being of the county depended on the workers spending their wages. In an attempt to lure the migrants downtown, merchants even hired crop dusters to airdrop reams of advertising leaflets over the camps. On Saturday nights, 3,000–4,000 workers thronged about the main street, filling the grocery and drug stores, dress shops, jewelers and theaters on one side of the tracks and crowding the gin joints, pool halls and gambling dens that lined Chinatown on the other side. Ten whorehouses scattered about town competed for traffic.

And yet as this dependence on an underclass grew, so did the petulance and chauvinism of the white community. It was as if they had come to resent the very backs on which their wealth had been built. Rudy saw its ugly face the night he accompanied his brother and sister-in-law to the county hospital. Aurora was about to give birth to her and Roberto's first child. "I was crying because Aurora was crying so hard in pain," Rudy recalled. "I told the nurse, 'Please take care of her,' and she said, 'No, we have to give her a bath first. You Mexicans are all dirty.'"

Fred Brooks, the bus driver, was another in a foul mood. Following the 1932 harvest, the camp had closed once again and Rudy, along with a handful of children of permanent workers, had returned to the school in town—the same one Jimmy Boswell attended. As Rudy boarded the bus and took a seat near the front, he could see Mr. Brooks's face in the big rearview mirror. He was working the plates of his false teeth, a kind of sour tonguing over.

"Mexicans in the back," he shouted.

Not sure they had heard him correctly—Mr. Brooks had always seemed a rather nice man—Rudy and his friends remained in their seats.

"Where I'm from, Mexicans and niggers sit in the back," Mr. Brooks repeated. He had come from Iowa.

The bus was quiet except for the nervous snickers of a few white kids. Rudy knew that silence wouldn't do. If anybody was going to speak out, it would have to be him, the son of a Boswell crew boss. "I'm not going to sit in the back, Mr. Brooks, just because I'm Mexican."

"You sit in the back or you and your friends can all walk," he replied.

Rudy and the others never took the bus again, choosing instead to walk the three miles from camp to school. The encounter left him shaken and full of doubt and shame. Part of him wanted so badly to be accepted by the children of the big growers that he never took his mother's tacos to school again. With his lunch money, he bought white bread, butter and milk or sometimes skipped lunch altogether, saving his dime for a packet of agate marbles. Once he saved enough dimes that he went knocking on Jimmy Boswell's front door to ask if his class-mate's shiny red bicycle might be for sale. It wasn't.

The Boswells were the one grower family that Rudy admired. It had less to do with the way they transacted business with his father, which was generally fair, than with his friendship with Jimmy. Bill Boswell's youngest child treated Rudy as an equal.

Mr. Bill himself visited the camp frequently to drink the beer that Rudy and his father home-brewed in the midst of Prohibition. Rudy washed and brushed out the bottles, put a pinch of sugar in each bottom and stuck a hose in the vat where a mix of malt, yeast and water fermented. He filled and capped each bottle and placed a dozen in a wet gunny sack in a cool spot under the cabin for his dad and Mr. Bill and constable Dan Leary to wash down the night. Mr. Bill would always give Rudy a nickel—big money for a kid in those days.

When the union activity began testing loyalties and sundering relationships in the days leading up to the 1933 harvest, Rudy told himself that his father and Mr. Bill would find a way around the troubles. But the beer drinking soon ended and the nickels dried up, and a strange feeling of precariousness fell over the Castro home and the camp as a whole. Rudy didn't know it then but his brother, Roberto, had begun to meet each night with a group of pickers and union organizers, and they were laying the groundwork for a cotton strike of crippling proportions.

THE UNION ORGANIZERS FANNED OUT from their base fifteen miles outside Corcoran—a tiny, fly-infested space where the only office equipment was a single rickety typewriter, and the side of a soda-cracker box served as a makeshift message board. The growers took one look at the Cannery and Agricultural Workers Industrial Union and expressed no worries, some to the point of smugness.

History, after all, was on their side. No union had ever won so much as a toehold in the California cotton fields, and this union, judging from its nerve center in Tulare, hardly seemed capable of changing the record. The Boswell Company, for one, had recovered from its near death experience two years earlier, and the Colonel was so confident that the union's bid would amount to nothing that he invested in two new forty-horsepower diesel engines. It was a major upgrade, and the company boasted that it was about to embark on a "heavy cotton ginning season."

As late as October 1, a few weeks after the pickers had returned to the camps, farmers throughout Kings County were sharing in that optimism, crowing that their workers were perfectly satisfied with the going rate of sixty cents for every 100 pounds of cotton picked. Any union seeking $1 for that same sack of cotton didn't deserve to be taken seriously.

Days later, however, it became clear that the union, threadbare as it might be, was not going away. When labor leader Pat Chambers gave the call to strike on October 4, pickers walked off the job by the scores. Whole crews deserted the fields and never once looked back. At the Hansen ranch and the J. Y. Petersen ranch in Corcoran, 200 pickers— their entire workforce—went out.

At the Boswell camp, things seemed calm until word reached the workers that one of their fellow pickers, Eliador Hernandez, had been struck by a car while walking down Whitley Avenue. The driver, a cotton grower named Harry Glenn who also owned the local telephone company, didn't even bother to stop and check on Hernandez, whose leg was badly broken. Writhing in pain, Hernandez watched car after car pass him by until he crawled to the side of the road and a pedestrian finally flagged down help. Back at the camp under a cottonwood tree, Maria Castro cleaned and basted the wound with a mix of herbs she had boiled, and fashioned a splint out of the tree's limbs.

For a day or so, with the help of Mateo Castro, the Boswells managed to get a few of their fields picked, and Mom Kate did her best to preserve a sense of normalcy. The weekend after the strike began, she laid out her lace tablecloth and celebrated husband Bill's forty-first birthday by preparing a beautiful three-course dinner for their bridge club. But nagging in the background was an awareness that the strike had already spread. Some 12,000 men and women all over the southern end of the San Joaquin Valley were proving their allegiance to the union. In a matter of days, the whole industry would shut down cold.

Fearing that thousands of acres of mature bolls were about to go to seed, Bill Boswell howled at Mateo Castro to regain control of his camp. Castro assured Boswell of his own loyalty, but there was nothing that he or anyone else could do about those who had gone over to the union. Indeed, the strike had divided the Castros, father pitted against son. Among those who now stood steadfast with the strikers was the family's oldest boy, Roberto.

The deep desire to keep the fields union-free was propelled by a simple set of economics. Between picking and other tasks like chopping and weeding, field labor accounted for nearly 45 percent of the total cost of production; nothing else came close. To drive down the expenditure, the biggest growers had done exactly what they were determined to prevent the workers from doing: They had banded together to maximize their leverage. Each year since 1926, a group called the Agricultural Labor Bureau of the San Joaquin Valley had met to decide the amount that would be paid for picking and chopping cotton. Sponsored by six county farm bureaus and chambers of commerce, the rates set by the ALB had always remained ironclad across the valley. Before the strike, workers had shown little stomach to resist the bureau. And those farmers possibly inclined to pay a bit more were hamstrung by their financiers—the cotton gins—which insisted that they stick with the ALB diktat.

On October 5, the farmers of Kings County stormed their labor camps and issued an ultimatum: Return to the fields and start picking cotton immediately for sixty cents—or get the hell out. The workers refused and the growers made good on their threat, tossing migrant families from the only housing they knew and heaving their possessions after them, until the highway was strewn with their meager effects: work clothes, rusted cooking utensils, a few clucking chickens for the most fortunate. "The people did not know what to do or where to go," Roberto Castro remembered years later. "They were under the sun, without water, without anything."

For the moment at least, the farmers had gained the upper hand. "We are going to rid Kings County of all the strikers and strike agitators," exclaimed L.D. Ellett, manager of the Pacific Cottonseed Corporation gin in Corcoran and chairman of a committee representing the growers.

Orchestrating such an expulsion wouldn't be so easy, however. The workers scooped up what belongings they could and formed their own settlements. Out of the mud of Tulare Lake, on a barren four-acre lot at the edge of Corcoran, grew the biggest of these ragged colonies, a place known affectionately among the strikers as "Little Mexico City." Within a week, 3,000 workers had wedged their way into Camp Corcoran. They put up a barbed wire fence, fashioned a crude water-supply system and arranged their tents and lean-tos in a chockablock but orderly grid. Then they named the streets for Mexican towns and heroes.

Even as Roberto Castro threw in his lot with the union, Boswell supervisors had asked him to keep tabs on the strikers. That way, peace restored, he'd waste no time bringing them back to the fields. As the strike wore on, however, Roberto felt his loyalties turn ever more in the union's direction. His father may have been able to justify trying to keep labor in the company's clutches for a mere sixty cents, but Roberto wanted no part of it.

He was all too aware that while cotton prices had declined since the onset of the Depression, the wages of the pickers had fallen even faster.

In the months leading up to the strike, there was actually some hope of recouping those pay cuts because the cotton market had advanced after President Roosevelt signed the Agricultural Adjustment Act. Yet from the growers' perspective, the offer of sixty cents was already a concession that they were doing better; they had paid only forty cents the year before. The sixty cents was also superior to what pickers in other states were earning. But many workers in the valley thought that they deserved more, and in Roberto's mind the battle line was now drawn: It was rich versus poor, *padron* versus peasant.

Rudy felt caught in the middle, but he could hardly blame the workers. Thirty motorcycle riders from the California Highway Patrol had roared into town, and on top of a cotton trailer two blocks from the railroad crossing they had mounted a machine gun pointed in the direction of the strikers' camp. "They were ready to mow down these strikers, all because they had the audacity to ask for more money," he said. "The attitude was, 'You're supposed to be out in the fields whenever we deem it necessary.' To me, it was an injustice. Even at that age, I knew that was wrong."

Strike sympathizers donated apples, beans and potatoes to Camp Corcoran, and it became Roberto's job to distribute what little food there was. The meals were cooked in a communal kitchen and two schools for the children were set up behind the barbed wire. A nine-member central strike committee kept order and a squad of sentries, many of whom had fought in the Mexican Revolution, patrolled with guns and crowbars. From a covered stage emblazoned with the words "Circo Azteca," entertainers stoked the spirits of the strikers with song and dance. In a bizarre coincidence, the Mexican circus had encamped on the same piece of ground days before the strikers showed up, and the performers gladly stayed and played.

Still, conditions at Camp Corcoran were grim, unforgettably so for writer and union champion Ella Winter, the wife of muckraking journalist Lincoln Steffens, who recounted a visit made one night. "We stumbled along in almost pitch-darkness among tents, people, burning oil stoves, refuse . . . smells of greasy cooking, stale fish, damp clothes and sewage. An occasional dim lantern lit up the mud we struggled through; there were small muffled sounds that gave the feeling of a crowded city, though we could not see the crowds. A sick baby wailed, and I had a sense, in that darkness, of people waiting, waiting."

The waiting was itself an achievement, as Camp Corcoran developed into a center—and symbol—of worker solidarity. With thousands of strikers housed together, the union could "maintain complete discipline," noted the union president, Pat Chambers. The close quarters, he added, safeguarded the bulk of his troops from being "picked off like strawberries" and thrown in jail or hounded out of the valley by the farmers.

Previous strikes by the union, including walkouts just weeks earlier in the vineyards of Fresno and Lodi, had been crushed. Chambers and his comrades had learned a lot from these setbacks, as well as from a few earlier successes. For instance, union strategists now encouraged the rank and file to participate in crucial decisions, and each group of cotton pickers—not only the Mexicans, who accounted for 95 percent of the strikers, but also the small communities of whites, blacks and Filipinos—was made part of the process. Chambers himself didn't merely plan strategy and then hightail it for a hotel along Highway 99 or back to union headquarters in San Jose. The unflappable Irishman lived among the farm workers, and they protected him wherever he went.

At the same time, most of the workers had little affinity for communism, and Chambers and other strike leaders took care not to flaunt their Marxist beliefs. Caroline Decker, a blond wisp of an organizer who had considered herself a communist since she was sixteen (she was all of twenty-one during the cotton strike) was a fervent ideologue, "prepared to give my life" to the party. Still, she preached doggedness instead of dogma, and her stirring speeches became one of the union's most potent weapons. The gin operators, she declared, have "proceeded to exact two pounds of flesh from the unfortunate cotton picker. . . . The strikers are entitled to hold out for more than 60 cents."

Despite their sound game plan, union leaders worried how long they could persevere. Private donations of milk and foodstuffs to Camp Corcoran and other union strongholds helped, but much of it never reached the strikers' kitchens. Sheriff's deputies seized truckloads of meat and vegetables, preferring to see the rations rot than the strikers fill their stomachs. "These were not fighting people," Roberto Castro said. "They were just asking for a bite to eat. That is what the people were asking for."

The night that Ella Winter visited Camp Corcoran, "a frowsy woman came out with a bundle, a tiny wizened baby whose face was almost black." When one of those in Winter's party realized the child was dead, evidently from starvation, she shrieked. The Mexicans stared back with blank expressions, doubtlessly wondering, said Winter, "who these strange people were, commiserating all of a sudden with a wretchedness they had always lived in."

The farmers had no shortage of ploys. When they failed to buy off union leaders, they arrested more than 100 pickets on vagrancy and other charges. They brought in strikebreakers where they could and tried to turn white worker against black and Mexican against Filipino. They threatened the Mexicans with deportation. Mostly, they got belligerent.

Even Kings County Sheriff Van Buckner, who stood on the growers' side, had "a hell of a time" controlling their appetite for violence. One day, he ran into a seething gang of farmers who told him that they were headed straight to Camp Corcoran to "kill all the Mexicans." Buckner finally shamed them out of their plans by remarking how much they sounded "like a bunch of young sports going to see how many ducks they could shoot before breakfast." Billy Boswell Jr., Jim Boswell's older brother, got into the spirit of things himself. Hoping to put a scare into some of the union men, he and a couple of farming buddies tossed a noose over a tree and gave them thirty minutes to get out of town.

Even though Mateo Castro stood strong with the company, his failure to convince his workers to stay and pick, capped by his inability to rein in his own son, had earned the disdain of Bill Boswell Sr. Forces may have moved beyond the control of the likeable crew boss, but failure of that sort couldn't be abided. So Boswell fired Castro, and the family was forced to find quarters in a labor camp run by a smaller grower. "What transpired between my dad and Mr. Bill, I don't know," Rudy said. "But my dad had a job to fill the camp and the workers just wouldn't come."

As the strike wore on, Colonel Boswell chose to stay behind the scenes, letting L. D. Ellett speak for the farmers. Still, he had a way of making his feelings known. At one point, he called up J. W. Guiberson, a member of Ellett's growers committee, to give guidance on dealing with the strikers. "You ought to have the guts to go in and clean them out," the Colonel said. "The cotton industry in the San Joaquin Valley will never be worth anything unless you do."

Blood was spilled, at last, on October 10, 1933. Twenty miles away from Corcoran, along Highway 99 in the Tulare County town of Pixley, about forty growers pulled up to where Pat Chambers had just addressed a throng of strikers. "This is no time for a backward step," Chambers had said, his wiry body perched on the back of a truck bed. "We must fight and show the farmers a solid front. We will match the farmers with their own violence. Let them start something—and we will finish it."

The talk was tough but empty. After Chambers spoke, the strikers moved across the road to a red brick union building, and the farmers brandished pistols and rifles. W. D. Hamett, a leader among the rank and file, begged the armed men not to shoot into the union hall. "There are women and children in there," he shouted. Hamett, an itinerant farm worker from Texas by way of Oklahoma, had a magnetic presence and a booming voice that he'd sometimes unleash on Sundays from a Church of Christ pulpit. He was six-foot-two with a barrel chest, and everyone called him Big Bill.

When the first of the advancing farmers shoved a rifle toward Hamett, he grabbed it and pushed it to the ground. As Hamett and the farmer wrestled in the gutter, a shot rang out and hit another man, Delfino D'Avila, a Mexican consular representative allied with the strikers. "They got me," he sputtered, his last breath rushing out of him. "Bill, you better run."

The strikers screamed and fled. Hamett ran all the way through the union building and escaped out a back fence. The farmers let fly with a fusillade. "Let 'em have it, boys!" one of them yelled. Another striker, Dolores Hernandez, was then cut down.

That same afternoon, near the town of Arvin in Kern County, more shots were fired after a five-hour stare-down between 250 pickets and thirty growers carrying guns. A third striker, Pedro Subia, was killed there. In addition, twenty workers were wounded in the two melees.

As night fell, one of Corcoran's own—Clarence "Cockeye" Salyer—began to make his way home, blood on his hands. He had been with the farmers in Pixley that day, and his thoughts turned to the Colt .38 Special he had used to fire on the strikers. He couldn't be 100 percent positive that it was his bullet that had slain Dolores Hernandez. But he was pretty damn sure—and wasn't about to take any chances.

A nervous man might have tossed the gun into one of the ditches or canals that ran between Pixley and Corcoran. But Salyer wasn't a nervous man. He was a practical man. He pulled into the driveway and walked inside his ranch house, clutching his .38. His wife, Cordie, grew anxious over the trouble that her husband had brought home. "My mother was all upset," recalled Fred Salyer, Clarence's son. But Clarence himself "was as calm as they come. . . . He just didn't want any ballistics expert to check his pistol."

So he headed out to the blacksmith shop on his farm and stoked the coal forge. Fred, then ten years old, went with him, cheerfully turning the crank for his dad. "It was fun for me," he remembered. The two of them waited until the hearth was red-hot, and then Clarence dropped in the gun. The cops would later come calling, but Clarence played dumb. A handful of growers who had gone to Pixley would be arrested, but Salyer wasn't among them.

Almost seventy years later, Fred Salyer would sit in his office, a photograph of an old Mexican labor camp on the wall, and recount the day that his father shot somebody in Pixley. Rumors of the murder had been floating around for decades, but Fred had never before spoken about the incident beyond his most intimate circle. It was a remarkable story—not just because of the way it underscored Cockeye's cold brutality, but because of the place it held in the pages of California history. And yet Fred talked about the incident as matter-of-factly as if he was talking about the weather. The way Fred saw the world, Pat Chambers was nothing but "a Commie agitator" bent on "putting cotton farmers in the area out of business." He didn't say it, but you got the distinct impression that Fred didn't have much feeling for the pickers who lost their lives in the strike. It was a war, and people die in war.

As for the gun, Fred recalled how he and his dad stood there until it disintegrated in the hellfire of the forge. "I watched it go up in sparks," he said. And then Fred Salyer—a man who liked to insist that nothing he or his family did was particularly noteworthy—gave away a little secret. After all these years, he had kept the forge, and he planned on making sure that after he died, it wound up in its rightful place: the museum in Tulare.

AFTER SOME INITIAL FOOT-DRAGGING by local authorities, eight growers were indicted in connection with the Pixley shootings, while an investigation into the Arvin killing continued. Valley justice being what it was, Pat Chambers was also sent to jail for allegedly inciting a riot.

Any growers who believed that the carnage would scare the workers into submission were sorely mistaken. Instead, the strikers bonded over the slaughter. "Our comrades died, but we promise to continue until we win or we die!" read a sign hanging in the union's Tulare offices.

The murders turned the cotton strike of 1933 into a cause célèbre. Actor James Cagney and other stars sent money to help the workers. Later, when the big-screen gangster had the chance to meet Caroline Decker at the Carmel home of Lincoln Steffens, he tried to teach the young labor leader how to tap dance.

Others found artistic inspiration in the cotton workers' fortitude. Langston Hughes wrote a play titled *Blood on the Cotton*, which starred a character named Jennie Martin as Caroline Decker's alter ego. But it was deemed too unwieldy and overwrought to be performed.

> Sheriff: "I have 100 men deputized in this county to protect law and order."
> Jennie: "And aren't we entitled to a little of that law and order then?"
> Sheriff: "You're an agitator, that's what you are."
> Workers: "Viva la huelga! (Long live the strike!) Viva Jennie!"

No American writer used the backdrop of the Central Valley to greater effect than did John Steinbeck, and the combat in the cotton fields provided him with a trove of material. Many details from the '33 strike were plopped right into his novel *In Dubious Battle*. Big Bill Hamett likely served as the model for the muscular London, who led the rank and file in the book's fictitious apple strike. Jim Casy, the lapsed reverend in *The Grapes of Wrath*, may also have been based on Hamett.

In the days following the slaying of the cotton pickers, the suffering intensified. Besides the lifeless infant that Ella Winter had seen, as many as eight other strike babies died from malnutrition. Sheriff Buckner prepared to close down Camp Corcoran as a menace to public health, though the local district attorney admitted that it "was just about as sanitary as other cotton camps," which the growers had always run.

With the situation fast deteriorating, the federal government leaned heavily on the farmers and the union to call a truce. President Roosevelt didn't like big labor any more than he liked big business, and his New Dealers saw it as their mission to bring the two sides together. As a first step, the government compelled both the union and the farmers to consider the input of a three-member fact-finding commission. Two days of panel hearings held in the town of Visalia lay bare just how horrible the lives of the pickers really were. "We have no water, lights or toilets," one field worker testified, going through a litany of wants that was typical for California's cotton hands. Another picker stated that she hadn't been able to afford shoes for her eight children since 1930.

Boswell's senior man in the valley, Louis Robinson, reminded the fact finders that times weren't great for farmers, either. "If the 60-cent picking rate is raised," he said, "the grower would lose what little he has." But others challenged that view, contending that many cotton growers would be willing to raise their wages if only finance companies such as Boswell would stop dictating such extortionate terms.

On October 23, the fact-finding commission recommended a compromise that was being privately urged by the feds: a new pay rate of seventy-five cents but no recognition for the communist-led Cannery and Agricultural Workers. Both sides found the deal hard to swallow, but the lead federal negotiator wouldn't let up. He threatened to exclude the growers from New Deal farm programs, and told the strikers that they'd be cut off from government relief, their main source of sustenance for the past two weeks. The seventy-five-cent settlement was promptly ratified, and the pickers returned to the fields.

The twenty-four-day strike proved to be a milestone, an important precursor to Cesar Chavez and his United Farm Workers. Its scope and duration and the monetary gains it secured—an estimated $1 million in additional pay—were unprecedented. Paul Taylor, a University of California economist whose firsthand report on the strike became the standard roadmap for generations of scholars, saw it in no less than cataclysmic terms. It had revealed the caste system within California's rural society, he said, the way the "faulting of the earth exposes its strata."

But for all that, the lives of most cotton workers weren't materially changed. Sam Darcy, the Communist Party leader in California, acknowledged that even if the union's demand for $1 in picking wages had been met, it would have been enough for "one square meal a day" and no more. By the spring of 1935, the Cannery and Agricultural Workers had withered away. Eight union leaders, including Pat Chambers and Caroline Decker, were convicted of criminal syndicalism after the longest trial in California history. They languished in prison for two years. As for the farmers indicted in the Pixley shootings, all eight were ultimately acquitted. No one was ever fingered in the Arvin murder.

In Corcoran, most people seemed eager to forget all about the strike, adopting a code of silence that Al Tapson found very odd. Tapson moved to town to teach school not long after the strike was settled, and on his very first meeting with the principal, he was warned point-blank "not to discuss the recent trouble." It was "very difficult to get people to say anything" about the strike, Tapson recalled. Once, when he tried to talk with the principal's wife about the communists' role in the labor movement, her husband shushed them right up. "There was an obvious split in the town," Tapson said. "It was literally the old story of one side of the tracks against the other."

Even some of the Mexicans acted as if they forgot. They asked Boswell to buy them baseball uniforms, and the Boswell Bears suited up as part of a new cotton pickers league. The Bears won the 1939 crown, thumping the Hanford All-Stars in the championship game, 3–1.

In the years to follow, Rudy Castro, whose young eyes had seen so much at the Boswell labor camp, was among those who'd airbrush away the more painful parts of history. Despite his father's firing, Rudy would discount any enmity between the Boswells and his family, stressing that whatever grudge the Boswells had, it didn't last. By the 1950s, both Mateo and Roberto had come back to work for the company, and Rudy himself remained there for forty-five devoted years, emerging as one of the Boswells' more valued supervisors and biggest cheerleaders.

Well into his retirement, approaching eighty and a bit infirm, Rudy rested in his small, dark house in Corcoran and limned the past with the positive. The brother he so loved and admired, the one who had walked over to the union side and opposed Boswell, was long dead.

"I can't tell you how much I love the Boswells," Rudy said. The Boswells are "the nicest, most compassionate, most intelligent, beautiful people in the world." If you coaxed him enough, he'd remember a few of the uglier moments, like the time the company put the hog pens right next to the labor camp when he was a boy. "But I hate like heck" to even bring that up, Rudy said. "I don't know if you can detect any bitterness. I try not to be bitter. There's a better day coming tomorrow."

"Goon-Squad Tactics"

I N T H E F A L L O F 1939, Alvin Odle said good-bye to Oklahoma and set out for California in a Chevy full of stolen gas—his two sons in the backseat and his wife beside him, nine months pregnant. Odle was a big brute of a working man, six feet two and all muscle, mean and head-strong with a heart that wouldn't quit. Once he got sour or sweet on a notion, it was nearly impossible to tell him otherwise. His wife, Pearl, thought it might take a few trips back and forth on Highway 66 to con-vince him that California—and not Oilton, Oklahoma—was the prom-ised land. But one day, a few years after that first move out west, he walked into their Corcoran farmhouse with his pocket full of cotton cash—"Okie rich" as his son recalled it. He informed Pearl that they'd be making a sixth trek back home to Oklahoma, this time for good, and she let him have it. "There's the road, Alvin," she said. "Feel free to take it. Just me and the boys ain't following." So Odle was cured of Oklahoma, at least until he made his first million dollars in the lake bottom years later and headed east again.

Why he pined for the cotton and oil fields west of Tulsa was a mystery to his wife and boys. A good part of Oklahoma, it seemed, had been blown west by the Dust Bowl in the 1930s. If Odle's region of the Sooner state had been spared the spectacular dust storms that blackened the panhandle sky, there was still plenty of drought and economic depres-sion sowing deep misery. Working the oil and gas fields first light to dark might bring home a few dollars, but not much more. And while the country outside Tulsa possessed forests of oak, hickory and cedar, little good they did when the sawmills weren't hiring. Odle knew all this but deep down he still hurt for home. No matter how many years he spent in California, the dirt under his fingernails was the dirt of Oklahoma.

He had grown up on 160 acres of homesteaded land so hilly and rocky that it could barely support his father and mother and siblings. So in

1933, after he and Pearl married, the twenty-two-year-old Odle left home and became an Oklahoma tenant farmer, growing forty acres of cotton, corn and peanuts, and raising hogs. His only help during the years on the farm came from a mule so ornery he named him Hitler. His neighbors in Oilton didn't know what to make of Odle. He seemed not to grasp that the odds of his lease were stacked against him. No matter how hard he toiled, the way the landowner had cut the deal—two-thirds crop to Odle, one-third crop to him—sealed his fate. Before he ever planted a single seed in spring, Odle had already said good-bye to his profit. Even so, they admired his spirit. He could do more work than just about any human they'd seen, and he was such a physical specimen— strong and swift and tireless—that he seemed to be cast in the mold of Jim Thorpe, the immortal Indian athlete who grew up on a reservation just down the road. Watching Odle plow the hardscrabble earth of middle Oklahoma, they wondered who would break down first, the farmer or his one-eyed mule.

"Hitler was a big mule with a big head. He'd just as soon bite you as kick you," Fred Odle, his oldest son, recalled. "But if you harnessed him from the blind side and got him in the field, nothing could stay with him. He and my father were quite a team. One day Hitler developed a fistula, a pressure sore on top of his shoulder. The vet came out and tied him down and started working on it. Hitler was in a lot of pain and he grabbed this guy with his teeth. He took a big bite and wouldn't turn him loose. He almost crippled that guy. They finally broke the guy free by breaking an ax handle over Hitler."

After seven years of tenant farming, the mule and an old Chevy were the only things Odle owned. Heading for California that first time, he sold the mule and still found himself short of cash to pay for the gasoline. In the middle of the night, he trespassed onto the grounds of the Drip Gut refinery outside town and cut a tiny hole in a gas line, siphoning enough raw fuel, drop by drop, to cover the entire 1,500-mile trip across the desert. The Odles passed Hoot Owl Hollow, the house next to the graveyard where Fred and his younger brother Jimmy were born, and waved good-bye to Oklahoma.

They made it to Indio in California's Coachella Valley before Pearl gave birth to Billy, the third of four sons. They took a puke-green army tent in a filthy labor camp filled with vegetable pickers from Oklahoma and

Arkansas, Texas and Missouri. Like the Odles, the workers had been chased out of the American heartland by drought, a dreadful economy and the ill effects of the Agricultural Adjustment Act—the federal government program to reduce farm acreage. "Gypsies by force of circumstance," Steinbeck called them. All told, more than 900,000 of these Bible Belt refugees rattled over the borders of the Golden State between 1930 and 1950. Into the San Joaquin Valley the more destitute of their ranks came—"ditch bank Okies"—seeking better wages. They ended up changing the culture with their tent-show revivals and charismatic healings, their strong backs and red necks and virile racism.

The cotton camps of Corcoran reflected the shift. The Depression had stemmed the flow of Mexican migration to the United States and sent many *campesinos* back across the border. Urban relief agencies in Los Angeles and elsewhere also embarked on an intensive repatriation program, and while farmers continued to recruit workers from Mexico to help with the harvest, many were ping-ponged between the two nations three and four times. By the late 1930s, it was no longer necessary to have much Mexican labor around at all. Suddenly there were plenty of Dust Bowlers willing to do the work.

"I was six years old and they gave me a five-foot long gunny sack with a strap on it, and I started picking cotton," Fred Odle said. "And every year the sack got bigger until I was carrying a fourteen-footer and picking 420 pounds when I was twelve. We moved from Boyett's labor camp to Forrest Riley's labor camp but the places were all the same. Terrible. It was shacks we lived in, and the bathrooms you couldn't stand to go inside of 'em. You ended up finding a place outside. Wintertime there'd be winos, so many winos, and I don't know how they stood it, except that they were drunk."

There was String Bean and Booger Red and Frisco Sid and Reno Ralph and Fresno Fred and a woman named Flournoy who watched over her son, a five-year-old they called Duck. He dipped snuff and chewed tobacco, and his raspy voice turned the air blue with cuss words that confounded boys twice his age. The winos would literally cotton-pick themselves to their next drink. With a canvas sack in one hand and a bottle of Gallo in the other, they'd take a swig, screw the lid back on and throw the bottle a good distance down the row. Then they'd pick extra fast just to reach another swallow.

Wages had inched up and cotton pickers were making as much as $1 for every 100 pounds of fiber picked. The growers felt a certain kinship with the Okie field hands, and those who showed any initiative were quickly promoted to tractor driver or gin operator, jobs that typically took years for a Mexican or black to achieve. To Fred Odle's eyes, this wasn't fair, the idea that Mexicans and blacks weren't capable of operating machinery. He saw that the growers had been shaped by their upbringing in the South, and they had turned Corcoran into a California version of a Southern town, only infused with the Wild West. "It was booming, like you see maybe on television or in Alaska or something. There were Mexican labor camps and black labor camps and white labor camps for miles around. People would go to town with a couple thousands dollars in a paper bag, and they'd spend it as they walked down the street. R. H. Curtis owned a car lot and he sold something like twenty-four cars each week."

On good days, the Odles stuffed an incredible 1,300 pounds of cotton into their sacks. Pearl kept up with her husband row by row, picking 350 pounds to Alvin's 650. The boys grabbed 300 of their own until their hands turned bloody. Each night, their mother would rub Campho-Phenique into their fingers and palms in hopes of turning them leathery. Eighteen dollars a day quickly added up, even accounting for Alvin's gambling and bad head for business. He was a sucker for all games of chance: poker, greyhounds chasing rabbits, and badgers fighting dogs— the white camp's version of a cockfight. His son never forgot the sound of those pitched battles. On Sundays, with most of the camp too exhausted to go to church, the men would pick a cabin and break out all the windows and chop the door in half. This way, they could better gather around and see the action inside. They'd take a badger no more than thirty pounds and the meanest 100-pound mutt they could find and throw them into the cabin together, one and two dollar bets riding on the outcome.

"I don't know if you've ever seen a badger fight, but pound for pound there's nothing tougher," Fred said. "They got big claws and the doggonest teeth. Their hide is real thick and loose so they can slip away. And they naturally hate dogs, like a big coon, only tougher. Oh Lord, it was like a couple of lions fighting. The dogs could never find a way to hold on to the badgers, and after a while they'd just quit, tired out. Finally they brought in this one big dog that learned how to kill 'em. He would grab a badger and push it down on the floor and crush him, crush his back. Those badgers got to the point where they wouldn't even fight him."

Riding one of his high times, grower Elmer von Glahn had built a track on the outskirts of town where he staged Sunday races between quarter horses and mules and just about anything with four legs that could run. Once a month, he converted the track into a ring and held boxing matches pitting the toughest workers from each ranch. No one in Corcoran could last more than a round or two with Alvin's brother Sam, who had learned to box in the CCC camps and, legend had it, once rode the rails to Chicago to spar with Joe Louis.

Sam Odle weighed 220 pounds and was strong as a bull and fast. After a long string of easy victories, he had gotten so complacent that when they brought in the so-called champion of Texas, a man who stood six feet four and weighed 230 pounds, he decided to buck hay the day of the fight. "Uncle Sam took him too lightly, I guess. He bucked hay all day and fought this guy at night. They were just like mules kicking, and he was doing real good. He was whipping this guy. But anyway, this guy broke my uncle's jaw in about four or five places. He had run out of gas and couldn't move. And this guy nailed him."

That first year in Corcoran, Fred learned how to use his own fists after his father caught him running home from school one day along the ditch bank, nearly peeing in his pants ahead of a group of Mexican boys.

"If you're running from them tomorrow, I'm gonna give you the whuppin' of your life."

"But they're wild Indians, Dad. Savages."

"Those ain't wild Indians, you fool. Them are Mexicans. From Mexico."

That night in bed, Fred told himself he's rather be beaten up by six Mexicans than whipped by his father. The next day, he stood his ground and knocked down one of the bullies, and they never pestered him again.

For the most part, the different racial groups kept to their own camps, heeded their own labor contractors and picked their own fields in different parts of the lake basin. The lines had been drawn in such a way that, sixty years later, each group would claim that they alone had chopped and picked the lion's share of Corcoran's cotton in the 1930s, '40s and '50s. This compartmentalization, not to mention the ebb and flow of their numbers, stymied union organizers. By the late 1930s, the talk of unions centered not on the fields, but on the gins where the cot-

ton was processed and baled. One way or another, the fight over unionizing the cotton kings wasn't quite over.

THE SIX GINS inside the Boswell processing plant clattered
along in their usual way until the strangest of sounds settled upon the
factory floor. Silence.

"What's the idea?" asked ginner Oliver Farr as each machine, without
warning, was turned off and the engines whirred to a halt. It was a Friday morning, less than a week before Thanksgiving, in 1938.

"They'll tell you about it outside," replied Bill Robinson, a plant foreman who had flipped the switches quieting Farr's equipment. "It's about
your union."

Union organizers—this time from the Cotton Products and Grain
Mill Workers, Local 21798—were back in Corcoran. If they were looking
to stir things up, they had picked the right place: a cavernous plant that
formed the industrial heart of the Boswell Company. Farr, a vice president of Local 21798, moved with caution out a side door and into the
bale yard. A pack of sixty employees, loyal to Colonel Boswell, swarmed
like dogs.

At a glance, Farr didn't seem much different from the faces that surrounded him. A big, broad-shouldered man hired by Boswell in 1936 after arriving from Altus, Oklahoma, he wore overalls and boots, and his
hands bore the calluses and grease smears of somebody well ac-
quainted with a hard day's work. But anyone looking closer would have
noticed that Farr and a handful of other Boswell employees were sporting something extra on their denim this day: an American Federation of
Labor button. It was the first time that any of them had ever worn union
regalia on the job.

"I want to know about your damn union," shouted a worker in the
crowd, now clustered around a cotton wagon.

"Well," Farr said, girding for a fight, "what about the union?"

"The company doesn't want your union here, and I don't see why you
fellows should turn again' the company you're working for."

One of the voices demanded to know the identity of the president of
the local. Farr replied that it was his friend, Lonnie Spear, a ginner at
Boswell for ten years. Spear was standing a few yards away, and the

crowd closed around him. He tried to back them off with reason, talking about the still shaky economy and the need to cut their twelve-hour workday into eight-hour shifts. This would ensure employment for more people.

The mob wasn't appeased. "Throw them out," Rube Lloyd, a maintenance supervisor, yelled. "Let's throw them out."

Three workers grabbed Spear and locked his arms while another pushed at him from behind. A fifth worker lunged at Farr, shouting, "I'm going to put you out." Farr stiffened his six feet four, 200-pound frame and managed to back the man off, but Spear couldn't wriggle himself free. Panting from exhaustion, Spear was dragged to the manager's office where a chorus of workers demanded that the union men be fired.

Both sides were told to cool down and ordered back to work. Farr, Spear and several other union workers gathered around one of the gins and tried to sort out what had just transpired. As they talked, plant foreman Robinson came right at them again.

"What are you going to do, Lonnie?" Robinson said. "It seems as though the boys aren't going to work with you. You should go home. That would be my advice."

"As a foreman, will you tell us to go home?" Farr asked.

"No, not as a foreman, but this is my idea, that you had better go home." And so they did, fourteen of them in all.

The forced exit of the union men highlighted the complexity of labor relations at Boswell. The fracas that morning couldn't merely be reduced to management versus the rank and file. True, the Colonel and his executives were ardent antiunionists, but most of those who actually pushed the AFL members out of the plant that day weren't Boswell bosses. They were fellow hourly workers.

Only five months before, as the union began to organize at Boswell, it must have seemed an auspicious time to rally under the AFL banner. The traditionally hidebound federation of craft unions, which in recent years had been out hustled by John L. Lewis's more nimble Congress of Industrial Organizations, was finally beginning to rejuvenate itself. Besides, those setting up Local 21798 thought they had a pretty fair case to make: The wages and conditions at Boswell were inferior to those that had just been negotiated at Anderson Clayton.

The Colonel insisted that he was paying as much as he could afford. But ginners at the Corcoran plant earned just fifty cents an hour, well under the average hourly wage for U.S. manufacturing workers of sixty-three cents. And those with less important jobs—sewing cottonseed sacks, for instance—often received as little as forty cents an hour. "A man can't make a living at it," groused Stephen Griffin, another of the union faithful at Boswell.

Oliver Farr, who was first exposed to the unflinching pragmatism of AFL President Samuel Gompers during a Texas rail strike in the early 1920s, tried to persuade his Boswell coworkers that all men should bargain their services. Otherwise, said Farr, it's "slave labor." Still, the majority of Farr's coworkers were suspicious. "We're one happy family," plant employee Frank Gonders told an AFL organizer who had come to Corcoran in July to set up the local. The union had sent out more than thirty invitations to its first meeting, but only a half dozen or so men showed up. A few attendees, like Gonders, openly questioned the need for an AFL presence in Corcoran.

Enlisting with the union took guts. To recite the AFL oath meant defying the town's power structure, which painted its union opposition in the most patriotic hues. "I stand for American institutions and ideals," said J. B. Boyett, a leading Kings County farmer. Although the Communist Party wasn't involved in the tussle at the gins, its sponsorship of the Cannery and Agricultural Workers five years earlier had left a stain on all of organized labor. Boyett's rhetoric dripped with the hackneyed phraseology of the best Red-baiters. He labeled union activists "subversive elements" and their techniques "un-American."

Most of those inside the Boswell plant didn't need a lot of convincing to share Boyett's view. The Dust Bowlers, even though they were just scraping by, didn't see unionization as the answer to their problems. As a group, they were strongly anticommunist and had sat through too many evangelical sermons not to know that political action diverted from personal salvation. Some were unwilling to join a union that included blacks or, as in the case of Local 21798, Mexicans. Beyond that, the migrants from the Southwest tended to possess a keen streak of independence. "Rugged individualism and collective action," one interpreter of the period has noted, "do not mix well."

Some of the union's complaints also rang hollow. Farr, for one, griped that shifts at Boswell routinely lasted twelve hours and sometimes up to sixteen hours—a much longer day than a man ought to work. But many at the plant liked the schedule, grateful for a way to puff up their paychecks. And as taxing as the hours were, Boswell's managers demanded no more of the shop-floor workers than they demanded of themselves. Gordon Hammond, the plant supervisor, was the undisputed leader of this cult of drudgery. He got to work at 5:00 A.M. every day, seven days a week. At 9:00 P.M., having taken time off only for a hurried lunch and a hurried supper, he could still be found overseeing the gins and the oil mill. That's when his brothers, all of them foremen, would sidle over and ask if they might be excused. "What," Gordon would say, "quitting in the middle of the day?"

It didn't help the union's cause that tough times endured. While the worst of the Depression had passed, the so-called Roosevelt recession still gnawed at the nation. Boswell had managed to bolster its balance sheet since the early 1930s, but the cotton industry as a whole still found itself saddled with huge crop surpluses and falling prices.

Meanwhile, the influx from the Dust Bowl had brought added economic strain across Central California. In Kings County alone, the population rose by nearly 40 percent during the 1930s, and the area two miles west of Corcoran came to be called "Oklahoma City." Relief rolls bulged, and health authorities fretted about the spread of tuberculosis. Thousands of anguished county residents pinned their hopes on Ham and Eggs, an initiative that promised to provide every unemployed Californian over age fifty with $30 each Thursday for life. While the pension plan's popular name evoked the full-belly comfort that so many hungered for—crowds would sing, "Ham and Eggs and glory, glory hallelujah!"—critics derided it as fiscal madness.

In November 1938, Ham and Eggs failed at the ballot box, leaving many in the valley feeling blue. Crooned a "ramblin' cotton picker" originally from Texas and now living near Bakersfield:

> We were all so happy
> And thought we were riding high
> But they knocked the whey out of Ham and Eggs
> And if I stay out here I'll die
> So I'll go right back to Texas
> And stay there all my life
> I hope to meet with some pretty little girl
> And I'll ask her to be my wife

I'll try to make her happy
I'll love her all the time
But I hope she is a millionaire
'Cause I ain't got a dime

Fallowness and flood added to the despair. The Agricultural Adjustment Act of 1938 had required a sharp reduction in cotton plantings, which compounded the glut of Okie labor and held wages down. What's more, the union couldn't have picked a worse weather year. Floodwaters had inundated 100,000 acres of lake-bottom land, driving down cotton production at Boswell from 47,000 bales to 10,000 in one year's time. The Corcoran plant, employing half its usual workforce, was down to ninety men.

For most of the workers, the decision not to go with Local 21798 proved an easy call. Who really wanted to make friction, especially for a cause that many had consigned to the radical fringe? Most plant workers were only too thankful to have a job in the current economic climate, especially one that allowed them to work indoors. At the end of the day, only twenty-nine Boswell employees joined the AFL—the fourteen who had been purged from the plant, plus fifteen others.

But the union men at Boswell weren't about to roll over. They'd leave their mark on the company, and on Corcoran, before they got through.

IN THE SHADOW OF THE BOSWELL PLANT, two men sat and waited. Their jalopy said "AFL picket car" on the side, but they didn't holler "scab" at anyone. They didn't scuffle with the scores of Boswell workers coming and going. They didn't even try to pass out union literature. As picket lines go, this one appeared pretty innocuous.

Yet it was causing big trouble for the company.

Ten weeks had passed since the union men had been accosted at the plant. And every time a truck rumbled up with supplies or came out loaded with cotton or seed, the union men stepped out of the car and asked the drivers to stop shipping goods to and from Boswell. Those who belonged to the International Brotherhood of Teamsters—their AFL brethren—readily obliged. When the company tried using nonunion truckers, dock workers at the Port of Los Angeles wouldn't touch Boswell's "hot cargo." Eventually the company had no choice but to start sending its cotton by rail, disrupting its normal distribution system.

"Goon-squad tactics," the Colonel said of the picketing. He wasn't the only one irate. With their cotton tied up in the Boswell yard, local farmers felt that their own livelihoods were being imperiled.

The growers of Kings County had come a long way since the 1933 strike. They had put together their own chapter of the Associated Farmers of California, a group dedicated to trouncing organized labor through a blend of political savvy, subterfuge and sheer terror.

On the political front, the Associated Farmers pursued three primary aims. The first was to promote antipicketing ordinances throughout California and motivate local law enforcement agencies to oppose union strike activities. The second was to encourage the arrest and prosecution of labor leaders under the state's criminal syndicalism law. The third was to force strikers back to their jobs by making sure that they weren't eligible for public relief, an objective captured by the blunt slogan "No work—No eat." Bankrolled by a Who's Who of the state's biggest agricultural and industrial companies, the Associated Farmers maintained a patina of respectability. But its county chapters weren't content with just shaping public policy. They also employed labor spies and kept secret files on suspected communists. And when all else failed during a union walkout, members resorted to baseball bats, tire chains and pick handles.

On the morning of January 30, 1939, some 200 Corcoran-area farmers met on Clarence Salyer's ranch and loaded themselves into fifty cars. Driving to the Boswell plant, they surrounded the two union men on duty that day, Stephen Griffin and Elgin Ely.

Lloyd Liggett, who farmed 200 acres of cotton, flung open the pickets' door. "You ought to be ashamed of yourself out here on this picket line, as good as the company has been to you," Liggett said, glaring down at Griffin. "They just can't stand this, and we're not going to stand for it."

"Listen, Lloyd, if I am violating the law, why don't you go get the law?"

"No Steve, you're not violating the law," Liggett answered. "But we're not going to wait on the law."

Forrest Riley, a 300-pound farmer with the worst case of bowed legs, then spoke up. "Boys," he said, "we are not going to have this goddamned AF of L in Corcoran." The crowd started to roar: "Turn the car over. Take them out. What are we waiting on?"

"No," Liggett said, looking at the rabble, "the boys are going to leave." Swinging around again toward the pickets, he told them, "You'd better be getting out of here, and don't come back."

Salyer, who just six years earlier had gotten away with murder, again ordered the pickets to move, but Ely said they couldn't go anywhere because the starter on the car was broken. In an instant, someone yanked off the AFL sign and tossed it into the backseat. The farmers then began to push the car, and it rumbled into gear. Several other union men who had been alerted to the ruckus arrived, and they directed Griffin and Ely to retreat to town.

"Don't stop in town," Salyer bellowed. "Get plumb out of the San Joaquin Valley. We don't want you here." The union men scattered, and the farmers drove up and down the street, blowing their horns in triumph.

That night, 300 farmers returned to Salyer's ranch to feast on barbecued pork cooked in giant kettles and listen to a series of speakers talk about the labor unrest in their midst. "I remember from my history books when the embattled farmers took their stand at Lexington," intoned Harry Lee Martin, former president of the Los Angeles–based Better American Federation, the shadowy publisher of such titles as *The Red Menace*. "The farmers are going to be a major part in the saving of this country for our grandchildren . . . but only if they take the flag in one hand and a club in the other and say, 'These two go together.'"

Then grower Bob Wilbur got up to speak. The union had the farmers "right where they want you," he told the assembled. If the AFL boycott at Boswell went on, Kings County might not be able to get its cotton to market. Then Wilbur raised an even more horrifying specter: Perhaps the AFL's attempt to make inroads at the gins was really a backdoor way to organize the fields. Those sitting under the stars at Salyer's ranch knew full well that farm workers had a tough time unionizing, in part because they hadn't been afforded the same federal labor law protections as industrial workers. But if enough AFL men occupied the plant, Wilbur suggested, they'd be in a position to refuse to gin all cotton not picked by union hands.

Over the next few days, it was the union's turn to make some noise. The AFL filed a complaint with the National Labor Relations Board, a body that by the late 1930s had made its union sympathies clear. AFL officials also reported the farmers' bullying to the new governor, Culbert Olson. Elected just weeks before the Boswell workers were kicked out of the plant, Olson was the first Democrat to occupy the governor's office in California in forty years, and he was extremely chummy with organized labor.

An early February meeting at the capital with the governor's executive secretary did little to calm things, however. Roger Walch, the district attorney for Kings County and an unabashed supporter of the farmers, questioned whether it was even appropriate to allow such a small minority of workers to picket. The Boswell employees were "entitled to choose for themselves," he said, and it was obvious that most of them didn't want the AFL around.

"We're going to picket in your county whether you like it or not," A. H. Peterson, an AFL organizer, growled at Walch. "If we have to bring in 1,600 men we will."

"If we have to bring in 16,000 we will," added a labor spokesman from Fresno. "If you want civil war, we'll have that, too."

"You better put your finger down and quit pointing it at me," Walch told Peterson.

Once the governor's staff restored order, the union and the farmers were asked to hold a public forum in Hanford to work things out. Olson's office also requested that the union stop picketing until the NLRB completed its case. Walch cautioned that if the AFL sent reinforcements to the Boswell plant, "bloodshed" was likely. "You're not dealing with San Francisco or Los Angeles," Walch said, "but with the San Joaquin Valley where highhanded methods of picketing won't be tolerated."

The Hanford forum didn't resolve a thing. The farmers walked out of the Civic Auditorium en masse and left behind a written statement asserting their right to "grow, harvest and market our agricultural products freely and unhampered." The union announced that picketing would resume right away.

Hobbling his way to the picket line, rotund Forrest Riley handed the union men an editorial from the Tulare newspaper. The warning was clear: "The San Joaquin Valley farmer is not a yokel or a serf," the paper averred, "but the sturdiest of free-born American citizens, peaceful when his rights are not interfered with, but a lion in strength and fury when abused or aroused."

T̲HE̲ B̲OSWELL̲ G̲IN̲ S̲TRIKE̲ N̲EVER̲ T̲URNED̲ B̲LOODY̲ like the one in the fields in 1933, but it got plenty vicious. Words, not guns, did most of the damage. The most brutalized victim was a frumpy mother of four named Margaret Dunn.

In February 1939, Dunn's daughter Dorothy was seen chatting with an AFL organizer in the presence of the Boswell pickets. Dorothy had been introduced to the man the week before when she was escorting Drexel Sprecher, an attorney for the NLRB who had come to Corcoran to investigate the scene. A student at UCLA, Dorothy had by chance ridden the bus with Sprecher from Los Angeles, and the two had flirted on the way to the valley. The only reason she was now talking to the AFL official was to find out when Sprecher might next be in town. She had a bad crush, and she wanted to see him again.

It was all quite innocent, really, the way Dorothy explained it, but no one seemed to want to hear any of that. The buzz was that her mother, Margaret, an operator at the Corcoran Telephone Exchange, was using Dorothy to shuttle communications from union leaders to the boys on the picket line. Before the Dunns knew what hit them, they were the outcasts of Corcoran. Rumors flew that a petition was being circulated among farmers and businessmen demanding Margaret be fired. A friend told Dunn that Bill Boswell had pledged to get her axed "if it was the last thing he did."

The ax fell a few weeks later when Harry Glenn, the owner of the telephone exchange—the same Glenn who had run over a farm worker during the '33 strike and refused to stop—dismissed Dunn. She had worked for him, devotedly, for fifteen years. Her honor impugned, Margaret was hysterical by the time her husband got home that evening. A little later, John Dunn went out and confronted Harry Glenn.

"What have you got against my family?" asked Dunn, a foreman for Standard Oil. Dunn would later maintain—and Glenn dispute—that their conversation that night went something like this:

D̲UNN̲: "I thought the problems at Boswell had died down since January."
G̲LENN̲: "Oh no, it is getting worse. They have told me that your daughters are carrying messages from your wife to the pickets."

DUNN: "Listen, all this gossip about my wife and daughters is a bunch
of bull. You know that."
GLENN: "There's not much I can do. Boswell is threatening to ruin my
business if I don't get rid of Margaret."

Such a threat couldn't be taken lightly. Boswell was the telephone exchange's biggest customer, and the company also financed Glenn's cotton crop. Glenn owed Boswell $30,000.

Margaret Dunn filed her own complaint with the NLRB, but as the
weeks passed she began to have doubts about going forward with her
case. She still hoped to win her job back and, in the process, salvage her
family's reputation. But she agonized over whether she really wanted to
air her mess in front of the federal government, not to mention every
scandalmonger in Corcoran.

From the start of the NLRB investigation, a key question was whether
Riley, Salyer and the other growers had routed the Boswell pickets on
their own or under the auspices of the Associated Farmers. For Margaret
Dunn, this aspect of the case had its own complications. J. B. Boyett, the
local president of the Associated Farmers, was close to her family. Getting the NLRB embroiled in everything "was going to cause an awful lot
of hurt," Boyett told Margaret, "friend pitted against friend." With
Boyett's assistance, Margaret wrote a letter to the NLRB trying to withdraw her complaint.

It was too late. On May 18, 1939, a hearing was called to order at Corcoran's American Legion Hall in the matter of J. G. Boswell Company,
the Associated Farmers, the telephone exchange and the union. The
case marked the first time the NLRB accused the Associated Farmers of
violating the Wagner Act, the groundbreaking labor law that spelled out
the rights of workers to organize.

In many ways, it was an ominous moment for a company like
Boswell to go on trial. Just weeks earlier, *The Grapes of Wrath* had been
published, shining a spotlight on what John Steinbeck characterized as
shameless profiteering among Central California's biggest growers, all at
the expense of the common man. Not surprisingly, the Associated
Farmers and other agricultural organizations attacked *The Grapes of
Wrath*, as did many Dust Bowlers who were offended by the way Steinbeck portrayed them. Bill Camp, the man who had given birth to the

modern cotton industry in California, called the book "obscene" and "vile" and presided over its burning in Bakersfield.

For the growers, however, *The Grapes of Wrath* was only the first shot in a full-scale bombardment. Three months later, Carey McWilliams's *Factories in the Field* came out, tracing the "land monopolization" among California's rich and powerful since 1850, and the "harsh repression" of farm labor in the state since the Wheatland hop riot of 1913. "Today," wrote McWilliams, "some 200,000 migratory workers, trapped in the state, eke out a miserable existence, intimidated by their employers, homeless, starving, destitute." His mutinous solution: The "present wasteful, vicious, undemocratic and thoroughly antisocial system of agricultural ownership in California" should be "abolished" and replaced with some type of "collective agriculture."

McWilliams—the son of a Colorado cattle broker, one-time corporate lawyer and literary protégé of H. L. Mencken—was more than just a nuisance with a pen. In January, Governor Olson had named him head of the state's Division of Immigration and Housing, and he used his office to try to impose a "fair" wage structure on the cotton industry. In late 1939, the Associated Farmers tagged McWilliams "Agricultural Pest No. 1 in California, worse than pear blight or boll weevil." By then, many growers were denouncing a conspiracy between McWilliams and Steinbeck, certain that the timing of their books couldn't be mere happenstance. "There is, I regret to report, one thing wrong with the conspiracy theory," McWilliams retorted. "I never met John Steinbeck."

The publishing world continued its assault on Big Ag in 1939 with *American Exodus: A Record of Human Erosion*. The book was a collaboration between photographer Dorothea Lange and her husband, Paul Taylor, the University of California economist who had documented the '33 cotton strike. In their journeys, Taylor and Lange had retraced the Boswell family's very own footsteps. Toward the beginning of *American Exodus* they tell of a plantation in Greene County, Georgia, which was abandoned in 1924 because of the boll weevil, and near the end of the book they show pickers emptying their sacks into a cotton wagon just outside Corcoran.

As the NLRB hearing went on, the case against Boswell came down to one man's word against another's.

The officers of Local 21798, Lonnie Spear and Oliver Farr, testified that the Hammonds had tried to dissuade them from bringing in the union long before they were expelled from the plant. The Hammonds swore that wasn't so. Another union man, Evan "Coon" Powell, said from the witness stand that Gordon Hammond had asked him to penetrate Local 21798 and be a snitch for the company—an accusation flatly denied. There was good reason to believe that Hammond held sway over Powell. He knew Powell's family in Georgia and had once helped Coon get out of a jam after he racked up too many debts playing "joker poker" and "California draw." Powell wasn't the most credible witness, however. The attorney for the Associated Farmers got him to admit that, besides being a gambler, he was also a big boozer and had been convicted of stabbing a man back in Georgia.

Yet the most scintillating testimony revolved around Margaret Dunn. According to her boss, Harry Glenn, Margaret wasn't fired from the telephone exchange because of anything to do with the union controversy at Boswell. Rather, he said, she was let go because other operators "were complaining very bitterly about her drinking." Another telephone company worker, who seemed genuinely concerned about Dunn, said he understood that she nipped from a bottle of port wine because of a stomach ailment that made it difficult for her to get "nourishment" any other way. Dunn's doctor, he said, had evidently prescribed the wine as medicine.

Glenn also testified that, while the union flap hadn't led to Dunn's ouster, Boswell had registered one complaint about her. Apparently, Glenn told the hushed hearing room, she was "running around nights with Fred Galusha," the manager of the local Anderson Clayton gin. And because she was handling Boswell's "intimate business calls over the phone, they objected" to her carrying on such a relationship "with a competitor."

After a month of finger pointing and the very public trashing of Margaret Dunn, the small-town soap opera came to a close. There was no clear winner. The boys from Local 21798 managed to bring out Boswell's antipathy for the AFL, but the company undercut the union's version of the facts more than once.

Associated Farmers representatives had claimed, fairly convincingly, that they hadn't instructed anyone to scare off the Boswell pickets. But it was equally clear that many of those involved in the affair were proud members of the Associated Farmers, and that Boswell was the largest financial contributor to the group in the county. Carey McWilliams would even cite Boswell's ties to the Associated Farmers as he decried the "vigilantism in California agriculture."

In 1941, the NLRB ruled that thirteen members of Local 21798 had been unlawfully locked out by Boswell and ordered them reinstated with back pay or put on a preferential hiring list. The telephone exchange was similarly ordered to rehire Margaret Dunn. For the opponents of the state's farming establishment, the ruling was reason to rejoice. "Fascist Farmers Lose Again," proclaimed *The Rural Observer*, an information clearinghouse dedicated to pumping out diatribes against the Associated Farmers. It called the case a "significant defeat" for Big Ag. The matter eventually caught the interest of Senator Robert M. LaFollette Jr. as he stretched his historic civil liberties investigation into the factories and fields of California agriculture. His committee, authorized to probe the tactics of union busters, subpoenaed Boswell's phone records and corporate files.

But in the long run, despite the NLRB ruling and the negative national spotlight, the farmers managed to fend off the union. The AFL never got rooted at Boswell, and in time it just faded away. Neither did organized labor accomplish much more in the fields. The CIO tried to get Okie and Mexican cotton pickers to go on strike, handing out fliers with an emotional pitch in the fall of 1939: "Two Men Killed in Pixley Exactly Six Years Ago . . . Murdered by the Bosses . . . Their Deaths Must Not Have Been in Vain!"

Not long before he sat down and penned "This Land Is Your Land," Woody Guthrie came to the big wooden union hall in Arvin and sang to a packed house:

> I ain't gonna pick your 80-cent cotton
> Ain't gonna starve myself that way.
> Gonna hold out for a dollar and a quarter
> Will they take us back again?

In spite of the union's plea, only about a hundred field laborers opted to strike in Kings County in 1939. Even Bill Hamett, the rank-and-file leader of the '33 walkout who had seen the growers shoot dead his friend in Pixley, had turned in his union card. Big Bill wasn't a stooped worker anymore. He had become a labor contractor, hiring out crews of

Okie pickers to California's cotton giants. If Hamett had any qualms about going from hero of the proletariat to boss man, he didn't show it. Years later, he told his grandson Roy that he had aligned himself with the communists in 1933 as a matter of survival, not out of any attraction to their hard-core beliefs. "He would have made a deal with the devil to provide more for his family," said Roy.

In time, the complexion of the fields would change again, from heavily white to almost exclusively Mexican and black, and union activists would need to find a new language and a different approach. The Okie cotton pickers moved on to better jobs, and their children turned up their noses at the fields.

Alvin and Pearl Odle and their four boys would travel back and forth to Oklahoma for the next decade, until Pearl put her foot down and took over the family finances. Alvin worked his way up from picker to foreman to farmer in Corcoran. He got his big break with the death of his boss, Forrest Riley, who put a clause in his will that Odle be given 2,500 acres of leased land for one year, free of charge. What a year it was! Great weather, a fine stand, a heavy boll set and a worldwide cotton shortage that sent prices through the roof. On top of everything, the government paid a fat subsidy. By 1972, Alvin Odle was a millionaire with a fourteen-room castle in Corcoran and a heart that ached for Oklahoma. Every summer thereafter, he went back home to go fishing.

River of Empire

KENNY BATTELLE STOOD ON A LANDING at the mouth of the Tule River and looked out at his new speedboat nodding up and down in the green waters of Tulare Lake. The vessel strained at its anchor, as if it couldn't wait to get going, and Battelle was mighty eager to ride the waters himself.

Even for a longtime Corcoran resident like Battelle, the sight of the old Yokut lake was mesmerizing. It had come back this time after a hiatus of thirteen years—a dry spell three times longer than ever before—and when it decided to return, it moved at an incredible pace. The lake mowed down ripe fields of barley and wheat, uprooted giant sheds of corrugated tin and scattered for miles the wooden houses of farm workers. The big, wide levees that had stifled all the smaller spring melts since 1924 were no match for the snowpack heavy on the shoulders of the Sierra this season. For weeks, the runoff had roared down the mountains some sixty miles east of where Battelle's gleaming boat, the *Betty Claire*, was now moored on this late morning, April 11, 1937.

Battelle, who had named the new craft after his two daughters—twelve-year-old Betty and eighteen-year-old Claire—wasn't an experienced boatman. A skilled farmer who also installed irrigation pumps, he knew a great deal about water, but that was of the tamed variety. Tulare Lake, now covering what had been tens of thousands of acres of farmland only a few months before, was a different story. Battelle had bought the boat in Los Angeles the previous week, figuring he might need a way other than truck or tractor to inspect the 4,000 acres he owned and leased in the lake bottom. One of his properties had flooded back in March, and while the levees protecting his prized ranch were holding for now, his days had become filled with watching and worrying.

This trip, though, was purely for pleasure, a maiden voyage for his wife, Esther, their two girls and their son, twenty-one-year-old Kenneth

Jr. Battelle told Esther and the kids to each bring along a companion, so that nine people would be packed into a boat that comfortably seated six. Besides the Battelles, there was Kenny Jr.'s buddy Albert Swisher; Claire's best friend, Doris Butcher; Betty's schoolmate Maxine Elliott; and Maxine's mom, who was Esther's guest.

A small man with stooped shoulders that barely filled out his overalls, the forty-five-year-old Battelle was known throughout Kings County as "the barracuda" because of his go-for-the-throat instincts at the poker table. As he surveyed the water that had poured into the basin, he beamed at his new plaything. It was a beauty, all right: a sleek twenty-two-foot model that shined white and red with chrome trim and mahogany inlays, and a 125-horsepower inboard motor that rumbled in its belly.

It was a sun-drenched Sunday, and Battelle wanted to set out across the lake before noon, before the wind kicked up out of the northwest and the whitecaps began to buck. But he had left the key to the boat at his house and sent Claire and Doris to retrieve it. The two girls, home from college for Easter weekend, hurried so they could rejoin the others.

No one was more raring to go than little Betty, a blond, blue-eyed tomboy who was thrilled at the thought of an adventure. She hadn't even been born the last time the Sierra snowmelt was heavy enough to bring the lake back. She wanted to ride shotgun, right next to her father, and hold on to the windshield. If anyone betrayed a little apprehension about the trip, it was Kenny's wife, Esther. She didn't know how to swim, and she had an intense fear of the lake. During the 1916 flood, when Kenny Jr. was just a baby, she made her husband move the family out of the lake bottom to a dairy ranch miles away. When the lake rushed back in March, friends invited her to go see the waters but she couldn't find the courage.

This time, her husband assured her that the boat was safe. When he bought the craft, he had taken it out along the Southern California coast and turned it sharply, and it had held steady all the way. So Esther reluctantly agreed to go along, cooking up a big batch of fried chicken for a picnic lunch.

Claire and Doris returned with the key, and Kenny Battelle started the motor. He gave the boat a trial spin and then came back to pick up his passengers. As the boat pulled away from the Virginia Dredger dock, they could see that Tulare Lake was teeming. Mallard, pintail and cinnamon teal ducks glided overhead with geese and blackbirds. And in the trees lining the banks, loons were mocking the almost warlike mobilization of farmer and heavy machine. The birds had built their nests out of the very grain that the farmers were trying to hurry to harvest. Flocks of pelicans, their gullets twitching wildly, swarmed the lake's surface.

People had forgotten how ruthless the lake could be. During it's thirteen-year absence, some farmers had grown complacent—sure that nature had somehow gone soft on them—that they had built equipment yards, gins and even cabins right in the lake bottom. Now they were forced to watch from the levee banks as these buildings floated away like bath toys. Around the periphery of the lake, they maneuvered dredges, tractors and earthmovers, desperately trying to fortify the muddy mounds defending the surviving fields. Whenever a levee gave way, the water rushed forth in a mad rivulet, overwhelming the land and expanding the lake's boundaries. As the water reached a new patch of dry land, puffs of dust blew skyward.

Sandy Crocket, who by 1937 owned a lot of farmland right in the lake basin, would long remember the night that he and his dad watched the water run over the top of a levee and destroy a substantial portion of his grain crop.

"How many acres do you have here?" asked Crocket's father, who was visiting Corcoran from Canada.

"Well," Crocket answered, "in this one block I have 8,000."

"Do you expect to see that all flooded?"

"Yes," said Crocket. "I've got other land in back of the big levee here that we are going to try to save."

Being an accountant, Crocket's dad started doing the math straight away: 8,000 acres at so many sacks to the acre, with a market price of so much per sack. "If my calculations are correct," he said, as the water filled the field, "that amounts to over a million dollars."

"Yes, sir," Crocket said, "that's right."

"You mean to tell me you're going to take a loss of a million dollars, and you're just standing here?"

"Papa," said Crocket, "what do you expect me to do? There would be no use to me jumping up and down."

Kenny Battelle let out the throttle of the *Betty Claire* about three-quarters, and as he neared the Kings River, just a short distance from the levee shore, he turned the boat to the left and it began to dip. He tried to right it and the boat responded for a second before it continued to veer left, getting caught in its own backwash. In a flash, the boat flipped and everyone went headlong into the water. It wasn't very deep—only about seven or eight feet right there—but the water was roiling and it was icy cold.

The stern sank and the prow poked out of the water, just enough boat for Battelle and his son to cling to and attempt a rescue. Nobody was wearing a lifejacket; there were several onboard, but they were heavy and uncomfortable. Battelle had ordered better ones, but they wouldn't arrive for another week. The elder Battelle shed his overalls and shoes and swam in the direction of the screams and shouts. He spotted his wife—she had been knocked unconscious but had somehow popped to the surface—and he and Kenny Jr. pulled her out of the water and onto the upside down boat.

Doris Butcher, who also couldn't swim, went under—once, twice, seven times in all, managing through sheer will to stick her head out of the lake again and again for gulps of air. Finally, when Doris thought she couldn't last any longer, Battelle swam out and grabbed her, leading her back to the boat. Mrs. Elliott couldn't swim either, but she remembered reading a story about a drowning man who had saved himself by staying calm and floating on his back. She floated for several minutes until Battelle senior and junior carried her to the boat, where Claire was also hanging on for life.

The three others were nowhere to be found. Albert Swisher had helped one of the survivors reach the boat and then was heard saying, "No, no, I can't." The son of the town librarian, he had suffered a broken neck during a diving accident four years earlier and had never fully recovered. Now weakened, he sank beneath the waters that also swallowed up twelve-year-old Betty and her friend Maxine.

The other six held on to the *Betty Claire* for almost an hour. By the time two onlookers heard their screams and paddled out, the forty-five-year-old Esther was also dead. Her watch had stopped at 12:15 P.M., just a few minutes into the trip. Her lungs hadn't taken in any water, but she had evidently succumbed to a blow to the head or perhaps heart failure. It took a team of farmers plumbing the murky lake bottom with grappling hooks to find the other three bodies.

An inquest was held, but nobody would ever know for sure why the boat overturned. The Battelles maintained that the rudder was strained and the propeller shaft damaged when the boat was dragged from the levee into the lake. Word around town was that Kenny Battelle's inexperience was responsible; folks said he never should have steered the boat into its own wake. He paid for all the funerals, and the town closed for business that day out of respect. He tried months later to get back into the boat as a kind of healing. He even removed his deceased daughter's name from the side and rechristened it the *Miss Claire*. But the boat held too many bad memories, and he took it out one last time and then got rid of it. For the rest of his life, he kept the receipts for the four coffins tucked away among his personal papers. "One minute we had our whole family," said Claire, "and the next minute it was all torn up."

By May, the lake blanketed more than 60,000 acres. But over the summer, between evaporation and the farmers siphoning for irrigation, the water level gradually began to drop. By the fall, most of the lake had retreated, and growers looked forward to a bumper grain crop. "Apparently," explained Corcoran chronicler Mae Belle Weis, "the farmers felt that thirteen dry years out of the past fourteen constituted pretty good odds," and so the area was seeded anew as if the flood was merely a one-time blip.

It was a lousy bet. In December, a huge winter storm ripped through the southern Sierra Nevada. The Kings River flowed at a record rate of 80,000 cubic feet—the equivalent of sixteen freight cars full of water racing by—every second. In some spots, eight inches of rain splashed down in just twelve hours. More heavy rains fell in January and February.

For the Okies camped along the ditch banks, the flooding that followed was catastrophic. It left them wet, hungry and out of work. Eighty-seven people throughout the valley died. The only consolation was that the disaster made front-page news, and the desolation of the agricultural migrants was seared into the public's consciousness for the very first time, stirring feelings of compassion toward the laborers and disillusionment toward the growers. An emergency relief committee distributed blankets, clothing, candy and tobacco to the sopping masses. The federal government also assisted them, setting up a grant-in-aid program that supported 50,000 Okies.

In due course, the government would extend a hand to the big farmers of Tulare Lake in the form of flood control dams on the Kings, Kern, Tule and Kaweah rivers. But for now, the growers were on their own, and the battle was tough. By March 1938, even before the spring runoff,

Tulare Lake again stretched over 40,000 acres—16,000 of which had been planted to grain. In April, four-foot waves chewed holes through the growers' levees, even though Boswell and others tried feverishly to reinforce them. In early June, the Boswell Company saw some of its land swamped by the freshet.

For the Colonel, the flood was the least of his troubles. Just weeks before, his darling Alaine had gone into the hospital for a hysterectomy. There were complications, and she never made it home. The premature death of his fifty-one-year-old wife—the Colonel's true love, the only one who could tamp down his outsize ego—was crushing. "If I were told that I should die quite soon," Alaine had written before she passed,

> I would not choose to live so differently.
> Only I'd ask to cut once more, a rose
> In its perfection in the amber mist;
> To hear the smoky owlet's throbbing note
> In the arroyo's dusk . . .
> To press my face close to the vivid grass
> Divinely perfumed,
> Then to turn and look into the sky
> So vast, so bright, and say, serene and cool,
> 'I go from whence I came,
> Back to the earth and sky and sea,
> Dust unto dust!
> So be it, and—amen.'

The Colonel paid off Alaine's last few purchases—a lace evening dress from I. Magnin, a blue hat from Bullock's, a novel set on the coast of Maine called *Dawn in Lyonesse*—and carried on the best he could. But the house in Pasadena, once so alive with her poetry and gardens and magnificent sense of style, was now empty and sad. "If I had to fight on living under that roof," said Alaine's brother, Fred, "I think I should go mad."

Back in the Central Valley, Boswell and the other farmers struggled to save what cropland they could. The people of Corcoran had plenty to distract them from the crisis at their doorstep. A women's basketball team, Olson's All-American Red Heads, rolled through town and beat the boys from the local fire department, 29–22. Eight-foot-eight Robert

Wadlow, billed as the "tallest man in the world," also paid a visit, padding around the Corcoran Mercantile Company in his size thirty-seven shoes. And a new movie house—aptly called the Lake Theatre—opened with a showing of *Mother Carey's Chickens,* starring Walter Brennan. But for most people, the biggest drama of all was taking place out on the real lake, where the grain harvest had turned into a frenetic race—farmer versus flood. Even when a levee broke, giant mechanical reapers were left in the fields, bumping along just ahead of the spreading deluge until, at the last instant, they were plucked out and quickly drafted into service somewhere else.

For at least a few growers, the lake's return was a blessing. Elmer von Glahn, the one-man boom-bust cycle who liked to play his accordion in the fields, had just started farming on his own again a few years earlier, planting grain south of the lake. The location was unbeatable; he was far enough from the basin in 1938 that his crops never got wet, but he was close enough to have all the free irrigation water he needed. His empire took off from there. "That water was the making of Elmer von Glahn," said trucker Sam Crookshanks, who shipped a mountain of burlap barley sacks for him during the 1938 harvest.

The Colonel also engineered a deal after the flood that was so shrewd people were still talking about it decades later with a sense of disbelief. He bought the Cousins ranch while it was under water, and then turned around and sold off the water at $10 an acre-foot, raising so much money that it paid for the land itself. While others focused on what a hazard the floods were, the Colonel was smart enough to recognize that the water was really a big benefit, especially among farmers who often found themselves pining for more sources of irrigation. Even the Corcoran newspaper, while dutifully reporting on the devastation wrought by the floods, grasped the silver lining. Noting that growers had long been forced to drill wells as deep as 2,000 feet to irrigate their fields, the editors made clear that the current conditions were no reason to be upset. "Water in Tulare Lake is what we need and have been praying for these many years."

Still, with the water showing no sign of letting up, many farmers felt more like losers than winners, at least for now. Levees fell in quick succession, and one large tract of land after another yielded to the lake: the El Rico District, Reclamation District 749, the Goldberg District, the Consolidated, the Lovelace—some twenty districts in all. More than

140,000 acres were eventually submerged, and for some inhabitants of Corcoran, the waters crept precariously close to town. At 223 square miles, the lake had gotten so large that airplane pilots who flew too low over the region risked losing sight of land, and a few became so disoriented that they thought they had strayed over the Pacific. The crop damage in the lake basin was ultimately put at $7.5 million (equal to nearly $86 million today). And some lakebed farmers were so shaken by the flooding that they looked to shift their operations out of the area.

The lake would also claim more lives. In July, grappling hooks found the limp body of Ed Bradley, fifty-six, about twenty feet upstream from his car, which had tumbled off a narrow levee and into Cross Creek. Robert Scutt, thirty-two, drowned after his boat hit a hidden obstacle on the flooded George Smith ranch, pitching him into a ditch. A search team heard the frightened screams of two duck hunters whose boat had overturned in the lake, but by the time they could get out there, Ray Dutra and Frank Costa were gone, too.

By the end of 1938, a rejuvenated Tulare Lake had changed the way the farmers thought about their land. For more than two decades, they had discussed putting a dam on the Kings River at a site in the Sierra foothills called Pine Flat. But in those early years, it was too little water, not too much, that had prompted calls for the dam. The idea was to build a grand reservoir that would store enough water to ensure that the growers could irrigate their crops without hassle, even during times of drought. Now, the damage to the fields and the loss of life created a new imperative, the logic behind it essentially the opposite of what had been argued before. A dam in the name of flood control should be built, the farmers said, and it was up to Uncle Sam to come in and get the job done.

BOSWELL EXECUTIVE FRED SHERRILL brimmed with hope as he reached the post office building in downtown Denver, home of the U.S. Bureau of Reclamation, and stepped past the majestic sculptures of mountain sheep just outside the 18th Street lobby. It was a fitting entrance. For more than three years—ever since the floods of 1937 and 1938 had kindled the notion of erecting Pine Flat Dam—the bureau had locked horns with a rival agency, the Army Corps of Engineers, over which one should supervise the project. The internecine struggle had gotten so nasty, in fact, that President Roosevelt felt compelled to intervene and try to bring the warring federal fiefdoms to a truce. But things had only gotten nastier.

For Sherrill and his boss, the Colonel, the preferred choice was clear. Like most growers up and down the Kings River, they wanted the Army Corps to build the dam. Since 1936, the corps had been charged with assembling flood control structures throughout the country, including potential projects in Central California. But the bureau, whose mission since 1902 had been to "reclaim" the arid West and through its waterworks make the land viable for family farms, saw Pine Flat in a different light. Its studies indicated that the primary advantage of the dam would be the enhanced irrigation of farmland—with flood control as a secondary benefit—and bureau officials contended that Pine Flat thus fell squarely under their purview.

Some of the distinction was nonsense. The way the corps and the bureau had conceived the dam, it would cost roughly the same regardless of which one built it: about $19.5 million, with half of that to be paid by the farmers. But there was one crucial difference between the two agencies' plans. The corps promised to give the farmers virtually total say over the use of water in the Kings River; the locals, in other words, would run the show. The corps would only get involved to prevent flooding. The bureau's blueprint, however, would grant the government much greater control. The bureau envisioned harnessing at least some of the river for electric power—a pet interest of Interior Secretary Harold Ickes, the old curmudgeon to whom Reclamation officials reported. And while bureau higher-ups and even Ickes himself insisted that they respected the farmers' long-held water rights, many worried that in years when the snowpack was light and there wasn't enough flow in the river to go around, the bureau might elect to produce megawatts over cotton and wheat. "We are afraid," said Sherrill, "to take the chance."

In short, Boswell and the other farmers liked the corps plan because it let them have it all. The government would come in and underwrite a big chunk of the dam and then step aside and permit the "free market" to operate in peace. That this "free market" was to be greased with government subsidy was an inconsistency that the growers didn't like to talk about. "It shakes down to a platform: Get out and give us more money," Bernard DeVoto, a trenchant observer of the West, once wrote.

Farmers throughout the region, he believed, were always "demanding further government help" and then "furiously denouncing the government for paternalism, and trying to avoid all regulation."

The fight between the corps and the bureau over Pine Flat Dam has been much discussed by historians over the years. Many have generally accepted Secretary Ickes's view that it was the corps—"spoilsmen in spirit," he called them—that defied its own commander in chief, "invaded the jurisdiction" of the bureau and worked "hand in glove" with the big farmers, all to the detriment of the public. "The Army Engineers are trying to grab off this project," Ickes complained to the president in May 1940. And Ickes's assessment certainly was valid—to a point.

In truth, the long-running affair was much more complicated than either Ickes or other observers have made it. The real story of Pine Flat Dam was one of shifting alliances and squabbling among the farmers themselves as they tried to decide which horse to back. And there was plenty of hypocrisy and doublespeak on all sides—Ickes's included.

At the center of the whole mess was none other than J. G. Boswell and his political guru, Fred Sherrill. Even though both men thought it best for the corps to build the dam, the Colonel and his well-wired lobbyist believed that simply disregarding the bureau—as some farming interests along the Kings were inclined to do—was a hopeless strategy. Secretary Ickes, after all, already had convinced President Roosevelt to back the bureau over the corps. For Boswell, standing firm until the corps might prevail could well mean losing the dam altogether. "Our flood control project is still in a coma," Sherrill had written a friend in February 1941, six months before his trip to Denver. The farmers on the river had been prepared "for a full fight," he explained, "but we went down for the count, and all I can see so far is the benign countenance of F.D.R. and the irascible countenance of 'Harold the Lovable.'"

Shortly before arriving in Denver, Sherrill had gone to Washington to discuss the matter with a senator who had served as chairman of the Committee on Irrigation and Reclamation. The outlook for the corps, Sherrill found, was far from promising. "Between Ickes and the president," he said despairingly, "they have already shut the army out." As for the combative Interior secretary, Sherrill couldn't resist adding: "It seems Ickes has a war going on as many fronts as Hitler."

And so, with the prospects dim for getting Pine Flat Dam swung the corps' way, Sherrill had come to the Bureau of Reclamation's Western headquarters to explore the possibility of working something out. Now headed into the third and final day of talks—July 30, 1941—he was optimistic that maybe, just maybe, the bureau as dam builder wouldn't actually be so dreadful. The bureau's top man, Commissioner John C. Page, turned out to be extremely accommodating to the farmers' position.

Page's acquiescence shouldn't have been all that startling. Unlike the man he worked for, Harold Ickes, Page cared almost nothing about politics. Indeed, the two were almost as different as they could be. Ickes was a one-time Chicago newspaper reporter, erstwhile Bull Moose Republican and Roosevelt's hatchet man, and he exuded passion in everything he did: from the way he guarded his turf in overseeing the Public Works Administration, to his tireless defense of minority rights, to his affair with the "sexually alert" woman identified in his memoirs only as Mrs. "X." His underling, Page, was unemotional and unassuming, a straight-home-after-work type who drank maybe one Scotch a year and hardly ever cussed or, for that matter, even raised his voice. "He said 'Damn!' once when he hit his thumb with a hammer," recalled his daughter Milly, "and that was a big event." The consummate engineer, Page reserved his enthusiasm for one thing and one thing only: building stuff, gigantic stuff. Before going to Washington, he had helped to oversee the construction of Hoover Dam, which, in subduing the Colorado River, had used up enough concrete to pave a highway stretching all the way from San Francisco to New York.

Pine Flat was to be nowhere near that enormous, but it was still an ample project. If giving in here and there to the local growers would help get it built, that was fine by Page. In some areas, much to the farmers' delight, Page agreed to go even "further in our favor than we had suggested," noted Sid Harding, an engineer for the Tulare Lake Basin Water Storage District who had accompanied Sherrill to Denver. Said Sherrill: "I was promised practically everything."

By the time Sherrill was ready to head back to Los Angeles to update the Colonel on Pine Flat, Page had signed off on a memorandum for Secretary Ickes that was to be used as the basis for formal contract negotiations between the bureau and the farmers. Those negotiations were still bound to be tough—the farmers didn't want to put as much of their own money toward the dam as the bureau was insisting on—but the agency's whole tenor had now shifted. "It is very unfortunate," Page wrote to Ickes, "that distrust of the bureau and its policies has arisen."

Page's memo included five key points that added up to one thing: Local farmers would have far more control over the river and the reservoir than the bureau's initial plan had implied.

Then Page made the most remarkable concession of all: The bureau wouldn't challenge the growers—at least not Boswell and the others right in Tulare Lake—on the size of their spreads. Under reclamation law, irrigation water from bureau projects was supposed to go only to farms that were 160 acres or smaller—"family farms" in the Jeffersonian tradition. For big guys like Boswell, already the world's largest cotton grower, enforcement of the law would have meant no less than the breakup of their empires. But acreage limitation wasn't something that the bureau had ever pursued with any vigor. And Page advised Ickes that in terms of Pine Flat Dam, adherence to the law was "probably unwise" and "might well be amended" for those in the lake bed. His justification: Even if the dam stanched most of the flooding, the occasional torrent could still wipe out any small farmer who tried to make a go of it there. Only big operations, with abundant financial resources and enough land to shuffle floodwater around, could make it through the rough times in Tulare Lake.

By September 1941 Ickes had formally approved Page's memo. "You may be sure," he wrote the Kings River Water Association, that when the farmers are ready to negotiate a contract with the bureau, "provisions embodying the recommendations of the Commissioner will receive my approval." Over the next year, bureau officials consulted with Harding and other local engineers on further studies of the Kings River area, and the Colonel regarded the alliance as a welcome development. Or at least it was preferable to what he believed to be the alternative: utter paralysis.

Having landed in Corcoran at the start of the thirteen-year dry spell, the Colonel had never seen Tulare Lake spring to life until the flood of 1938 and, naturally, he lived in fear of its next awakening. It came in 1941 and again in '42, as a brief dry cycle surrendered to powerful rains and cascading Sierra snow. Once more, tens of thousands of acres in the basin were engulfed, and the lake's return became a focus of local life. As crews drained highways in the lake bottom to open them to traffic, people took up the sadistic sport of "dry-land fishing," clubbing the wriggling perch, crappie and carp that were left behind. The latest run of floods also drove packs of rattlesnakes from their levee lairs to higher ground, terrifying town residents. Even Walter Winchell took a respite from his verbal strafing of the Nazis to inform his radio audience about the onrushing waters around Corcoran.

Obtaining the dam—and obtaining it quickly—was now the Colonel's top priority. With President Roosevelt continuing to push for the bureau to build Pine Flat, "the old army proposal is as dead as Hector's pup," he reasoned. "While I hold no brief for Ickes and his department, I do believe it is the only way to get the dam."

The outbreak of World War II had raised serious doubts about whether money for Pine Flat would be found anytime soon, but word from Washington was hopeful even on that score. If the growers and the bureau could come to terms on reservoir storage, "I think I can get an initial appropriation to cover initial construction, not after the war is over, but right now," Republican Congressman Bud Gearhart of Fresno advised the lake farmers in the summer of '42. He was confident that "a tremendous demand for power on the Pacific Coast within the next very few months" could help him sell the dam as an indispensable tool of war.

Yet many growers weren't willing to engage with the bureau, Page's assurances and Gearhart's advocacy notwithstanding. For those farming upriver, flooding wasn't nearly as big a worry as it was in the lake bottom. They wanted Pine Flat built not so much for flood control, but because the carefully timed releases from the dam would let them irrigate their fields, orchards and vineyards with far greater precision than ever before. And for that, there was no real hurry. They could afford to wait until their allies in Congress had found a way to give the project over to the corps—the agency, they believed, that would be willing to give them the sweetest terms on their water-storage contract.

By the middle of 1942, then, the situation had become completely topsy-turvy. The upper-river farmers, who coveted Pine Flat because it would improve their irrigation practices, wanted the flood control agency to construct the dam. Lake bed farmers like Boswell, who wanted the dam built as rapidly as possible for flood control, lined up behind Page and the irrigation agency.

As the months passed and the bureau and the corps slugged it out on the bureaucratic front, the shadow war between the two groups of farmers also grew hotter. "The attitude which Mr. Boswell has apparently taken . . . concerns me considerably," cautioned H. L. Haehl, an engineer working for the upstream growers. If the Colonel keeps siding with the bureau, "cordial relations between all the water users on the river" may "become impossible."

Despite the warning, Boswell's men continued to cozy up to the bureau, and they made no apologies for it. To hold out hope that the corps was going to build the dam "in the face of presidential oppositions and directions to the contrary," Sherrill wrote to Haehl in September 1942, was nothing but foolish. Anyhow, Sherrill told him, "with Colonel Boswell, good faith means good faith. If he feels committed to negotiate with the bureau . . . he is going to negotiate on this basis—not with his tongue in his cheek and the thought in his mind that we just go through the motions and go back to the army."

By the spring of 1943, the growers in Tulare Lake were more determined than ever to see the Kings River—and eventually the Kern, Kaweah and Tule—get dammed. The lake was now full of water for the third straight year, and all the while, the government was urging area farmers to grow more food and fiber for the war effort. It was no simple task. Not only had the flooding stifled production, but with their regular crews of Okies now signed up for the service or employed in the state's shipyards and aircraft factories, growers faced a labor shortage. The Colonel pressed the government to send reinforcements, but cotton wasn't at the top of the list of crops eligible for emergency workers, and he came away feeling like he might as well have asked Mother Nature to kindly stop dropping so much snow in the mountains. "You can't fight the two G's—God and Government," the Colonel said.

For a time, the local high school tried to ease the crunch by busing students to the fields so they could pick cotton. The women of Corcoran pitched in, too. Kate Boswell went out one day during the 1942 harvest and stuffed fifty pounds of bolls in her sack in three hours. Then she went home, "prettied herself up and came to a missionary meeting," Mae Belle Weis reported in her *Corcoran Journal* column. "I think that is a very good record."

In the end, the federal government did bring in Mexican laborers as part of its new *bracero* program. Soon they were supplemented by hundreds of German and Japanese prisoners of war who, under armed watch, were ordered to pick 150 pounds of cotton a day for a small wage. The Germans proved fitful, as more than a dozen tried to escape Corcoran—all of them unsuccessfully. The Japanese were more compliant.

They "feel the disgrace of having been captured," said one of the guards at the Boswell POW camp. "As far as they are concerned, they are dead." One Japanese prisoner was so despondent he hanged himself in the fields. His body was found dangling from a tripod used to weigh cotton sacks.

Like many small communities, Corcoran swelled with pride during the war years. Each week, the *Journal* printed an honor roll listing the names of the town's fighting men and women. Elmer von Glahn's airport—which had been opened in the spring of 1941 as a throng of 4,000 jitterbugged to trumpet legend Joe "Wingy" Manone—was turned over to the military for pilot training. A cadre of businessmen, including Bill Boswell, led the local war bond drive under the rallying cry "Back the Attack." Residents rationed sugar, meat and truck tires and waited for their husbands and sons and brothers to come back home. Jimmy Boswell, who had left Stanford University to serve as an army private in the South Pacific, was wistful about the possibility in a letter that he wrote from the trenches:

> I've only been out here a few months, but already Corcoran and its fine people seem like a dream. Out here we can think of the little things that seemed unimportant at the time. There's a lot of things I'd like to do—ride a bike, kick a football, go down to Johnny's for a Coke. . . .
>
> Strict censorship prevents me from saying anything about my setup out here. However, I will say that it's a mighty tough life. I don't think Hollywood's idea of the South Pacific will meet the approval of these fellows when they get back home. I'm well and happy and excited. I wouldn't take a million dollars for some of the experiences I've had. . . . I'm certain that if I come through OK, I'll have learned more than I ever could at college. . . .
>
> It's been a long day, and I'm ready for a little shut-eye, so I'll say so long for now.

In the summer of 1943, with 112,000 acres of Tulare Lake land under water, the Colonel's men demanded action. Flood conditions "make it imperative that relief be provided as soon as possible," Harding, the engineer, told a top bureau official that June. The feuding between the bureau and the corps, he added, was "entirely out of place." The only important thing was that the dam be built. Whether the assignment was given to the bureau or the corps was a "federal administrative question" that "should be decided without requiring extensive public debate or forcing local interests to take sides."

That said, the Colonel and his boys had already cast their lot. Boswell's point man in the valley, Louis Robinson, even wondered whether it was time for the lake basin growers to cut their own deal with the bureau, freezing out the upper-river farmers and the army. Such an open breach never occurred. But the lake growers did go so far as to tell the bureau that they supported coordinating the operation of Pine Flat with the agency's main plumbing unit in California: the Central Valley Project. A leviathan system of interconnected canals, dams and hydroelectric plants, the CVP had broken ground in 1937 and would eventually range over the entire length of the valley, impressing one latter-day observer as "the most mind-boggling public works project on five continents."

"It has never seemed reasonable to me—in fact, I do not think it good management—to have the army construct the Pine Flat project" with the bureau still at work on the CVP, said Robinson. For the rest of the year, the Colonel and his men continued to push this line, confident that they'd be able to iron out an agreement with the bureau. Then one morning in early February 1944, everything came unglued.

The turning point was a congressional hearing in Washington. At the moment, only the bureau had been granted official authorization to work on Pine Flat, owing to a provision in the reclamation law that gave the Interior secretary almost unilateral power to launch a dam-building project. The bureau had been unable, however, to secure any funding for the job. The Army Corps, meanwhile, was still seeking step one: authorization to construct Pine Flat. And now the House Committee on Flood Control was moving along a bill that would give the army such sanction.

At the start of the hearing, a colonel from the corps made the case for why Pine Flat was predominantly a flood control—not an irrigation—project, and why the army should therefore be the one to oversee its construction. The new Reclamation commissioner, Harry Bashore, argued the other side. Bashore, a thirty-seven-year bureau veteran, had taken over the agency just six months earlier, after John Page had gotten sick and stepped down. Then, in the midst of his testimony, Bashore went off on a subject that caught Boswell totally by surprise. If the bureau built Pine Flat, he said, any farmer receiving water from the project would be forced to abide by the 160-acre limit—even those in the lake bottom.

"We are concerned with the little people," Bashore told the committee, and by breaking up large land holdings, new farmsteads can be established for the "soldiers coming back from the war." Besides, he added, some control should be brought over big growers who "come to the federal government" seeking to have projects built that will allow them to "perfect their water rights."

By the second day of the hearings, the committee riveted its attention on this new, unexpected theme. Two California congressmen on the panel, Alfred Elliott and Norris Poulson, grilled Bashore on how the Bureau of Reclamation could champion the 160-acre limit when it would so disrupt the valley's economic structure. Bashore's response was simple: Restricting the size of farms that acquire water from a bureau dam was the law of the land, and it had been that way for more than four decades. "The limitation on holdings is not something that was concocted by a bunch of fuzzy-brained professors with Phi Beta Kappa keys and whatnot," Bashore said. "That was put into the reclamation law by the Congress in 1902." Then he added: "I want to make it plain that I do not believe in invading a man's rights at all, but I think it is for the best interests of the country at large that we have as many people own farms" as possible.

For Boswell, Bashore's statements were the ultimate act of betrayal. Everything the company had been working toward with the bureau for nearly three years was instantly and irreversibly damaged. The 1941 memo that Page had negotiated with Fred Sherrill in Denver—and Ickes had blessed—lay in tatters. Page had given "to me as solemn a commitment as any man ever gave to another that the land limitations provision had no application," Sherrill said years later, his rage still palpable.

The farmers, heretofore at odds over which agency should build Pine Flat, were a united front after Bashore's performance. The corps promised not to shackle the growers with any acreage limits, and now the upriver crowd and the lake farmers were both 100 percent behind the army spearheading the project. Short of that, even the status quo looked better than what the bureau had in mind.

"If what we can get is only going to be on a 160-acre basis then we don't want any dam at all," the Colonel said. "We would prefer to go ahead as we are."

THE IDEA THAT SMALL LANDHOLDINGS are more virtuous than vast estates is as old as the Bible itself. "Woe to you who add house to house and join field to field," warns the Book of Isaiah, "till no space is left and you live alone in the land."

The notion that the yeoman farmer is far nobler—and far better equipped to safeguard democratic ideals—than somebody who oversees a giant agricultural realm has its roots in the earliest days of the republic. "The small landholders," wrote Thomas Jefferson, in a rebuke of the landed aristocracy back in Britain, "are the most precious part of the state."

The principle that 160 acres—and not a smidgen more—is plenty for a man to prosper on was espoused by no less than Abraham Lincoln, who sealed the number into law when he signed the Homestead Act in 1862. "I am in favor of settling the wild lands into small parcels," he said, "so that every poor man can have a home."

Despite such lofty sentiments, convincing people to confine themselves to small farms has been a tricky business, at least in the two states where the Boswells have resided. The antebellum plantation, like the one that Johnson Boswell had in Georgia with his thirty-three slaves, held no regard for the Jeffersonian dream. Neither, of course, did the cotton empire that Johnson's grandson, the Colonel, founded in California.

After years of rampant speculation and land monopoly by the likes of Henry Miller, the reformers finally got a tool to create small farms in the irrigated West. In 1902 Teddy Roosevelt signed what one prominent historian has called "the most important single piece of legislation" in the annals of the region. The National Reclamation Act was the inspiration of the "irrigation crusaders," a band of Progressives who believed it was incumbent upon the federal government to build great water delivery systems beyond the 100th meridian and make small farms flower across the bleak deserts of outer America.

The vision of redeeming a region that received so little rain had been described more than a decade earlier by John Wesley Powell. The indefatigable scientist had lost his right arm to a Civil War Minié ball and gained widespread fame when his expedition careened down the uncharted Colorado River in 1869. Powell later oversaw the mapping of the West as head of a U.S. geological and geographical survey, and by 1890 he had devised an irrigation plan that reaffirmed the small farmer as the backbone of American democracy. The way Powell saw things, the rivers of the West should be conquered and its deserts made to bloom. But at the same time, Western settlers should each receive no more than eighty irrigated acres—an amount that would allow 1.25 million families to have a decent farmstead.

The 1902 act doubled Powell's prescription to 160 acres, a quarter of a square mile section of land. Though somewhat arbitrary, the 160-acre threshold had grown out of one of the first laws passed by Congress, the Land Survey Ordinance of 1785, and over the years the figure had come to define the archetypal "family farm."

Backers of the new reclamation law were confident that the 160-acre constraint would provide a potent antidote to the land barons of the Central Valley and elsewhere. Senator Francis Newlands of Nevada, the driving force on Capitol Hill behind the legislation, was one who predicted that the 1902 act would shatter the "existing land monopoly, which has been so bothersome in California" in particular.

Newlands expected that most of the land set to be irrigated by the newly formed Reclamation Service (it became the Bureau of Reclamation in 1923) would be culled from the public domain: the millions of western acres owned by the federal government. But private property also was eligible to be served by reclamation projects—so long as these holdings stayed true to the 160-acre limit. Any "excess lands" had to get their water from someplace else or remain dry—or be split apart and sold off.

From the start, however, there were difficulties in implementing the acreage limit. Although most of the areas benefiting from reclamation projects were in compliance with the law, large landowners in certain locales treated the 160-acre ceiling as a joke. In Arizona's Salt River Valley, enforcement was so spotty that a 1916 review board found that it "closely verges on fraud."

Part of the problem was that the 1902 act and a later version passed in 1912 didn't have the teeth necessary to bust up big land holdings. The Reclamation Act of 1914 tried to remedy that, but this law was as feeble as its forerunners. Finally, in 1926, Congress got tough. Those with more than 160 acres were required to sign contracts with the government in which they agreed to sell off their excess lands before receiving any project water.

Even then, there was little desire among bureau officials to crack down on flouters. When Harold Ickes first asked about the law in 1934, his minions at the bureau told him that acreage limitation had been "a dead letter for years" and that it was best to "let sleeping dogs lie." Like John Page, these were bureaucrats intent on building monumental structures, not pursuing social policy. In 1938 and then again in 1940, bureau officials retreated still further from the 160-acre limit. They abetted Congress as it exempted three projects—one in Colorado and two in Arizona—from the law.

But Secretary Ickes soon hardened his position. Two men were largely accountable for this new attitude. The first was Walter Packard, a New Dealer who, in a provocative 1942 study, argued that "the existing patterns of large-scale production possess characteristics which are basically unsound" for society. Packard spread the gospel to Paul Taylor, the roving Berkeley professor married to photographer Dorothea Lange. Taylor was a natural ally in any battle against big farming, but when Packard first mentioned the 160-acre limit to him, Taylor wasn't quite sure that he had heard his good friend and neighbor correctly. "I asked him if he meant that ought to be the law," Taylor later recalled. "He replied, 'Oh, it *is* the law.'"

By late 1943, Taylor was obsessed with seeing the 160-acre limit upheld. Hired on as a consultant to the bureau's electric power division, he had direct access to Ickes on the issue, and he came to believe that the secretary could be counted on as a "staunch and effective supporter." So, too, could President Roosevelt, who proclaimed that reclamation projects shouldn't benefit "the man who happens to own the land at the time," but should give a lift "to *The Grapes of Wrath* families of the nation."

In March 1944, the newly energized defenders of the 160-acre limit found their position under heavy assault, following a bold sneak attack. Alfred Elliott, the California congressman, had attached to a rivers and harbors bill a little noticed "rider" that would exempt the Central Valley Project from any acreage limits. Elliott had "slipped it through . . . at the last minute," Taylor remembered. The Democrat's amendment "hit the floor with only a few hours of warning."

Elliott was a corn and alfalfa grower from Tulare County, the site of some of California's king-size farms, but he himself wasn't one of the behemoths; he owned only about forty acres and ran a small dairy. Nor was he insensitive to the concerns of the less fortunate. Once, when tensions flared at the county fair because blacks and Mexicans were trying to dance in the same hall as whites, the six-foot-one, 200-pound Elliott snatched the microphone and shouted: "This dance floor is for all of Tulare!" That display of courage, along with a continued commitment to deliver relief to his constituents of color, made Elliott a hero in Tulare's black community.

On acreage limits, though, Elliott had become a shill for the big growers, his defense of their holdings downright inflammatory. The Interior Department, he said, was "trying to socialize agriculture and force Communism upon the people of the San Joaquin Valley." Caught totally off guard, Interior Department officials could do little but watch Elliott's amendment pass the full House.

The fight then turned to the Senate, where California's Sheridan Downey took up Elliott's—and Big Ag's—cause. Even in a place celebrated for second acts and midlife conversions, the path that Downey had taken to reach this point was truly amazing. He was born and raised in Laramie, Wyoming, the son of a congressman who once proposed decorating the inside of the Capitol with frescoes depicting the story of Jesus. In 1913, at the age of twenty-nine, Downey made his way to Sacramento, where he practiced law and endeavored to get rich quick. All of his crazy schemes, including one to import elephants and monkeys into the United States, were flops. Mired in debt by the late 1920s, Downey began self-publishing pamphlets jam-packed with economic nostrums. Among them was a call for a radical redistribution of income. "I abhor two great American institutions as undemocratic and as destructive to national welfare: our vast fortunes and our professional politicians," he wrote. "I hope the coming generation will eliminate both from the American stage." Even Paul Taylor said Downey's leanings were "almost more liberal than I found easy to take."

In 1934 Downey ran for lieutenant governor as the less famous half of Upton Sinclair's End-Poverty-in-California ticket. Then in 1938, having stumped for the Ham-and-Eggs pension plan, he was elected to the Senate as a Democrat. The next year, he pushed to bring Senator La Follette's civil liberties investigation to California—the probe that helped expose the Associated Farmers and scrutinized the activities of Boswell and others. Though some regarded him as a crank, the last thing anybody could have guessed was that Sheridan Downey would one day be accused of fronting for California agribusiness. But that's exactly what happened. The longer Downey served in the Senate, the more conservative he became. Fighting off the 160-acre limit, which he now called "grotesquely inappropriate," was the ultimate expression of his new-found philosophy.

In the spring of 1944—with the dispute over acreage limits now generating headlines across the country—Colonel Boswell embarked on a two-pronged lobbying campaign. The first order of business was to try to ensure that the Army Corps built Pine Flat Dam, and Harry Bashore's Bureau of Reclamation didn't. To that end, it was vital to make an all-out push on the flood control bill working its way through Congress. But if the flood control bill failed, and the bureau wound up wresting control of Pine Flat and integrating it into the Central Valley Project, it was essential that the Elliott rider undoing acreage limits pass. That was the safety valve.

On the morning of May 10, the Colonel reached Washington by train and checked into the Mayflower Hotel. His aim was to spend the next few days winning over undecided lawmakers the old-fashioned way: by employing a little race baiting.

THE CONSPIRACY THAT THE COLONEL FEARED—or at least professed to fear—was one that he thought southern senators would find particularly shocking. "Rumors are circulating freely here," he said, "that the CIO and the New Deal are going to transport hundreds of thousands of Negroes to California from the South," presumably to take over the 160-acre farms that would become available if the large land-holdings of the Central Valley were dissolved. Many of these blacks, the Colonel noted, couldn't vote in their home states because of Jim Crow laws. But once they moved to California, he said, their "places in the South" are going to be "filled by European refugees, mostly Jews"—a plot designed "to keep the New Deal and radical labor in office."

"I believe this should be explained to a lot of our Southern senators," said the Colonel, "and if they can be convinced this is true, I am sure they will do what they can to help us from being Mexicanized or social-ized by certain vicious groups."

The Colonel's theory was absolutely preposterous—the nuttiest thing he ever put down in his many pieces of correspondence that would survive the years. And there is no evidence that anybody in the Senate or anywhere else ever took seriously the scenario he laid out. Yet the Colonel's half-mad tone did accurately reflect how threatening he found the "160-acre provision that the Reclamation Bureau is trying to ram down our throats." It was obviously going to be "a tough fight," he said, "and we have got to do everything we can" to win.

The Colonel was not without some advantages. One of the biggest was that the company's second-ranking executive, Fred Sherrill, was in a splendid position to make certain that the growers' concerns were adequately heard. He was now sitting right inside the Army Corps of Engineers.

Sherrill had started thinking about hitching back up with the service from the moment the United States had been thrust into World War II. "What I am doing" in the cotton business "isn't as much help to the country in its present extremity" as signing up with the army would be, Sherrill told a friend, just days after the Japanese had bombed Pearl Harbor. The Boswell Company "would undoubtedly get along all right without me, and in the final analysis will probably have to."

It made perfect sense, really. By its nature, cotton was the ideal industry from which to draw men who understood how to deliver massive quantities of raw material and, because of the crop's global reach, also knew a thing or two about the way the world worked. The most eminent of the cotton men who joined the government was Will Clayton, the chairman of the old Boswell rival Anderson, Clayton & Company. In 1944, Clayton was named assistant secretary of state and would go on, in the words of his biographer, to be "a participant, catalyst, and in some cases prime mover" of several momentous "acts of statecraft," including the Marshall Plan for Europe's post-war recovery. Sherrill's job at the corps—coordinating the purchase of lumber used in the war—wasn't nearly as illustrious, but it was important enough that he earned a Distinguished Service Medal.

Notably, Sherrill's position had nothing to do with building dams. He also took care to insulate himself from possible conflicts of interest, declining Colonel Boswell's offer to make up the difference in his salary. "I don't want to have to testify that while I was serving my country, I drew at best more than 60 percent of my income from a cruel and heartless corporation, that thus undertook to keep me on a string," Sherrill wrote

to Leslie Groves, an old friend from West Point who arranged his commission in the corps. "Not for an instant do I think that Colonel Boswell
or anyone connected with this firm would ever ask anything improper
of me while in the government service. I am determined, nevertheless
. . . to have no reflections, by conjecture or innuendo, cast upon the
army or the office of the Chief of Engineers, upon the J. G. Boswell Co. or
Colonel L.R. Groves."

Even if Sherrill was too honorable to do anything corrupt, there were
undoubtedly worse things for the Colonel than having his lobbyist stationed inside the corps command center at such a pivotal time for the
company.

Even before he went back into the army, Sherrill had used his con-
nections to try to drum up a little government business for Boswell,
which had seen its biggest market—Japan—vanish because of the war.
"What's the use of having friends if you don't use them!" Sherrill had
written to Groves in October 1941, about a year before Groves took
charge of the A-bomb-building Manhattan Project. With top-quality
cotton in short supply, Sherrill told him, the army's clothing branch
might well want to know about "high-grade California cotton, shipped
by reliable merchants (of which we are one). . . . If it is not asking too
much, I should appreciate your calling this matter to" the appropriate
officer's attention.

Once he got to Washington in May 1942, Sherrill concentrated his energies on the job he had been hired for, and by the time the war was over
he would supervise the buying of some 23 billion board feet of timber—
enough wood to build nearly two million 2,000-square-foot houses. But
he did find spare moments to aid the Boswell cause in California. Sherrill had been in the army for all of a few weeks, for instance, when the
Colonel asked him to poke around Congress on a tax question that
could affect the company. In the fall, the company called on Sherrill to
help dislodge a "bottleneck with respect to Mexican labor" in the fields.

Mostly, though, Sherrill's extracurricular activities focused on one
thing: Pine Flat Dam. "If you can find time to do a little work on this
matter," Louis Robinson told Sherrill, "you will do the lake interests in
general and Mr. Boswell's interests in particular a real service." As the
events of 1944 played out, Sherrill became Boswell's eyes and ears on
both the flood control bill and the Elliott rider.

By the summer, Elliott's amendment to gut the 160-acre limit set off what *Business Week* magazine described as a controversy of "national proportions." Sherrill sent the Colonel regular updates on the legislation and coached him on how to lobby for its passage, while Paul Taylor, working on behalf of the Interior Department, angled for the amendment's defeat. Many years later, Taylor would "still remember the hot June days and the hot pavement as I went from place to place in Washington," rounding up support for the 160-acre limit. Among those he recruited to adopt resolutions and blitz lawmakers with appeals were the American Federation of Labor, National Catholic Charities, the American Legion and National Grange, which asserted that the "welfare of the nation" hinged on the preservation of the "family farm."

Despite Taylor's fervency, the Elliott rider was still intact as the rivers and harbors bill emerged from a House-Senate conference. Impelled to make a last stand for small farms, Taylor turned to on an old college classmate: Senator Robert La Follette.

La Follette understood instinctively what was at stake. Five years earlier, his civil liberties investigation had picked up on many of the same concerns that Taylor was now flogging. "The unchecked force" of big agriculture in California, the La Follette Committee had found, is "rapidly eliminating the prosperous 'family farm,' the institution around which our rural democracy has functioned." Now the Wisconsin senator was more than happy to marshal a defense of the 160-acre limit. "If you press" for the Elliott rider, La Follette warned Downey, "I will deliver a three-hour speech on the floor of the Senate against you"—an indictment in the form of a filibuster. He even asked Taylor to draft the speech deprecating Downey, but it was never needed. Intimidated by La Follette, Downey backed off, and the entire bill bogged down in the Senate. The Elliott rider died with the Seventy-eighth Congress. "We stalled the gravy train!" said a gleeful La Follette.

There wasn't much chance to savor the victory. The other measure being sought by the big growers, the flood control act, had even more momentum behind it. Despite the unambiguous wishes of the president—Pine Flat Dam should be "built by the Bureau of Reclamation and not by the Army Engineers," Roosevelt said in May, reiterating his long-held stance—the House Committee on Flood Control was on track to legislate the reverse.

Critics found it outrageous that the Army Corps was so keen on going along with the committee and frustrating the president's plans. It's "insubordination," San Francisco writer Robert de Roos and Harvard University professor Arthur Maass wrote in a *Harper's* magazine article that blasted the Army Corps' "arrogance" and denounced the "very cozy" relationship between the corps and a group of senior congressmen. It was as if, Wallace Stegner would later add, the corps "got its finger into the bung of the pork barrel . . . and nobody has ever been able to get it out."

Such condemnations aside, there were a few plausible reasons to ignore FDR on this one. Most compelling was that, while other bureau facilities sent mostly government-owned water to farmers, every molecule of water to be channeled through Pine Flat Dam was already in private hands—a point that John Page's bureau had itself acknowledged back in 1940. Given the extent of the farmers' water rights, "there is no need whatever . . . for control by the bureau," said Charles Kaupke, the Kings River water master and a leading spokesman for the growers.

Congressman Elliott, a member of the flood control committee, also functioned as a mouthpiece for the farmers, repeating the now-familiar mantra that Pine Flat's main contribution to the valley would be to stem flooding, and it was consequently the corps' responsibility to build it. The principal purpose of Pine Flat "is not irrigation," Elliott told a Senate hearing. "It is not power." The Interior Department, he continued, "has become one of the greatest hindrances to agriculture in the state of California. . . . They have sold themselves down the river" by pushing a "socialistic program . . . that Mr. Ickes has in the back of his mind."

Harry Bashore and his assistants at the bureau countered that not only did their studies show irrigation to be the chief benefit of Pine Flat, but the dam was an integral part of a valleywide irrigation network revolving around the Central Valley Project. Boswell's own executives had articulated essentially the same vision less than a year earlier, but now their faction expressed shock—shock—that anybody would promote such a plan.

As the year went on, the bureau tried to have Pine Flat stripped from the nationwide list of projects included in the flood control bill, and to have pre-construction work on the dam funded under the Interior Department's budget. But Elliott and the Kings River farmers were able to get the bureau's budget request nixed. And by late 1944 it was apparent to everyone that the flood control act was going to be passed with Pine Flat Dam still a part of it and the Army Corps in charge of construction.

Nevertheless, the bureau wouldn't admit defeat. Congress might authorize the corps to build Pine Flat, but Ickes and Roosevelt remained resolved that reclamation law—and the 160-acre limit—would apply. The secret was to insert language in the flood control act that spelled all this out. The big growers could get their dam built by the corps, but if they drew a drop of water from Pine Flat, they'd have to break up their land holdings. Unfortunately for Roosevelt and Ickes, the language their subordinates inserted into the bill contained a huge unintended loophole:

> Hereafter, whenever the Secretary of War determines, upon recommendation by the Secretary of the Interior that any dam and reservoir project operated under the direction of the Secretary of War may be utilized for irrigation purposes, the Secretary of the Interior is authorized to construct, operate, and maintain, under the provisions of the Federal reclamation laws, such additional works in connection therewith as he may deem necessary for irrigation purposes.

As Ickes would soon come to realize, the key words in this jumble of a sentence—contained in section 8 of the final bill—were "such additional works." Was Pine Flat itself an "additional work"? Or did Congress somehow mean for the 160-acre limit to apply only to "additional works" beyond Pine Flat Dam?

After some more discussion on the Senate floor, Ickes grew confident that the lands served by Pine Flat Dam would indeed be held to the 160-acre limit. And he urged the president to approve the flood control bill. In Ickes's eyes, section 8 amounted to a "sweeping defeat of the California interests who oppose . . . the land policies of the administration."

He shouldn't have been so exultant. Regardless of what had been said on the Senate floor, the prose was sufficiently opaque to keep teams of lawyers poring over it for many years, parsing section 8 like Talmudic scholars studying some divine text. Nineteen months after President Roosevelt had signed the flood control act into law, even Senator John Overton, the chairman of the flood control subcommittee, felt compelled to ask the bureau for a clarification. Overton believed there was still "an uncertainty as to . . . what application would be made of the

160-acre limitation." In the late 1950s, nearly a decade and a half after the act had passed, the U.S. attorney general was still trying to sort out what section 8 was all about.

More immediately, the hazy language gave the growers the excuse they were looking for. "We are advised," the Kings River Water Association said in one letter to the bureau, "that this project contains no additional works bringing it within the scope of section 8."

Or, to put it another way: Acreage limit? What acreage limit?

On April 2, 1945, President Roosevelt signed a War Department appropriations bill, giving the corps funds to begin preliminary work on Pine Flat. The president did so "reluctantly," however—and emphasized that he'd try to avoid dispensing any more money to the army for the dam.

No matter what the flood control act said, Roosevelt was sticking to his conviction that Pine Flat was primarily for irrigation and should be linked to the Central Valley Project. Accordingly, the president said, he planned "in the near future" to recommend to Congress that it take the dam away from the corps and transfer it to the bureau.

He never got the chance. Ten days later, Roosevelt was getting ready to eat lunch at his home in Warm Springs, Georgia, when he looked up at his butler and said, "I have a terrific pain in the back of my head." Then he collapsed, the victim of a cerebral hemorrhage. Alice Rothery, Colonel Boswell's secretary, remembered well what it was like when the news reached Los Angeles:

It was 2:50 Thursday when the word of Roosevelt's death came to us . . . and it just didn't register at first. JGB was across the hall. . . . I went over there and told Mr. B., and he just said: "Roosevelt dead." . . . Nothing more. The "boss" went to his room without another word, and when I went in later he was sitting, looking out of the window, and his face was very red. He seemed very subdued when he went out, and said "good night" so quietly.

By evening it had sunk in, and you can imagine that I felt as though I had lost someone in the family, and I have heard others make the same remark. I have wept a bit every day since, and judging from appearances, I have not been alone. JGB's eyes looked pretty red the next day (perhaps not from that, but he avoided looking directly at me, I noticed). As a matter of fact, he hadn't said anything against Roosevelt, to my knowledge, in months. I don't know why.

Whatever sentimental feelings the Colonel may have had for FDR, business was still business. Within a couple of weeks, he and Fred Sherrill were trying to figure out where the new president stood on Pine Flat. Sherrill's take was that Harry Truman "leans definitely to the congressional point of view." And because Congress had passed the flood control act "in spite of presidential pressure" to back the bureau, "I believe the new president will decide in favor of the Corps of Engineers."

It was not one of Sherrill's best pieces of political analysis. Two months after Roosevelt died, Truman informed Secretary of War Robert Patterson that, regardless of the way the farmers and the corps were interpreting section 8, he was hewing to his predecessor's position. The flood control act, he said, "clearly established the intent of the Congress to support and maintain the principles of the federal reclamation laws." And it should be the bureau—not the corps—that negotiates a storage contract with the growers.

Truman, like FDR, intimated that the corps might not even get to build Pine Flat—the flood control act be damned. Both the army and the bureau, he noted, were developing comprehensive plans on the major rivers of the Central Valley. And until those reports were completed and sent to the White House for review, Truman said, "no commitment" should be "made to anyone as to when or by whom construction will be undertaken" on any project.

Despite the president's directive, the corps told Congress in the spring of 1946 that it was ready to begin construction of Pine Flat, and it set out to get more funding. The Senate appropriated $1 million for the job. President Truman, escalating the game of chicken, then turned around and impounded the money. He said he'd release it under two conditions: first, that the corps and the bureau reach agreement over how to allocate the cost of building Pine Flat—apportioning a certain amount to irrigation and a certain amount to flood control and, second, that the bureau negotiate an adequate repayment contract with the farmers covering the irrigation portion of the dam. (By law, flood control was free of charge.)

That summer, administrators from the two agencies and four engineers representing the local farmers gathered at the Army Corps' offices in Sacramento. Distrust abounded over four days of meetings. Both sides did agree that Pine Flat's price tag had shot up from the original $19.5 million to $33.5 million, a reflection of certain changes in the design, such as larger spillways, and general cost increases. The farmers, though, said no more than $10 million of that should be allocated to irrigation; the bureau indicated that up to twice that amount would be appropriate. The corps, trying to be Solomonic, suggested that the costs get split right down the middle: $16.75 million for irrigation and an identical sum for flood control.

The gulf in the calculations stemmed from the inherent biases that each side had going into the process. The bureau wanted to allocate the costs of the dam by using a formula that measured what the water in the Pine Flat reservoir would ultimately go toward. The answer: mostly irrigation. The farmers, however, preferred a formula that ostensibly measured where the dam's economic returns would come from. The answer: mostly from flood prevention. The whole exercise was laughable. Depending on the methodology you picked, you'd get whatever answer you wanted. Any attempt to settle things, needless to say, went nowhere.

A few weeks after the Sacramento meetings, the Kings River farmers presented to the bureau a list of fifteen conditions that any contract with the government would have to contain in order for them to sign it. The growers insisted, for instance, that Kings River water could never be diverted outside the immediate area, and that Pine Flat wouldn't be made part of the CVP. The big precondition, though, was number 7: "Any so-called acreage limitations shall not apply" to those being serviced by the dam.

Michael Straus, who had taken over for Harry Bashore as Reclamation commissioner, referred sardonically to the farmers' list of demands as a "15-point catechism." But the bureau did finally respond to the growers' ultimatums in October 1946 and, on the all-important acreage-limitation question, the agency seemed to offer Boswell a fair bit of hope. While acreage limits "will be followed on the Kings River" generally, the bureau said, "the excess land provisions, in their present form, may not apply appropriately" to those in the Tulare Lake bed.

All of a sudden, the promise that John Page had made to Fred Sherrill in 1941—and Harry Bashore then dashed two years later—looked to be

back on the table. And that was just the beginning. In a stunning rever-
sal, bureau officials in California eventually drew up a plan to exempt
not only the disaster-prone lake bottom, but the whole Kings River area
from the 160-acre limit. Their rationale was that "it would be a matter of
extreme difficulty to delineate" where the flood risk that precluded
small-scale farming began and the other parts of the river ended "without being arbitrary and without working hardship upon individuals in
the transition zone."

The most stalwart defenders of the 160-acre limit weren't surprised
that a Straus-led bureau was so willing to knuckle under this way. Despite his training as a newspaper editor, Straus was known to be notoriously loose with the facts. Independently wealthy but a devoted New
Deal liberal, he was also pompous and a total slob. "The characteristic
Mike Straus pose," said Floyd Dominy, who followed him as commissioner, "was for him to plant his feet on his desk, almost in your face,
and lean back in his swivel chair flipping cigarette ashes all over his
shirt. At the end of the day, there was a little mound of ash behind his
seat. He was an uncouth bastard! He carried one white shirt with him on
trips. I remember one night when Reclamation was throwing a party,
and a cub reporter came by and asked me where to find Mike Straus. I
just said, 'Go upstairs and look for the guy who reminds you of an un-
made bed.'"

It wasn't that Straus didn't care about the 160-acre limit. He just didn't
care about it enough to expend any political capital on it, especially
when that might cut into his real agenda: building dams—lots and lots
of dams. "There was nothing on earth that gave Mike Straus quite as
much boyish, exuberant satisfaction as erecting dams," said one water
historian. He believed in "dams for dams' sake." And in his eight years as
commissioner, "he would become responsible for as many water projects as any person who ever lived."

To be sure, Michael Straus talked like he was the great guardian of the
160-acre limit, vigilantly "defending the historic policy," as he himself
put it. And Straus wasn't one to endorse an outright abandonment of
the 160-acre law; that was as politically risky as backing it in earnest.
Rather, he sketched out for Congress a step-by-step guide for keeping
the acreage limit on the books—and then evading it.

Straus called the gambit "technical compliance," explaining with a
wink that this wasn't to be confused with "spiritual compliance." For ex-

ample, Straus told how a big farmer could subdivide his land into 160-acre parcels, sell them off to relatives and employees and then lease everything back. Straus also promoted the idea that a big farmer could escape the acreage limit by paying the government for all his irrigation benefits in advance, before the law required him to sell off his excess land—a ten-year window. This contrivance—seen as very questionable within the bureau itself—was a sop to the richest farmers, and it would ultimately become the loophole that Boswell seized on.

As the proposed exemption for the Kings River area made its way through the bureau hierarchy in Sacramento, Paul Taylor did his utmost to head it off. In a blistering, seventeen-page appeal to the bureau's regional director, Taylor argued that, while excusing the farmers from the 160-acre limit might speed along the agency's contract negotiations on Pine Flat, caving in would be a "grave mistake." The cost, said Taylor, would be no less than "sacrificing the main purpose of the reclamation law." Taylor admitted that there were some complexities in applying the law in Tulare Lake and elsewhere on the river, but he said that these problems were "less critical than they are made to appear."

The big farmers themselves, Taylor pointed out, had stated repeatedly that they wanted Pine Flat built for flood control only if they were guaranteed the right to receive the water stashed behind the dam whenever they requested it for irrigation. As those in the lake bottom liked to say: "We'd be no better off getting dried out than we would by being drowned out." But to Taylor, this arrangement gave the farmers the upside of reclamation—precisely choreographed releases of water from a government dam—while allowing them to dance around what they saw as the downside: the 160-acre limit. When all was said and done, Taylor persuaded the regional director to hang tough on the Kings. Boswell was foiled again.

Negotiations on allocating the cost of Pine Flat weren't going much better. In the fall of 1946, the Army Corps was still valuing the irrigation piece of the dam at $16.75 million and the bureau was fixed on $20 million. But Sid Harding, the Tulare Lake engineer, told Fred Sherrill that he remained doubtful that more than $10 million was "justified" or "can be sold" to the farmers. "I am still expecting to see the day when the Bureau of Reclamation can be put in its proper place," added Harding, who was one of the negotiators for the growers. "Until then, we have to stay in there trying."

Sherrill didn't appreciate such bluster. He now worried that the dam was taking too long to get under way, and the farmers were coming off as intransigent. Unless they were willing to move off their $10 million figure, he feared, the White House might simply inflict the bureau's $20 million asking price on them—or else "there will be no project" at all. Sherrill then proposed a way to break the logjam: If the locals and the corps would come together around a figure of $13 million or so, he said, "the Secretary of War will be in a greatly strengthened position" to prod the bureau away from the $20 million mark.

In early 1947, Congressman Elliott telephoned Louis Robinson in Corcoran to announce that a breakthrough had finally been reached. The deal, which Elliott had played a lead role in brokering, traced the lines that Sherrill had mapped out. The farmers and the bureau would continue to work out an agreement for repaying the irrigation costs of Pine Flat, but the final amount wouldn't exceed $14.25 million. Meantime, the army would build the dam and operate it for flood control "without hindrance" from the bureau.

On February 18, President Truman released the $1 million he had impounded for the project. At long last, the building of Pine Flat was a certainty. The 160-acre limit, however, was not.

THE TORTURED SOUNDS OF TROMBONES AND TUBAS from five high school marching bands playing the "Star Spangled Banner" were still ringing in folks' ears when Charles Kaupke, the Kings River water master, took the podium to begin the groundbreaking ceremony for Pine Flat Dam. It was May 27, 1947.

Kaupke—who bore a passing resemblance to the politician he hated most, Harry Truman—had surely wondered whether he'd ever see this day. He had been the watermaster on the Kings, gauging and divvying up its flows, since 1917. Along with Sid Harding, he was one of the growers' representatives in their haggling with the bureau. He had trooped to Congress many times, testifying about the need for the dam.

On this day, though, the political sparring and interagency feuding seemed distant. The start of construction on Pine Flat, Kaupke told the 2,000 people present, was "the most important event in the history of the Kings" since farmers had first tapped the river. Then he introduced the dignitaries on the dais: officials from the Army Corps, local and state politicians and the leaders of the Kings River Water Association, including Boswell's Louis Robinson.

Soon Colonel E. H. Marks of the corps addressed the assemblage. "The Kings has had its own way," he said, "and the farms and cities in its path have prospered or suffered at the whim of the river." But the 429-foot-tall structure that "will rise a few hundred yards upstream from here . . . will end the hazard of the disruption of business and community life during and after floods."

The main speaker, California Governor Earl Warren, stood in the shadow of the granite mountain walls, stared out at the river and pronounced this "one of the happiest days of my life." Warren, who grew up in Bakersfield, told the crowd that "no matter how long I live," Central California "will always be home to me." Pine Flat Dam, he said, will bring "prosperity and safety" to "hundreds of thousands of people in this valley."

Finally, Colonel Marks turned to the governor and said, "I would like to present to you the key that will unlock the resources of the Kings River." With that, Warren inserted the key into a switch box and turned it, sending a radio signal to a group of hard hats high on the north side of the canyon, about a half mile away. One of the men pushed a plunger connected to forty charges of dynamite, and the face of the south canyon wall exploded in a dirty cloud. The crowd waited for the sound, but for about five seconds there was none. Only quiet. Then a thunderous crash swept down the river, and a mushroom of smoke billowed and spread. Another racket ensued—this time, clapping and shouts of hurrah.

Yet not everybody cheered. The bureau still had its own designs on Pine Flat, and within weeks of the groundbreaking, Fred Sherrill was complaining that the agency had started "flooding the country with propaganda, the principal argument now being that the speculative land barons of California are standing in the way of obtaining farms for soldiers."

Such recriminations would long characterize the contract talks between the growers and the bureau. Year after year went by—1947, 1948, 1949—with still no agreement on how much the irrigation portion of the dam was worth, whether the 160-acre limit should apply or how to resolve any number of other claims and counterclaims. The bureau didn't even present the growers with a draft of a contract until 1950. When the agency did, the farmers promptly rejected it. What followed was an endless series of angry letters, hostile meetings, tense phone calls, spiteful accusations, taut nerves and all-round huffing and puffing. The bureau also made another attempt to integrate Pine Flat with the Central Valley Project, but that bid was stymied, and the congressman who carried the bureau's legislation, Cecil White, paid the price during the next election. "We knocked him flat," said one official of the Kings River Water Association.

Pine Flat was completed in 1954 at a final cost of $41.6 million, seven years after the groundbreaking and almost a full decade after the flood control act had been made into law. The Army Corps then dammed the other three rivers that fed Tulare Lake—the Kern, Kaweah and Tule—by 1962. And still there was no contract with the bureau. Interim agreements did allow for Pine Flat to function, but it wasn't until 1963 that a final repayment pact was signed—and only then with the recognition by both sides that the critical issue of acreage limitation couldn't be decided.

Instead, with Congressman Bernie Sisk serving as the go-between, the bureau and the farmers agreed to punt that question into federal court in a "test case," where the rancor, like the river itself, would flow on and on.

Blue Blood

"Boy," the Colonel shouted, his Georgia twang echoing down the hall, "come on in here."

The Colonel's stepson, Warren Williamson, traipsed through the twenty-room Tudor mansion that he now called home, passed the elevator and entered the library. He was seventeen years old, and everybody knew him as "Spud," a nickname that had stuck since he was a baby and fairly resembled one. Just months earlier, in January 1945, the Colonel had married Spud's widowed mother, Ruth, the daughter of Harry Chandler—the man *Time* magazine said was "more responsible than any other" for the building of Los Angeles. While some relatives on the Boswell side of the family considered the Colonel impossibly bigheaded, Spud took an immediate liking to his new dad. "He could charm the bird out of a bush," Williamson remembered years later. "He had that old southern way."

As Spud walked into the library, the Colonel shuffled over and shut the door behind them. Then he settled into an overstuffed chair near the fireplace, which provided an ideal backdrop for the monologue about to take place. Among the ecclesiastical figures embellishing its seventeenth-century gesso facade was one whose legs sprawled open profanely, while another clutched a bottle of wine. "Your mother is worried about you," the Colonel said, as Spud's disquiet grew. "She wants me to give you the birds-and-bees lecture.

"Well, I don't know anything about the birds and the bees," Boswell went on. "But back in Georgia, when my mother felt I needed that sort of education, she called in my Uncle Luther, and so I'm going to tell you what Luther told me." A Greene County farmer and member of the Georgia legislature, Luther Boswell had been hailed as the "sage of Penfield," and Spud was about to learn why.

"Boy," the Colonel began, "you can mess around with one of those little debutante girls, and you'll most certainly wind up at the altar with her daddy's shotgun in your back. Or you can fiddle around with a married woman until you raise your head off the pillow some night, and you'll be looking down the barrel of a .38. Or you can go to the red-light district, and when you come away from there, you'll most assuredly have some terrible social disease. Boy, of the three options, always take the third."

The Colonel paused and then continued: "Now you know everything I do about the subject." Spud was speechless.

At age sixty-three, the Colonel had reached the top of his career. In the San Joaquin Valley, the Boswell name was synonymous with immensity and power. And in downtown Los Angeles and the exclusive enclave of San Marino, where he and Ruth had recently bought their 12,000-square-foot house, the Colonel was a pillar of a community known for its uncompromising conservatism. He may have been a southerner, but he "was not a 'rebel,' much less an unreconstructed one!" joked Bill Munro, a professor of history and government who sat with the Colonel on the board of the California Institute of Technology.

The Colonel had, of course, achieved high social standing even before he met Ruth. And now, their marriage further solidified his position, for it knotted together two great California empires: the giant of industrial agriculture and the most elite clan in Southern California.

Ruth was one of six children of Emma Marian Otis and Harry Chandler, the Horatio Alger of the Pacific. Born in Landaff, New Hampshire, Chandler journeyed to Los Angeles in the early 1880s, while still a teenager. The warm, bone-dry climate offered him an opportunity to salve his lungs, which had become badly infected after he accepted a dare from a Dartmouth College classmate and dove into an ice-covered vat of starch. Chased out of one flop house after another because of his hacking cough, Chandler landed as a fruit picker in the San Fernando Valley, where he recovered from his illness in the hot sun, slept under the stars and through hard work and thrift pocketed a few thousand dollars. Chandler took his small fortune and bought newspaper circulation routes (which in those days were sold independently of the papers themselves), and soon he crossed paths with that most ruthless of capitalists, *Los Angeles Times* owner General Harrison Gray Otis. Otis and Chandler could well have gone to war. Instead, they conspired to drive a

Times rival out of business, and Otis hired Chandler as his new circulation manager. Sometime later, he was made business manager. And in between, the up-and-comer won a promotion of a different sort: He married the boss's daughter.

Chandler took over for his father-in-law as publisher of the *Times*, but he was never a newspaperman in the manner of his ink-stained arch-rival, William Randolph Hearst. Rather, he was a businessman who happened to run a newspaper, and he used the *Times* to boost the city and, by extension, his huge portfolio of holdings. The *Times* coddled Chandler's friends and attacked—or, more devilishly, ignored and thus silenced the voice of—his enemies. It was said that the paper's editorial-page philosophy through the 1930s and '40s could be summed up most simply: "Think of what is good for real estate."

Dubbed "California's landlord," Chandler bought up and subdivided property in the Los Angeles County communities of Arcadia, San Gabriel and Baldwin Hills. At one time, he owned upwards of 862,000 acres along the Mexican border (where he had his own gigantic cotton operation until the early 1920s). He controlled the 270,000-acre Tejon Ranch at the southern edge of the San Joaquin Valley and acquired the 340,000-acre Vermejo Ranch in New Mexico and Colorado. And not least, there was his deal in the San Fernando Valley with General Otis, Corcoran founder Hobart Whitley and the other members of the Board of Control, who reaped untold millions when Owens River water came splashing down the Los Angeles Aqueduct.

Land was but one piece of Harry Chandler's kingdom. He helped organize Western Air Express and, with a group of friends, launched Douglas Aircraft. He drilled for oil and supplied the thrust behind the development of the Los Angeles Harbor, the L.A. Civic Center and the Hollywood Bowl. He was the brains behind one of the city's most famous icons: the fifty-foot-high HOLLYWOOD sign perched up in the hills. So intricate was Chandler's web of dummy corporations and secret trusts that the *Saturday Evening Post* noted "that nobody, with the possible exception of himself, has ever been able to count them."

Chandler died in September 1944 at the age of eighty. Four months later, the Colonel slipped out of his office (by then located in the H. W. Hellman Building in L.A.'s Spring Street financial district) and drove with Ruth to Yuma, Arizona—the place where it had all started for Boswell as a cotton merchant. The pair eloped. To his good friend John

McWilliams, "the marriage was perfect," and the Colonel himself quipped that he had always wanted to stick his feet "under a rich widow's table." Ruth's first husband, a well-known attorney named Frederick Warren Williamson, had died of a heart attack in 1942.

Nobody remembers exactly how the Colonel and Ruth met, but they moved in similar circles in Pasadena, and it's quite conceivable that some mutual friend introduced them. (The Colonel and Harry Chandler surely would have known each other through the California Club and other downtown business fraternities, but they weren't intimates.) Fifteen years younger than the Colonel, Ruth was an attractive woman with pronounced features, blue eyes and hair that had grayed prematurely. She was about the Colonel's height, kept fit, and was as much a stickler about good posture as he was—a nod to the time when she taught physical education at Pomona College in the early 1920s. "It was 'Sit up,' 'Sit up' all the time," recalled Spud's sister, Sue.

Ruth was quiet, shy even, but anyone who took her for a pushover was dead wrong. She had opinions, firm ones, and in later years she would stand her ground against her firebrand of a sister-in-law, Dorothy Buffum Chandler, as the family tangled over the direction of the *Times*. Ruth "was a very strong person," said Al Casey, who was president of the newspaper's parent company, Times Mirror. "You weren't going to snow her."

His first wife, Alaine, had been gone for more than six years when the Colonel remarried, and there is no indication that he had dated anyone between Alaine's passing and the time he began courting Ruth, showing up at her Arden Road residence in his green Buick to whisk her off to dinner or the theater. When things got serious, the Colonel—not a tender man, never all that comfortable with children, given that he had none of his own—did his best to reach out to Ruth's kids: Spud, Sue and their younger brother, Norman. "He'd roll in and say, 'Everybody got your homework done? Let's go to the movies,'" Norman remembered. "And away we'd go to the movies—sometimes without our homework being done."

The Colonel wasn't smitten with Ruth the way he had been with Alaine. But he had starting telling people that he was "getting old and was lonesome," and Ruth was a marvelous companion: smart (with a Stanford and Wellesley education), polished (she was also a stickler when it came to diction and grammar), and possessing a sharp sense of humor. The Colonel "loved Alaine," said his nephew, J. G. "He had the greatest respect in the world for Ruth."

Considering her pedigree and the imperious reputation she had in some quarters, Ruth could be amazingly down-to-earth. When she visited Georgia, she liked nothing better than tooling around Greene County in a rusted pickup truck with a jammed front door that belonged to the wife of one of the Colonel's boyhood friends. Once, while dropping by Boswell's Arizona ranches, she and the Colonel were told that the worker housing had fallen into disrepair. It was Ruth who expressed concern, tramping through the dust to see the tumbledown cabins for herself, while the Colonel stayed in the car, angry that the issue had even come up. She then pushed the Colonel to tear down the old wooden structures and put up concrete ones.

Ruth may have seen a little of her late father in her new husband. Although they were nothing alike physically—Harry Chandler, a hearty six feet two, would have towered over the Colonel—they shared a deep affection for accumulating land, a loathing of organized labor and an intuitive grasp that water was the secret to unlocking California's riches. Harry Chandler eschewed the sorts of sybaritic trappings that the Colonel had cultivated a taste for, but both of them shunned alcohol, and they were also alike in another, more basic way: Each carried himself with an exceptional degree of certitude. Neither was afraid to say what was on his mind, and to do so unequivocally, vehemently, with an aura of being in complete control. Spud Williamson, when he had hit his seventies, liked to say that he had known two truly great men in his day: his grandfather, Harry Chandler, and his stepfather, J. G. Boswell.

Yet from the time he married into the family, the Colonel couldn't help but feel, on some level, that he had to prove himself. It was hard keeping up with the Chandlers, or at least the ghost of Harry Chandler, who at his peak had reportedly become one of the twelve richest men in the world. "No Easterner can understand what it has meant in California to be a Chandler, for no single family dominates any major region of the country as the Chandlers have dominated California," David Halberstam wrote in *The Powers That Be*. In the East, it would take "a combination of the Rockefellers *and* the Sulzbergers to match their power and influence."

The Colonel made it clear that after he married Ruth, he wasn't going to live in her house on Arden Road, though there was plenty of space there. That might make him look like some arriviste or, worse, that Ruth was paying the bills, which she wasn't. "I don't want to be a kept man," the Colonel said. Thinking about it years later, Jim Boswell remarked that for the Colonel, the "accumulation of money" had always been a kind of obsession. "And then when he married Ruth it became a real fetish," he said. "He was going to outdo the champion." Spud Williamson recollected overhearing the Colonel bluntly telling his mom: "I can hold my own at the bank or any other place with those folks from the *Los Angeles Times*." At that stage in the Colonel's life and at that point in the Chandler dynasty, Williamson added, "I don't think there was any question but that was true."

Nevertheless, it was probably a good thing that Corcoran's name hadn't been changed to "Otis," in honor of Ruth's grandfather, as had been contemplated back in 1905. Setting up shop in a town that saluted the Chandlers' forebear might have been too much for the Colonel to take.

THE COLONEL TRAVELED THROUGH TWO WORLDS—the Chippendale-furnished opulence of San Marino and the gritty, mud-splattered environs of Corcoran—and from time to time, Spud went along for the ride.

At first, Spud couldn't figure out where they were going; because of the Colonel's brogue, he kept looking on the map for a town called "Cochran." Once he pinpointed the place, the drive there was nerve-racking, at least as long as his stepdad was behind the wheel. The Colonel, who would forgo his chauffeur for these father-son trips, turned out to be a serial tailgater. "We'd come around the curve, and there would be a big truck laboring along in front of us, and I'd think we were going to go right through the back end of the thing," Williamson recalled. "I mean, your knuckles were white, and he'd just be chattering away. I don't know if it was a lack of depth perception or he had always driven that way. I don't think he ever had an accident, but he could scare the liver out of you." Spud assumed the driving whenever he could.

After trundling over the mountains and into the valley, the Colonel would spend the day inspecting field and gin, and Spud would tag along. The teenager was intrigued by the business, and he loved hearing stories about how the Colonel—this self-described "boy from a Georgia cotton patch"—had put together such a colossus. Still, Corcoran wasn't on anybody's scenic tour of California, and Spud learned from the very first just how ugly and smelly a factory farm and cattle-feeding outfit could be. As impressive as it was, he said, "it was the kind of place where you were thinking, 'How quick can I get home again?'"

And what a home it was. The house that the Colonel and Ruth purchased at 1155 Oak Grove Avenue had been completed in the late 1920s for an oil heiress named Emery, and she had spared no expense. Spread over ten acres, the estate was designed by Myron Hunt, the Pasadena architect whose panoply of credits included the Rose Bowl and the Huntington Art Gallery. The detailing was exquisite, from the mullions on the windows and the long stone balustrade outside to the oak-paneled walls and regal, front-hallway staircase inside. It was this very staircase, its banister hand-sculpted by teams of Italian carvers, that Barbra Streisand (playing Fanny Brice) would ascend in the 1968 film *Funny Girl,* calling out to Omar Sharif (Nick Arnstein): "Ah, Nick, it must have cost a fortune. . . . Anyway, it's the perfect house for a millionaire."

While the Colonel was living at Oak Grove, Elizabeth Taylor and Van Johnson showed up to make a movie with an ironic title, given their host's aversion to liquor. *The Big Hangover* was about a war veteran whose allergy to alcohol made him drunk at the most inopportune moments. Several years later, when Ruth's son Norman was stationed in France during the Korean War, he and some other soldiers were watching Dean Martin and Jerry Lewis romp across the big screen in one of their buddy movies, and the Oak Grove mansion popped up again. Norman was floored, but kept his mouth shut. "When you're in the army," he said, "you don't sit there and go, 'Hey, look guys, there's my house.'"

Ruth took charge of the entertaining at Oak Grove—affairs that ranged from the magnificent to the ridiculous. When Pat Boswell, Walter's daughter-in-law, once visited the house, Ruth came in carrying silver trays, only to reveal that sitting on them were piles of peanut-butter-and-bacon sandwiches. "There's just one thing about Ruth," the Colonel said, a tad embarrassed. "She ain't what I'd call a good feeder."

Ruth was responsible for decorating the mansion, but the grounds were the Colonel's province. The gardens at Oak Grove were so beautiful that a glass slide of them would one day be archived at the Smithsonian

Institution. They had been laid out originally with the assistance of Florence Yoch, a celebrated designer who directed the landscaping around Tara on the set of *Gone with the Wind*. Yet even Yoch's touch wasn't southern enough for the Colonel, who proceeded to plant magnolias here and there. The Colonel's chief gardener, Shintaro Tanaka, was a magician with tree and shrub, and Boswell found much pleasure in walking his estate, always formally attired, his dog Jeb (for Jeb Stuart, the Confederate general) at his heel. After he had been at Oak Grove for a while, the Colonel discovered one imperfection, at least in his mind: The property lines were uneven. So he bought out parcels from his neighbors, determined to have his borders neat and orderly, like his fields of cotton. For the Colonel, "there was no such thing as having too much land," said Spud Williamson. "You only wanted more. You never wanted less."

Some members of the Boswell family, unable to hide their contempt, started calling the Colonel's house "Buckingham Palace," and thought that he and Ruth were acting, in the words of one nephew, "very pleased with themselves." Bill's wife, Kate, had no patience for the Colonel or Ruth's airs. At one family gathering, the Boswell nephew was given the job of keeping Kate and Ruth "from going at each other's throats." The reason for much of the animosity was simple: Ruth and the Colonel were beyond rich, and Kate always felt as if she and Bill hadn't been given their fair share. "There was that contrast," said Bill and Kate's daughter Jody. "She didn't have what they had."

Back in Corcoran, Kate and Bill lived for years in the same house on Whitley Avenue, a house she totally hated. "I said, 'Bill, sell it. I'll go anywhere. Just get me off this main highway with my children running out in the streets, with the bicycles, tricycles and all the animals and somebody crying all the time. I can't stand it.'" After enough pleading, they moved to the other side of town and rented a little cottage from, of all people, the Salyers. "It was a cute little place," said Kate. "It was like living in a doll house." All along, though, Kate thought of this as a temporary stop, and she urged Bill to get busy building them their own house—her dream house. But J. G. for some reason discouraged the idea, and Bill always deferred to his brother. "I don't know," said Kate. "Every time he told Bill to do something, by golly, he'd do it." Finally, Kate got so fed up that she went out and bought a new place on her own, and then worked up the plans to refurbish it. She even had a decorator come up from Pasadena to help her. "And so I got my bath and big closets and dressing room and everything," she said. But she would always resent how long it took and how the Colonel had stood in her way.

Like her mom, Jody Boswell found the Colonel insufferable. Not long after Alaine had died, the Colonel took his niece on a South American cruise, and nothing was quite the same after that. The hostility stemmed from a party in the Colonel's stateroom, where Jody heard him talking to one of his guests. "Yes, she is lovely, isn't she?" the Colonel said. "You know, she's my ward." Jody had been aware of the deal made in 1925, in which her father, Bill, became the Colonel's employee and the Colonel agreed to compensate his brother, in part by providing the best education possible for his children. But never in her wildest dreams did Jody imagine that the Colonel regarded her as his "ward" because of the arrangement.

"I want to settle this whole thing right now," Jody snarled at the Colonel later that night, unable to hold her anger any longer. "I have a mother and father, and I am not your ward. . . . If I'm your ward, you explain that to me. I always thought this was part of my father's salary." The Colonel was so stunned by her tirade, Jody later said, it was the only time that she ever saw him at a loss for words. Finally, he managed a weak rejoinder: "Well, I, I, I, don't know. It just came out. Why, I didn't mean it."

"You sure did mean it," she said hotly. After that blowup, Jody and the Colonel retreated into what she called an "armed truce."

Not everyone in the Boswell clan thought ill of the Colonel, and occasionally he demonstrated a flicker of warmth. In 1948, his older brother Herbert's daughter went off and wed a sailor. Virginia Boswell was only sixteen at the time, and Herbert was so upset that he threatened to have the marriage annulled. Virginia's sister, Louise, wrote to the Colonel, and he responded by sending Virginia $1,000, along with a letter telling her that his mother, Miss Minnie, had gotten married at seventeen and everything had turned out just fine. It'll be that way for you, too, he assured her. He also gave Virginia and Louise a fair bit of stock in the Boswell Company. "I want you girls to have a good life," he told them.

The Colonel and brother Walter also remained close, and Walter was a frequent houseguest at Oak Grove, where the two of them would sit for hours and cackle about the old times in Georgia. Lots of friends were around as well, including John McWilliams, a Pasadena investor (and the father of celebrity chef Julia Child), the Van Nuyses (scions of the old grain-farming family for whom the San Fernando Valley town was named), Merrill Lynch executive Lyman McFie, and Robert Millikan, who headed the California Institute of Technology.

The Colonel had been a supporter of Caltech since the late 1920s, and as time went on, the school became increasingly important to him. Some of the biggest stars in science had graced the institute's Pasadena campus over the years—Einstein, Oppenheimer and Pauling among them. And then there was Millikan himself. A Nobel laureate in physics, he was "next to Einstein . . . undoubtedly America's most public figure in science," as one historian has noted. "When Millikan spoke, the country listened." Caltech's board of trustees had always been composed of luminaries in their own fields: law, business and medicine. In late 1946, the Colonel joined his brother-in-law Norman Chandler, Herbert Hoover Jr., attorney John O'Melveny, copper-mining magnate Harvey S. Mudd and other VIPs on the board—a coronation of sorts for Boswell as one of the real lions of Los Angeles.

As the 1940s wound down and the era of McCarthyism dawned, the Colonel worried that the nation's academic institutions were at risk of being "contaminated" with faculty members promoting "Communism, socialism and so forth." Even a place like Caltech, he believed, was vulnerable, leaving Millikan to assuage the Colonel's fears that the institute might become "a menace to, instead of a bulwark of, free enterprise." "Nothing can possibly happen in that direction," he told the Colonel, "unless everything that the trustees . . . tried to weave into the warp and woof of Caltech is pulled out by the trustees themselves. . . . There isn't one chance in a million that they will ever do that."

No academy, though, was more in step with the Colonel's free-market thinking—and his concerns—than Claremont McKenna College, which was established in 1946 (and where Spud Williamson matriculated). The school was led by George C. S. Benson, a former army officer and Northwestern University professor whose best-known book, *The New Centralization*, questioned the concentration of power in the federal government. Benson and the Colonel talked frequently about the need, as Boswell put it, to "clean out the 'pinkos'" from university lecture halls. And in 1948, the Colonel pledged to give Claremont McKenna $50,000 for a James G. Boswell Professorship of American Economic Institutions. The holder of the post was to be someone "in sympathy with the

fundamental principles of the American system" of constitutional government and capitalism. By 1950, Benson was revising the college cur-
riculum so that all seniors would be required to take a course from the
Boswell professor. "We are going to see that the graduates of one college
in the country have a clear-cut conception of the values of American
political and economic heritage," Benson told the Colonel.

The Colonel dismissed Ruth's philanthropic endeavors as "goose
feathers and chicken feathers," but he liked to give generously himself,
and his donations went far beyond the ideological. An $85,000 gift
funded a new dormitory—Boswell Hall—on the Claremont McKenna
campus, though the Colonel agonized that having his name on the
building might attract the one thing that the Boswells have always reflex-
ively resisted: the public eye. Benson tried to calm him, informing the
Colonel that the college publicity director who, conveniently, was "also
the *Times* correspondent for Claremont . . . feels we can carry out your
desires to the letter. We will not announce the name of the dormitory
when we announce construction. . . . We will then simply call it Boswell
Hall in the fall when the students are assigned to it. The only paper which
is likely to inquire closely as to who the Boswells are is the *Claremont
Courier,* and I think they can be satisfied with a very brief story."

In 1947 the Colonel started his own charitable foundation with Ruth
and Walter as directors and an asset base anchored by some $3.5 million
in Boswell stock. (By 1999, the James G. Boswell Foundation boasted assets of $62.4 million, while granting some $2 million annually to mostly
educational and medical causes.) Much of the Colonel's largess benefited folks back in Georgia, where he created several loan funds for university students and bestowed $50,000 for the perpetual upkeep of the
Penfield cemetery. He also handed over $240,000 to build a much
needed rural hospital, named for his mother, in Greensboro. The
Colonel followed that up with a $300,000 hospital endowment. The re-
lease of any funds, though, was contingent on those living in Greene
County also contributing. "Colonel Boswell would match anything we
would do, but he said that it breaks down something in a person's character to just give them money," explained Arthur Stewart, who served as
one of the first administrators of the hospital.

Yet as munificent as the Colonel was, and as lavish a life as he led at
Oak Grove, he remained frugal when it came to company spending.
Debt financing was still deemed an abomination. "I do not think it wise

for this company to have outstanding commitments over any period of years," he said. The Colonel was tight with his workers, too. In September 1946, troubled by inflation, he agreed to give everyone a 10 percent increase on their first $3,000 of salary. But he also stressed that the rules of physics would apply to the pay raise: What went up was sure to fall back down. "There will come a time," the Colonel wrote in a memo, "when the spiral of prices will go in reverse, and you must realize that the company will be forced to make corresponding reductions in salaries at that time."

The Boswell Company is "primarily interested in protecting its employees with respect to the high cost of living," he reminded them, "and not the cost of high living."

"CALIFORNIA'S COTTON RUSH." That's what *Fortune* magazine called the boom under way in the San Joaquin Valley in the late 1940s. More than two decades had passed since J. G. Boswell and other pioneers brought cotton to Central California, and now their efforts had reached full flower.

The number of valley acres planted to Acala cotton more than tripled to 935,000 between 1944 and 1949, and within a few years that figure would approach 1.5 million. California soon surpassed Alabama to rank fourth among cotton states, trailing only Texas, Mississippi and Arkansas. Cotton now topped citrus as California's most valuable field crop, and its value rivaled that of motion picture production out of Hollywood. *Fortune* also credited cotton with stabilizing the "San Joaquin Valley's economy, which formerly was dangerously dependent upon the market price of table, raisin and wine grapes." So taken were valley residents with cotton's advance, when a local newspaper poll asked them to name "the 10 greatest living men," they mentioned George Harrison almost as often as they did Dwight Eisenhower, Douglas MacArthur and Pope Pius. Harrison's claim to fame? He was the senior agronomist at the U.S. Agriculture Department's experimental cotton farm in Kern County.

Nor was the recognition of California as a cotton powerhouse limited to locals. Traditionally, as Boswell executive Fred Sherrill pointed out, the state had been "identified with gold and the Golden Gate, Los Angeles city limits, Hollywood, smog, bathing beauties and similar tidbits of the cheesecake of life." But "the fact that California is a cotton state has finally penetrated . . . the cotton centers" of the world. The symbolism was as striking as the statistics. In 1949 the National Cotton Council chose Suzanne Howell, a University of California coed from Bakersfield, as its Maid of Cotton. That March, the council held its annual convention at the Los Angeles Biltmore—the first time the get-together had ever been convened in California. And *Newsweek, Time* and *Life* all joined *Fortune* in noting the ascent of cotton throughout the valley.

Not that everything was rosy. The industry, for instance, remained under fire from liberal commentators. Cotton "has been treated as another fabulous California success story, another 'gold strike,'" Carey McWilliams wrote in February 1949. "But as one penetrates the fog of ballyhoo, one sees that increased cotton production is aggravating the tensions in California's agricultural economy." He called the previous season's picking rate of $3 per 100 pounds "scarcely a living wage." And he said that "from every section of the valley came reports of filthy camps, racketeering contractors, gambling, prostitution, company stores, liquor concessions and the rest. It is an old, old story, and nothing in it has changed except that the scale has been greatly magnified."

Yet even a detractor like McWilliams couldn't help but recognize, however grudgingly, that by accelerating the industry's shift from South to West, California's cotton men had achieved something awesome, something historic. The valley farmer is, "above all, a smart operator," McWilliams acknowledged. "He combines a western toughness and taste for gambling with a remarkable willingness to experiment with technological innovation, and he finds in California the funds for huge capital investments. His energies and his discoveries are shaking the South's cotton empire to its foundations."

To those around the Colonel, it seemed like the perfect time for him to begin thinking about retirement. But he wasn't ready to go just yet. Although Boswell was starting to show his years, looking pale and weary and thin, he "gets a little panicky and withdraws into his shell at the thought of anyone assuming any part of his responsibilities and prerogatives," his secretary, Alice Rothery, observed in June 1945. Still, the Colonel was a practical man, and he knew that the day was approaching when he'd have to make way for new blood. His first choice to take over the company was Bill and Kate's oldest son, Billy Jr., who was invited to live in Pasadena and attend UCLA. A bit later, the Colonel took Billy on a trip aboard the *Empress of Britain*, hoping to instill in the then-twenty-three-year-old a sense that he was his rightful successor.

Billy was a hellion, though, and the thought of joining the Boswell Company and playing by the Colonel's thick set of rules left him cold. "Oh, Billy was smart," said his friend Verdo Gregory. "But he didn't give a rat's ass." What's more, Billy had visions of starting his own cotton operation, a business not affiliated with the one that the Colonel had already put his stamp on. Billy would say that if the Colonel did it "on his own, I can do it on mine," recalled his mom, Kate. Whether Billy might have changed his mind and taken control became moot when the Colonel caught him running a bookmaking operation out of the company's offices. As Billy told the story, the Colonel was furious and kicked him out the door with just $35 in his pocket. Kate turned around and lent her boy $7,500 to lease his own land. The Colonel, chagrined, never knew how he paid for it, but Billy's wish was fulfilled: He was now farming solo.

Billy was a pretty good farmer, and his assistant, Nacho Gomez, was even better. Over time, however, alcoholism crippled Billy. His drunken escapades around Corcoran—often in cahoots with the man he called "my scoutmaster," Cockeye Salyer—became the talk of the town. One local legend had it that when the bartender at the Brunswick Pool Hall wouldn't serve him any more vodka after last call, Billy refused to be denied. He whipped out his checkbook, bought the saloon on the spot and then poured himself another drink. Billy, who owned the Brunswick for seven years, insisted that this wasn't how he came to purchase the place, but many others swore by the story. Billy did concede that he had some trouble getting his liquor license once he bought the Brunswick. That's because on the application, he was required to mark down if and when he had ever been arrested. "Too numerous to list," he had written, recalling all the times he had been hauled in for drunk driving. He eventually got the license anyway.

With Billy out of the picture, the Colonel had few places to turn for an heir apparent. The Colonel's brother Walter was even older than he was. Brother Bill was eleven years younger, but he had his own beer binges to contend with, and overseeing the company from big-city L.A. was never his bag. Besides, Bill had grown disenchanted with the way things were going at Boswell. A number of managers had cut side deals for them-selves, using company equipment to farm their own land and otherwise double-dipping through silent partnerships. Bill had repeatedly warned the Colonel about the problem, but he turned a blind eye, unwilling for some reason to confront those engaged in such activities. One former Boswell employee speculated that the Colonel realized he didn't pay that well, and so he was willing to live with a certain amount of graft to prevent "management unrest." Whatever the Colonel's reasoning, Bill got so disgusted that he retired from active duty with the company in early 1951 and began devoting his time to other things. He helped start a rodeo in Corcoran, pulling in thousands to watch some of the best rop-ers and branders in the West, and he ran his own cattle ranch in the foothills east of Porterville.

As for the next generation, the Colonel's stepson, Spud, was too young and not really right for the business, and Walter's boys wanted nothing to do with it. Then there was Jim Boswell, Bill Sr.'s younger son and Billy's little brother. He had joined the company in 1948 as a cowboy in the Arizona feedlot. He was only in his mid-twenties, fresh out of col-lege and, at least at this juncture, couldn't imagine taking over for the Colonel. "I didn't even think about it," he said. So, almost as a matter of default, all expectations fell on someone outside the family: the Colonel's long-time lobbyist, Fred Sherrill.

Sherrill had returned to Boswell at the conclusion of World War II, af-ter three-plus years inside the Army Corps of Engineers. It wasn't the friendliest homecoming, however. Despite Sherrill's fine work on Pine Flat Dam and myriad other issues, the Colonel couldn't find it in himself to welcome him back graciously. In May 1945, Sherrill had told the Colonel that, despite an offer to head a lumber industry association, he was intent on rejoining Boswell and "charting a course which . . . will lead to a day when I can relieve you of some of the load you have car-ried." The Colonel's reply was not what Sherrill hoped to hear. The com-pany, the Colonel told him, wasn't in a position to match the $25,000 salary that Sherrill was seeking and could unquestionably command elsewhere. With its stock so thinly traded, Boswell is "forced into a very conservative position," he wrote. "This would not be the case in some of the larger businesses with their stock holdings scattered, which places them in a position where they can better afford to bid for high-class executives.

"We would all hate to lose you," the Colonel added. "On the other hand, it would be a source of great satisfaction to me and to the other of your friends here to see you 'go places.' I think you are eminently fitted for the position offered to you, and the lumber business will be very fortunate to have you."

Yet, as was often the case with the Colonel, what he didn't say was as important as what he did say. And because of Alice Rothery, the Colonel's secretary, Sherrill was able to steal a peek into the unsaid. "JGB had me read the letter back to him after dictating it, and then worked it around," Rothery told Sherrill in a cover note that she typed on a scrap of paper and sent along, sub rosa, with the Colonel's missive. "He had me leave out: 'On the other hand, should this or anything else not develop, I am looking forward to having you come back to your old position with me. I have always looked forward to having you back with me.'"

As Sherrill continued to negotiate with the Colonel and entertain other offers as well, including one to become president of the Federal Reserve Bank of Boston, Rothery's secret notes provided invaluable insight into the boss's state of mind. There was no hint of romance in the correspondence between Rothery and Sherrill, and it's not entirely obvious why she took such a big risk to help him. "I should not want this letter, if it were returned for any reason, to fall into anyone's hands other than my own," Rothery wrote, as she passed on the salaries of other Boswell employees so that Sherrill would have a better sense of what he should shoot for. What is apparent is that she had far more regard for Sherrill than she did for the Colonel, and her loyalties followed suit. "I can't see anyone else capable of taking JGB's place, and I think he must know it," she said.

Later came this: "This afternoon JGB asked me if he had ever had a reply from you to his last letter, and I said, 'No, I don't think so.' . . . Then he dictated the enclosed, asked me if I could get it right out, that he was going home, so I rushed it out. When he signed it and handed it to me, he whistled a few bars. Whistling to keep up his courage? I noticed that he stayed in his office quite awhile after signing the letter, in spite of his telling me he was in a hurry. At times he seems quite pathetic to me—I don't know why—but I think he is worried that you will not come back—and I am convinced he wants you back."

By June, Sherrill was satisfied that the Colonel really did want him to rejoin the company, and he also believed that it was the perfect moment for the Colonel to step aside. "The nation needs his services more than the company does," Sherrill told Rothery. "He should be in the State Department with Will Clayton. We need people in there with the broad background of knowledge J. G. has, with his ability to weigh the factors in the human equation and figure out the answers." However, Sherrill also understood that getting back to Boswell on his own terms wouldn't be easy. The Colonel was bound to keep dickering if only because he knew no other way. "Nothing is "an 'open and shut' deal with him," Sherrill said. "The very love of trading makes it necessary for him to haggle almost over a ten cent store purchase just to get the most out of life."

Tiring of all the back-and-forth, Sherrill tried an alternate route in August: He attempted to buy out Boswell or, more precisely, get someone else to buy out Boswell and then put him in charge. That someone else was Chicago industrialist Henry Crown, who had turned piles of sand and gravel into one of America's greatest fortunes. During the war, Crown had taken a break from his job as chairman of Material Service Corporation to serve as a procurement officer in the Army Engineers, and he and Sherrill struck up a close friendship. Crown had founded Material Service in 1919 with his brother, Sol, and their cement (or "sment," as Henry called it) helped build many of the landmarks of the Windy City: the Loop subway system, Wacker Drive, the Merchandise Mart, the Civic Opera House and the original McCormick Place. On their way to becoming billionaires, the Crowns would one day own everything from defense contractor General Dynamics Corporation (where Sherrill was put on the board of directors) to the Empire State Building in New York. But late in the summer of 1945, it was the Boswell Company that Henry Crown had his eye on. After an overture from Crown's bank, the Colonel allowed that he "was interested in selling anything he owned," Crown reported. "We realize that this is a little vague, but at least it indicates that he is open to inquiry."

The buyout wasn't to be, however. Exactly why it fell apart isn't clear. Crown later told the Sherrill family that he had been set to meet the Colonel to hash out an agreement, but he wasn't able to keep the appointment, and the Colonel backed off after that. Sandy Crocket, the old Tulare Lake farmer, had his own version of why things unraveled. He claimed that Sherrill brought him in on the negotiations, with the idea that he would be part owner. He said he even flew to Chicago to meet with Henry Crown. Crocket recalled that he then went on to Los Angeles, where Boswell decided "to check up on Crown," ordering Sherrill to get a bank reference. "The bank wired him back: 'If Crown wants to buy California, sell it to him because he can pay for it.'" The bank's reply made the Colonel feel so emasculated, according to Crocket, he became "infuriated" and "refused to sell the company." The Colonel also flirted with selling the company to Merrill Lynch, but in the end that went nowhere as well.

Eventually Sherrill did return to his old job at Boswell, and at first things were not much different than they had always been. The Colonel was as domineering as ever. One of his employees, Glen Wilson, took pains to carefully position the boss's hat rack in a way that everyone in the office could see when his fedora was hung there. It was a signal—a warning, really—that the Colonel had arrived, "and they had better watch it for the rest of the day," recalled Wally Erickson, who was Boswell's outside auditor in the mid-1940s. John McWilliams had once admonished the Colonel for "popping off unnecessarily and belittling those who worked for him," but he concluded that his friend was "just made that way, and there is nothing anyone can do to change him."

By late 1949, though, the Colonel had at last started to slow down, mostly due to a painful case of the shingles. And for Fred Sherrill, the future finally seemed to hold some promise. The Colonel asked him to assist in the drafting of his will, an exercise that suggested the boss was genuinely grateful to those who had helped him build the empire. A third of his Boswell shares would go to Ruth and her children. A third would go to his brothers Walter and Bill and their children. And a third would go to Sherrill and a handful of other top employees—enough to make them all extremely wealthy men. It might not have been as good as if Henry Crown had bought Boswell, but for Sherrill it seemed like all he had worked for, all he had put up with, would one day pay off big.

THE COLONEL REACTED to his ongoing bout of shingles—an acute viral inflammation affecting the skin and nerves—the way the rich often do when they suffer from an illness: He threw money at it. His $225,000 donation to Caltech for the James G. Boswell Foundation Fund for Virus Research was intended to spur the discovery of treatments for viruses "that cause diseases in man," particularly shingles and pneumonia. Unfortunately for the Colonel, science works more slowly than commerce, and the biologists at Caltech weren't able to give him any relief.

At Oak Grove, meanwhile, the Colonel discovered a tonic of his own. The Colonel's doctor, Larry Williams, recommended one drink a day, and the man who until now had been utterly abstemious started downing an old-fashioned in the library every evening before dinner. The Colonel never developed a taste for alcohol, and he made sure that his tumbler contained more than its share of sugar, bitters, maraschino cherries and orange slices to mask the bourbon. Yet he got to liking the ritual—and the buzz—well enough that if his wife was a little slow in mixing his cocktail, the Colonel could be heard bellowing through the mansion: "Hey, Ruth, what's going on 'round here? The liquor's flowin' like glue."

In the summer of 1952, the Colonel suffered a heart attack that confined him to his bed, and he faded in and out of consciousness. Dr. Williams asked the Colonel's nephew, Jim Boswell, to move into the bedroom next door to the Colonel's and be available to him, bringing his uncle his mail and keeping watch for him at the office. "Three o'-clock in the morning or noon the next day or sometime, he would become lucid and start talking," Jim recalled. During one of those fits of clarity the Colonel sent for Dr. Williams, his lawyer Joseph Peeler, John McWilliams and Ruth.

"I want to do two things," he announced, sitting up in bed. The first was to change his will so that more of his wealth would be left to his nephew Jim. The second was the real bombshell: Jim, he said, would become Boswell's next president. Jim had gone from the feedlot to running the whole Arizona operation, and his part of the company had recently outearned its California counterpart. The Colonel was now persuaded that Jim, though only twenty-nine, had the right stuff.

Jim protested the first order, telling the Colonel that they could sort out his will when he got better, but he didn't balk at being made president. In the few years since he had joined the company, his interest in the business—and his confidence—had soared. At the same time, the Colonel had divulged that he didn't think Fred Sherrill was up to the presidency after all. Sherrill had proven himself nonpareil as a lobbyist, but the Colonel questioned whether his business instincts were as good as his political ones. Deep down, the Colonel feared that Sherrill couldn't be counted on to make the right decision when it came to cutting a deal. "Fred Sherrill is a scholar and a gentleman," the Colonel told Jim, "but his yes and no isn't worth a damn."

Even at the end, the Colonel's southern roots held fast. He had always insisted that everyone at the office address each other by their last names—"Mr. Robinson, that file is over here," "Thank you, Mr. Curtis"— and he practiced the same formality with the nurses who looked after him around the clock at Oak Grove. Except for one. He greeted the night nurse as Mary.

"Why do you do that?" Ruth asked him.

"I've never said that name," the old Confederate replied, "and I'm never going to say it." Mary was Mary Sherman—a surname that she happened to share with the marauding Union general.

J. G. Boswell died at the dawn of the 1952 cotton harvest, on September 11, at the age of seventy. His body was cremated and his remains were buried along with Alaine's at an Arlington National Cemetery gravesite that he had cared for with characteristic fastidiousness. When Fred Sherrill was back East with the Army Corps, the Colonel had directed him to head over to Arlington and make certain that the plot was well sodded and the pores of the large granite headstone were kept dirt free. As he was laid to rest, the Boswell board proclaimed that the Colonel had been "responsible as much as, if not more than, any single individual for the establishment of the cotton industry in the Pacific Southwest." The *Greensboro Herald Journal* praised his "keen analytical mind" and "wise counsel." The Colonel's obituary in the Corcoran paper highlighted the "many communities" that had profited from "his philanthropic acts." At Caltech, the trustees all stood in the Colonel's honor as a tribute was read: "Few men have had the good fortune to serve their day and generation as fully as Colonel Boswell has done. He is perhaps best described as a very fortunate and successful self-made man—self-made because he had developed within himself the qualities of character, mind and heart which made all who knew him well trust, admire and love him."

But for some of those closest to the Colonel, the eulogies struck a false note, for they left out another part of his legacy—one that Fred Sherrill, already reeling from having been passed over for company president, would soon discover. The will that he had helped the Colonel to prepare wasn't the one that had been filed with the probate court. Instead of leaving a third of his Boswell shares to Ruth, a third to his brothers and their children and a third to his top employees, the Colonel left it all to Ruth—the one person who needed the $2 million in stock (more than $12 million in today's currency) the least. Sherrill and the other executives got nothing. As for Walter and Bill, "I do not make provision for my brothers . . . for the reason that during my life I have generously provided for them or otherwise helped them," the Colonel declared in the document. The Boswell side of the family was so outraged, some members vowed to contest the will in court. But Jim threatened to withdraw as president if they took such action, and everybody simmered down.

Fred Sherrill's family turned bitter, too. His oldest son, Peter, began referring to the Colonel as "the great exploiter of men." Fred himself was more charitable. Many years later, when his granddaughter asked him about being double-crossed, he would only say this: "They spell it S-h-e-r-r-i-l-l, but it's pronounced 'Sucker.'"

PART THREE

SUMMER

It wasn't much good having anything exciting like floods,

if you couldn't share them with somebody.

—A. A. Milne, *Winnie-the-Pooh*

Summer

THE SEASON BEGAN IN APRIL and ended in October, a summer measured not by the calendar but by the pick and pull and squeeze and crush of the valley's unbroken harvest. Every third vehicle that whooshed past the oleanders that divided Highway 99 was a big rig and, more often than not, they were hauling away some piece of the valley's lavish bounty. Even accounting for the puffery of the county agricultural commissioners in their annual crop reports, the summer harvest was a grotesque wringing out of the land.

From the orchards came 1.5 million tons of apricots, plums, nectarines and peaches as well as 614,000 tons of almonds, walnuts and pistachios. From the vineyards came 4 million tons of grapes to eat and make raisins and to mix with their snooty cousins from Napa and Sonoma. From the fields of row crops came 400,000 tons of cantaloupes, 515,000 tons of onions and 6 million tons of tomatoes. From the dairies came an astonishing 12 billion pounds of milk and cream. On the road between Hanford and Corcoran, past the large banner affixed to the cotton trailer that read "Caution: Abortion Increases Your Risk of Breast Cancer," was a sign that said even more about the place: "Cheese and Kiwis, make a left."

The valley turned mad beneath a summer sun that didn't let up for six months. There was no hour when the big rigs didn't run. A truck driver heading one way would sometimes fall asleep and come barreling across the highway. If you were driving the other way, the oleanders were tall enough that you never saw him coming. Suddenly the pink flowers parted, and a trailer spilling one crop or the other stood right before you. Every summer, the harvest took its human toll, and no one paid a higher price than the farm workers from Mexico killed in traffic accidents or crushed by machinery or doused with pesticides. What seemed especially cruel, if not ironic, were the *campesinos* who died on their way home from work, their vans smashing into trucks that carried the very crop they had picked that day. Amid summer's frenzy, no one much noticed the bodies. Like the fruit, they moved from field to box.

Then in the early morning darkness of August 9, 1999, came the Dodge van packed with fifteen farm workers heading home after a grueling night in the tomato fields. The driver had no license and the passengers sat on homemade wooden benches barely bolted to the floor. California law, in a bow to the urgencies of the harvest, required no seat belts. They had worked from 6:00 at night until 4:30 in the morning, ten men and five women, picking tomatoes by mechanical harvester on the graveyard shift at Terra Linda Farms. During the long drive home, they had somehow forgotten to turn on the headlights. On a dark narrow road outside the town of Five Points, not too far from the Kings County line, a big-rig driver had pulled over to take a nap. He was in the middle of a U-turn when the van emerged out of nowhere and slammed into the trailer's side. Thirteen of the farm workers, ages seventeen to fifty-four, were killed on impact. Hours after the crash, the crushed bodies of the driver and a front seat passenger remained wedged in the metal.

The tragedy didn't halt the harvest, not even for a day, but it did make national news. State senators and assemblymen went on TV vowing to change the law that exempted labor contractors, the farmers' middlemen, from providing seat belts in their vans. The *Fresno Bee* took the unusual step of assigning a reporter and photographer to follow the bodies on the 2,000-mile journey back home.

Magdalena and Carlos Florentino, sister and brother, had come all the way north from Lindero, a tiny village tucked in a lush green valley in the Mexican highlands. It had been nearly two years since Magdalena said good-bye to her daughters, ages seven, eight and eleven. She could outwork any man in the fields, and that included Carlos. That first month in the San Joaquin Valley, she sent home $1,200. She tried to do the same every month that followed, sometimes managing only $800 and other times outdoing herself with $1,250. The money went straight to her mother to buy clothes for her daughters and to pay their way to school. "Magdalena left for the same reasons Carlos left," one sister remarked. "She did it for her kids."

Alejandro Norberto had crossed the border three years earlier, his mother greasing the way by paying a coyote. "I brought him into this world and, in a way, contributed to him dying," she said. "I shouldn't have taken out that loan. His death is partly my fault." He was accompanied to the San Joaquin Valley by his good friend, Serafin Hernandez, a boy with an old man's soul who would sit alone for hours on the hilltops of Lindero, playing his banjo and harmonica. "He was the best son I ever had," said the seventy-two-year-old grandmother who raised him. "He told me he didn't want me to be poor anymore. He wanted to go to the other side to find work so he could send me money. I told him that as long as I have God in my heart, I don't need money."

Some nights, she said, she woke up to the sound of his music. It was so real that for a precious moment she wondered if the accident—and not the music—was a dream. "The only comfort I have is knowing that in a few years I'll be buried in the ground next to him," she said. "We'll be together again."

The front-page story in the Sunday *Bee* ran beneath the headline MEXICO IN MOURNING. It was accompanied by a large photograph depicting Carlos Florentino's grieving widow and their four small children trudging along the three-mile path to his funeral. Three weeks later, the California Senate passed a bill that required farm labor vans to be equipped with seat belts and set aside $1.75 million to fund ten highway patrol officers to follow the harvest and inspect the vehicles.

The bill was passed on the same September day that a crop duster sprayed a crew of workers building a levee in Five Points, near the same spot where the thirteen *campesinos* had died. The crop duster was trying to kill an infestation of aphids in a cotton field, and the nearby levee workers were suddenly overcome with headaches, nausea and chest pains. Hugh Taylor, the owner of Hugh's Flying Service who had been dusting cotton fields for thirty-four years, said the workers were making it all up. "There was no way they got sprayed. People see an airplane and say, 'Oh my gosh, I've got a headache.'" The deputy agricultural commissioner downplayed the incident. "Normally, things like this are short-term," he said. "We would expect symptoms to be acute, but they're very recoverable. I don't think we have anything real serious."

THE COTTON HARVEST, 835,000 tons of fluff and 873,000 tons of seed, was still three months off, but Mark Grewal, The Wild Ass of the Desert, could already smell something big in the air. The turn from spring to summer brought out Grewal's competitive side, not that the farmer who swore to bleeding Boswell blue showed any other side. We were standing next to the Tulare Lake canal that cut across the basin at a diagonal, one of the few places where it was easy to glimpse Boswell land side by side with the land of a neighbor.

"Remember back in spring when I told you that first week or two was the most important? If you stuck the seed in the ground and missed your moisture zone, the plant would come up late and it would dog you through the season? Well, look at our cotton and look at our neighbor's cotton. This is a Boswell field, and that's a Gilkey field."

The Gilkeys had missed their moisture zone. Their first irrigation—whether too light or wasted on ground not worked up properly—failed to hold. It took them a week to discover the mistake and do something about it. They hurried in sprinkler lines and watered the seed from above, but this turned the soil hard and crusty, a more difficult push for the baby plant. The Gilkeys managed to get a stand, but it was late and patchy.

"They lost only a week, but look at the difference," Grewal said. "Our field is a three-bale-to-an-acre field. I guarantee you they won't make two bales to the acre. That's a $500 an acre difference. All because of timing, speed. I feel bad because the Gilkeys are good people. But if they keep having cotton crops like that, they're not going to be in the cotton business for long."

Tulare Lake had a way of bedeviling even the best of growers. Killer salts heaped up on the fringe, and the soil ran thin where the water once deposited beach. What gave Grewal fits, though, were the eleven different soil types that marbled the rest of the land, some fields moving from clay to loam to sand in the same 640 acres. Variability on a small scale was a problem. But when it played out over 140 square miles of cotton, the fluctuations complicated every big decision Grewal had to make. The soil dictated the variety of cotton, the frequency of irrigation, the amount of fertilizer and even the timing of the pick.

Grewal made no concession to the small farmer when it came to knowing his weak and strong ground. What missed his eye from the window of his truck or the vantage of the company helicopter came through clear in the charts and maps of each harvest. Sensors inside the snouts of the pickers measured the fiber's flow, revealing where a three-bale-to-an-acre field suddenly dropped to two bales. This is where Grewal and his five district managers concentrated their efforts, deciding how best to fix the weak spots. If thin ground happened to run far and wide, the decision was simple: They pumped up the whole field. But if a weak spot popped up here and there, the call for a hardier cottonseed or a boost of fertilizer made less sense. Precision agriculture, at least on the scale that Boswell did it, sometimes meant living with a patch of bad ground.

"The Chamberlain ranch, the Stevenson ranch, the Homeland ranch—that's the land on the perimeter," Grewal explained. "It's our harshest ground, our least mellow, the land that requires the most water. If you visualize the lake, that's where the beach was." He then turned and pointed to the low ground where the Kings River met the Tule and the levees were built wider and taller. "Over there is the middle of the lake. The Basin, the El Rico, the Cousins. That land's got so much pop that you have to be careful. The cotton wants to grow too fast. On that land, you've got to slow things down."

Whether the problem was salt or a blight called verticillium wilt or too much nitrogen and not enough zinc, farming Tulare Lake began with grooming the soil. In late fall, right after the harvest, a fleet of John Deeres, each one outfitted with a dozen metal shanks that looked like the blades of the grim reaper, clawed so deep into the earth that they called it "ripping." A good rip reached down twenty-eight inches into the soil, turning over a layer of clay and breaking up the fibrous roots from a season of hay or wheat or safflower. Mothballing a field for winter required not only brute force but also the soft touch of high-tech. To get the perfect tabletop finish, Grewal's men set up lasers on tripods and used global positioning satellites to shoot beams of light that raised or lowered the tractor's bucket to the same exact inch. The flatland became flatter.

The action in spring had been no less fussy. All across the basin, 450-horsepower Caterpillars dragged lines of discs that whipped the soil into a fluff known as "fines." It rolled off the back end like a surfer's wave, and to the layman's eye it looked as if nothing more needed to be done before planting. But this was Corcoran, and no matter how many times the big Cats lumbered back and forth, Grewal knew that hard chunks of clay were being left behind. Turning the last clods into fines was the job of the harrow, an implement that beat the earth with four-inch spikes. Once the pummeling was done, it was time to seal in the moisture from the winter rain and prevent the kind of drying out that could vex planting. A team of tractors, each lugging a giant metal grate called a "float," made one final pass. The float looked like the device groundskeepers dragged across the infield during the seventh-inning stretch of a baseball game, only ten times bigger. Boswell had light floats and medium floats, and floats for the heaviest soil.

"They don't talk about floating and sealing in the bible, do they?" Grewal asked. He was referring to *Cotton*, the 832-page book on the origin, history, technology and production of the fiber edited by two professors from Texas A & M. The hairs of Texas cotton, Grewal reminded us, were one-eighth of an inch shorter than the fiber grown in Tulare Lake. Texas cotton was too coarse for anything better than blue jeans. Boswell cotton, meanwhile, was a unique variety of Acala known for the strength of its fibers. It went into fine bedsheets and shirts and dresses for L.L. Bean and Land's End.

"A lot of our techniques aren't covered in the bible. Here in the lake basin, we had to invent our own way of doing things. We plant our cotton on the flat ground, not furrows. Our tractors, as you can see, are like military tanks. They run on rubber tracks, and the wheels never dig into the ground. That way we don't compact our fields. Compaction erases everything we're trying to do out here. We're trying to turn the clay into silk."

Grewal drove us to one of the worst patches of ground Boswell farmed as a way to illustrate his point about the before-and-after miracles of tilling. "This is desert. I mean nothing wants to grow out here but tumbleweed. But we worked the ground and reconditioned the soil and now look at this cotton. It's pretty, isn't it? We happen to be growing this for seed, not fiber. It's all going to Greece."

The most fertile parts of Tulare Lake were a kind of Club Med for cotton. The problem wasn't getting the plant to grow but getting it to stop growing. Each fall, Boswell and the other lake farmers sprayed tens of thousands of gallons of defoliants to kill the leaves. This way, the mechanical pickers would have only fiber to gather up. But the defoliants shut down the plant for only so long. New green shoots kept pushing until the day each December when, picked clean, the cotton fields were plowed under.

The stress that came to a Boswell field was, for the most part, stress that Grewal and his men induced by playing with water. In the name of

good cotton farming, they waited three to four weeks between irrigations in summer. Once they decided to make their move, it took a week
of day and night pumping to move all the water from ditches to fields.
Each irrigation, Boswell pumped 100,000 acre-feet of water onto its Tulare Lake lands, drawing on rivers and state canals and the ground 1,000
feet below. This was enough water to furnish the needs of 200,000 families for a full year. As we zigzagged across the basin, it was easy to tell the
fields—right down to the row—where the water had run. Plants fresh
from a soaking stood tall and fresh; plants still awaiting their midsummer shot wilted in the July heat.

Grewal considered irrigation a game of brinkmanship, taking the
plants to the edge and then nursing them back. He drove from field to
field measuring the recovery time, the period between the water hitting a
particular row and the plants in that row moving from a state of wilt to
taut. "If it takes three or four days to go from droopy to firm, that means
we waited too long between irrigations," he said. "We stressed the plant
too much." Good stress made bolls and bad stress knocked them off.
Sometimes bad stress was beyond the farmer's control, like a cold snap
in late spring. Because the cotton plant aborted plenty of bolls by nature,
the last thing Grewal wanted to do was encourage this impulse by deny-
ing water for too long. He pointed to a field in the midst of recovery. What
looked like pure wilt to our eyes—bad stress—appeared to Grewal as the
first sign of rally, the first hint of taut. "That cotton right there looks perfect. If you come back tomorrow, it will be standing tall. Those plants are
loaded to the max. I've got six bolls already on that first branch."

Every summer, Grewal pursued the same end: fields that showed the
perfect balance between vegetative growth and fruiting growth. The
main stem of the cotton plant shot straight up, and if things were going
right with irrigation and weather, the first six or seven branches were
vegetative. This was the leaf base, the carb load, needed to turn flowers
on the higher branches into bolls and bolls into fluffy cotton. A stingy irrigator who induced too much early stress would find his field thrown
into a tizzy of fruiting. Bolls might appear on a plant only four of five
branches tall, far too early in its maturation. The plant, thinking its days
were numbered, was producing offspring right and left. Unfortunately,
it would never develop the big leaves needed to capture the sunlight to
feed its babies.

On the other end of the spectrum was the sloppy farmer who watered too often and fertilized too much. His fields stood a good chance of turning rank, all leaves and little fruit. A rank plant waited eight or nine branches up the main stem before it began to set bolls, far too late. The branches sprouted such wide wingspans that they shaded out the plants in the next row over. Without direct sun, the bolls aborted en masse.

In a Boswell field, the plants appeared perfectly spaced, and the sun had no trouble getting through. Flowers in these fields had a 90 percent chance of setting into fruit. Other growers were lucky if they hit 70 percent. What made this success rate even more amazing was that each flower stuck around for only a day. It opened at dawn and closed that same night, and if its spiny pollen failed to make the trip from anthers to stigma—by bug or by breeze—it was never heard from again.

"Branch seven or eight," Grewal said, "that's where I want my first fruit to appear. And I want the flowers to set close to the main stem because those bolls mature the fastest. Early maturity means early pick. You can tell when a farmer has irrigated his cotton right. The space between the branches is two inches all the way up. I'm looking for a plant that has seven vegetative branches, eighteen or nineteen fruiting branches—twenty-four to twenty-six branches bottom to top. I want something that doesn't grow any taller than my waist, a canopy that's not too wide and not too narrow. That's the ideal cotton plant."

As much as Grewal hated to admit it, 90,000 acres of cotton found all sorts of ways to defy him. No matter how many soil samples his five district managers took, or how finely they tuned their nitrogen to phosphate ratios, or how hard they struggled to avoid soil compaction, the plant had its own mind. One week things looked great, and the next week a field had spun out of control. Suddenly the perfect two-inch space between nodes had doubled to four inches and the plant, unable to rein in its vigor, was kicking off bolls.

When Grewal faced such a field, when he needed to slow things down in a hurry, he didn't hesitate reaching for the same bottle of purple liquid—the same fix—that lesser farmers reached for. They called it Pix, a growth regulator applied to the leaves that almost immediately slowed the rate of cell division and shifted the momentum away from vegetative growth. Eight ounces of Pix per acre turned the rank into the fertile.

Grewal wanted us to know that Pix, or mepiquat chloride, didn't hurt the birds or the bees and quickly metabolized into carbon dioxide once it hit the soil. Still, he didn't like taking short cuts; he considered chemical farming a bow to defeat. When he did pull the trigger, he said, he tried to stay away from the multiple and heavy applications used by other cotton farmers. "We're not big users of Pix or any other chemicals for that matter. Farming is a two-end game. You can produce all the cotton in the world, but it means nothing if your chemical costs are through the roof. No sizable cotton farmer does a better job of eliminating chemicals than we do. We're damn near organic on a lot of fields."

Organic cotton was an ideal that few small farmers, let alone the nation's biggest farmer, had ever attained. On its face, Grewal's claim was close to laughable. In Tulare Lake alone, Boswell was spending $30 million a year for fertilizers and pesticides. It took three different chemical suppliers to cover all of the company's needs, and that didn't count the $4 million paid each year to Lakeland Crop Dusting for fuel and pilots. Watching the sleek planes scream back and forth across the basin, shooting five or six different blends of chemical defoliants on the fields below, the last thing that came to mind was the bedraggled organic farmer wielding his net, trying to estimate his bug counts and praying that the batch of beneficial insects he had just unleashed in his field would outduel the bad ones.

Yet if it was true that no cotton grower had done more than Boswell to embrace the wonders of petrochemicals, it was also true that no other grower had spent more time and money trying to find a way around them. Each year, Boswell turned over 10 percent of its land to experimentation, to varieties that promised greater yields and shorter growing seasons and hardier immune systems to ward off disease and thus reduce the reliance on chemicals. The company had spent $20 million teaming up with Ciba-Geigy and Dow Chemical trying to genetically engineer just such a super seed. At the same time, its thinking on pests had shifted. The old notion that you sprayed on a given calendar day, no matter what the bugs in the fields were doing, had yielded to an approach called integrated pest management. At its heart was the recognition that you could outfox the bad bugs by creating a refuge in the fields that nurtured the good predatory bugs.

Grewal and his men knew that the pests threatening their cotton fields roosted in the weeds of neighbor farms and rode in twenty or thirty miles on the northwest winds. A swarm of migrating lygus, a dreaded bug with a neat yellow triangle on its back, went straight for the young bolls, working its menace from both ends. Its mouth pierced the boll's inner wall with a series of puncture wounds that erupted into wartlike growths. Its ass dropped a trail of toxin-laced feces that carved black pits into the outer wall. In a matter of days, the bolls shriveled and turned brown and dropped as if a killer frost had descended in June. Lygus, however, had a weakness. It happened to love to drink from the sweet fields of green alfalfa almost as much as it loved to partake of cotton. The beauty was it could pierce and suck all it wanted while doing little damage to the alfalfa.

So instead of bombing the cotton at precise intervals each spring and summer, Grewal and his men came up with the idea of erecting a wall of alfalfa around the rim of the basin, trapping the lygus miles shy of the cotton. The beguiled bugs, it turned out, were content to live within the confines of their 20,000-acre refuge for weeks at a time. When the alfalfa needed to be harvested in summer and late fall, Grewal simply left behind enough strips to give the rousted bugs a place to land. In years when cool weather held down the pest population, the Boswell wall could mean a whole season of cotton without pesticides.

"People are always talking about 'thinking outside the box.' Well, J. G. was doing that long before I got here. He encourages us to be leaders and motivators. What farmer in this country brings in John Wooden to a company meeting to talk about the thirteen national championships he won at UCLA? Or Roy Clark, the guy from *Hee Haw*? Or Senator Orrin Hatch? J. G. wants us to break the mold. The five farmers under me are going to be better than me someday. They're all college-educated agronomists and licensed pest control advisers. They understand the systems and they can farm anything. These guys are going to kick my ass, and I welcome that day."

For all his brashness, Grewal saw himself as part of a continuum. It wasn't unlike a pianist who traced his instruction back to the greats— my teacher is Lorenz and his teacher was Arrau and his teacher's teacher was Liszt. Grewal had descended from a long line of Boswell men who had changed the way cotton was bred and grown and ginned. Most everything he was doing had its roots in the innovations of Leslie Doan and Audie Bell, the legendary Boswell farm managers who ran the fields for almost forty years.

Doan was a budding soil scientist at the University of Arizona in 1942 when he caught the eye of Colonel Boswell, who was trying to move from barley to cotton but couldn't get the plants to grow. He accepted the Colonel's offer to work in Corcoran and barely got his feet wet before Patton's Third Army came calling. He was stationed in Belgium when his commander discovered that Doan, who looked everything like his English father and nothing like his Mexican mother, could speak Spanish. This was enough for the counterintelligence branch to whisk him away, and he began shadowing Russian spies and Nazi war criminals through Europe. It was a seven-year detour, and when he returned to Corcoran with his German bride, the war had brought the miracle of petrochemicals to Boswell. The Colonel's men were throwing fertilizer on everything, paying no mind to the rate or the timing of the applications. Boswell had the sensibilities of a merchant and a ginner, not a farmer, and his 33,000 acres looked it. The wheat was the wrong variety for the clay soil; it fell over each time they watered. Sugar beets, alfalfa and safflower were all perfectly sick. A bumper crop of barley was the only thing holding up the farm.

Doan teamed up with Bell, a brilliant agronomist who talked root and leaf physiology day and night in an Okie twang. He had grown up in Delano and liked to say that he graduated from Fresno State on a Thursday and went to work for Boswell the next Monday. He and Doan left nothing to chance. Before they applied a single bag of fertilizer, they took soil samples from every field and measured the nutrient loads. For some mysterious reason, they couldn't get the cotton to mature in the Cousins district, what by all rights should have been one of the best pieces of land. They tested for nitrogen, iron, phosphate and potash looking for a culprit. It got so vexing that "Cousinitis" became a term in the West for a cotton plant that for no good reason came up stunted. Then a flood struck and they noticed that wherever fish had washed up and decomposed, the cotton plants flourished. The missing element, they surmised, was zinc. By spraying it straight on the leaves—a teaspoon for every gallon of water—they turned 7,000 acres of wasteland into gold.

In time, Doan and Bell would confront an even bigger curse, the same one that had wiped out ancient Mesopotamia. Years of irrigation were pushing to the surface a toxic brew of salt and selenium. On some fields, the alkali had crusted so thick that it looked from a distance like fallen snow. They worked from six in the morning to seven at night devising a system to "wash the soil" and reclaim these fields. The first step, an early version of ripping, employed metal shanks to shatter the top layer of clay. The goal was to change the soil's permeability so that a good irrigation—or a good flood—would wash down the salts three feet below the surface. This was deep enough so that when the main root did hit the salt zone, the plant had the size to survive it.

The second step was a bit of alchemy. Doan and Bell learned that a molecule of sodium could be liberated from a particle of clay. The trick was sulfur. The yellow crystals acted as a kind of scouring pad for alkali soil. Yet it wasn't enough for the sulfur to simply split up the bond between salt and clay. They had to move fast to marry the clay to a new partner, or it would reattach itself to the old salty one. Doan and Bell found the perfect mate in gypsum, the calcium bled from rock and used to make plaster of Paris. Gypsum not only blocked salt's return but it made the structure of clay more porous. As a test, they threw a shovelful of the white chalk onto a puddle of water that wouldn't drain. By the next afternoon, the hardpan had become a sponge; the water was gone. This was enough for Doan and Bell to begin applying 100,000 tons of sulfur and gypsum to the fields each winter. Because each crop sucked something different out of the ground and put something different back in, they fine-tuned a rotation system that alternated cotton with wheat and cotton with alfalfa and cotton with safflower. By the late 1950s, their method of turning badlands into productive ones was such a success that Jimmy Boswell, the nephew now in charge, set about seizing an empire.

From farm to gin to shop to lab to executive suite, Boswell stocked his company with managers who needed no push from above, men who were among the finest in their fields and whose interests went a good ways beyond agriculture. If they came off the farm, if their old men had kicked the hell out of them, taught them how to drive a tractor when they were seven and work all day in the sun and irrigate at night, and then sent them off to the university, well all the better, Boswell liked to say. "The esprit de corps was magic," said one company insider. "J. G.'s leadership skills were incredible. He could rally people behind him to march through the fires of hell. You go down the list and those guys were all studs. They were all can-do men."

Walter Bickett, the company finance chief, grew up on an Idaho sheep farm and had a knack for taking on big land deals. Boswell didn't let his "golden gut" accept or reject an offer without Bickett's pencil first working over the numbers. Rice Ober was a cotton marketing whiz who had graduated from Occidental College in Los Angeles with a degree in economics and then spent time in Costa Rica as part of Nelson Rocke-feller's Good Neighbor Policy. Len Evers, who started out as a book-keeper, had a photographic memory and such a keen appreciation for the bottom line that it was all but impossible to slip anything past him. Evers didn't pretend to be an agronomist or a chemist. What he knew were the costs—right down to the nickel—that went into each aspect of the operation. Whenever Boswell got fixed on a wrong-headed notion, it was said, only Len Evers could right him.

There was Paul Athorp, the equipment man whose belly hung so low that Boswell's wife, Roz, used to tease him about sending out a search party for his pee-pee. He started with the company as a ditch digger at age seventeen, ferreting out leaks along a 105-mile stretch of irrigation pipe, outworking even the Mexicans. He ran the Dairy Avenue equip-ment shop and was considered a genius when it came to designing plows and compactors big enough to match the land. Trying to beat the 1983 flood, Athorp and his men worked around the clock for forty-five days erecting fourteen massive carbine pumps that made the Kings River run backward.

There was Arvel "Vandy" Vandergriff, who held twenty-eight patents related to ginning technology, and H. B. Cooper, who held a doctorate in plant genetics and was hailed on the cover of a Land's End catalog as the father of California's best cotton varieties. And there was Stanley Barnes, the "civil engineer," as his license plate boasted, who was known to search through obscure texts and turn-of-the-century hydrology re-ports to find historic pretext why Boswell acreage deserved more water in drought and less water in flood.

Of all the old Boswell managers we tracked down, Barnes was per-haps the most peculiar. Long after retiring, he had preserved his color markers in a Tupperware container he labeled "air-tight." To show us the intricate waterworks that served the Tulare Lake basin, he pulled out a crisp map and began marking the rivers in blue and canals in pink and levees in yellow, serpentine colors everywhere. "I don't know why I'm doing this," he said. "You're never going to understand it." He had turned his home into a museum of the old West. A framed display of various types of barbed wire hung in a living room decorated with Native American rugs and three brands of western saddles, and his library held books on Geronimo, the Hopis, the Navahos, the Mayas and the Hapsburg Empire, 1526–1918. He seemed all bottled up with dry facts, an engineer's engineer, until he began to tell the story of the 1969 flood. His daughter, Janet, who played clarinet in school and guitar in church, had come home one day troubled with gossip. Her classmates were saying that her father had manipulated the wheeling of water to screw all the neighbor farmers. "Listen, honey," he told her, "when I leave this house every morning, I'm no different a person. I have the same integrity, the same honesty, the same values outside this house as I do inside. I'm the same man." At some point, he recalled, he looked across the dinner table and saw his wife's coffee cup vibrating and realized he was pounding the table as his daughter, the brightest of four children, the one who got to his heart the most, was nodding her head yes. A month later, as he and his wife headed out the door for a bridge game, Janet suffered a massive brain hemorrhage and died in his arms. The next Monday, floodwaters rising and levees breaking, Barnes went back to work.

If Boswell's men had one mandate from the top, it was a commitment to innovation and cutting-edge technology. As early as 1962, the company installed data processing equipment, an unheard-of step for even a big farmer in those days. Boswell began packaging its cotton not in jute, a practice that predated the Civil War, but in plastic swaddle. This way, buyers could see exactly what they were getting: pure cotton, not fiber heavy with shipyard sweat. By delivering the bales straight from its gins to the textile mills of the Carolinas, Boswell killed off a whole industry of middlemen who grew fat parking cotton. The company didn't let the feelings of gin manufacturers get in its way, either. Processors such as Boswell had a choice of installing one of two gins, either a Lummus or a Continental. Some swore by Lummus and others swore by Continental, but only Boswell found a way to marry the virtues of both. Salesmen for the two manufacturers were aghast the first time they stepped inside the new Boswell gin and glimpsed a Lummus feeder perched atop a Continental stand. It was like an audiophile who had mixed and matched stereo components to get the best sound. Said Ross Hall, the gin boss: "We Boswellized it."

After the company's embrace of DDT proved a disaster—it was killing all the good bugs even as the bad bugs were becoming resistant—Doan and Bell bought a kitchen mixer and poured in hundreds of pesky boll-worms that had died from a virus. They mixed the green goo with water and sprayed it on twenty acres and, poof, the healthy bollworms started dying, too. Problem was, they had no way of mass duplicating the virus. It was biotech in a blender—twenty years before its time.

Such innovation, whether bubbling up from field or lab, wasn't the only reason the company had a leg up. One of the perks of having the boss sit on the board of General Electric in the early 1970s was that Boswell became the first farmer in the country to gain access to satellite technology, using a Landsat to inventory its fields. All these years later, a twelve-inch satellite receiver sat atop each eighteen-row planter, taking signals on how many seeds to drop and how far apart to drop them. Once a driver lined up a planter, no hands were required to guide the wheel. Drivers finished whole burritos between east and west turns.

Every few years, another gee-whiz advance made its debut, from six-row pickers to robots that inspected tufts of fiber at the gin. Because the lake bottom was so flat and vast, it became the ideal place for the big tractor companies to partner with Boswell and test the durability of their newest models. After forty years, the upshot of all this innovation and technology was plain. The cotton Boswell produced and picked was, in the view of Jockey International president Howard Cooley, even better than what its customers contracted for. The uniformity of the land, the consistency of the picking and ginning machines, the steadi-ness of the workforce meant the first bale was nearly identical to the 10,000th bale. "We don't even have to check it," Cooley said.

Now that same bent for innovation was driving Mark Grewal to change everything. First to go was the premium Acala cotton that had dominated the lake bottom for half a century. Grewal was looking to re-place it with a fancy breed of Pima whose longer and more delicate fibers commanded twenty-five cents more per pound in the market-place. It didn't matter that Pima's longer growing season meant subtle changes in farming. It didn't matter that Pima demanded the building of a new cotton picker, one that kept its fibers knot-free and enabled Boswell to compete with the hand-picked Pima of Egypt. It didn't even matter that Pima required a whole new approach to processing, that its cotton peeled off the seed so easily that Boswell had to transform its old sawtooth gins into roller gins—at a cost of $8 million. It didn't matter because Grewal already had taken the first steps to convert 90,000 acres of Acala into Pima.

He was equally intent on seeing vegetables rise where only cotton and grain had grown before. Grewal had taken a tiny experiment in tomatoes, onions and lettuce and turned it into a twenty-six-square-mile operation irrigated by drip and an overhead sprinkler system that moved across the fields on giant metal wheels. It was hard not to marvel at the reach of his revolution. As we drove back to Corcoran under the July sun, more than 700 acres of onions were being picked, bagged and shipped to Costco, and cannery tomatoes, 10,000 acres in all, were making the turn from green to pink. This didn't count the 20,000 acres of alfalfa fields being harvested not for hay but for planting seed, or the wheat and barley already bagged, or the safflower now aflame with orange flowers. The biggest cotton farmer in the world was fast becoming the biggest diversified farmer in the world.

"People laughed at me when I came here and started talking about converting to Pima and planting vegetables. 'It won't happen, Grewal.' Even J. G. thought I was crazy. Well, we're going to be largest tomato grower in the state next year. And look at those onions. Primo. It doesn't get any better than that.

"And nobody in the world touches us on seed alfalfa. Nobody. Take a look at those alfalfa bales right there. There must be 500 of them lined up in that field. We averaged 1,000 pounds of alfalfa seed per acre. Unheard of. That's a net of $13 million—on the seed alone. We slammed it."

BACK IN TOWN, J. G. Boswell was sitting in the lanai across the street from Corcoran's one McDonald's, sipping a double Jack Daniels over rocks and plotting the next changes to a company he had now taken a firm grasp of once again. He had flown in from Idaho the day before and greeted us with a warm punch to the shoulders. "The tape-recorder boys," he called us. Gone was the charade that he had come back merely to assist his son, J. W., in a few cost-cutting chores. As he crushed ice in the kitchen and poured us a drink, he began to tell the story—proudly, it seemed—of how he had recently demoted J. W. and cut his salary right in front of the other board members.

"Sometimes you have to kick the mule in the ass," he explained. If there was a less humiliating way to get the job done, he couldn't be bothered. He had grown weary of what he called his son's "paralysis by analysis," watching him hide behind his secretary and staff and not even answer his own phone. After four straight years of downward earnings, J. G. said he had no choice but to step in. It was time to return to the seat-of-the-pants decision making that had always served him and the Boswell empire best. "I'm not going to try to tell J. W. how to run his life, except get it right," he said. "This company comes first—before the feelings of my son. . . . I was patient and then I got impatient."

If his son didn't have the right stuff, at least not yet, J. G. was willing to accept some of the blame. He never groomed him for the job, never even imagined him working for the company. "My father never prepared me for this job," Boswell said. "The only advice he ever gave was the day I went off to war. He told me to keep my head down. That was it. I do regret not doing more of that with my son. But running this business isn't that complicated. You sit around with a bunch of guys you trust, and you get a feeling. Then somebody has to stand up and say, 'Let's go.'"

To hear the father tell it, J. W. had pestered his way into the company. He came looking for a job after college, only to hear his dad tell him that he wasn't "hirable." When J. W. persisted, Boswell decided to send him to Geneva to work with the company's agents selling cotton to the Eastern bloc. He figured the kid might last a few months and come back home with a new career in mind. "He learned how to drink vodka twenty-four hours a day and work for forty-five minutes," J. G. said, not bothering to hide his sarcasm. J. W. eventually got serious and taught himself the business from the cotton plant up. Then in 1984 his father surprised the cotton world by naming J. W. as successor. He got the big office and the title, but the old man wouldn't let go. Managers from the different departments in Corcoran, Pasadena and Australia would compare notes of the early morning and late night phone calls from J. G., supposedly retired and golfing and skiing in Sun Valley. "Do you hear from the old man much?" a governmental official in Australia asked the local Boswell manager. "Oh no," he replied. "Not more than two or three times a day."

They could see that J. W. wasn't a comfortable leader. He had built his management style on consensus, the very thing his father detested. When he blew the bugle, one old-timer said, no one followed him. Still, they winced as they watched the father dress down the son at meetings and even insult him in front of his mother. Nothing seemed to sum up their relationship better than the fact that they hunted, fished and hiked together, and yet when Junior decided to get married, it was to a girl his father never knew existed. "One day he tells me he's getting married," Boswell recalled. "I went around the desk and shook his hand and said, 'Who's the lucky gal?'"

Back in command, Boswell required little more than a pencil, scratch pad and telephone to run the company. He had scribbled out a two-year plan, a list of nine goals that included firing the board, clearing the decks of debt and tripling earnings. To shore things up, he decided that the company needed to trim 30,000 acres of land from its Corcoran operation. Not prime ground, mind you, but the ground at the edge of the old lake. "Anybody can buy land," he said. "It takes a good man to sell it."

Nowhere on his list of things to do was building a dairy, much less five dairies clustered on one Boswell ranch that would rank among the largest milk producers in America. But quietly, R. Sherman Railsback, Boswell's chief of operations, had put together a plan to sell the 7,000-acre Chamberlain ranch to dairy farmers fleeing the sprawl of Southern California. Railsback was looking to lure 55,196 Holsteins over the mountains. This was more cows than people who lived in Hanford, Kings County's government seat, population 41,000. Boswell's cow town, at full occupancy, would produce 7 million pounds of waste each day, the equivalent of a city twenty-six times the size of Hanford. Its residents would be a far cry from the contented cows roaming California's green hillsides in the TV ads exclaiming that "Great Cheese Comes from Happy Cows." These Holsteins, like thousands upon thousands in Kings and Tulare counties, would squeeze in and out of concrete stalls and kick up immense clouds of dung that hung in the summer air.

Railsback insisted that the company wasn't going into the dairy business. Boswell was merely acting as a developer, using its expertise to acquire land permits and partner up with dairymen who were tired of battling suburbia and wanted a place where the buffer between city and farm was miles wide. Developing land was nothing the company hadn't done before, and with great success. Besides Sun City, the retirement community outside Phoenix, Boswell had developed EastLake, a master-planned village in eastern San Diego County. It also had developed Interlocken, a high-tech business park outside Boulder, Colorado, that featured a hotel, conference center and twenty-seven-hole golf course.

A community of dairies, even one the size envisioned by Boswell, was hardly a big leap. Much of the land would remain planted in corn, alfalfa and wheat to feed the cows and soak up the waste. A cheese plant to process the milk and to take a bite out of the county's double-digit jobless rate would also rise from the ranch. "We want this site to be as environmentally sensitive as we can make it," Railsback said. "Because of our size and our name, if we in any way try to cut corners, they're going to be down on top of us."

That the project had soared over every hurdle in Kings County, where numerous staff members and planning commissioners and elected supervisors had worked for Boswell or had family who worked for Boswell or did business with the company, was no great surprise. It looked to be a sure bet, greased in that subtle way that Boswell often greased things, and then the unexpected struck. The state attorney general's office, alarmed that the valley was approving giant dairies with the sole requirement that a farmer complete only a simple form, filed a lawsuit. The AG was seeking to impose a basic environmental review process on the counties, even as Boswell was paying for an extensive review of its own. The intervention of the state's top cop had the effect of strengthening the hand of environmentalists. They were quick to point out that the valley was about to overtake Los Angeles as the worst place for smog in America. No matter the assurances from Railsback, the dust and gas from tens of thousands of Holsteins would only degrade the air more. One feisty group, the Center on Race, Poverty and the Environment, rode in from San Francisco and vowed a long fight.

While Railsback and others tried to deal with turmoil surrounding the dairy, J. G. focused on another task: finding an investor who was willing to buy a chunk of Boswell Company stock. He had no choice in the matter. The company had carelessly run afoul of certain federal rules governing the Colonel's old charitable trust—yet another misstep that J. G. blamed on his son. Now it had to unload a bunch of shares to get back into compliance. J. G. set his sights on Robert Rowling, a Texas oil billionaire and hotel magnate who had helped Boswell develop the Omni at Interlocken. Rowling was J. W.'s age, an evangelical Christian who had quit a fledgling law practice to join his daddy's oil company. There, he demonstrated an amazing ability to pick up valuable assets at bargain prices. Now he was eyeing Boswell's business, and he found himself completely charmed by the old man. "They don't print them like him anymore," Rowling said, calling J. G. "a throwback, a padron, a heavy, a legend, an icon."

Boswell invited Rowling for a tour of Tulare Lake, and the Texan couldn't quite believe the precision of the fields and gins. He also couldn't help but notice Boswell's historic rights to the Kings River, as well as the proximity of all this liquid gold to an aqueduct that could shoot the water straight to the beak of Los Angeles. As Rowling and Boswell stood outside the company hangar at the entrance of town, they cut a deal. The man who once remarked, "Everything I own, really, God owns," would invest some $20 million in America's biggest farmer.

Tucked away in Pasadena with all the trappings of a corporate president's digs—the big desk and brown leather sofa and Remington sculpture—J. W. couldn't understand why his father had gone with Rowling. He didn't trust him, and he didn't appreciate his father's decision to give the investor a seat on the board. "His daddy threw him the keys," J. W. said of Rowling. "My daddy didn't throw me the keys." He feared that Rowling had merely played to his father's ego, and that he was trying to take over the company as a water grab. His father, though, wouldn't listen to him. "I've been put out to pasture," J. W. said. The way he saw things, every time he wanted to jump in on one matter or another, his father shot him down. "Why don't you let so-and-so handle that," J. G. would tell him. J. W. said he had enough office work to keep busy, but he was now president in title only—a fact that had become so obvious to his secretary, Carol, it brought tears to her eyes.

J. W. acknowledged that his father had never been motivated by money, but he had a deep need to hog all the glory. Truth be known, J. G. was taking credit for implementing plans that J. W. and his team had put together years earlier. When J. W. first began developing Interlocken, he remembered, his father called him a fool because the project looked so iffy. Now J. G. was referring to the Omni as the hotel he had built. "My father is a bullshitter," J. W. said. "He can rationalize anything.

"He has stepped on everybody in his life. But he's not going to ruin my life. I have a job to do and that's keeping this company going. Every day that goes by, the more he needs me. But he'll never admit that."

At some point in the conversation, J. W. caught himself and tried to soften his words, but there was too much bitterness. He recalled an incident twenty years earlier when he was newly married and the entire family was vacationing in Idaho. They were having a drink in a cowboy bar in Sun Valley, and J. W. decided to take a photo of the clan. The flash went off, and a guy drinking at the bar freaked out. He started screaming how it was taboo to take a picture inside a bar, and he grabbed at the camera. J. W. said he was scared as hell and didn't know what to do. Just as he and the guy were about to come to blows, J. G. suddenly swept around and put his maimed hand on the back of the guy's neck and started to squeeze—real, real hard. The guy backpedaled across the bar, and J. G. kept clinging to his neck with those three fingers, all the while whispering in his ear. He finally let go and the guy bolted straight out of the place.

J. W. said it was one of the most embarrassing moments of his life—his daddy having to rescue him. After some time had passed, J. W. managed to work up the nerve to ask his father what he had whispered to the guy. Apparently he had persuaded the jerk that if he laid a hand on his boy, he would break both his legs.

It had now gotten to the point that J. W. was thinking about raising outside capital to make a play for the company himself. But to do so would require taking on J. G. in open warfare. And that would not only destroy the family, but he knew he stood a good chance of losing. "My dad taught me everything I know," he said. "But he operates by fear and intimidation. He loves fucking with people's minds. He's not going to give up the reins. One of his friends once told me: 'Your dad is a very competitive man. And you're the last person on the planet for him to compete with, and you're on his turf.'"

Boot Hill

AMERICA'S ETERNAL QUESTION came to the San Joaquin Valley a little late. It arrived by way of a small item tucked inside an early issue of the *Western Cotton Journal and Farm Review.* It appeared not in big bold headlines but in seven plain paragraphs that ran beneath a story about the wonders of certain soils. After singing the praises of sandy loam, the *Journal* took a poke at the question that had haunted the South for the better part of two centuries. What should be done about the Negro problem in cotton?

Colored folk were coming to the San Joaquin Valley to raise cotton, the *Journal* reported, and they were finding plenty of white landowners willing to rent them small patches. "You all will have your nigger program here if you don't watch out," a grower visiting from the South was quoted as saying. The *Journal* suggested that the larger question of "big versus small" agriculture was really a question of "white versus black." The Japanese, Mexican and Hindu pickers had each come and gone without any illusion of growing cotton. But these blacks, fresh from a South where tenant farming had become another form of bondage, wanted something more from the land. They actually believed that the middle of California held the promise of a new Dixie and that they deserved a piece of cotton paradise.

"Are we going to make this into a 'ten-acre-and-mule' country? Are we going to have Negroes as neighbors on all sides of us in our agricultural districts?" The *Journal* then let its readership of growers, ginners and cotton merchants know exactly where it stood. "There are two ways to keep these people from coming to California, by the owners not renting land to them and by the finance companies refusing to finance them."

As it turned out, the article, which appeared on the eve of the Great Depression, fretted for no reason. The specter of a Negro landed class never came to be in the valley. By the late 1930s and early '40s, a new question posed in even more urgent tones now hung over the booming cotton fields. Who was going to pick all of this fiber? The pipeline from Mexico had all but dried up, in part because the peasants were still unsure what to expect from America's farm economy and in part because the growers remained wary of a class of laborers known to strike for better wages and working conditions. During the Depression years, more than 100,000 Mexican guest workers were returned to their native lands. The Japanese cotton pickers had pooled their resources and bought vineyards and fruit orchards. As soon as the war was over, they would come back from internment camps in the desert to once again raise their crops. By the 1950s, many of the white Okie pickers had moved up the same ladder, graduating to tractor dealers and truck drivers, foremen and farmers. What was needed was a new migrant, one as desperate as all the others had been. Who better than the Negro farm hand, an old friend who didn't need to be taught a thing about cotton?

The growers, especially those from the South, were willing to believe that any Negro with the gumption to come west was a Negro not plagued with the idleness of so many sharecroppers back home. Besides, the thinking went, they wouldn't be sticking around long enough to create a "nigger problem." A mechanical cotton picker already had been tested in the Corcoran fields, and Jim Boswell would soon own forty-one of the machines, as well as a train car filled with a new product from American Cyanamid, a chemical that actually singed the leaves off the cotton plant. By burning the green, the machine could pick the white unfettered. If the big operators could squeeze a decade of toil out of the Negro—enough time to develop an even more efficient machine—the sight of workers trailing twelve-foot canvas sacks would be forever erased from the fields.

Not every grower held to such a calculating view. Many in the lake bottom were simply responding to the urgencies of the land and had no idea how the westward journey of black sharecroppers might end. Businessmen as longsighted as Boswell did have an inkling of the extent to which mechanization would upend the lives of the 10,000 black, white and Mexican pickers who swarmed the lake bottom each fall. But not even a Boswell could foresee just how much Kings County and Corcoran—from the corner drugstore to the corner liquor store—would be changed by the absence of this workforce.

And so the call went out in the mid-1940s to the black cotton pickers of the South. Among the first to jump at the vision of sunny California was an Oklahoma bootlegger named Robert Parker. A small, bug-eyed man with a corkscrew body, Boots Parker had ferried white lightning from Idabel to Broken Bow during the Dust Bowl. Now he was about to take the wheel of what would become a modern-day underground railroad, stealing blacks off the plantations of the Bible Belt and carting them west to the cotton fields of Tulare Lake. Over the next few years, Parker would make the 1,500-mile journey on a road to freedom known as Route 66 nearly three dozen times, transporting hundreds of families to a promised land a long cry from Chicago or Philadelphia or New York City.

The son of a sharecropper, Parker had heard the stories about the Golden State while growing up in the far bottom corner of Oklahoma. Grapes as big as jade eggs and watermelons the size of small boats and cotton fields that didn't quit. Row upon row, mile upon mile, it was all there for the picking in a giant valley in the middle of California. He was barely nineteen but already could taste the bitter that came with his father's land. There was a reason they called it the "four steppes of hell," a swath of plantation country where the Red River braided Oklahoma, Texas, Arkansas and Louisiana. Long before Parker came rumbling through the night in an old Ford schoolbus, black sharecroppers hopped from one steppe to the other, looking for a place that met their dreams. Texas was considered the depths of hell. Next came Louisiana and then Arkansas. If you had a chance to leave any of those states, you had better take it. Oklahoma, the farthest up from the devil's fire, was considered the least brutal.

Least brutal was the way a black man measured things. *Least brutal* to Parker meant forty-one lynchings marking the life span of Oklahoma as opposed to several hundred lynchings in Georgia or Mississippi or Texas. *Least brutal* happened to be a tradition passed down from Oklahoma's Indian heritage, the Choctaw and Cherokee blood that ran through Parker's own veins.

The first black settlers had arrived in Oklahoma in the 1830s by way of a great exile of Indian tribes booted out of their ancestral lands and forced to march the legendary Trail of Tears into the Sooner state. The Five Civilized Tribes, as they were known, made the trek with their Negro chattel and black freedmen. The Cherokee and Choctaw erected plantations and dispatched their slaves—many of them half black, half Indian—to work in the cotton fields and on the steamboats that traversed the Arkansas and Red rivers. Try as they might to mimic the southern plantation boss, Native Americans tended to be more lenient slave masters. Their definition of a freedman was fluid enough that some of the more famous gunslingers and lawmen who roamed the Oklahoma frontier—Cherokee Bill and Bass Reeves—were a mix of black and Indian.

Such was the lure of the Indian Territory that when the Civil War ended, blacks from the bordering states kept pouring in to Oklahoma. They stayed one step ahead of the Ku Klux Klan by building the all-black towns of Taft and Boley and Vernon. Out of the dust of Tulsa rose the most successful black financial district in the nation, a community of stately houses, churches, shops, theaters, banks, newspapers, doctor's offices, jazz clubs and two of America's finest Negro-owned hotels. So much money flowed up and down Tulsa's Greenwood Avenue, a place that Booker T. Washington dubbed the "Negro Wall Street," that even the shoeshine man at the Palace Building dangled twenty-dollar gold pieces from his watch chain.

And then in the spring of 1921, it all came burning down in the worst race riot in America. Thousands of marauding whites laid siege with gun and fire to the heart of Greenwood, killing 100 blacks and wounding 800 more. Oklahoma had become such a tinderbox—all that tolerance had turned the black man uppity in the white man's eyes—that almost no spark was needed. When it was all said and done, the only spark anyone could think of had come a few days before. A nineteen-year-old black delivery boy named Dick Rowland had lost his balance in a downtown elevator and nudged up against a seventeen-year-old white girl named Sarah Page. The *Tulsa Tribune* trumpeted the smoldering ruins: "BLOODSHED IN RACE WAR WILL CLEANSE TULSA."

Boots Parker and his twin brother were born seven years after the Tulsa race riot in a town sleepy by comparison. Idabel sat at the edge of Quachita National Forest, and everything south to the Red River was flatland and cotton. The Frisco line divided the town in two, east side for whites and west side for blacks. The schools were segregated and government jobs were the province of whites. The restaurants and cafés, trying to capture the Negro dollar without driving off white patrons, set up little dinettes in the back room for blacks. Except for a sawmill where

Parker's father, Herman, bought scraps of stove wood to sell to other blacks, there was no industry in town. Blacks worked as domestics and gardeners and, if the terms sounded right, they could be enticed to take another stab at the tenant farms along the river.

Sharecropping and tenant farming were supposed to be two different institutions, the sharecropper with one foot still mired in slavery and the tenant farmer with one foot on land that was only a big crop away from being his. That big crop almost never came. Not for Herman Parker or any of his three sons or any other black farmhand they knew in Mc-Curtain county. The white man owned the land, owned the implements, owned the seed and often owned the mule. Whatever supplies the sharecropper desired were to be purchased at the little country store across the field. It, too, was owned by the boss man. If he didn't have the cash to cover his slab of salt pork and beans, the sharecropper could count on the landowner fronting his purchases at an interest rate ranging from 15 to 40 percent. This advance was known as the "furnish." As part of the terms, the sharecropper was obliged to give up at least a third of his crop, while the tenant farmer agreed to pay a rental fee for the land. Crop or cash, though, it made little difference at settle time. One by one, sharecropper and tenant were called into the landowner's office each January, sat down before the great ledger book and given the same sorry news: "It was a good year, Herman. You worked hard and did real fine despite the price of cotton being low. You owe me 12 bucks is all."

Boots Parker said he saw his father bite his tongue so many times that he began to wonder if the tongue was like the inside of the hand, so thick with dead skin that a man no longer noticed. "Plantation life ain't nothing but a halfway penitentiary," Parker said. "You couldn't leave. My daddy was a good man but scared to know a different way. I couldn't even get him to walk into a front door in no place unlessen he was asked." Even in Idabel, far from the Tulsa fires, a black man knew his place. It had been thirty years since a Negro transgression had roused the white community to take up a lynching. That lynching in the spring of 1916 happened to be the second lynching in the town's history—and the last. A thirteen-year-old white girl named Edna Martin said a black man tried to rape her while she was picking wild onions in the woods outside Wright City, thirty miles northwest of Idabel. The sheriff arrested a black man who had served time in Arkansas for a similar rape, but this case never made it to trial. During the preliminary hearing at the Mc-

Curtain County courthouse in Idabel, a mob of 500 whites grabbed the suspect from deputies and strung him up to a beam on the second-floor balcony. He dropped twelve feet in the air before the rope came up taut and he tried to grab up enough slack to save himself, at least for a few seconds. It took just that long for someone to pull out a six-shooter and riddle him with bullets. They let the body hang for three days before the sheriff brought him down and buried him in a shallow grave near the Little River. "Never come to Oklahoma to commit such crimes," read the last paragraph of the story in the *McCurtain Gazette*. "A mixture of Arkansas and Texas blood won't stand for it."

Boots Parker and his twin brother grew up barefoot in a hovel on the squalid side of the black ghetto, a place with no name except the one that whites had given it: Niggertown. They watched their mother prepare to give birth to their baby brother and the black midwife call their father into the room and lay down the most awful of choices: save his wife or save his child. "Let my son live," he remembered his mother shouting. Three kids with no wife, and they never heard their father utter a mumbling word. Boots got so sick of being hungry and having sorghum molasses shoved down his throat that the smell of it dogged him through life. As a young man, he realized that if he stuck around Oklahoma long enough, the debt to the white man that got passed from his grandfather to his father would get passed to him. So in the summer of 1946, like so many other black Okies chasing a myth, he hopped into the back of a truck with cardboard sides and a canvas top—a Negro prairie schooner—and headed west.

He had a job lined up before he got there, signing on with Gus and Walter Irons, two white brothers who were going into the business of relocating black sharecroppers to the San Joaquin Valley. Gus was a rodeo man who had moved out to California and built a labor camp in the Tulare Lake basin. He had promised several white cotton growers that, come fall, the camp would be filled with black families eager to work. Walter Irons was the sheriff in McCurtain County, a big man in starched khakis and a white Stetson who had taken on the role of recruiter. He was guaranteeing safe passage to any black farmhand in lower Oklahoma who wanted to escape his debt and move west.

If Gus Irons had a streak of bandit in him, Sheriff Irons was beloved by the 8,000 blacks of McCurtain County. The trust went back to a bloody incident years before he ever joined the sheriff's office. A black field hand had talked his landowner into letting him farm an acre of cotton on a back corner. Every day after work, the black man tended to his patch and raised such a fine stand that when picking time came, the landowner told him he'd be taking half. The field hand protested, and the landowner and his sons beat him unconscious in front of his wife and child. "Go fetch Mr. Walter," the wife screamed, and her son ran a full mile to reach Irons. By the time Irons arrived and carried the bloody sharecropper to his truck, the sheriff and a deputy were blocking the way. "You ain't taking that nigger anywhere," the sheriff said. "Set him back down." Irons knew the nearest hospital was seventy-five miles away, in Paris, Texas, and time was running out. "Over my dead body," he told the sheriff. He laid the man down in the back of his truck, got in and drove across the Red River. A few miles short of Paris, the man gurgled once and died. Word got back to Idabel's black community what Walter Irons had done. No one ever forgot it.

"When my daddy became sheriff, he took me everywhere in his car and that included the black district," recalled Marcia Fulkerson, Irons's youngest daughter. "We'd drive across the tracks to have dinner at one of daddy's black friend's house and the whole neighborhood would come out shouting, 'Mr. Walter is here. Mr. Walter is here.' They would shake his hand and hug him, and he had a big smile on his face. They all loved him. It wasn't fear love. It was true."

With Walter Irons in his corner, Boots Parker had the perfect setup. The sheriff would tell blacks to gather at the courthouse each Sunday with a suitcase and $35 a piece, the one-way fare to California. If Parker's bus happened to fill up, the families signed on with Bubba Lee or Cowboy Williams or Mozell Stokes, labor contractors who pulled up in big flatbed trucks outfitted with homemade benches. They drove all the way to California like that. To those who stayed behind, it looked as if their friends and neighbors were heading off to one big picnic, food baskets and all. "They were 'going to California,' that's all you heard. They were 'going to California,'" remembered Luke Etta Hill, whose family never left Idabel. "I was a kid and they were so excited that you wanted to leave with 'em."

Parker never encountered a problem in McCurtain County, where Sheriff Irons ran the show. Trouble started when he had to cross state lines and pick up blacks in Arkansas and Texas, two states that didn't take kindly to sharecroppers ditching debt. To sneak them out, he'd set up rendezvous points in the piney woods and under bridges at three and four in the morning. "I made the journey into hell thirty-two times," he said. "You weren't allowed on the plantation. I had to steal them off. Arkansas was the worst. If you got caught with a load of blacks before you crossed state lines, they'd take the families back, threaten them, put them in jail and burn up the damn bus."

It wasn't until they came upon the New Mexico–Arizona border, 1,000 miles into the trip, that they felt light enough to start cracking jokes and singing old spirituals. From there it was on to the California desert, then Los Angeles and finally the San Joaquin Valley. Minnie Patterson, a sharecropper's daughter who grew up in Carthage, Texas, had imagined California being all sorts of things. What lay before her wasn't one of them. "Oh my goodness," she muttered to herself. "This is not the place for me." She recognized the cotton, of course, and figured out what were grapes and peaches and plums. And she could smell the big dairies a mile before the cows ever came into view. But those fat red orbs with thick hard shells, the ones hanging like Christmas bulbs with ruby red berries inside, she had never seen before. This was the valley of the Pomegranate King, as well.

They turned onto a sandy road full of chuckholes where families— black, Mexican and some whites—were living in tents and Union Pacific boxcars and little stoop houses made of lettuce crates and grape stakes. One gypsy family had taken up in an abandoned bus that read destination "Guam." Each labor camp sat on the outskirts of town, a safe distance from the good citizens of Corcoran, Hanford, Tulare, Bakersfield and Fresno. One camp went by the name of *Las Moscas*, home of the flies.

Patterson and her husband, Willie, noticed right off that the rivers weren't rivers here but deluxe irrigation canals that parceled out the snowmelt from mountaintop to cotton field. The canals didn't stop where the labor contractors had set up their $2-a-month tents that the Negroes called "old commons." The irrigation water flowed right on by. The only trees standing were clusters of eucalyptus that the first settlers had planted with visions of harvesting their oil and turning the wood into railroad ties. The trees, though, turned out to be a particularly poor variety. Like the patches of alkali, they were good for nothing, not even midway shade.

Minnie Patterson decided that first night she wouldn't be staying.

What kind of land have you brought me to? she asked her husband. Same shacks, same fields. Driving three miles to fetch water with a milk pail. Reading Scripture by kerosene lamp, frying catfish by kerosene stove. You might as well have kept me hitched to the plantations of east Texas. She wanted a home, nothing fancy, in the civilized city. A tract house up the road in Fresno or Bakersfield would do. But Willie Patterson kept pounding nails and boards onto their crooked little hut, and the black folks kept trickling in from Oklahoma and Arkansas, Texas, Louisiana, and Mississippi. They had done something no other blacks had done. They had followed a trail that didn't head north and didn't exactly veer to the mythical West either. Unlike millions of other southern blacks who threw their fate to the city, they were looking to keep alive their rural souls, right down to the cotton picking. Here in a valley that had no end, they could be free from Jim Crow and the half-empty cotton sacks of sharecropping. They had found a new South.

Between 30,000 and 40,000 black Okies arrived in the valley in the years after World War II, 7,000 of them settling in the Tulare Lake basin. Their migration west happened to come at the same time the federal government was implementing the Bracero program, allowing farmers to import seasonal guest workers from Mexico. From 1944 to 1954, the number of Mexicans toiling in the fields of California jumped from 36,000 to 84,000. Because there were enough crops in the San Joaquin Valley to sustain seven or eight months of work, the black Okies, like the Mexicans, hopped from harvest to harvest within a 150-mile radius.

They thinned out the rows of cotton in May, dug up onions and potatoes in June, picked grapes in July and August and returned to the cotton fields in October. Back home, the only water for farming was the water that came out of the sky, and the cotton grew scrawny. Out in the lake basin, with its maze of dams and dikes and ditches subduing four rivers, the fields were nothing but white. Cotton to the left, cotton to the right, cotton in front and back. Standing in the midst, field meeting sky, cotton meeting cloud, it was enough to make your head spin. It took only a little walk down the row to make 100 pounds.

Evelyn Grigsby, whose father made shoes and sugarcane syrup back in Marshall, Texas, had migrated to Corcoran early enough to pick cotton with the German prisoners of war interned in the lake basin. She and some of the other women wore kneepads so when their backs gave out they could continue picking on all fours down the mile-long rows.

"It ain't like it is now," she said. "You picked cotton until February. You picked until you met the plow." The one store near her camp didn't sell black-eyed peas or collard greens, only dry beans that Grigsby managed to boil while she worked in the fields. She learned to kill the stove at just the right time by putting in only so much kerosene on her way out the door that morning. "That way, when I came home, I wouldn't have to do anything but cook some corn bread."

Some mornings, when the fog hovered too thick and the cotton drooped wet, the call would roll across the field: *Cotton too heavy.* This became an invitation to light a fire in an old tire, spread your sack down at row's end and shoot craps until the fog lifted. Even when it cleared, some of the men might only fill a sack or two, enough to get a cotton ticket and return to the game. By lunch time, they were too filled with jug wine to be any good at picking. Juanita Noble, who was raised by her black Choctaw grandmother in Hugo, Oklahoma, caught the gambling and booze bug herself. "I'd pick enough to get a cotton ticket and then I'd lay that ticket down in a game of tonk. If I lost, I'd pick a little more and come back with another ticket. The games might last two or three days right there in the fields. I'd go home with a few hundred dollars without ever picking."

Beulah Handsbur was introduced to the game of dare between picker and farmer as soon as she and her mother arrived from Okemah, Oklahoma, in the summer of 1947. Just like the man overseeing the Mexican pickers, her crew boss had rigged the scale so it began at five pounds in the negative. Each time she picked 100 pounds, she got credit for only 95. And just like the Mexican pickers, she and the other blacks devised a way to counter the rigged scale, throwing in wild melons and other junk, adding back the cheated pounds—and then some. "Did I feel bad? No," Handsbur said. "I had no husband and seven kids in school. I had clothes to buy and food to shop for. The good Lord knew I had to make it."

Some camps offered a classroom with a part-time teacher, and other camps were close enough to the public elementary for the children to walk to school. On weekends and holidays, they were expected to work alongside their parents in the fields. In the bigger families, a crew of six older kids could bag more than 1,000 pounds—$30 a day worth of tickets, which the merchants in town treated like cash. Children too young to pick learned to make the fields a playground. On cold mornings, they stood with their rumps catching the heat of the kerosene-lit tires, and the rubber would sometimes pop and sizzle onto their pants. Come spring, when the girls began wearing dresses, the boys could spot the scars on the back of their legs. The grownups, picking their assigned rows plus a "snatch row" shared in the middle, sometimes lost sight of the children. More than once, a kid wandered off only to be found floating dead in a canal. One small boy got tired of waiting for his parents and climbed up the ladder and fell asleep in one of the trailers filled with cotton. By the time they discovered him gone, the trailer had been hauled to the gin. He died a horrific death, chopped up inside the saw blades, body parts pressed into a cotton bale.

Frankie Allen, one of twelve children born to an Oklahoma share-cropping family, said his father would lay down an ultimatum on their long drives out to the lake bottom: He had to pick 200 pounds a day or face a whipping. He found out right away that 197 pounds wasn't 200. "My Dad was tough, but he wasn't cruel. Osie Allen was his name. We came out from Wewoka, Oklahoma, 'cause he heard money growed on trees in California. We didn't have running water. We had buckets sitting on a bench outside our shack. If we were still asleep at cotton-picking time, daddy got a dip of that ice-cold water and threw it in our face. 'Wake up. Wake up.' We had to survive."

If the weather was nice, they'd gather on Sundays at a park and play baseball and slow-cook ribs and chicken under mulberry trees. They formed a Colored All-Star team that played the Boswell Bears and other squads in the local Cotton Pickers League. Osie Allen was a catcher and could throw a guy out stealing second without ever getting up from a squat. Just a flick of his wrist, his son Frankie remembered. "My Dad was an excellent baseball player. He could have went pros. He stood five feet ten and weighed about 170 pounds. He was lovely. All his fingers was broken. One of his hands was like two of mine. He'd thump my head and it would hurt worse than a whuppin'. Never went to a doctor or nothing. He didn't drink, didn't smoke. Had no vice till the day he died.

"Then something just snapped. One day he picked up the shotgun and shot my momma in the back. Then he sat himself down on that same bench that the cold water sat on. Laid the shotgun across the bench, tied a string around the trigger and just pulled it. I didn't see him do it. I saw the neighbor burying his brains in the backyard."

If the California that greeted them didn't exactly live up to their vision of the Golden State, most figured they had plenty of time to make it right. After a season or two, the careful ones had saved enough cash to replace their walk-in tent with a shack. Each season, if things went well, they nailed on a few more feet of board or bought a piece of alkali all by itself and set down a wooden boxcar. Some pickers figured if they had to live in a shack, the shack might as well be back home in Oklahoma or Arkansas. So they hightailed it east. They might return to California, they might not. During the San Joaquin Valley earthquake of 1951, the fields shook so hard that many migrant families left for good. "Every aftershock that came, the Negroes would throw down their hoes and hit the ground and start shouting and praying, 'Oh Lord, if you just save me, I won't ever drink that wine again,'" recalled Emma Wilds, who helped her brother Gus Irons run his labor camp. "One woman ran out of her cabin so fast that the screen door caught the sleeve of her dress and ripped it clean off. Her husband was yelling, 'Come back here, you don't have no clothes on.' She ran for a mile naked as a bird. You could hear her hollering back, 'Naked is the way I came into this world and naked is the way I'm leaving.'"

Boots Parker, who became a fix-it man for Irons when he wasn't running blacks out Route 66, returned to his shack one day to find his wife and four children vanished. "She kept saying 'Let's go back to Oklahoma,' and I kept telling her, 'I ain't got shit in Oklahoma.' I guess she couldn't take it no more. I came home from work one day and they was gone. No warning. No nothing. Not even a note. The big trunk was missing. No clothes left but mine."

HOWARD AND GERTHA TONEY and their nine children came westward like a stutter. From the northeast corner of Texas, they moved to Abilene and from Abilene they moved to Fredrick, Oklahoma, and from Frederick they landed in the Arizona desert in a labor camp just a quarter mile from where the Japanese were corralled. It was 1943, and the Nisei and their children were still two years away from returning to their homes in California. Black sharecroppers, responding to the ads of "Cotton Pickers Wanted," were two steps ahead of them.

Gertha Toney liked to boast that she came from a family of slaves raised up in the white man's house. Her elders could all read and write, and three of her aunts on her father's side were schoolteachers. She and Howard had married in Naples, Texas, when she was only sixteen, but they had the good fortune of farming land that belonged to a righteous white man. He asked for only a third of their cotton and a fourth of their corn, and they managed to leave Texas free and on their own, "owing nobody nothing." Their journey west came in such fits and starts that Gertha gave birth to five children in different towns along the way. Her "stopover kids," she called them.

"There were lots of us stopovers running about," said Dorothy Toney, the middle child, four siblings in front and four in back. "I was born before we got to Oklahoma. My brother Bobby, who came after me, is an Okie. We were all born in different states because my dad was afraid he was gonna get lynched in Texas and he had married that woman, my momma, who was always protesting about everything."

They had a 1936 Dodge that Howard Toney bought for $160 and sold for $200, and he plunked down the cash on eleven bus tickets to Arizona. They stayed five good years and might have made Arizona home if not for the itching that came thanks to Gertha's brother. He was one of thousands of southern blacks who had migrated to Oakland and San Francisco to build ships for the war. Tossed to the wind by peace, he decided to stay in California and follow the harvest inland, picking grapes in Selma and cotton in Corcoran. He made it all sound so good to his sister and brother-in-law that they decided to join him.

Howard Toney, tall and dignified and kissing a perpetual Lucky Strike, believed a man's word was sacred. Friend or relative, he would be on you with all four claws if he ever caught you lying to him. He figured the watermelons really didn't grow the size of small boats in the valley, but he could see there was enough truth in his brother in-law's crowing about Corcoran to call it home. He knew how to drive a tractor and jerry-build an electrical system, and he didn't like stooping in any cotton field. He settled the family in side-by-side tents in a labor camp west of town and found work as an irrigator. Gertha and the older children picked cotton on weekends, sometimes hitching a ride to the fields with Mr. Pumphrey, who was known as a sleeping man. The morning fog would lull him to slumber and suddenly they'd find themselves on the wet side of a levee. Most times, Mr. Pumphrey would awaken before they got too far down the embankment and he'd deliver them to a field full of dew at seven sharp. "There was every color and every kind out

there, even a one-armed man who said a bear bit it off, and tramps picking cotton, too," Gertha Toney said. "The thing I remember most is the
chuck wagons with catfish and hot links cooking right in the fields. You
could smell the Louisiana hot sauce in the air."

Dorothy Toney, who was ten years old when the family arrived in the
basin in 1949, couldn't believe her eyes that first morning out. "I was told
that there used to be a lake filled with water out there, but now it was
just cotton. Not a building or a house or a tree as far as you could see.
Just cotton." On foggy days, when the horizon broke into thirds—cotton, fog and then sky—she wondered if the stories about some varmint
in the lake bottom, a monster as big as the Loch Ness eating people
alive, might be true. "Oh man, you should have seen it. Sometimes in
the middle, just above the tule fog, you'd see these cranes flying over,
these big long birds that looked prehistoric. They were headed for whatever water remained. And sometimes the men would leave the cotton
fields and go to where those cranes were and bring back water."

Gertha was a large woman who didn't care much for cooking, especially inside a tent that Howard tried to make homey by building a wood
floor over the cold damp earth. He'd come home from irrigating and tap
dance on that floor, and she'd complain that he was keeping her barefoot and pregnant. Out of a sense of guilt, she figured, he took over the
cooking chores. About the only dish she still fixed with a smile on her
face was her "fist" biscuits. She got up at five in the morning and made
them from scratch, yeast and all, creating so much racket in the kitchen
that it sounded as if she was throwing the pots and pans. She figured if
she had to get up that early, everybody else had to get up that early, too.
Just as soon as the oil would sizzle, she'd plop the balls of dough into the
frying pan and push down on each one with her fist. They came out
sweet and golden brown with the folds of her knuckles still on top. If the
noise didn't wake up the young ones, the smell of the biscuits did.

Gertha taught Sunday school and went to Wednesday night Bible
study, and she and Howard, a church deacon, loved to belt out gospel
songs after dinner. They taught their children a reverence for work and a
respect for the rod. Gertha could fashion a switch out of anything, but
she preferred the young branches of a cottonwood tree—not too thin
and not too thick, with just the right whip. She kept both eyes fixed on
her four daughters. There were parts of the camp, gamy parts lit by a
candle's flicker, off-limits to the children. L. K. Griggs, the black man

who ran things, was a pimp and a gambler still a decade shy of becoming a preacher. His wife kept a few girls to service the migrant men. "It was a frightening time for my mom because she had these four daughters living in that camp area where these men would come to drink and gamble and mess around with these ladies," Dorothy Toney recalled. "There was all this commotion going on all the time, and we were hearing things we never should have heard."

Johnnie Lou, the oldest daughter, was sixteen and built large like her mother. She might as well have been a frail thing, though, for all the days she spent sick in bed. While the rest of the family was out picking cotton or potatoes in Bakersfield, Johnnie Lou stayed home and helped with the cooking. Griggs had a brother-in-law named Paul Jackson who was just as wild as he was, and he began sneaking over to the Toney's side of the camp near the woodpile to visit poor Johnnie Lou. Before her senior year at Corcoran High, she became pregnant with his child and had to drop out. Gertha turned crazy with the news. She marched over to the district attorney's office and filed charges of statutory rape. Jackson was picked up and put in jail, and Johnnie Lou, a single mom, was put on public assistance. It shattered everything Gertha wanted to believe about rural life, making her question why they ever passed over Chicago or Detroit and chose the tule sticks.

"The reason we didn't follow all those blacks to the big city is because the big city was dangerous," she explained. "The rural was safe for my children with a whole big space for them to play and learn how to do things. One year in the camp ruined all that. It just tore us up. She was the oldest daughter and she was lacking just five months from graduating from high school. Jackson had a wife right there in the camp. In fact, his wife and my daughter gave birth at the same time. And those two little girls ended up going to school together and you could hardly tell them apart. They looked just like twins."

The last thing Gertha Toney needed was a traveling salesman banging on the door of their tent, promising to flush away her ills. His dusty Plymouth had been outfitted with retractable shelves filled with hundreds of sundries—from coconut oil shampoo to vanilla extract, from a teat balm for dairy cows to pie filling that tasted like it came off the backyard peach tree. He introduced himself as Edwin Matheny and said he lived just down the road in Tulare. While he worked as a door-to-door pitchman for the Furst-McNess company, his real passion was buying and selling cheap land to give Dust Bowl families, white and black, a shot at the American dream. He had heard good things about Howard Toney, that he was a hardworking man with his gaze straight ahead, and he wondered if he might be interested in an acre of land on the north side of town. He was selling it for $400—$25 down and $15 a month.

The price seemed fair, and so did the interest rate, but Toney needed a little time to make it work. He took a job in a sugar beet field, chopping weeds with a short-handled hoe, to come up with the first payments. After signing the paperwork in Tulare, he returned to the camp to pick up one last item: the wood floors he had built inside both tents. Then he drove his wife and nine children to their one-acre spread and set down the floors on a patch of Kings County that belonged to no other man. Like the rest of the great valley, Corcoran was using restrictive codes in real estate deeds to lock out people of his color. Out here on the fringe, though, nothing stood in his way but a field of tall dry weeds. His children turned the weeds into swords and began dueling.

That summer, as Gertha and the kids cut grapes in Fresno, Howard spent all their savings at the Farmers Lumber in Corcoran. Over the base of the wood floors, he began building the family a house. It was four rickety rooms, one right behind the other, with no running water and no bathroom, a "shotgun house" where you could see the light from the front door all the way to the back door. Yet for a family coming in from a long and dirty raisin harvest, it was just about perfect: a real home made of real wood with a kitchen and separate bedrooms and a tarpaper roof that would sit beneath the shade of a cottonwood tree. The children joked that by the time the tree was big enough to yield its whuppin' sticks, they'd be long gone.

The Toneys had graduated three more children from Corcoran High when Edwin Matheny showed up again, this time without his sundries. In a matter of a decade, he had turned the alkali scrub on the fringe of several cities—land he had bought for a pittance—into little country settlements for black families. The old cotton pickers would come to his office clutching Karo syrup cans filled with silver dollars. He would offer them such reasonable terms that years later, in the wake of his death, the black Elks Lodge in Tulare would hang his prized elk's head as a tribute to Matheny.

The traveling salesman in his industrial shirt and tie and gabardine slacks had become a millionaire by buying houses in the path of highways and moving them to his subdivided land. Howard Toney bought two of Matheny's houses that day and planted them next to his own shack. He rented out one house and let his younger son live in the other, while he and Gertha stayed put. Enough black families had followed in their path that the forty acres of Matheny land outside Corcoran had been given a name by white folks: Boot Hill. Some blacks took the name literally—that they had been "given the boot" to the outer limits of town. Others thought it was a reference to ghosts, black ones, in Corcoran's midst. Not until a decade later, when Howard Toney decided to do something about the colony's lack of drinking water, did Boot Hill get an official name. He and his neighbors had grown weary of lugging milk pails to the same spigot in town. As he sat down in the summer of 1964 to ask the federal government for money to help them connect to the municipal water system, Toney struggled to find the right words. He knew that a neighborhood that took its name from an Old West cemetery didn't exactly convey the right feelings. So he chose to begin his letter this way: "We the people of Sunny Acres. . . ."

THE BUSES CAME UP FROM LOS ANGELES each harvest carrying black men who weren't looking to settle down in Corcoran any longer than it took to cover a gambling debt or make the law forget. One glance and a lot of farmers shied away from hiring them. Not Forrest Riley. Ten miles outside town, in a camp with a plantation store, he erected enough tents and shacks to house more than a hundred black workers. He never met a black man he didn't believe belonged in the fields. They were bred for cotton picking, he liked to say, and he'd pluck them right off the street corners or straight out of the joint, the rougher the better. If they showed a polish that made him second-guess his first inclination, he'd direct them to his family spread along the Tule River. It was fancy living by ranch standards, with furniture from Barker Brothers in Los Angeles and clothes from Bullocks and I. Magnin. There, he put them to work as his "house Nigras."

Riley was born on a Ventura County ranch, a third-generation Californian on his mother's side, but he could "out-South" the boys from Georgia and Virginia any day. With his first crop of grain in the 1930s, Riley struck it rich. He loved to tell the story of how the kid with a terrible case of rickets went on to farm 20,000 acres and ride his palomino atop the most dazzling silver Spanish saddle in four Tournament of Roses parades. The rickets had left him with the legs so bowed he had trouble walking. It didn't help that he weighed 300 pounds. He was so fat that he paid his daughter a dime every night to pull off his boots. Because he couldn't mount his palomino the usual way, he trained the horse to spread its legs until it got low enough for him to climb aboard. Even then, he needed a lift from Bozo, the black ex-con with the razor-slashed face who tended to Riley's every need. Bozo ironed his clothes and shined his shoes and kept him stocked in Planters peanuts and RC Colas and, when the boss got that little twinkle, he'd find a Nigra girl for a quickie in the fields.

"Daddy was good to his colored help, but they weren't equals and they didn't want to be," said Frances Nanasy, one of Riley's three daughters. "Our cook was a Nigra. Suzie. I don't think she had a last name. And her son, Willie, did the gardening. Bozo was my dad's colored boy. He had his own quarters and he took care of daddy. He went into town every once in a while and ended up drunk and on the rough side of the tracks."

The rough side of the tracks was where a handful of blacks who didn't call Boot Hill home had settled, some as far back as 1915. They had built the First Baptist Church with the help of Reverend Isaac Pearson from nearby Allensworth. The town was founded by a former slave who had been the highest-ranking black officer in the Union Army. For a brief moment, Colonel Allensworth's colony stood as the only place in California built, financed and governed by blacks. They eked out a bare living before a shortage of water and an excess of salt did them in. By 1922, they were gone, scattered to Corcoran and other hamlets.

Within shouting distance of First Baptist sat the honky-tonks and whorehouses run by Corcoran's black godfather, Willie Montgomery, one of the few African Americans who had a steady job at Boswell. The Strip was the first thing you saw when you entered town on Whitley Avenue, the black bars lining the north side and the Mexican bars lining the south side: King of Aces, La Frontera, La Paloma and Mexicolindo. The Chicken Shack was a huge Quonset hut with a stove and big red letters that read "Community Kitchen." It was all a contrivance, a cover for the little plywood booths where five to fifteen black prostitutes, depending on the season, hopped from cot to cot. Business was so bold and brisk that the cops, no matter how much boodle Boss Willie was willing to pay, were forced to shut it down. A farmer ended up buying the hut and feeding his calves in the same plywood stalls.

The action simply moved over to the Corcoran Hotel and Simon's Busy Bee, and the 604 Club, which was owned by a black woman named Stella who had movie star looks. She was a hooker who had made good, and her back room concealed some of the biggest dice games in town. The highlight of the night was seeing Pretty Miss Pete, who was nearly as beautiful as Stella, get drunk and do a striptease on one of the club's pool tables.

Fred Odle, the son of the farmer who never stopped pining for Oklahoma, was a cop working the 7:00 P.M.–3:00 A.M. shift during Corcoran's wide-open days. The chief of police asked only two things of young Odle: patrol the Strip on foot and keep the peace and no more. Odle took this to mean that the big growers wanted their workers to enjoy a little R&R. He didn't need to hassle the prostitutes and gamblers. His job was to make sure the cotton pickers' dollars stayed in town and to keep them from killing each other with the knives they used to cut grapes.

He learned right off that certain black men prized by Riley and Cockeye Salyer were more protected than others. One worker known as Snake killed three men by Odle's count and never served a day in jail. Snake got his nickname because his bottom teeth had been knocked out and his tongue, which had a huge growth on it, darted in and out when he talked. Likewise, nobody messed with Big Six, a six-foot-six bulwark whom Riley recruited off the streets of L.A. because he had grown tired of cotton pickers challenging the crew boss who weighed their cotton. He put Big Six on the scale and never heard another complaint. "I've never seen anyone bigger or stronger in my life," Odle said. "And not an ounce of fat on him. He would have been an All-Pro anywhere at anything if given the chance. You know the song 'Leroy Brown'? Well, Big Six fit that perfectly. The baddest man in the whole damn town. He was a smart guy, too. He'd clean up on weekends and go uptown and gamble all night, and he always seemed to win."

Some of the black families along Pickerell and Gardner grew tired of living beside the Strip. Their Sunday strolls to First Baptist came lined with empty muscatel bottles and men fighting about who owed whom from Saturday night—and even a dead body or two from an argument turned knife fight. They wanted the slow and quiet life that the Toneys and other black families enjoyed on the far end of town, even if it meant having to fetch their own drinking water. They made the trek to Matheny's office, picked out a highway house and bought enough land to raise a few pigs and chickens.

There were no churches in Boot Hill but plenty of preachers. A half dozen families gathered on Sundays in the home of Thelma and Bedela Tillis until a new church could be built. St. Paul's Missionary Baptist, Corcoran's rival black parish, opened its doors on October 28, 1951. Over time, it would draw so many parishioners from First Baptist that some Sundays the old church couldn't even muster up a choir. But even out here, protected on all sides by the fields, the sins of the Strip still found them. Two church stalwarts, Seamon Hill and Millie Walker, began an affair in front of the whole congregation. Hill's wife, Editha, was in the dark until her husband moved in with his lover. She seemed to hold the church partly responsible. Six months after their separation, she stashed an unloaded .38 caliber revolver in her purse and took a seat in the back pew.

She didn't think she had the guts to go through with it until she locked eyes with Millie Walker, fat and smug. As Reverend W. B. Smith wound up his sermon, she went to the restroom and loaded the six bullets into the .38. She could see Walker chatting so nicely with Reverend Smith and Howard Toney. Without a word, she walked up and fired two shots point-blank. The first bullet struck Walker in the chest and, as she slumped to the floor, Hill fired a second bullet into her back. Gun in hand, Hill then strode to the front of the auditorium and faced her husband. "This is all your fault," she muttered. He ran and she fired two more shots, missing both times. Hill later told police that she took the gun to church only to scare Millie Walker into letting go of her husband. That look on her face, though, changed everything.

THE TWO GRAD STUDENTS LOOKED LOST that summer day as they walked the trails connecting Boot Hill. Seeing them move from field to field, residents wondered what business they had there. Unless you were working for Matheny or hauling a hog to Bunk Cole's slaughterhouse, a white person didn't have much reason to venture to Boot Hill. But this young couple, a skinny guy in horned-rimmed glasses and a girl in a taffeta dress with a toothy smile, seemed a thousand miles from home. As it turned out, it was a bit farther than that.

Anne Mathers and Chuck Chomet had met at Oberlin College in Ohio. She was Scotch-Irish, the daughter of a socially conscious banker and school secretary. He came from a family of Viennese Jews who escaped the Holocaust, the son of a physician and a housewife turned socialist. They fell in love during their third year at Oberlin, a liberal college founded by abolitionists, and headed to the University of Michigan for graduate studies. Mathers was pursuing a master's degree in social work, and Chomet was specializing in public administration. They had married in May 1964, on the eve of coming to California to work as summer interns for the Migrant Ministry program. Corcoran was to be their honeymoon.

The Migrant Ministry, funded in part by the National Council of Churches, was so committed to righting social wrongs that even some church people considered it a commie front. Cesar Chavez was making noise in the valley grape fields and, with the help of a Migrant Ministry nun named Sister Anne Harrison, his National Farm Workers Association (a forerunner of the United Farm Workers union) had signed up more members in Corcoran than any other place. The Chomets found themselves captivated by the idea of reaching out to the Mexican farm hands who were still crossing the border in significant numbers, even as the Bracero program was being dismantled. As a Jew, Chomet made it clear he wanted nothing to do with the proselytizing that crept into the ministry's message. The head of the program honored his request, giving the young couple the freedom to roam Corcoran and find their own area of interest. A week or two later, they stumbled on Boot Hill.

"I remember feeling a little stunned at first. We had no idea there were African Americans in Corcoran," Anne Chomet Petrovich said. "It was unbelievably hot and they would sit outside these board shacks on mattresses waiting for a little breeze, talking and laughing. They had no drinking water. It was that summer, with a little push from Chuck and me, that they decided to do something to change the situation."

That first day out, the Chomets kept hearing the name of the Toney family. If the colony had an elder statesman, it was Howard Toney. What electrical systems were wired to the houses or attached to makeshift wells, he had built. He was now working for a millionaire named Hovannisian, the biggest buyer and seller of highway houses in Tulare and

Kings counties. Gertha was putting on a white dress and cleaning houses and ironing clothes for Corcoran's schoolteachers. Field hand to domestic, this was how things progressed in the valley for a black woman of her era. Meeting the Toneys for the first time, the grad students knew then how they would spend the rest of the summer. The federal government had recently initiated the War on Poverty program. Each state commanded a share of money for projects and few were more worthy than bringing water to an enclave of black cotton pickers on the outskirts of Kings County.

During one early meeting, as the talk inside his living room grew excited, Howard Toney wondered if the Chomets might write a letter to Paul O'Rourke, the man who headed the War on Poverty effort out of Governor Pat Brown's office in Sacramento. The couple, heeding the first lesson of the Migrant Ministry—to empower by showing and not doing—said they would type the letter, but Toney would have to write the words himself. "He was a very gentle and dignified man, and I found myself awed by his presence," Petrovich said. "We had this secondhand Smith Corona, and I will never forget him standing over us, dictating the first words of that letter. 'We the People of Sunny Acres. . . .' It gave me chills. They had taken the language right out of the preamble of the Constitution."

If they could convince Corcoran to bear some of the costs of extending water lines to Sunny Acres, the state and feds might act more quickly. So the Chomets accompanied Howard Toney and other community leaders to city hall to make their case. The city administrator assured them that he understood their plight, but the three-quarters of a mile they lived outside the city limits was too great a distance to bridge. Corcoran would be willing to hook up Sunny Acres to the municipal water system if residents could secure a state or federal grant to install the lines. For a group seeking the city's support to win just such a grant, it was a meaningless offer.

The Chomets left city hall disappointed but not entirely surprised. Barely a month had passed in their summer internship but they had already begun to see Corcoran as a company town that had less and less use for its black field workers. Seven years earlier, more than 60 percent of the cotton grown in Kings County was still picked by hand. Now, practically all of it was picked by machine, idling tens of thousands of Mexican and Negro workers. The black Okies tried moving to other crops, but they found themselves outdone by the migrants from Mexico, the vast majority of whom had crossed the border without legal documents. The Mexicans were willing to work longer hours and for cheaper wages.

The way the Chomets saw it, they had come West, all right, but it wasn't a West that college or the Migrant Ministry had prepared them for. They might as well have landed in Texas or Alabama, for they had entered a realm of the South where racism had embedded itself into everyday life. The bigotry might have glowed a little more benignly in the California sun, but that was all. "I had this romantic idea of the family farm, but the San Joaquin Valley was a very feudal society," Petrovich said. "The black families had been cogs in the wheels of these huge corporate farms and now they were being pushed aside."

Jim Crow, California-style, was born in the 1920s and '30s, when the KKK controlled local boards of supervisors, police departments and school districts all across the valley. The cheery news of the White Knights of the Invisible Empire filled the pages of the *Corcoran Journal* next to updates about the bowling league and a cattle plague. "Ku Klux Klan Will Hold Picnic July 4th," a 1931 headline read. A basket lunch would be followed by a sack race and nighttime cross burning. When a state roster of the KKK revealed the names of valley law enforcement and the chairman of the Kern County board of supervisors, there was no run for cover. "I am proud to be associated with some of the best citizens who are members of the Ku Klux Klan," the board chairman said. In the 1940s and '50s, as the Klan finally receded into the fields, still more southern racial baggage rode west in the mass migration of whites from the Dust Bowl states.

Some children of black Okies, those who graduated from local schools and wanted no part of the fields, applied for jobs as city and county workers. Their reception was not unlike the one that greeted Ruby Hill at the Delano Police Department in 1961. The star high school running back had entered the police academy in Bakersfield without a hitch. Even though he was the only black candidate, no one had bothered to whisper a word of discouragement as he made his way through the six-week training program. He earned his certificate and returned to Delano, his hometown in Tulare County, to apply for an opening. The police chief, a big-bellied white man with a Sooner twang, sat Hill down in his office.

"How long have you lived here, Ruby?"

"I've lived here all my life, sir."

"Well, in all that time, have you ever seen a black policeman in this town? It can't happen here, Ruby. It isn't going to happen."

Hill shook his head but kept his cool. "I got a family. I need a job."

"Well, if it's a job you need, I can help you with that."

"What kind of job?" Hill asked.

"A garbage man," the police chief replied.

Hill hauled away Delano's trash for eight years, all the while applying for police openings in Kern, Tulare and Kings counties. The response through the 1960s never wavered. He was either one-third or one-half an inch too short or his muscular frame too stout.

Out in the fields blacks were invisible. No one took a second look. In town, on their way to eat out or shop or take in a movie, their color all of a sudden registered. Hiett's Drive-In served the best malt shakes in Tulare County. Residents would hook a tray to their car windows, and the pretty girls in roller skates would glide on out to take their orders. Unless the customer happened to be black. Then the girls would suddenly turn blind and the customer would have to walk up to the take-out window and shout his or her order through the glass. The treatment didn't differ at the Plunge, a boardinghouse on the south side of Delano that barred blacks from its swimming pool well after other parts of America had begun to integrate. Or at the Curb Inn, a popular eatery that refused to seat blacks. Or at the big roller rink near the Kern County fairgrounds.

Even if city leaders were so inclined, the institution of "separate but equal" wasn't needed in Corcoran. Blacks knew enough to keep to their side of the tracks and hardly ever challenged a white business owner by seeking access. Corcoran wasn't like Visalia, a town across the highway where white racists pounded their chests. The white power structure in Corcoran did it slick, in the words of one black resident, their bigotry just below the surface. Every Wednesday at midnight, the Lake Theater had a single showing of a "Negro movie." At the Corcoran Theater down the block, no signs directed blacks to the far side of the aisles, but that's exactly where they sat. Throughout the civil rights era, the black Okies went about their business paying little mind to the protests that were wrenching the outside world. Their rural isolation had somehow dulled them to the upheaval of the times. The Toney children and their friends, who grew up in the 1950s and early 1960s, would recall not a single march or protest against the discriminatory practices of local government and businesses.

"These counties didn't care nothing about civil rights," said Lealga Fortson, a Texas sharecropper's daughter. "We knew it and we didn't push the envelope. I don't know why we swallowed it. Maybe it had to do with the fact that we had a strong sense of our own self back then. We didn't need government programs or affirmative action. We depended on each other, neighbor to neighbor. I remember we had hayrides and cookouts. We went to a place of our own called The Roadside where our parents danced all night and us kids stayed outside roasting weenies. You could hear the music and see their shadows moving inside the barn. We had our own thing and to hell with what the whites had. That was our attitude back then."

The Toney's sixth child, Janie, who graduated from Corcoran High in 1953, said she put on blindfolds and wandered through everything. "Coming from Texas where blacks and whites took up different spaces, I didn't give a thought to racial attitudes in Corcoran. About the only thing I remember is a girl in one of my classes getting up to give a book report. She kept referring to 'those darkies,' and our teacher stopped her and said, 'We don't talk that way here.'" Her sister Dorothy, the family rebel who graduated in 1957, saw a lot more. She was a skinny girl with big bald spots covering her head, a condition she blamed on the alkali bathwater that bubbled up from the well her father dug. She was forced to wear the clothes that her mother brought home from her domestic jobs, hand-me-downs from the kids she went to school with. She said she could feel them staring at her as she walked down the halls.

She fought back by subverting the status quo any little way she could. When she took a seat in the Corcoran Theater, she plopped down in the middle aisle—white folks land. One weekend night, Dorothy Toney found herself caught in a melee that broke out between white and black students. The theater had decided to show a provocative double billing—the film debut of Sidney Poitier in *No Way Out* followed by Henry Fonda in *The Grapes of Wrath*. Poitier played a doctor at a large metropolitan hospital who failed to save the life of a white hoodlum wounded in a robbery. The dead man's brother accused the doctor of letting him die because he was white. His racist rants climaxed in a full-scale urban riot. It was the first movie Dorothy Toney ever saw that confronted racial hatred in America, and she sat there transfixed. Each time a white character uttered a racial epithet or blacks were portrayed talking in ghetto slang, whites in the theater began to snicker and parrot the word "nigger." Then came the showing of *The Grapes of Wrath* with its earthy depiction of a white Okie family caught in poverty's vise. Black students in the audience couldn't pass up the chance to give it right back. "We started laughing at those rednecks on the screen the same way they were laughing at us. And they didn't like the tables being turned one bit," Dorothy Toney said. "Blacks and whites started tossing food from one side to the other. It started with light food and then it got heavier and heavier. Sometime in the middle of the *Grapes of Wrath*, it broke out into a full fight. I was a skinny little black girl with no hair, and I just ran."

Nothing happened in Corcoran without at least a wink from the Boswells or Salyers. The city operated so firmly in their grasp that it only made sense that race was one more matter whose pulse they dictated. If there was any doubt about this—the Boswells, after all, tended to be influence peddlers of the shiest order—the two farming families left their fingerprints all over an election in 1962 that proved to be one of the city's nastiest. It focused on a Migrant Ministry worker who had preceded the Chomets, a New York seminary grad named Sue Carhart who had arrived in town in 1959. Carhart did her best to fit in. She collected clothes for the poor and organized a sewing club for women and a vacation Bible school for labor camp youth. The Presbyterian church the Boswells attended was so supportive of her do-good formula that it helped pay her salary. Then the director of the Migrant Ministry, embracing a brand of social activism that would evolve into full-scale support for Cesar Chavez's union, informed Carhart of a new mission: forcing Kings County and Corcoran to build low-income housing for Mexican farm workers. Carhart, a heavyset blond with a radical bent, was more than happy to oblige.

She began registering Mexicans to vote and gathering signatures for a ballot measure to require Kings County to tap into federal funds for low-income housing. The measure ended up passing in 1960 by a comfortable margin, but local officials still refused to seek any federal housing grants for the poor. Provoked, Carhart decided to fight back. With three spots up for grabs on the Corcoran City Council, she joined up with local Democrats and formed a slate to run against the candidates favored by Boswell and Salyer. It was a rare show of public defiance against the cotton kings, and the campaign fast grew ugly. The Carhart candidates—a storekeeper named Mendes, a crop duster named Quaadman and a physician named Smith—were hardly left-wingers, much less socialists. But five days before the election, on the morning of April 5, 1962, the citizens of Corcoran awoke to an advertisement in the *Journal* that took up three full pages and called for a "Vote to Prevent City Socialism."

The ad was the handiwork of the Good Government Committee, a group headed by Boswell vice president Jim Yost, Salyer American vice president Everette Salyer and Faye Gilkey, the matriarch of Gilkey Farms. A vote for the Carhart slate, it warned, would amount to a raid on the city treasury and would lead to municipal bankruptcy—all for a bunch of "pie-in-the-sky" giveaway projects. Nowhere did the ad mention low-income housing for minorities, but it didn't have to. After weeks of shrill editorials by Percy Whiteside, the *Journal*'s publisher, it was perfectly clear what "giveaway projects" meant. "Tender-footed ministers and social workers" are bent on deploring everything good about our community, Whiteside wrote. They know nothing about the "truly great progress" Corcoran has made to provide better housing, schooling, day care for children, water, sanitary and health services.

Whiteside, who was born in Tulare and ran the newspaper in nearby Lemoore, had come to town two years earlier to take over the *Journal*. He didn't let the fact that he was broke stand in his way. Whiteside knew the *Journal* had never written an article critical of the big growers in forty years. Given that the Boswells, Salyers, Gilkeys and Hansens already owned the paper in spirit, he figured they might as well own it in shares, too. So he put together a partnership in which the Boswells and Salyers bought a 60 percent stake and the other grower families controlled the rest. Soon after, Whiteside began driving around town in a brand-new Edsel and penning editorials that praised Corcoran for having the "second tallest skyline between Los Angeles and San Francisco." This skyline consisted of a dozen grain elevators ten stories high.

With Whiteside's doomsday pronouncements, it was a wonder that even one of Carhart's candidates, Dr. James Smith, ended up winning a seat on the council. "We have breached the cotton curtain," the physician crowed, vowing to carry on a fight to defeat two other grower-backed councilmen in the next election. "We will represent all of the people impartially." Two months later, Carhart was gone. Boswell's minister at the local Presbyterian church had yanked his financial support from the Migrant Ministry. With no one willing to pay her salary, Carhart had no choice but to leave town.

The Chomets, who had arrived two years after the election, soon found themselves facing the same threat. At first, they endeared themselves to Whiteside by lugging gallons of paint to spruce up the shacks of Sunny Acres in a citywide beautification campaign before Corcoran's Golden Jubilee celebration. Now their scheming to bring water to black cotton pickers was seen as more of the same "socialist poppycock." At one point, they were pulled aside by locals and reminded of the fate that befell Carhart. "We were told not to get too involved because the people before us were run out of town," Anne Chomet Petrovich said. "For the most part, though, the attitude of white Corcoran was politely condescending."

Like a lot of good southerners, the Boswells and Salyers showed a downright folksy side in their one-on-one dealings with blacks. There was no shortage of stories of down-on-their luck blacks who had a serendipitous encounter with Bill Boswell or Cockeye Salyer and came away with a job offer or a loan that never had to be repaid. "If my truck broke down or I needed a couple thousand, Mr. Salyer would never do nothing but reach into his pocket and give me the money," said James Hatley, a black man who hauled products for both Salyer and Boswell. Willie Johnson lost his cab company and landed a job cleaning offices for Salyer American. Famous for firing workers, Cockeye never once threatened Johnson, not even after the janitor caught the boss naked in the back office with his mistress. At home, the Salyers drove off one black maid by throwing around the word "nigger," but other maids seemed to understand that they meant no harm by it. "You know I was raised by a black woman," Linda Salyer, Cockeye's granddaughter, said. "Her name was Lola and if we ever knew her last name, we forgot it. The first time I heard the word 'nigger' was from my grandmother's mouth. I asked her a question and she said, 'Well, go ask the nigger in the kitchen.' She adored her nigger, Rosy Ramble. She was four feet wide and three and a half feet tall and she slept with a pistol under her pillow."

Far more pertinent and damning were the employment records of the two farming giants dating back a half century. Only a few of the blacks who migrated west to the Tulare Lake basin ever climbed into the ranks of tractor drivers, irrigators or gin operators. Gambling and booze had something to do with it, but neither Boswell nor Salyer made any genuine attempt to recruit qualified blacks. Although Fred Salyer would flatly deny it, the word in Corcoran's black community was that Salyer American made a practice of stamping the applications of black job seekers with a big "C" for colored, and then rejecting nearly all of them.

Boswell and Salyer could afford to be choosy. The mechanical cotton picker had industrialized the fields even as Mexican migrants continued to flock to Corcoran. Most of the peasants had come without any legal status, hardly the kind of workers to gripe about wages or long hours. Lorenzo G. Luna, who brought out his family from the Rio Grande Valley in 1958, had no trouble finding a job at Boswell. He grabbed his hoe and by the light of a kerosene lantern began irrigating the cotton fields at three in the morning. His children and grandchildren would go on to work for the company as well, each one climbing a bit higher up the Boswell ladder. Those first few years, he earned $1 an hour. His children remembered a Thanksgiving working all day in the fields and coming home to the strangest turkey. It was laid out on a big platter, and as they moved closer they could see that slices of Spam, interspersed with slices of hard-boiled egg, ringed the entire perimeter. In the middle sat a giant mound of refried beans.

Just how far down the pecking order the black cotton picker had fallen became clear to a visiting journalist one afternoon in the early 1960s. Ron Taylor, a reporter for the *Fresno Bee* who later worked for the *Los Angeles Times,* was the only newspaperman to chronicle the struggle for water in the black enclaves. He was sitting on the porch in the settlement of Teviston when he heard an old black preacher walking toward him, muttering "God damn geese, God damn geese." Taylor asked the preacher what he meant by his lament. "God damn geese," he said. "I lost my job chopping weeds in the cotton fields to a bunch of God damn weeder geese." A number of farmers in the lake had taken to employing birds to cleanse their fields of weeds; the flocks of geese, it turned out, would nibble away at the Johnson grass, Bermuda grass and other invaders but leave the cotton alone.

It was in those years of the late 1950s and mid-1960s that the camps would close and the streets of downtown Corcoran would empty. Publisher Whiteside would become fixated with the notion of bringing a bowling alley and golf course to Corcoran, even as Nick's Liquor Store and McMahon's Furniture and J. C. Penney and Thurman's Stationery shut their doors. The Chomets would return home that summer to finish their graduate studies in Michigan. Howard Toney and a few of his neighbors would soon form the Sunny Acres Water District and, after receiving a $46,000 federal loan grant, the city would agree to meet them halfway. The barrio section on the outskirts of town would be annexed into Corcoran, but Sunny Acres would forever sit outside the city limits. Gertha Toney would praise the Lord that the machine had come and wiped clean man's labor from the cotton picking fields. But she would watch the children of the black Okies, her daughter Dorothy among them, take the highway going the other way—to the jobs of Los Angeles. There, Dorothy Toney would hear about the civil rights movement and a preacher named Martin Luther King for the first time. For the next forty years, her mother and a handful of her neighbors would hang on to their little spot in the sun. Like the cotton in late winter, they would somehow escape the machine's plunder, still clinging to the land.

The Stud

THE RAIN POURED DOWN through a canopy of ponderosa pine and white fir, a cold summer storm, and Jim Boswell was growing wetter and more chilled by the minute. The two men hiking with him—each one two decades his junior—had the sense to bring along hooded ponchos in case the weather turned on them, as it often did this time of year. Boswell, though, fancied himself stronger and tougher than just about anybody he knew. Beyond a bedroll and a stash of his much loved Jack Daniels, he was disdainful of packing a lot of gear. The only thing protecting him from the elements was a plastic trash bag, with a slit cut in the bottom for his head to poke through. And it was doing a pitiful job of keeping him dry.

It was 1980, and the trio had been tramping for several days across the High Sierra, twenty to twenty-five miles at a clip, ascending and descending thousands of feet, always to the same rhythm. Boswell, pushing age sixty, raced out on the trail, way ahead of the others. "His whole deal," said John Grant, one of his companions, "was to bury you." Eventually Grant and his buddy Chris Reynolds found Boswell perched atop some rocky ledge with his feet hanging over the side. "Where have you been?" he asked, and then, to rub it in, he stared at his watch.

Practically everyone who had ever accompanied Boswell into the great outdoors had some variation of the same experience: "He hiked my ass into the ground." "He left me in the dust." "He skied me into the mountain." The ailing little Georgia boy with the asthmatic wheeze had, as a man, turned into a bona fide stud. Boswell's friends took to calling him "the Cheetah" because he was so hard to catch. Even before he quit smoking Marlboros in the early 1970s, taking long drags that made his voice a little gravelly, he was a wellspring of energy. After the '69 flood, Boswell had treated several of his executives to a fishing trip on the Dean River in British Columbia. The furious pace he set to reach the campsite nearly killed Stan Barnes, the company's top water engineer. The other guys finally divvied up Barnes's pack to help him make it up a steep ridge. Boswell didn't care; he just stood at the top of the trail, yelling for everybody to go faster, faster, always faster. "Come on! Come on! Let's go!" At least the laughs were worth it. On that trip, Boswell decided to drink his Jack Daniels from an old body lotion bottle that his wife, Roz, had given him. But the smell of the stuff, called "Intimate," never came out, and the boss was left choking and spitting.

Now Boswell decided it was time for another hike, and the two stooges he brought along were Grant and Reynolds, both of whom had worked for the company at different times. They knew going in that Boswell would make them tote the heavy stuff—the bottles of wine, the butterfly lamb chops, the thick cuts of steak—and then he'd get back and tell everybody that they just couldn't keep up. In Boswell's version of events, Grant said, you were sure to be cast as the "whining weakling." They didn't let the name-calling get to them, however. If you wanted to hang out with Jim Boswell, a guaranteed good time, you had to accept the ego and machismo, the hyper-competitiveness, as part of the bargain. It didn't matter what it was: gin rummy, bridge, squash, racquetball, golf, dominoes, shooting ducks, shooting craps, fishing, stacking wood, drawing cards to see who would wash the dinner dishes—Boswell was bent on winning every time, and usually he did.

Boswell long ago had befriended John Grant's father at Doyle Springs, the mountain retreat, and when "Pop" Grant died in a 1958 tractor accident, Boswell had stepped in as his son's mentor. Boswell resolved to teach the boy—a notorious screw-up in school—a little discipline. So he stuck the fourteen-year-old at the end of a hoe in the Arizona sun and made him chop cotton. "I had a bunch of other men in my life who weren't going to take me on because I was a problem," said Grant, who over the years would stay for long stretches with Boswell and his wife at their Pasadena home. "Jim Boswell put me to work, and I needed that."

As the rain fell that day and the hours passed, the three men made their way to Mountaineer Creek, a tributary of the Little Kern river, where they huddled under a tarp and built a fire. Boswell, though, couldn't seem to get warm. He shivered uncontrollably. He vomited. And he started to become disoriented—"really, really goofy," as Reynolds described it—a symptom of hypothermia. Boswell peeled off his sopping clothes, and Grant and Reynolds rubbed him down to raise his body temperature. Then they shoved him into his sleeping bag. Slowly he began to thaw. Grant would later attribute Boswell's recovery to three things: the bedroll, the campfire and a more-than-ample helping of Jack Daniels. Grant still cherishes the photograph he took of Boswell sitting there that night, bug-eyed drunk, flipping him off with the stub of his middle finger.

When they got up the next morning, perhaps expecting to find the man moving slower than usual given the ordeal he had been through the previous day, Reynolds and Grant could only smile. Boswell was already gone, a message left behind: "If you lazy bastards are going to sleep in, I'll see you down the trail."

He kept right on going, too. In the coming months Boswell would join an assault on China's Mount Kongur, which at more than 25,000 feet was one of the highest unclimbed peaks in the world. A friend of Boswell's from Hong Kong was sponsoring the 1981 trip, and Boswell ultimately reached through the thin air above base camp to about 19,000 feet. He was "a very gutsy trekker," recalled Chris Bonington, the renowned British adventurer who led the climb. Though Boswell was "the oldest member of the entire expedition," Bonington pointed out, he was also "one of the fittest."

A few years later, on his sixtieth birthday, Boswell would tackle the 211-mile John Muir Trail, from one end to the other. A decade after that, bum knee and all, he'd head out on the same trail again, this time to celebrate his seventieth birthday. He'd march some 200 miles over fourteen days before becoming so ill he was forced to abandon the journey. He left the trail only after his son, J. W., and a doctor badgered him into doing so. Even then, Boswell would swear that he had been fine, that it was J. W. and the others in his party who were too feeble to stick it out. "I didn't get sick," he'd say. "Those wusses got sick."

He vowed that if he reached eighty, he'd do it all over again, charging past everyone, just as he had done his whole life.

EVEN AS A BOY, there was a part of him that detested his uncle, the little peacock who made his father tremble and his mother lie. He didn't want to believe he had been named after him. There were "Jim

Boswells" up and down the family tree, after all, and any one of them could have been inspiration for his christening. Growing up, Jimmy Boswell could see that accumulating wealth was a contest for his uncle. The way he figured it, he took after his father, who didn't give a hoot about money.

During those first years in California, his mother, Kate, made a game of sticking cardboard in the soles of his shoes. Her thriftiness became such a statement that his father laid into her one day about the hand-me-downs that made up the boy's wardrobe. "God damn it, Kate, I don't care if we're broke. I'm gonna buy this boy a pair of britches." Bill Boswell then took his youngest son to the Mercantile in downtown Corcoran and dressed him in a brand-new pair of jeans. Even after the company began to prosper and the hard times had eased, Bill and Kate Boswell maintained a modest household that bore no resemblance to the lavish lifestyle the Colonel led in Pasadena and later San Marino.

Whenever he came to town to rattle the cages of his top managers, the Colonel would have dinner at his brother's house, and Kate would go to her usual lengths to hide Bill's boozing. The Colonel would check the report cards of Jimmy and his older brother, Billy, and two sisters, Kathryn and Josephine, and put them through an arithmetic test, offering the winner all the loose change in his pocket. Then he would give a nod to his chauffeur and off he'd go to the fancy digs of the Tulare Hotel. "He never spent the night," Jim Boswell said. "He wasn't a warm man. I didn't love him. But I respected him."

As much as he wanted to be his father's son, he saw his formative years shaped as much by his imperious uncle as by his laid-back dad. At the Colonel's insistence, Jimmy left Corcoran in the ninth grade and entered an all-boys school in the hills of Ventura County. The Thacher School, founded in 1889, had carved out a reputation as a West Coast feeder for Yale, and what excited Jimmy most wasn't the top-notch classroom instruction but an emphasis on the lessons of the outdoors, hiking and rafting and riding horseback. Every freshman had to care for a horse, and Jimmy made the trip south with Peanuts, the bay his father picked up in Mexico. None of the other students brought a pony of his own, and young Boswell soon learned why. The school horses were experts at gymkhana, the sack and barrel races that pitted boy against boy. Peanuts, on the other hand, was a cattle pony. Every time Jimmy leaned out of the saddle to retrieve a sack, Peanuts thought he was getting ready to rope a steer. "He'd put his ass down and slide," Boswell said. "You couldn't take the cow out of that horse."

It took a while to get accustomed to Thacher, but once Jimmy made the emotional break from home, he never really returned. The four years of boarding school allowed him to detach from Corcoran so that for the rest of his life, as he moved between Los Angles and Phoenix and Sun Valley, he would say that home was a "long time ago." Looking back, Boswell said it never occurred to him that his uncle was grooming him to one day take over the company. Yet he lived with the Colonel and his Aunt Ruth during the summer, and after graduation the Colonel took his nephew on two long cruises, one to Panama and the other to the Orient to meet with cotton agents and merchants.

The Colonel made him dress each evening in a white dinner jacket and black tie and inspected his nails for dirt and his hair for cowlicks. On the trip to the Far East aboard the same American President's Line that delivered Boswell cotton worldwide, they ate at the captain's table every night. The Colonel caught a flu bug in Tokyo and had to stay behind for several days. By himself, eighteen-year-old Jimmy headed to Nagasaki, a trip he would never forget, for it came on the eve of World War II and he could see the glower on the faces of the Japanese soldiers in full regalia with swords. "At the train station, there was a huge gathering of Japanese in three concentric circles. They were wearing headbands and a guy in the middle was leading them in a chant. Then he yelled something and you could hear a pin drop. I was scared shitless."

He took a steamer to Shanghai and checked in to the Park Hotel, where a big band was playing in the grand ballroom on the top floor. As he ate dinner that first night, he watched twelve Chinese couples stroll in, the men in crisp tuxedos and the women in long silk gowns with high collars that showed off the grace of their necks. "The band struck up 'I'll Never Smile Again,' and I've never seen a prettier sight than those twelve couples gliding across the dance floor." He spent four days in Shanghai without the Colonel, four days walking through a city where history seemed to be happening right before him. Shanghai no longer belonged to China. Except for an enclave under the control of the United States and another under the sway of Britain, the entire port city was occupied by the Japanese. He heard a bomb explode outside the hotel and watched Japanese soldiers strip-search the American ambassador and his wife. He saw the wife in her bra and panties and looked away as the ambassador muttered, "Do as they say, dear." It was during those four nights, free from the hawk-eye of his uncle, that he met the daughter of a prominent American missionary and made love for the first time.

A month later, he was settling into dorm life at Stanford University, planning to pledge Alpha Delta Phi and win a starting job on the Cardinals baseball team. He would later repeat the story of how the war so depleted the squad that he, a mere freshman, earned a varsity letter in the spring of 1942, before the war came calling for him, too. But records and yearbooks at Stanford University showed that Boswell played only freshman baseball and soccer in 1942 and, in subsequent years, never participated in either sport again. Depleted squad or not, he didn't earn a varsity baseball letter. His fudging never went so far that he claimed to be a first-stringer. Rather, he recounted that a shortstop named Bobby Brown took his spot. This was no shame because Brown would one day become a World Series star for the New York Yankees, a cardiologist and president of the American League. Boswell, though, still found a way to best him. "Bobby Brown got second base, but I got his girl," Boswell would say.

That girl was Rosalind Murray, the pretty Stanford graphic arts major he'd later marry. She was the daughter of Feg Murray, an Olympic track star from Stanford who penned a nationally syndicated cartoon called "Seeing Stars." It was Hollywood's version of "Ripley's Believe It or Not" with a drawing and a little-known tidbit about a different movie star each Sunday. "The Murrays had a beautiful home off Wilshire Boulevard in Los Angeles, and each one of the children followed in their father's footsteps and went to Stanford," Bobby Brown recalled. "We dated for two years and then I went into the navy. We drifted apart, and she and Boswell got married. I knew him just a little, but it wasn't from baseball. I don't remember him as a baseball player."

In the fever of Pearl Harbor, Boswell enlisted with the army and found himself in the hold of a troop ship headed to the Pacific islands. At nineteen, he landed at Guadalcanal at the tail end of the famous battle. He came back shaky from his first patrol and the sergeant stuck a cigarette in his mouth and that was the beginning of a thirty-year, three-pack-a-day habit. They moved from jungle to jungle in the Fiji Islands; the worst of it was coming across the remains of six villagers skinned alive by the Japanese. "There is no in between when you see something like that. You want to kill every one of those fuckers." He rose to sergeant and returned stateside to train as an artillery spotter with the Piper Cubs, all the while taking correspondence courses through Cal Berkeley.

At twenty-three, discharge papers in hand, he reenrolled at Stanford and took a job at the James Allen & Sons meat-packing plant south of San Francisco. Allen was a giant Scotsman, well near 300 pounds, and he had gotten to know Boswell's father during trips to Corcoran to buy cattle. Jim and his fraternity brother, Pete Gadd, the boxing champ at Stanford, got out of bed at four in the morning and donned their long white coats to work in the thirty-eight-degree chill of a slaughterhouse where 300 hogs, 100 sheep and 600 cattle rode the trolley in and out each day. They'd carry on their shoulders a last delivery to Chinatown, making sure to knock off the hat of at least one old lady by swinging the big hindquarter just so. Then they'd hustle back to campus in the 1936 Ford that Jim had won in a craps game. "I was happy as hell pushing beef. I knew one thing. I wasn't going anyplace at Boswell. The last thing I wanted to do was work for J. G."

He was one course shy of an economics degree when his dad showed up with a problem that needed fixing. The cowboy running the Arizona ranch had contracted with a big beef shipper to finish feeding more than 1,000 head of cattle. Normally, this was nothing the company couldn't handle, but Boswell workers had made the mistake of mowing down the alfalfa fields the cattle were to graze on. Jimmy gave his dad a "don't-bother-me-with-this-now" look, and big Jimmy Allen, standing beside Bill Boswell, lit into the boy. "When your daddy says he needs you, you don't ask, 'Why?' You say, 'When?' Your job will be waiting for you when you get back."

Jim Boswell would never return to the Allen slaughterhouse. He told his father he'd take the job in Arizona on one condition—that he'd answer to him and not his uncle. The Colonel was none too pleased to hear this, throwing a fit in his Los Angeles office in front of his brother and nephew and secretaries. How dare a wet-behind-the-ears punk challenge his authority? But in the end, the Colonel relented and Jimmy, his course work in economics completed, headed to Arizona with buddy Pete Gadd at his side. He solved the alfalfa problem in no time and learned how to run the feed lot, cotton fields and gin. By 1949 he was calling all the shots in Arizona.

Even as the boss, he liked to think of himself as a simple cowpoke, straight-shooting whiskey each night on the same Phoenix barstool next to the Gill brothers, who once ran one of the largest cattle ranching businesses in the nation, spread across nine western states. Boswell, who wanted nothing more than to be one of the boys, earned his spurs in a most horrific way. One Sunday morning, astride his horse, he was roping cattle in an open-sided pen under a tin roof. He had landed the rope around the steer's neck and was trying to reel in the 1,200-pounder when the lasso somehow wrapped around a corner post. All of a sudden the slack was gone and it was a fight between horse and steer. The horse, no match, began to fall and Boswell tried to brace himself. He and the horse landed with such force that it crushed every finger in his right hand.

One of the cowboys saw the mess of blood and bone and threw Jimmy into his feed truck and drove him to a doctor in Litchfield. The doctor gave him a shot of morphine and bandaged him up and sent him to the Good Samaritan Hospital in Phoenix. For the next two days, he tried to outdrink every cowboy who came to visit him. The doctors, he said, had agreed to whatever amounts of whiskey and visitors he desired, all in the name of getting his blood to circulate. Unfortunately, gangrene set in anyway and two doctors came into his room to tell him that his whole hand had to go. They would amputate in the morning.

He said he got out of bed, put on his clothes, stole out of the hospital and hitchhiked to the airport. He didn't have a dime on him and didn't bother calling his wife. He said a black man who worked at the airport saw his condition and loaned him enough money to cover his flight to Los Angeles and a cab ride to the Good Samaritan Hospital there. As luck would have it, he said, a doctor who specialized in fixing hands— he had spared several Arizona cowboys from amputation—was on duty and outfitted him with a wire lattice that forced open his hand. Each day, the doctor snipped off more black flesh until his middle fingers were no more. At one point, Boswell said, he grabbed the scissors himself and started cutting away. They were able to stay ahead of the gangrene just enough to save the rest of his hand. "I didn't dwell on it. I can still bowl and swing a golf club, and I'm the only guy you know who can get away with giving the middle finger in a social situation."

The hand, which resembled a pig's knuckle, surely had been mangled this way. But the story he told about how it was saved sounded more than a little like myth. As with other Boswell stories, it was difficult to know what really had taken place. Friends and coworkers who were present would recall different, less grandiose, details. They doubted that Boswell sat up in bed and drank whiskey or that he sneaked out of the hospital in the middle of the night or that a black stranger paid his way to Los Angeles. Even Boswell would recount, in a different telling, that it was his Aunt Ruth Chandler, and not dumb luck, who found the doctor that saved his hand.

THE COLONEL'S DEATH IN 1952 was still sinking in when his nephew took a seat at the boardroom table of the company bearing his name. Jim Boswell was twenty-nine years old, callow and untested in the corporate world.

Fred Sherrill—the wounds of getting passed over for president still raw, the sting of being stiffed in the Colonel's will still fresh—settled into his seat, as well. He had decided to make one last bid for power. As Ruth Chandler called the meeting to order, Sherrill started right in. "My plan, madam chairman," he announced with his usual stentorian delivery, "is that we're going to build several new gins"—a vision that moved the company deeper into cotton processing and put less stock in farming the land.

Boswell had been blindsided—and nothing made him madder. "It was kind of like he was reaching over to have a little jab at me," he'd later remember.

And so Boswell jabbed back. "With all due respect," he said, his own voice rising, "that's not the way it's going to be." Sherrill blanched. The room fell silent. Had any board member harbored doubts that J. G. was the Colonel's rightful heir, they were vanishing fast. My god, he looked young sitting there, with that boyish face and a sinewy frame clearly more comfortable in ranching duds than a coat and tie. Yet there was something extraordinary about the way he commanded attention, the decisiveness with which he had batted down Fred Sherrill's overture. He had "this ability to hold people in the palm of his hand," said Bob McInerny, who was a friend from Stanford and later handled the Boswell Company's insurance. Everybody in the room now knew that he was their leader, and they all stared at him, wondering where in the world he was taking them. "We're going to become more of a producer"—a grower, a landowner, Boswell explained. "That will put us in a position where we can have flexibility. You can always sell the land or lease the land" if it doesn't work out, and adjust from there.

No one dared say another word until John McWilliams, the Colonel's dear friend and a long-time director, at last spoke up. "Ruth," he said, "I vote with Jimmy."

"I do, too," said Boswell's Aunt Ruth, who, while more than a mere figurehead, was content to let her nephew steer the company in whatever direction he liked.

And that was that. From then on, according to Boswell, Fred Sherrill was nothing but 100 percent supportive of him and his blueprint for the future. Shortly after the contretemps, the old intellectual even presented Boswell with a gift of ten books, including Hendrick's *Bulwark of the Republic* and Bowers's *Jefferson and Hamilton*, that he believed were must-reads for any well-rounded CEO.

Jim Boswell would play down his boardroom decision, insisting that it was more a spontaneous response to Fred Sherrill's effrontery than a well-considered scheme. "Honest to God," he'd say, "it's just something that came out . . . a knee-jerk reaction." Whether he actually had been that capricious, or more calculating than he'd ever let on, the timing couldn't be beat. Profit margins in the custom ginning and farm loan business were shrinking, the result of increased competition from cotton co-ops and other factors. If Boswell hadn't switched tracks, his hand would have been forced soon enough. "The change was strictly a matter of economics," said Len Evers, the payroll clerk who emerged as one of his closest advisers.

The path was now set: Boswell would put less emphasis on financing and processing cotton for others and, instead, buy up as much good cropland as possible beyond the 45,000 or so acres the Colonel had assembled. This basic strategy would shape the company for the next half century and, as it unfolded, strike some observers as nothing less than a modern version of Manifest Destiny.

Those initial years in the job were marked by one challenge after another. A huge flood in 1952 put some of Boswell's best farmland fifteen feet under water. In '53, a valley congressman named Harlan Hagen accused the company of rigging prices, prompting the Justice Department to investigate. That same year, more labor unrest hit the Boswell gins, an echo of the mayhem from the late 1930s. The International Association of Machinists, which had enjoyed great success in organizing the aero-space industry, was trying to penetrate the company.

Unlike the Colonel, Jim Boswell was no tightwad when it came to tak-ing care of his employees. He made a personal entreaty to his workers to reject the Machinists, reminding them that the wages he paid and the benefits he offered were better than anything around. "Interference from any union," he warned, would just be counterproductive. Yet in December, with union leaders promising that they could obtain more job security, the Machinists were voted in. The following June, a con-tract was signed.

Now the real fight began. The Machinists' view was that the Boswell Company had become a "union shop"—and everybody covered by the contract had to automatically join their ranks. Boswell, though, refused to go along. If anybody wanted to pay his dues and sign a union card, that was fine. But Boswell sure as hell wasn't going to mandate it. By mid-October, with the '54 cotton harvest under way, talks broke down. The union struck, dividing the workforce.

Those who crossed the picket lines were greeted with shouts of "scab" and, in the case of the black workers, "nigger." Some of the strik-ers wielded slingshots, hurling rocks at those who dared to enter the company gates. Cotton trailers, piled high with fiber fresh from the fields, were set ablaze. Trucks lugging cotton were slowed by sugar and honey in their gas tanks and tacks on the road. As Boswell executive Rice Ober and his wife watched a high school football game in town one Friday night, they could hear a boom in the distance. The Machinists had set off two sticks of dynamite on the Obers' front lawn. "Just big fire-crackers," said Machinists organizer Charles Jones, dismissing the episode as nothing serious. The next one could have been, though. Jones learned that some of the rank and file had squirreled away a big-ger load of explosives, which they were planning to use to topple the Corcoran processing plant. "What are you going to do after you blow the goddamn place down?" Jones asked the rabble-rousers. "Then you won't have a place to work at all." Confronted with this irrefutable logic, they reluctantly dropped the idea.

Families did get torn apart. Roberto Castro, who had helped lead the '33 cotton strike, supported the Machinists, while his younger brother, Rudy, ever loyal to the company, tried to talk him out of it. Years later, Rudy blamed his mother's heart attack on the tension and worry that the '54 strike caused. She felt Roberto was "fighting against me," Rudy said. "She'd cry because he was walking those picket lines." Maria Castro died that fall.

In due course, things settled down. By fortifying its levees, Boswell saved enough of its ranches from flood damage to squeeze out a profit of $1.3 million in 1952, and it earned even more money the following year. The company denied wrongdoing in the price-fixing case, and the Justice Department's antitrust inquiry fizzled. Even the union was defeated in the end. Boswell never backed down in his conviction that his wasn't a union shop, and he thwarted every effort to make it otherwise. When a federal mediator prodded the parties toward a resolution, Boswell's representative rang up the boss to get him to bless the agreement. He came back to the bargaining table, shaking his head.

"Do you want me to quote him exactly?" Boswell's man asked.

"Sure."

"He said to go screw yourselves."

In 1955 another vote was held. The Machinists lost this round and were decertified. "I always had hoped that we could do something there," said Jones. "We didn't have a chance."

Boswell pushed equally hard when it came to the regular routine, overhauling the antiquated financial systems and organizational structure that the Colonel had left in place. "I sometimes wonder," he said several years into the job, "if maybe I'm not too demanding in seeking the 'perfect' operation." Pete Gadd, who had been beside him since Stanford, was brought in as a manager in Corcoran—and then dismissed on grounds that Boswell would reveal to his Aunt Ruth:

> I spent the entire summer at Corcoran for a reason that I couldn't state to anyone, but it had to do with a very serious hunch on my part that Pete Gadd wasn't working out too well, and I had serious doubts as to his potential. The only way that I could satisfy my suspicion was to work with him and beside him and closely observe.
>
> To make a long story short, I have just let him go. As you no doubt know, this is particularly hard on me because we have been very close in school and during the war. Pete is honest and sincere and certainly works hard. The only fault I had to find was lack of judgment. This is always an intangible and boils down to a question of his judgment versus mine. . . .
>
> It is hard enough to take business problems home with you, but it is

unbearable to take personnel problems of this magnitude home with you. I'd rather face them and get them behind me.

Jim Boswell had withstood every test. But where he really proved himself, even more than in handling adversity, was in seizing opportunity. Time and again, it came knocking at his door.

"SANDY'S BEEN FUCKING ME."

It was the spring of 1956, and Sandy Crocket's partner of two decades, Casey Gambogy, had just walked into Jim Boswell's Los Angeles office with his wife, Esther, by his side. Boswell hadn't seen Gambogy in years but kept tabs on how he and Crocket had been picking up large tracts of land around the valley. Just eight years earlier, they had added Elmer von Glahn's huge Tulare Lake holdings to their own, giving them a grain and cotton fiefdom that now stretched across about 100,000 acres. They were easily the biggest thing around.

Yet all was not well, as Gambogy confided to Boswell this day. A large man shaped like a bowling pin, he squished into one of the chairs facing Boswell's desk and let loose about the way that Crocket seemed to be siphoning more than his fair share of cash from their operation. He couldn't prove it; he wasn't even sure how Crocket might be cheating him. But Gambogy, who controlled 40 percent of the venture, was convinced that something was amiss. He was fed up and, besides, his kidney stones had been bothering him. He wanted out.

Boswell asked Gambogy for a financial statement and then spent a few minutes soaking it in. Over the years, Boswell would demonstrate a tremendous gift for parsing facts and figures. Those who marveled at his deal-making prowess would invariably cite his uncanny instincts, that golden gut. He liked to talk it up himself, burnishing his image as a daring cowboy, the antithesis of the Organization Man. Yet while he did have a remarkable instinct for what would make money and what wouldn't—and he'd rely on that intuition to point him in the right direction—those closest to Boswell knew the real truth: He would backstop his gut by analyzing hard data. "Consolidated after-tax returns," "a P/E multiple of 15," "a nine-percent-plus post-merger value"—the memos that Boswell wrote were littered with the parlance of corporate finance. When he picked up a balance sheet, "he could grasp more information in an hour than most people could in a week," said Walt Bickett, who served as Boswell's chief financial officer. Len Evers witnessed the same sober skills at work. "Jim Boswell is a very misleading individual to a lot of people," he said. They "really thought Jim was floating on the surface. They never knew how deep his thinking was."

Boswell passed the document back to Gambogy and looked over at him. "I'll tell you what," he said, a portrait of the Colonel beaming down from the wall. "I'll give you a million bucks for your stake right now, no questions asked." It was a laughable figure, really, the lowest of lowball bids. Gambogy's interest was plainly worth ten times that much. Nevertheless, Boswell knew he was dealing with a desperate man. Gambogy accepted on the spot.

"Honey," Gambogy said, turning to Esther, "give me the stock certificates." She opened her purse and handed him the paperwork.

Boswell's finance men thought that the boss had lost his mind when he buzzed on the intercom and ordered them to write out a $1 million check and, please, hurry it in. They should have known better.

Just a year earlier, Boswell had received another unsolicited call, this one from a real estate broker representing two brothers named Allen, who owned a 37,000-acre chunk of farmland west of Tulare Lake. One brother had taken ill, and while he was recuperating on a cruise, the other took it upon himself to sell out. Boswell dispatched his top managers to scour the property, which was called the Boston Land Company, and they couldn't quite believe their good fortune. Not only was the seller anxious, but this was some of the best dirt they had ever seen. Boswell snapped up the place for less than $6 million. It was such a steal that by immediately selling off about a third of the acreage plus that year's crop, which Boswell had wangled as part of the original purchase, he recovered his entire expense. Lapsing into business-speak, Boswell crowed about his "zero-cost basis" on the deal. In plain English, that meant he had just gotten 23,000 acres of productive cotton land for free.

Now he was at it again, building the empire on the cheap. Boswell forked over the check and shook hands with Gambogy. Then, as soon as he had escorted Gambogy and his wife out the door, he phoned the guys that he knew would make his new partner, Crocket, most nervous: the accountants.

By the next morning, a team of CPAs led by Wally Erickson was nosing around in Crocket's books. Crocket contacted Boswell and offered to buy him out for $2 million; it was a chance to double his money overnight. Boswell declined. "Sandy," he said, "I think I'll stick around, play a couple of cards and see how it goes." As Erickson would remember it, there was never any concrete evidence that Crocket had been filching, and Crocket would deny that he had ever done anything improper. But the more Erickson crunched the numbers, the more he became persuaded that Crocket was indeed hiding something. "There were a lot of shenanigans," Erickson said later. By the fall—guilty as Gambogy had suspected or perhaps just weary of all the scrutiny—Crocket agreed to sell his stake, too.

It didn't take long to come to terms. Boswell had a much different style at the negotiating table than did the Colonel, the compulsive haggler. Some back-and-forth was inevitable, to be sure. But Boswell hated "pussyfooting around," as he put it, and he wasn't averse to paying a premium for something that he wanted to buy, or offering a discount for something he was trying to unload. If you were making a deal for the right strategic reasons, another 5 percent or even 10 percent, one way or another, wasn't going to count much in the long run. Boswell acquired Crocket's share for about $10 million, a transaction that took in tens of thousands of acres of land outright and leases on tens of thousands of acres more. Among these was a lease for Buena Vista Farms outside Bakersfield, a last remnant of "Cattle King" Henry Miller's million-acre realm and an important entrée into Kern County to the south.

The following year, after a nasty legal scrap, Boswell assumed control of the Tulare Lake Land Company, adding another 10,000 acres to the mix. In less than five years, and at a stage in life when many men are just starting to get serious about their careers, Jim Boswell had more than doubled his company's size, completing the greatest land-buying spree in California since that of Miller himself. He also had guided the company far past its traditional borders, procuring land, cotton gins and an oil mill in Mexico in 1955. J. G. Boswell wasn't just a farming company anymore; it was an international farming concern.

Boswell would always stress that he didn't go looking for sellers; they had somehow found him. It wasn't quite so simple, however. He may not have been predatory, but he was as opportunistic as they come, ready to pounce whenever he smelled the slightest opening. The slogan at the company's California ranching division, Boswell once joked, was "covet thy neighbor's property."

And yet he never let that hunger cause him to lose his way. As big as the company had become, as rapidly as it had expanded, Boswell kept sight of a few fundamental principles—ones that the Colonel had also strictly abided by. First off, he frowned on long-term debt, funding one land purchase after another with cash from the company's day-to-day operations. "Jim is really a very conservative person," remarked Evers, a trait that would prove critical for surviving—and even thriving—in times of flood or drought, or when the commodities markets tanked. Second, Boswell never bought land just to get bigger; each purchase fit into some grander plan.

Certain parcels were acquired because they carried with them a valuable "cotton allotment"—a right to grow the crop under a government program restricting overall production. Others contained indispensable pieces of plumbing. The takeover of Crocket & Gambogy, for example, provided Boswell with access to several key canal systems that it hadn't been able to tap in the past, allowing it far greater manipulation of the water flow in Tulare Lake. Boswell bought some land because it had additional water rights attached to it. During the Colonel's era, the mantra had been: "Get the water, and the land will come." Now, the next generation was flipping that philosophy on its head, buying up land that afforded it more control over the water. Most of all, Boswell bought acreage that promised to enhance its efficiency in the fields. The goal was to put together solid blocks of ground, not scattered holdings, so that they could be cultivated and harvested en masse, getting the most out of every worker and machine.

As Boswell's footprint grew, so did the angst of some of its neighbors. Playing the bully was nothing new for the company. In the early 1950s, for example, Boswell had pumped well water from an area north of Tulare Lake and threw it in a canal running south to its fields, leaving local farmers screaming that their groundwater was being hijacked. Boswell didn't even bother hooking up a power line so that it could pump quietly; instead, it revved up a bunch of diesel engines, keeping people awake at night. "There was a level of arrogance that was exercised from Los Angeles in those early years," conceded Stan Barnes, the water engineer. As time went on, Barnes said, he and other managers made a point of cooperating with the neighbors wherever possible, and the

company did band together with smaller farmers on a variety of mutu-
ally beneficial projects. Sometimes Boswell would agree to transfer water or repair a broken levee for a local farmer without any gain for itself,
just to be accommodating. Nonetheless, some would always see
Boswell as the 800-pound gorilla because it was the 800-pound gorilla.
As the company reengineered more and more land in and around Tulare Lake, it was unavoidable that others would be hurt. Floodwater that
used to stream one way had been redirected and was now coming
straight at them. Even in its most benevolent moments, Boswell was obviously going to look out first for Boswell.

The company's political reach was as long as one of its massive
canals. Over the years, one-time employees and close customers would
find their way onto the Corcoran city council, the Kings County board of
supervisors and the local planning commission, assuring Boswell a
warm reception whenever it needed to get something passed. The police department had a special bond with the company, too. More than
one man around town would confess that whenever he got pulled over
after having had a few too many to drink, there was one surefire way to
stay out of jail: Mention that you worked for the Boswell Company, and
you got a friendly escort home. Until the late 1950s, Boswell even had its
own personal lawman. Deputy Constable Gene Ely wasn't on the government payroll; he was paid directly by the company. After he retired,
Ely loved reminiscing about the "Blue Moon of Kentucky" case. "Some
guys stole 239 bales of cotton from Boswell," he said. "We couldn't figure
out how it was done or where the cotton went. I spent two months dig-
gin' . . . until finally a truck driver squealed. The bales had been taken by
some guys working for the company, slipped away to a compress in
Fresno and then sold. We started rounding up the guys who did it, and
when I went up to arrest one gin-yard worker, he was loading his family
into the car, ready to go back to his Kentucky home.

"He had been talked into the theft to get money for the trip. As we
drove to town to the jail, I had my radio playing in the car, and it started
blaring out the tune 'Blue Moon of Kentucky.' My guy started bawling
and crying. 'I'll never see Kentucky again.'"

Digesting all that land was difficult. Boswell managers, grappling
with how to deploy their people and equipment across such a huge expanse, had to learn to be master logisticians, not just farmers. Sluicing
the water over so much space was its own trial, a job without end. "The

irrigation system is just like Disneyland," said Glenn Jorgenson, one of Boswell's water engineers. "As long as there's imagination, it'll never be finished." The ginning operation also had to be modernized and reorganized so that it could cope with Boswell's sudden surge of cotton production.

At the same time, Jim Boswell had to make sure that all that output would find a home. Until now, most of the company's crop had been shipped overseas, but Boswell was increasingly concerned about competition from abroad cutting into his business in Europe and Asia. So he headed to the East Coast to try to convince the spinning mills there to buy California cotton. It was a tough sell. "As long as you're getting along good with what's growing right in your own area," said Duke Kimbrell, the chairman of Parkdale Mills in Gastonia, North Carolina, "why would you want to change for something that's more expensive?" But Boswell "kept banging away," as Kimbrell recounted, determined to prove to him and other domestic mill owners that the reason California cotton cost more was because of its superior quality. It took a while, but Boswell cracked them all: Parkdale, Fieldcrest Mills, Acme Mills, Spartan Mills, Textiles Inc.

Boswell would head down other avenues as the years wore on: aircraft manufacturing, oil and gas, cattle (leading the company to obtain another 400,000 acres for grazing in Oregon and California's Yokohl Valley), even the pig business. But at this juncture, in the late 1950s, Boswell and his men were focused on a single objective: being the best cotton grower in the world.

With one exception. In Arizona, Boswell grew houses.

By 1959 Boswell's Santa Fe and Marinette ranches outside Phoenix were becoming hopelessly expensive to run. The water table had fallen so low from decades of pumping that the wells could scarcely be used. The sprawl of Phoenix, with houses planted closer and closer to the fields, was making it more difficult to spray fertilizers and pesticides. The Farm Belt was giving way to the Sun Belt.

That summer, the Webb Company was scouting for land—lots and lots of land. Del Webb, chairman of the development and contracting firm, was already a legend on the American scene. He had built the

Flamingo Hotel in Las Vegas for Bugsy Siegel, hung out with presidents, partied with Howard Hughes and, in 1945, bought the New York Yankees, who won ten World Series titles during his twenty years as owner. Now his company was about to push into a new arena: retirement living.

Boswell had been alerted to the project by Bob Johnson, a Webb executive who worked out of Los Angeles and belonged to the 21 Fund, an investment club that counted Boswell as one of its members. Chatting over dinner one night, Boswell told Johnson that he might have some land in Arizona for sale, and after he got home, Johnson phoned Del Webb's right-hand man, L. C. Jacobson, to tell him the news. Jacobson planned to call Boswell back the next day but never got the chance. That morning, Boswell flew to Arizona and barged in on him.

"I understand you're interested in buying some land," Boswell said.

"We are," replied Jacobson, "but we're interested in rather large tracts—and you may not be qualified to provide that."

"You're probably right," Boswell said, deadpan. "All we've got is 10,000 acres."

Jacobson was a bit taken aback. "Well, let's go look at it."

Together they rode out to the ranch, and Jacobson could instantly see the possibilities. He suggested they return to the office and keep talking. On the way back, Boswell nonchalantly gestured toward more of his cotton fields along Grand Avenue. "Oh, by the way," he said, "there's another 10,000 acres over there, if you're interested."

By 2:00 A.M., a deal had been struck. It was classic Boswell. No lawyers, no investment advisers. Just two men and a yellow legal pad on which to scratch out some figures. The agreement called for the Boswell Company to sell its farmland, but the pact also made Boswell a 49 percent partner with Webb in the development of the retirement community. Each side put up an initial $50,000, and Boswell loaned another $600,000 to start up a golf course, recreation facility and shopping center. Jim Boswell used his seat on the Safeway board of directors to bring in a supermarket as the anchor tenant. But it was Boswell's business mind, as much as his money, that would have a lasting impact on the place. "He stayed on top of what we were doing," said Owen Childress, a Webb executive. "We'd come up with a new concept—a change in the financing plan or something—and Boswell would just pick it apart. Boy, he'd raise my blood pressure with some of the questions he'd ask, but I eventually learned that he was just boring in to find out if I really knew what I was talking about."

The initiative that Webb and Boswell were pursuing was risky. No one had ever created a village where senior citizens could come together and maintain "an active way of life," not on this scale anyway, and some geriatric experts insisted that older people would rather be with their families, not isolated in some out-of-the-way spot. "Financially," cautioned one report presented to Webb, "such a project would fail because of 'cannibalism' (more people would die than move in)." As Webb and Boswell pressed forward in the face of such skepticism, even some of their own men expressed misgivings. "How am I going to get a 30-year mortgage on a guy who is 65 years old?" Childress fretted.

On New Year's Day 1960, the newly named Sun City opened. The naysayers couldn't have been more wrong. A line of cars backed up for more than two miles along the highway, waiting to get in, and that first weekend 237 houses were purchased for $2.5 million. Within twelve months, 2,000 homes had been sold, defying management's original plan to peddle 1,700—over three years. More than a booming business, Sun City was a sociological phenomenon. *Look* and *Life* magazines featured it in photo spreads. Calvin Trillin spent ten days there and devoted forty pages to the experience in the *New Yorker*. It was Del Webb, not surprisingly, who became the big hero, pictured on the cover of *Time* magazine under the banner: "The Retirement City: A New Way of Life for the Old." But as Boswell would tell it, it was Webb's colleagues—himself included—who were the real force behind Sun City; Webb was merely the front man, the marquee name they needed for marketing. "I used to write part of Webb's speeches for him," said Boswell, "and they all started, 'When I visualized Sun City, standing there in the middle of a cotton patch. . . .' Hell, he couldn't find his way out there if he had to."

Sun City and its sister community, Sun City West, earned many millions of dollars for the Boswell Company over the years. Even after Boswell sold out to Del Webb decades later, the family remained a fixture on the landscape. Walter Boswell, who had overseen the Arizona ranches, had died in December 1952, just a few months after his brother, the Colonel, passed away. Now the hospital at Sun City would carry Walter's name forever, drawing his children, grandchildren, cousins, nieces and nephews to the dedication ceremony. As Walter's boy, James O., stepped to the podium, it wasn't hard for him to imagine what his dad would have thought about such a large gathering of the clan. "Jesus Christ," he could hear the old boy saying, "we're up to our ass in Boswells!"

Boswell, though, wasn't finished building the empire just yet. In 1963 he ventured across the globe into Australia. Two old farm-loan customers from Arizona had convinced him to take a shot at growing cotton in the state of New South Wales. Boswell put up the necessary capital while partners Jim Blasdell and Richard Rhodes ran things Down Under.

Blasdell was a real go-getter who insinuated himself with local politicians and other power brokers, and never met an obstacle he couldn't figure a way around. When he learned that government rules barred individuals from buying more than 400 acres of cotton land, he had his Australian accountant line up twenty-two "straw men," as he called them, so that he could string together thousands of acres and also obtain the water licenses needed for irrigation. "We broke every law in the state," he'd say with pride many years later, though the truth was that Blasdell merely bent and twisted the laws, taking pains never to actually run afoul of them.

Yet as clever as Blasdell was at working the system, he wasn't so good at working the land. To make matters worse, that first year of planting he and Rhodes ran into some of the heaviest rain ever seen in the area. Because they had not used any herbicides, a snarl of weeds came up with the harvest. By the time Jim Boswell stopped in to check on things, it was one fat mess. The cotton stood all of three inches high, while the weeds stretched to a foot.

Blasdell knew that the partnership was kaput; under their arrangement, Boswell had the right to take over some farmland in Arizona that Blasdell had put up as collateral, or assume 100 percent control of the Australian venture. Though things were in disarray in Australia, the hard-to-get land and water licenses were now in place. Boswell chose to go it alone. "Jim is pretty good at making deals where people don't have an option," Blasdell said.

Boswell's next move turned out to be one of the most significant in the company's history. He picked Jim Fisher, a young agronomist, to try to turn things around in Australia.

Fisher, who would one day become president of the Boswell Company—the only non–family member ever to hold the post—had grown up on a small farm near the San Joaquin Valley town of Strathmore and earned a master's degree at Fresno State. From the moment he arrived in the Australian hamlet of Narrabri, he displayed the traits that would fuel his rise. He was smart and tenacious, and he handled people well with a knack for making even the drudgery of a job somehow seem exciting. That first full year, Fisher planted 5,000 acres of cotton and nurtured them with the right diet of chemicals and water from the Namoi River. The result was more than two bales to the acre—enough, observed Fisher, that in a single season "we covered up all the red ink that the place ever had."

Getting the cotton to cooperate, though, was only part of the trick. Politicians and farm organizations were wary of a big American company encroaching on their turf, and Boswell had to win special parliamentary approval to obtain outright title to the land. A couple of local solicitors assured Boswell that this was all perfunctory, but others held out less hope. "They just sold you the Sydney Harbor Bridge," Tom Lewis, the minister of lands for New South Wales, told Boswell when he first came calling. Yet Boswell persevered and, with Lewis's help, overcame the hostility. The governor of New South Wales, Sir Roden Cutler, even paid his respects by driving into the Boswell fields in a black Rolls-Royce and taking afternoon tea there. Much to her displeasure, Fisher's wife, Lois, had to learn to curtsy for Sir Roden's visit.

In just a few short years, everything had fallen into place. "We looked out over the . . . plain," Fisher recalled later, "and Boswell said: 'How much bigger can you make this? Two, three, five times bigger?' And I said, 'Boswell, it about broke my pick getting this stuff done, what we've got here. Now you want me to go out there and do some more.' But that was the way he thought." In time, Boswell's Australian arm, known as Auscott, would become that nation's biggest cotton grower, swelling to more than 60,000 irrigated acres, while also getting into cattle and dry wheat farming. And thanks in no small part to Boswell's exploits, Australia would go on to meet its goal of becoming one of the world's top cotton exporters.

As the 1960s wound to a close, even some of Jim Boswell's own men couldn't quite believe the run that he had put together. During a span in which many other farming magnates went under, Boswell had become the Midas of the soil. He had tasted failure, as all businessmen do; his foray into Mexico, for instance, had been a bust. But for the most part, "everything he touched turned to money," said A. L. Vandergriff, who ran Boswell's ginning operation. Some insiders, watching earnings increase year after year, called it "Boswell luck." Everyone who was honest, though, had to admit that a lot more than just luck was involved.

JIM BOSWELL COULD BE A REAL SOB TO WORK FOR.

Not always, not even most of the time, mind you. When things were going right, you'd be hard-pressed to find a bigger-hearted boss, one more willing to listen to what even the lowest man in the organization had to say. "He didn't surround himself with yes men," said Jon Rachford, who joined the company in 1968. But God forbid a deal derailed on you or you didn't make your numbers—or you hadn't done your homework and you tried to fake your way through a presentation. Then came a dressing-down that you wouldn't soon forget. It wasn't that Boswell was a screamer. He sure could make you feel stupid as hell, though.

Sitting over lunch in Corcoran one day with Boswell and one of his top managers, you could literally see the fear building in the man's eyes as the boss dug into him. It wasn't so much what he said. Boswell always gave out plenty of zingers. It was the way he said it—the icy tone, the edge in his voice—that made it abundantly clear he wasn't too pleased with the manager's performance of late. At the end of the meal, when the man suggested that they drive out for a look at the fields, Boswell glared at him. "You go your way," he said coldly. "I'll go mine."

Fisher called it "getting the red ass." Wally Erickson's term was "being on the bubble." Somebody or another, it seemed, was always on the bubble. Boswell "didn't put up with any bullshit," said Bob McMicken, who ran the cattle business. "He was tougher than a boot."

He demanded exactness, whether it was in the fields, the gins or the conference room. If a meeting started at 10:00, you'd be wise to show up by 9:50, or you might be in trouble for shaving it too close. "The poor guy's probably gotten there a minute before ten or ten on the nose, and he still catches the red ass," said Fisher. Boswell never let up, either. He liked to get his people going on projects at the tail end of the week, just to make sure they didn't start winding down. "I give my big orders on Fridays," he once said. "That way, I keep 'em running."

He sometimes pitted his top people against each other, having two

The Boswell clan of Greene County, Georgia, in a photograph likely taken around the turn of the century *(courtesy of Emily Griffin Langford)*

Walter Boswell, J. G.'s older brother, in his army uniform. He served as an aide to General John J. "Blackjack" Pershing before going on to run the Arizona operations for the J. G. Boswell Company *(courtesy of Steve Boswell)*

The Penfield, Georgia, cemetery where many a Boswell is buried *(Library of Congress)*

Men gather during the establishment of Corcoran, early 1900s.
The figure standing second from left is Hobart J. Whitley, known as
"the Great Developer" *(UCLA Library, Department of Special Collections)*

Jack rabbit slaughters were a local version of the fox hunt for
Tulare Lake basin pioneers *(courtesy of Gwen Smith)*

Bonanza wheat harvest in the Tulare Lake basin, 1920s *(courtesy of Johnny V. Baker)*

Women demanding food during the 1933 cotton strike *(Library of Congress)*

The annual convention banquet of the California-Arizona Cotton Association, held at the Alexandria Hotel in Los Angeles, 1927. J. G. Boswell can be found on the left side of the table farthest to the left, third up from the bottom
(courtesy of the Western Cotton Shippers Association)

A line of pickets from the 1933 cotton strike *(Library of Congress)*

The strikers' tent city in Corcoran during the 1933 walkout
(Library of Congress)

"Cotton Pickers Wanted" sign perched in the San Joaquin Valley, 1930s
(Dorothea Lange, Library of Congress)

Hauling a cotton sack to be weighed
(Dorothea Lange, Library of Congress)

The Boswell Bears, the labor camp baseball team from the late 1930s
(courtesy of Lawrence Galvan)

Cotton pickers' shacks near Corcoran
(Dorothea Lange, Library of Congress)

A portrait of Colonel J. G. Boswell, late 1940s *(courtesy of Minnie G. Boswell Memorial Hospital, Greensboro, Georgia)*

A look at the 12,000-square-foot home in San Marino where the Colonel lived with his second wife, Ruth Chandler. Boswell relatives, thinking the couple acted a little snooty, came to call the place "Buckingham Palace" *(Architecture Design Collection, UC Santa Barbara)*

Elizabeth Taylor and Van Johnson star in *The Big Hangover* (1950), filmed at Boswell's estate. The mansion was used as the set for several movies, including *Funny Girl (Academy of Motion Picture Arts and Sciences)*

Fred Sherrill, the Colonel's political operative through the 1930s and '40s. Boswell had promised to leave Sherrill a big stake in the company, but the Colonel reneged, leaving his long-loyal lieutenant to remark: "They spell it S-h-e-r-r-i-l-l, but it's pronounced 'sucker'" *(courtesy of the J. G. Boswell Company)*

Alaine Buck, the Colonel's first wife and his true love, was a poet. She died young *(courtesy of Barbara Hamilton)*

The Colonel (center) and Ruth Chandler (top of the stairs, to the right) visit the Hotungs, one of Hong Kong's most prominent families, in 1948 *(courtesy of Theodore W. Lindabury)*

The Salyer family lines up its armada of farm equipment along the edge of a levee in the lake bottom, early 1950s *(courtesy of Johnny V. Baker)*

Clarence Salyer and his wife, Cordie. He had a mangled eye that wandered so far to the left, it earned him the moniker "Cockeye"
(courtesy of Verdo Gregory)

Jim Boswell as a toddler, along with his siblings and
his mother, Mom Kate *(courtesy of Jim Boswell)*

Two-year-old Jimmy Boswell and his sisters playing
in a mound of cotton, with a bale behind them,
1925 *(courtesy of Jim Boswell)*

Bill Boswell and his oldest son, Billy Jr. Billy was supposed to inherit the empire from the Colonel, but he was too much of a hellion to take the reins
(courtesy of Jim Boswell)

BELOW: The Colonel and his nephew, Jim Boswell, on a trip to the Far East. "I didn't love him," Jim Boswell would later say of his uncle. "But I respected him"
(courtesy of Jim Boswell)

ABOVE: Jim Boswell on horseback as a young man
(courtesy of Jim Boswell)

Old cars are lined up along a levee during the 1969 flood
to prevent the waves from crashing through. Jim Boswell
would claim credit for the scheme, though others
remembered it differently *(courtesy of Gwen Smith)*

Jim Fisher would rise to become president
of the Boswell Company, only to be fired
without warning
(courtesy of the J. G. Boswell Company)

Irrigation water gushing onto a Boswell cotton field,
which runs clear to the horizon *(courtesy of Jim Fisher)*

Jim Boswell and his board of directors, 1976 *(courtesy of Jim Fisher)*

Jim Boswell and Del Webb sign the land agreement that would turn old Arizona
cotton fields into America's first large-scale retirement community, Sun City
(Sun Cities Area Historical Society)

Jim Boswell during the cotton harvest with giant modules of cotton
lined up behind him *(photo by Matt Black © 1999)*

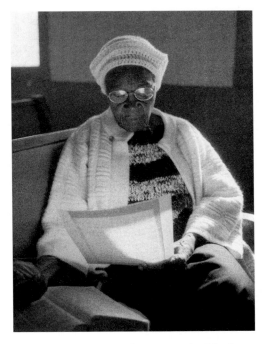

Hattie White, seated on a pew inside the
First Baptist Church of Corcoran
(photo by Matt Black © 1999)

Harvesting cotton *(photo by Matt Black © 1999)*

A lone Boswell cotton picker chugs across a field
(photo by Matt Black © 1999)

Jim Boswell camping the way he loves best
(courtesy of Jon Rachford)

managers tackle the same assignment. He'd tell Fisher, "Gee whiz, Jim, Yost has your cotton-picking deal beat all to hell." Then he'd turn around and tell Yost that it was Fisher who had his deal licked. More than anything, Boswell expected his men—and this was a fraternity; no women ever reached the inner circle or even the outer circle—to make decisions. It was a lesson he preached over and over: Practicing due diligence was fine. Scrubbing the numbers was well and good. He did plenty of that himself, after all. But dubiety was a sin, overstudying a situation the worst. Nobody survived who wasn't willing to step up and put a plan into motion. Buy or sell. Plant or fallow. It almost didn't matter what you did, as long as you could justify it and were willing to stick by it, see it all the way through.

That said, you'd also better be prepared to shift gears, and shift them fast. It wasn't uncommon for Boswell to send you scrambling in one direction and then, without any real explanation, order you to make an about-face and go clear the other way. Had he sensed a change in the market? Had he been made privy to some piece of information that you hadn't had access to? Or was this all for his amusement, a way to test your moxie? Often you never really knew for sure.

Because of his methods—and sometimes in spite of them—Boswell fostered an incredible sense of shared pride. There wasn't anything his men couldn't do—wouldn't do—for Jim Boswell and the Boswell Company. If their confidence verged on hubris, and at times even crossed the line, that was okay with the boss. You need that kind of attitude when you're farming the bottom of a lake. "It kind of reminded me of my time in the Marine Corps," said Brooks Pierce, who enlisted with the company in 1958 and went on to hold various supervisory positions.

Boswell was careful not to micromanage. He knew farming, but he didn't pretend to be an agronomist. He knew the way the water moved over his land, the basic dance steps, but he didn't make himself out to be an engineer. In fact, he relied so much on the expertise of his underlings, a few critics perceived that Boswell was disengaged, maybe even a little fainthearted. Those around Boswell knew that, while he definitely had his faults, this particular charge was a crock. "Oh, he's a delegator," said Fisher, "but he also wants to know what the hell you're doing." Barnes, the water chief, put it like this: "You never forgot who the leader was—at all times."

Boswell's desire to be a workingman's boss—he bristled at any sug-

gestion that he removed himself from the front lines in moments of crisis—occasionally led to a fish tale or two. Such was the case, it turned
out, with his story of grabbing a shovel and single-handedly fighting off
the 1969 flood while hatching the plan to line the El Rico levee with junk
cars. "Well, I'll be goddamned" was Jim Fisher's reaction when he heard
Boswell's version of events. "That's bullshit." According to Fisher, it was
he and a neighbor—not Boswell—who devised the car scheme over
drinks at the company lanai. "I don't even know if we told Boswell"
about the plan right off, he said.

Taking the glory and dishing out the brickbats, Boswell could be
rough on his top men. Yet he prided himself on the kindness he displayed to employees further down the line. "When I came in here," recalled Ross Hall, a senior manager who hitched up with the company in
the mid-1970s, Boswell "made sure that I understood . . . the culture of
those people who worked at the hourly level. He made sure that I understood what they had to go through, and what they lived for and
thought about and felt, and how important it was that they . . . were
treated with dignity."

Unlike the Colonel, who never bothered to speak to his laborers,
Boswell chatted up everybody as he made his way around Corcoran and
the ranches in Arizona. He knew them all by name: the irrigator, the
truck driver, the grunt piling bags of seed in the warehouse. And it was a
good bet that he knew their kids' names, too. "I don't care what group of
people he's with—rich, rich, rich or poor, poor, poor—he's one of them,"
said Harvey Ruth, who joined the company in 1946 and then befriended
Boswell when they were cowboys together in the Arizona feedlot.

"He's quite a guy. Very folksy," said Peter Munk, a real estate tycoon
who controlled the Sears Tower in Chicago and Watergate complex in
Washington and, in the '70s, developed resorts with Boswell around the
South Pacific. "There was no crap about class or background. There
were no servants, no butlers, no chauffeurs. He and you threw the steak
on the fire."

When it came to his Latino workers, Boswell showed an especially
soft spot, time and again pointing out their contributions to the company and Corcoran. He did not hesitate to promote the most qualified
Latinos, and many of them, such as Rudy Castro, would talk about the
boss with a reverence reserved for saints. But on the subject of blacks,
the most ticklish of questions for the scion of Georgia plantation own-

ers, Boswell was much more conflicted. There were times when he found himself in the middle of a joke or a story from the 1940s that he uttered the word "nigger." In the midst of the Watts riots, his son recalled, Boswell slammed down the morning paper on the breakfast table, fuming at his daughter, Jody, "It's you and your goddamn generation. It's all your fault." His company had a spotty record hiring blacks in Kings County. "We don't go out of our way to search out any diversity groups," he explained. And he would continue to insist that black cotton pickers played no significant role in the lake bottom. "We really didn't have any blacks here."

Yet if he was a product of his generation and his family's history, he could rise above those things, too. In the 1970s, Boswell became heavily involved in Up With People, a multiethnic singing troupe of youngsters who touted the ideal that music—in their case, saccharine-sweet music—could bridge class and culture. Boswell served on the Up With People board alongside Jesse Owens, and one day he invited the Olympic star to eat with him at the California Club in downtown Los Angeles. "It was pretty bold to have a black going to dinner" there—even someone so famous, said Blanton Belk, Up With People's founder and one of Boswell's closest friends. "At one point, Jesse said to him, 'Hey, Bos, do you know what you're doing?' All the waiters—Mexican, black and white—were poking their heads in the door to see. He was doing something that was breaking a barrier."

In the 1960s, Boswell led an effort to tear up the old labor camps around Tulare Lake, eradicating what he called "The Grapes of Wrath days," and then he backed home loans for his workers to make sure they could find a decent place to live. The hourly wages he paid, high for agriculture, also "pissed off a lot of the neighbors," said Barnes, but Boswell just said "too bad" and kept on paying. He looked out for his hired hands in other ways, as well. When his cousin Bill, Walter Boswell's boy, inquired as to whether his son might be able to get a summer job at the company after the 1969 flood, the answer was categorical: No way. "We have over 50 percent of our land at Corcoran still under water. . . . You can well imagine the frustrating time of attempting to provide some continuity of work for a tremendous number of key personnel with many years of dedicated service. I can't in all good conscience provide work for another, as this would mean one less job, for which there are a great many local employees available."

Still, some weren't satisfied that Boswell was doing enough, that the workers were being treated with the respect that they deserved—and the aspersions would make him burn.

At a cocktail party in the summer of 1969 Cynthia Jones, the wife of a close friend, walked up to Boswell and, bourbon and water in hand, laid down a direct challenge: "Why don't you be a hero and accept Cesar Chavez?" At the time, Boswell had a 500-acre table-grape operation in Arizona, the biggest in the state, that was being picketed by Chavez's United Farm Workers Organizing Committee. Jones, a self-described "limousine liberal" who had married a conservative investment banker, was the oddball in her social circle: a passionate activist who had been swept up in a number of progressive causes, including Chavez's grape boycott.

Many a CEO would have sloughed off a comment such as Jones's. That wasn't Jim Boswell's MO. "I'll tell you what," he told her, "I'll take you and your friends and fly you down to our Arizona ranch and show you the other side."

Before she knew it, Jones and three others—her minister from All Saints Church in Pasadena, Bill Rankin; Rankin's wife, Sally; and UFW supporter Ralph Hurtado—were being flown in the Boswell airplane to the company's vineyards. Boswell toured them around Sun City and treated them to dinner. He even stayed that night in the same hotel room as Hurtado, who was awakened around 4:00 A.M. by a stirring on the floor. He glanced down, and there was Boswell, muscling through his early-morning pushups.

Not long after daybreak, the visitors were allowed to go out and roam among the pickets, who were waving their black and red eagle flags. Because of his position on the Safeway board, Boswell was a juicy target for the union, and UFW leaders were quick to hand out fliers accusing him of exploiting the poor:

> The life expectancy of a migrant farm worker is 49 years. This man, Boswell, who donates land for a hospital in Sun City refuses to have health insurance for the people who make him rich, the migrant farm workers. This man who pretends to be civilized still allows labor contractors to steal wages from his workers.

Boswell, naturally, had his own point to get across. Perhaps his most persuasive argument was that the bulk of his workers were still picking grapes, unmoved by the union's call to strike. Only about fifty of Boswell's 500 crew members had walked off the job, and the UFW was never able to establish itself at the company. If Boswell wasn't paying his employees adequately, if he wasn't doing right by them, wouldn't a majority—not just a tiny portion—have gone on strike?

Boswell said good-bye to his visitors, giving each of them a lug of grapes to take home. No minds had been changed either way. "He genuinely felt he was showing us what a good job they were doing down there," said Cynthia Jones. Yet what she saw was "a very paternalistic relationship: 'These are my folks, and they're whistling a happy tune.' We were more convinced than ever that the farm workers needed a union." Bill Rankin agreed. "Boswell was considerate and respectful of his workers," he said, but it was "on his own terms." There's a difference, the clergyman added, "between charity on the one hand and justice on the other."

A few years later, the UFW returned to Arizona. Governor Jack Williams had signed a bill outlawing secondary boycotts and harvest-time strikes, and Chavez protested by fasting for twenty-five days. Shortly after the fast ended in the summer of '72—doctors monitoring Chavez had noticed an erratic heartbeat—Boswell's top manager in Arizona, Hank Raymond, sent the governor a shipment of grapes. The little note attached to it read: "The 400 American-Mexicans who are helping us harvest these non-union grapes join us in thanking you for signing the Farm Labor Relations Act."

Governor Williams, in turn, thanked Raymond for the delicious treat. "Give my regards to the pickers," he said.

In EARLY 1973, flush with confidence from his experience in Australia, Jim Fisher flirted with farming cotton on a whole new continent: Africa.

He had been invited to the Sudanese capital, Khartoum, for meetings with President Gaafar Nimeiri, who wanted the Boswell Company to lease 200,000 acres of land and feed its fledgling textile industry. Fisher flew out over the wedge-shaped clay plain between the White and Blue Niles, and was spellbound by what lay before him. The contours of the land were perfect for growing cotton. "You could have gone in there with a big ol' cutaway disk pulled by a D–8, land-planed the damn thing, put in your field, and you would have been in business. It was that simple, that easy."

Jim Boswell, though, wasn't so sure. He worried that the country wasn't stable enough politically, and just a few weeks later, his fears were borne out. Eight terrorists from the Black September cell of the Palestine Liberation Organization seized the Saudi embassy and killed three diplomats: two Americans and a Belgian. "Ah, Fisher," Boswell told him, "now I know what you wanted me over there for. You wanted me to get my ass shot off."

Back home, Boswell gobbled up still more land—though it wasn't so easy this time. In 1972 the company began negotiating to renew its lease at Buena Vista Lake with Henry Miller's heirs, when it discovered that competition had arisen from some of the descendants themselves: the Bowles brothers. They and another branch of the family, the Nickels, had fought bitterly over the spoils of their inheritance until, finally, fifty-three lawyers later, a San Francisco court had meted out Henry Miller's millions in the mid-1960s. More recently Henry Bowles and his sibling George had been arguing that Buena Vista wasn't being managed to its full potential. They believed that Boswell should have been farming sugar beets, melons and tomatoes instead of cotton. Boswell countered that the cropping plan was the product of "exhaustive studies," and that Buena Vista had been highly profitable over the past fifteen years for both Boswell as lessee and the Bowles-Nickels group as lessor. It was "inadvisable," Boswell asserted, "to change for the sake of change." Yet suddenly the Bowles boys were proposing that their own farming company take over the land. Boswell, they told their relatives, should be dumped.

In July a member of the family approached Jim Boswell to sort things out. As was typical, Boswell left little to chance, scripting the entire meeting in a memo to farming chief Les Doan and other senior executives. "Les," he wrote, "my opening question to you will be, 'Why do we have the present cropping plan. . . ?'" During the course of the day, Boswell instructed the troops, "we will emphasize accounting, computer programs, marketing and financial controls. . . . The Bowles brothers have apparently made a strong 'pitch' that they, with a newly formed nucleus of staff, can generate profits at least equal to ours. We know this is incorrect, but it's damn difficult to prove. Let's give it a try!"

And try they did, succeeding wildly. By October, the Miller heirs—including the Bowleses—had concluded that Boswell should not only keep farming Buena Vista, but they even agreed to give up ownership of the property. For the next year, Boswell worked to close the deal. One cousin, George Nickel Jr., also craved the land and put up stiff resistance. But Boswell won over the rest of the relations, and in the summer of 1973, Nickel was outvoted 7–1. "I think my family made a mistake," he said long after the merger agreement was completed. "Boswell knew people, though. He knew how to persuade them."

Six years after buying Buena Vista, the Boswell Company further beefed up its presence in the Bakersfield area, acquiring 18,500 acres around Kern Lake. The seller was Hollis Roberts, a six-foot-three, 250-pound son of the Dust Bowl whose dreams were as outsize as he was. Roberts had an aw-shucks demeanor that cloaked an impressive array of political contacts and the kind of fierce entrepreneurial drive that led him at one time to control 128,000 acres of land. Boswell, for his part, called Roberts "the best horse trader I've ever met."

Roberts had a little chapel attached to his office, "and after he made a deal," recalled Bakersfield attorney Bob Williams, "he took everybody in there, and they got down on their knees and prayed that it would all work out." But Roberts's prayers fell short. He got mixed up with a man named C. Arnholt Smith and hit hard times. Smith was a prominent San Diego financier who watched his U.S. National Bank collapse in 1973 and, six years later, stood convicted of state income tax evasion and embezzlement. Named in a series of lawsuits because of his ties to Smith, Roberts tumbled into bankruptcy.

By the early 1980s, when Jim Boswell found himself caught in the rain of the High Sierra, he stood as the biggest farmer not just in California, but in all of America. In Tulare Lake, only one other family—the Salyers—was even in the same league. And they had already begun to sow the seeds of their self-destruction.

Truce

THE GHOST OF HAROLD ICKES HAD BEEN SILENT for more than a decade when it rose up and threatened to rip the Boswell empire into hundreds of pieces.

No one saw it coming. A federal appeals court in San Francisco was now saying that the giant farms of Tulare Lake had to abide by the old reclamation law. They could be no bigger than 160 acres.

The New Deal–era fight over whether acreage limits should apply to Pine Flat Dam had been simmering all these years. Until now, though, things had gone the big growers' way. In 1963, after much dithering, the government had agreed to store water at Pine Flat and, meanwhile, leave it to the courts to decide the knotty question of whether all the farms served by the project should be capped at a quarter square mile. Then, nine long years after that case was first filed, a ruling finally came down. From his chambers at the federal courthouse in Fresno, Judge Myron Crocker had decreed that "reclamation law has no application to lands within the Kings River" area.

The Colonel couldn't have said it any better himself. The Bureau of Reclamation's thirty-year attempt to break up Boswell and Salyer and the other agri-giants of Tulare Lake had been soundly rebuffed. The farmers had won the suit—or at least the opening round—by hammering away on two basic points. Congress, their legion of lawyers had contended, never meant to apply the 160-acre limit to a flood control facility like Pine Flat Dam. What's more, they argued, even if that had been the intent, it didn't matter. The growers had already reimbursed the U.S. Treasury for that minor portion of the dam allocated to storage and irrigation benefits. And that payment, in and of itself, should have removed the manacles of reclamation law, they said.

That the farmers could buy their way out of acreage limits—in the case of Pine Flat, for an interest-free total of $14.25 million—had long been a contentious proposition. Through the Truman, Eisenhower and Kennedy administrations, officials at the highest levels had debated whether or not the policy was sound, flitting back and forth between blessing and denunciation. The lawyers for the growers, of course, highlighted the former. The extensive paper trail they presented to the court included, for instance, a 1952 memo from Reclamation Commissioner Michael Straus, who went so far as to call the payout principle "established departmental policy." Straus—the architect of "technical compliance," the nod-and-wink approach to reclamation law—had added that "it would be unfair" to keep negotiating with the Kings River farmers if their upfront payment wouldn't obviate the 160-acre rule. In July 1957, Interior Secretary Fred Seaton assured those on the Kings that "the Department continues to recognize and support the basic concept of reclamation law that full and final payment . . . ends the applicability of the acreage limitation."

Armed with such compelling documentation, the farmers had been confident that they would prevail in their "test case" against the bureau. And Judge Crocker hadn't disappointed. By paying back the government, he had ruled in March 1972, the farmers were "relieved of any . . . restriction" as to the size of their holdings.

Now suddenly, four years later, came the stunning news: The Ninth Circuit Court of Appeals found that Judge Crocker was totally in error. According to the court, Ickes, Roosevelt and land reformers like Paul Taylor had been right all along.

Reinterpreting everything from the history of the 1944 Flood Control Act to the Interior Department's policy regarding lump-sum payments, the appeals court decided that Congress had intended for the 160-acre limit to apply on the Kings. And even a $14-million-plus recompense from the growers wasn't going to change that. "The Kings River project is subject to the reclamation laws," Judge James Browning wrote in an opinion bursting with footnotes. Those drawing water from Pine Flat must "sell their excess land"—in Boswell's case, more than 88,000 acres irrigated by the Kings—or else stop receiving releases from the dam. It was the judicial equivalent of a sucker punch.

Just a month before, at a black-tie dinner celebrating the company's fiftieth anniversary, Jim Boswell had cheerfully thanked, among others, his corporate lawyers. "They keep us out of trouble," he told his seventy-five guests, as they feasted on beef Wellington and asparagus hollandaise. Now reeling from the April 1976 appeals court decision, Boswell took it upon himself to find the best attorney he could to appeal immediately to the Supreme Court. He settled on Simon H. Rifkind, a former federal judge who quit the bench in 1950 to take up a high-profile practice that included representing Jacqueline Kennedy Onassis and, undoubtedly more thrilling to Boswell, serving as a special master for the high court in a Colorado River water dispute. To bring Rifkind up to speed, Boswell hand-delivered to him at his New York office a slew of maps and charts and books, including a harangue against the Bureau of Reclamation by the late Senator Sheridan Downey, *They Would Rule the Valley*. Boswell also brought with him a company-made film that had been submitted as evidence in the Tulare Lake case: *Man, Land and Water.*

"The outcome of this case," Boswell reminded Rifkind, "will affect the lives and livelihoods of hundreds of people. Obviously, assistance to you has the highest priority for all of us."

Although those sentiments were surely true, some growers on the Kings were advocating an altogether different tack. They wanted the fight moved outside the courtroom. Their notion was to head right back to Congress; let the House and Senate pass a new law clarifying, once and for all, that the Bureau of Reclamation's 160-acre ceiling had no bearing on an Army Corps dam on the Kings River. Jim Boswell had been leery of pursuing this line in the past, and Rifkind assured him that the Supreme Court was bound to reverse the Ninth Circuit. The way the lawyer talked about it, the case was a slam dunk. "I hate to take your money," he told Boswell.

Still, with all he had built now on the line, Boswell wasn't inclined to buy into any lawyer's bluster. So when Fred Salyer requested a meeting at the Burbank airport to kick around a political strategy to go along with the legal one, Boswell decided to hear him out. He and Jim Fisher arrived in Burbank not quite knowing what to expect from their long-time rival in the lake bottom. Judge Browning's opinion had left the Salyers, who had more than 30,000 "excess acres," in their own tight spot. Maybe Fred Salyer's desire to meet was sincere and the two sides could work as a team to save their empires. But so much enmity had passed between the Boswells and Salyers, so much sourness had stuck in the craw since Jim and Fred were childhood pals fished out of a water tank by old Cockeye, that even meeting at an airport terminal was a leap of faith. The Salyers were known as the most disputatious clan in Kings County. They couldn't even see eye to eye among themselves.

EXCEPT FOR THE TIMES when Cockeye would get raging drunk on bourbon and drive his Cadillac into some irrigation ditch, headlights bobbing in the black water, the Salyers did a pretty good job of keeping a lid on things. Their bent toward privacy bordered on the pathological. Fred Salyer, for one, regarded the news media as "the worst enemy of the United States." He might have liked reading history books, but he saw no value in preserving his family's story. He was plenty proud of what he and his father and brother had built, but he was the last guy to brag about it. Like Jim Boswell, he intended to die with Salyer history on his pillow.

It began with Clarence and Cordie Salyer boarding the Santa Fe in southwestern Virginia and heading to Corcoran in 1918. They chose their destination on the word of Clarence's father, who had made the trip west in the late 1880s and never forgot the land of the lakes. They arrived in early December with "little money, no land, no job."

"The list of problems are interminable that have tested the ingenuity of the Salyer Family," read the company's ad in a 1964 edition of the *Corcoran Journal*. "Too much water, not enough water, a depression, a war, weather, prices, taxes, insects, rising costs, governmental interference and inadequate financing are a few."

If you knew anything about the Salyers, the list of obstacles was one glaring obstacle short. Weighing as heavily on the family as water or weather or depression or Uncle Sam's meddling was a feud that pitted Clarence against his two boys, Everette and Fred. By 1964 the only thing keeping the Salyers from an all-out court fight was a bizarre, twenty-page legal truce called "Operation Armistice." The compact was a long list of do's and don'ts and vows not to steal company records and take every step possible to keep the rancor secret and preserve détente. And yet it had never stood on shakier ground. The Salyers were one material breach from war.

"Termination of the Salyer Family Compact would be a grievous mistake," Walter M. Gleason, the attorney who brokered the truce, warned both sides. "No one will gain or win in such a terrible family war. To the contrary, all members of the family will suffer very real losses. The casualties, financial and personal, will be heavy and tragic."

The last thing Corcoran wanted to see was Cockeye and his boys go down by their own hand. If the Boswells were half-ass aristocrats who commanded the respect of townsfolk, the Salyers were loved for their beautiful foibles. They were pure hillbilly, and proud of it. "Blue blood? My God, there is no blue blood in us. We're a coal mining family," Linda Salyer, Fred's daughter, said. "We're right out of the hills. The hills of Virginia."

Everette was the firstborn, his father all over again except for two good eyes. He was a plain man, his closet full of the same tan trousers and white shirts, and he handled all the financing and dealt with the commodity brokers. You didn't have to know Everette very long to understand that he had an incredibly nimble mind. "Everette was one of the most brilliant men I ever ran into," said Thomas Keister Greer, the Salyer brothers' longtime attorney. "Fred had no deficiencies in his IQ either, but his judgment wasn't equal to Everette's."

Fred was one the valley's best pilots, an expert mechanic who could fix even the radar systems on his planes. He ran the farm with every bit the precision of Boswell. It wasn't enough that his fields were free of weeds. The irrigation canals had to be razor clean, too. The bales of cotton that Salyer shipped were pristine as well, carrying a tag that read: "Don't You Think I'm Pretty? At Salyer Land Company's gin in Corcoran, Calif., they pampered and cuddled me so I would look neat and clean. Please don't tear my dress with hooks and do keep me out of grease and dirt." Fred traced his fussiness to his mother, Cordie, a plump, stately woman who taught him that there were two kinds of dirt on a kid. Today's dirt was clean dirt. Yesterday's dirt was dirty dirt. She didn't tolerate any dirty dirt. She made Everette and Fred and their little sister, Virginia, take a bath every night so they didn't wake up the next morning dirty. Of the three, Fred was the most dogmatic. His nasally voice became indignant whenever he took aim at his favorite target: the pinkos in Washington who were bringing down the country by sticking their noses in places that were none of their business. "Fred makes John Birch look like a liberal," Jim Boswell said.

The brothers ran the company the best they could with a father who held most of the stock and let them know it each time they dared a new idea. If Clarence Salyer entertained any notion of one day handing over the company to his two boys, he hadn't bothered to tell them. And they hadn't bothered to ask. Ever since they were kids, they had walked in complete fear of him. Fred was there the day his father came home from the labor troubles in Pixley in 1933 with blood on his hands and a .38 Special that needed melting. He grew up wondering if the murder that day was his father's first or his last. "My father was a very violent and dangerous person," he would say years later. "If someone crossed him, goddamn he would hurt you. He had no rules. His weapon of last resort was his fists. Whatever was handy he used."

There was the time they were heading down the mountain in their Studebaker, Clarence and Cordie up front and the three kids standing up in the back. On a bend in the road, Clarence caught sight of a man waiting with a pistol in one hand and a flashlight in the other. The guy was up to no good, he figured, possibly a highway robber, and he punched the gas and went straight for him. "Dad was a fast thinker. He didn't hesitate. He tried to run him over and barely missed. And mother had a fit. 'Clarence, Clarence. You almost ran him over.' He looked over at her and said, 'What in the hell do you think I was trying to do?'"

There was also the time one of his workers, a half-wit named Chester, forgot to screw in the plug after changing the oil on a Caterpillar tractor. Later that day, the machine froze up in the fields, and a small investigation turned up the plug under a tree outside the blacksmith shop. Clarence fired Chester on the spot and assumed their dealings were done. But a few days later, Chester knocked on the front door of the Salyer estate, looking for a last check owed him. Chester was a good-size man, and Clarence happened to be injured and walking on crutches at the time. "You tore up that sonofabitch Cat," he yelled. "Hell if I'm going to pay you." Chester kept insisting, and Clarence picked up one of the boy's BB guns and smashed him over the head. He dropped like a side of beef.

"I hadn't seen a dead person before but he sure looked like it," Fred would recall. "There was blood everywhere, all over the carpet, and the BB gun was still in my father's hands, bent with its springs coming out. My mother came in the room screaming, 'Clarence, Clarence, you killed Chester.' It turned out he wasn't dead, but dad gave the same answer: 'What do you think I was trying to do?'"

To his grandkids, he was just one big mischief maker. Even when he managed to drive drunk and not land in a ditch, it was an adventure.

"We would be heading down Highway 43 and all of a sudden his false teeth would start bothering him," Linda Salyer remembered. "He'd pull them out of his mouth and toss 'em out the window. We'd watch them shatter along the road and start roaring with laughter. He did this all the time, and the dentist kept having to fix him up a new pair. I thought it was the funniest thing I'd ever seen. I didn't know at the time I was laughing at a drunk man."

No cop or sheriff in Kings County had the brass to arrest Clarence. If an officer caught him driving erratically, he'd take one look at his driver's license and chauffeur him back to the Home Ranch. The Salyer compound of emerald green lawns and fine ranch houses sat wedged between the family's towering grain silos and its 7,001-foot runway, big enough to land a 747. Clarence had come to expect a safe pass from the cops and fumed if they didn't personally deliver his Cadillac coupe, too. Why the hell was he paying Sergeant Ely $2,000 a month if he couldn't get a favor or two? The way Clarence saw it, the kid gloves from the sheriff and county inspectors befitted a cattle and farm baron who commanded 30,000 acres and directed a powerful water agency and the chamber of commerce.

No one could accuse Clarence of not giving back to the community, as he donated his time to the Boy Scouts and his prized rams to the high school kids at the 4-H club. Each October, on the eve of the harvest, the town's biggest shindig, the "Cotton Under the Stars" gala, was held on the parklike grounds of the Salyer cabana. The steaks were big and the bartenders poured with the unhinged hand of someone else's booze, and you could always count on a spectacle or two floating in the Salyer pool. At one gala, three flamingoes made of cotton glided across the water on a grass raft. Another gala saw Salyer launch a fifteen-foot scale model of a Mississippi riverboat. The event, which benefited Valley Children's Hospital in Fresno, brought out the best in Clarence. He charmed the ladies and held court over the men, and by the end of the evening, if he had thrown back one too many shots of Early Times under the eucalyptus trees, the trip home was a mere walk across a dirt path.

Where Clarence risked trouble were the nights he crossed over onto the turf of the California Highway Patrol, whose officers weren't nearly as solicitous as Ely's boys. He found out just how diligent the CHP could be a few nights after Christmas 1960 when two officers stopped him on the outskirts of Hanford. This is where his mistress, Gertie Smith, lived

in a house surrounded by a brick wall that Clarence, in piques of jeal-
ousy, liked to crash through. On this night, he refused to get out of the
car, refused to take a sobriety test, refused to provide his license or date
of birth. "I'm the biggest taxpayer in Kings County," he growled, words
slurred. He was dressed in overalls, and the work shirt pulled tight over
his fat belly smelled of booze and cigars. "If you want a fight, I'll give you
a good one right here." On the way to jail, Salyer kept muttering that
Governor Edmund "Pat" Brown, his close friend, would hear about it in
the morning. "You boys are history."

Whether Salyer made good on his threat after being bailed out of the
drunk tank a few hours later isn't clear. It is a good bet, though, that
Governor Brown did hear about it in the morning, and the backslapping
Irish pol was known to pull all sorts of strings for his buddies, especially
ones who raised the kind of dough that Salyer raised. On campaign missions to the valley, the governor liked to bring along his wife, Bernice, a
bossy woman whose constant demands annoyed Fred Salyer to no end.
It was Fred who flew the governor and first lady from farm spread to
farm spread collecting cash money—bags of it. Only one time did Fred
protest, telling his father he had enough of that "son of a bitch and his
pushy wife" and wasn't about to fire up the plane on a moment's notice
and pick up Clarence and the Browns in Sacramento. "I don't give a shit
what you think," his father shouted back into the phone. "That son of a
bitch is our son of a bitch. Now get your ass up here and do it."

Brown named Clarence to the State Board of Agriculture and gave
him an open invitation to barge into his Capitol office whenever something irked him, which proved far too often. The two men, to no one's
surprise, had a falling out over a water project in the early 1960s. Years
later, Brown fondly recalled the easy access Clarence enjoyed to his office, even if it meant sitting on the receiving end of one of his famous
tirades.

This willingness to forgive Clarence for his savage temper and rascal
ways was almost universal back home in Kings County. Every farmer,
tractor dealer and chemical salesman, it seemed, could tell a story or
two of how Clarence passed a bad check or ripped him off in a handshake deal or defrauded Uncle Sam by claiming crop damage for a crop
that never existed or extracted a transportation fee for government
grain that, in reality, never left his possession. The general attitude was
that if Clarence managed to get one by them, well, they only had them-

selves to blame. Everyone knew he had the guts of a bandit and pulled his best tricks on Sunday when people were soft from church or too much shuteye.

Tom Cherry, who operated a cotton gin up the road in Fresno, counted himself among the victims. Salyer used the old "damn-I-forgot-my-checkbook" trick to cheat Cherry in a tractor sale. They had agreed to the $6,000 price, and one of Salyer's men already was heading home with the Caterpillar when the boss began fumbling around for his checkbook. "I'll be back," Salyer promised Cherry. When he returned— three months later—he had a screwy smile on his face and a check for three grand. "He asked me, 'Isn't that what we agreed to?' And I said, 'No, Clarence.'" Salyer wouldn't budge, and Cherry finally caved in. "What was I to do? I thought, 'My God, I've been taken by that guy, too.'"

By the mid-1960s, Clarence's behavior on the home front was threatening to undo the fragile peace of Operation Armistice. He was now shacking up with Gertie Smith, splitting time between her house in Hanford and the Home Ranch in Corcoran. Cordie tried to hold her head high, but it was a scandal. Because the names "Gertie" and "Cordie" sounded so much alike, townsfolk greeting Mrs. Salyer would sometimes slip and call her by the name of her husband's lover. Lenders had grown tired of Clarence writing bad checks, his accounts drawn down to nothing because he had purchased another piece of land. He may have been a man of exceeding vision—the first grower, for instance, to devise a mobile fix-it shop so his machines could be repaired right in the fields. But as far as the banks were concerned, the line between Salyer innovation and Salyer mental illness had never been so thin. Credit would no longer be extended to Clarence's manic whims. The banks were shutting him off, agreeing to finance the company only if Everette and Fred were the ones running the show.

This drove Clarence crazy. He grabbed a shotgun and threatened to kill any employee no longer willing to take orders from him. Workers were caught between Clarence's hair trigger and Everette's pledge that anyone who obeyed his father would be fired. In the course of one argument, Clarence knocked down attorney T. Keister Greer, a six-foot-two-inch former marine, not once but three times. Clarence took a swing at Fred and mocked Everette for having a heart attack, vowing to induce a second coronary that would kill his oldest son. In the name of peace, Everette and Fred pushed the idea of buying 7,300 acres of farmland in the Sacramento delta, 200 miles to the north, and essentially exiling their father. Clarence called their bluff. He'd go to Victoria Island but not without his mistress and not without the standing threat that he would soon return to Corcoran to divorce his wife and move back into the Home Ranch with Gertie.

Operation Armistice now hinged on the hope that Clarence would busy himself with improving the new ranch and planting asparagus. Once he had a stake in Victoria Island, it was thought, he'd agree to part with all his shares in the company for clear title to the delta land. But Clarence had another idea. If his sons were so eager to disown him, going so far as to buy him a $3.5 million island, he'd stick them with the interest payments—hundreds of thousands of dollars. And if they were counting on him to be grateful and hand over the company he had built with his own shovel, they were the worst kind of pigeons. He wanted his island retreat and he wanted a voice in his company back in Corcoran, too. His daughter, Virginia, who controlled enough stock to make it difficult for her brothers, stood firmly on the side of her father. "Tell those two sons of mine to sue me," Clarence told one go-between. "They ain't got the guts."

THIS WAS THE SALYER CLAN that in the spring of 1976 was proposing to make peace with Boswell and join forces to wage a legal and political fight to keep the federal government from enforcing acreage limits in the lake bottom.

The Boswells and Salyers were hardly the second coming of the Hatfields and McCoys. Mom Kate and Cortie were bridge partners, and their children were playmates through grade school. When Cockeye needed money to expand his grain and cotton plantings in the early days, it was the Boswells who seeded him. But cordial relations between the two families ended, so to speak, where the levees began.

Before the flood of 1952, Cockeye had split the lake in half with his huge dirt dike. When the waters struck and Salyer stayed dry, Jim Boswell hopped into a rowboat and tried to blow up the levee. In 1963, when the Kings River farmers came together to wage an early battle against acreage limits, Boswell and the Salyers bickered over which attorney would best represent the farmers. Boswell pushed ahead with its choice, and the Salyers threatened to sue for $150 million. "We have no interest in bluff," Everette warned the Boswell board. "This situation is too serious for bluff."

The two families fought over the plumbing system that delivered water into Tulare Lake, and they fought over how to get rid of the water when it became too much. The most bitter of all these battles took place during the 1969 flood, when the dams upriver went from empty to full in twenty-four hours, gushing water onto grain fields and cotton land. Boswell and the Salyers adhered to the Golden Rule of Floods: Every farmer for himself. In the fight to keep dry, Jim Boswell even wiped out his brother, Billy, in '69. "His land was up on the Kaweah channel, and when the water was coming down there it blew out on him," Boswell said. "He came over looking for me and said, 'You flooded me out.' I said, 'Well yeah, my levee's higher than yours. I guess I did.'"

Even so, just as in war, there were certain conventions that no one flouted. One of these, at least in the eyes of the Salyers, was that flood-water from the Kern River should be considered different. Its rightful destination wasn't Tulare Lake but Buena Vista Lake to the south. In past floods, considerable attempts had been made to divert the Kern into Buena Vista. But in the flood of '69, Buena Vista Lake happened to be farmed by the Boswell Company. The last thing it wanted to see was a second basin planted with its crops go under.

Where the water should go had the makings of a titanic battle, and no one itched for war more than Keister Greer, the Salyer attorney who despised Jim Boswell. Greer was a transcontinental commuter who split his time between his big client in Corcoran and his home in Rocky Mount, Virginia, where his family was one of the first to settle in Franklin County, the moonshine capital of America. Greer was a big man with lots of hair and a handsome pink Celtic face, and he wore only London-tailored suits with a gardenia in his lapel and a bow tie. His critics were certain he had fallen in love with the regal pitch of his voice way back as a college debate champion. He carried himself not as a mere lawyer but as a barrister. "Sometimes in error but never in doubt" is how Stan Barnes, Boswell's water man, described Greer. "He was the goddamndest guy, stirring the pot every time he got a chance," said Jim Fisher.

With Greer leading the charge in '69, the Salyers tried to win support from the Tulare Lake Basin Water Storage District to divert the flow of

the Kern River into Buena Vista Lake. But the Boswell Company controlled the district's board with its six members, and the Salyer plan was
tabled, 6–4. The Salyers surprised no one with their next move. They
sued the water district, seeking to unseat the Boswell directors on the
grounds that the board's makeup was unconstitutional. Greer argued
that because only landowners were allowed to vote in water district
elections—and the largest landowner enjoyed the most votes—the
process violated the equal protection clause of the Fourteenth Amendment. He demanded a resolution based on that bedrock tenet of
democracy: one man, one vote.

Four years later, the case made it to the U.S. Supreme Court. In a 6–3
decision delivered by Justice William Rehnquist, Boswell prevailed. A
water district, Rehnquist wrote, has a "special limited purpose" and
therefore merits an exception to the one-man, one-vote standard. As for
Boswell having more voting power than any other farmer in the district,
Rehnquist took note of the fact that the company was also assessed
more for construction projects and the like. With weighted voting, he
said, the burdens equaled the benefits.

Unconvinced, Justice William O. Douglas issued a blistering dissent.
"The result," he declared, "is a corporate political kingdom undreamed
of by those who wrote our Constitution." Years later, the case would still
vex him. In his autobiography, Douglas recalled *Salyer Land Co. v. Tulare
Lake Basin Water Storage District* as one of the few opinions of that era
"that had the flavor of the philosophy of the robber barons. . . . This was
not a decision in the dim past of the 1880s and 1890s when the vicious
powers of the corporate world may not have been fully appreciated; it
was a decision handed down in a year of enlightenment—1973."

WITH THE ONE-MAN, ONE-VOTE DECISION fresh in every-
one's minds, Everette and Fred Salyer met Jim Boswell and Jim Fisher
around some benches outside the Burbank airport terminal. Fred Salyer
did most of the talking, as Fisher would remember it. His message was
straightforward: It was time for a truce between Boswell and Salyer—a
real, honest-to-goodness truce. With the government poised to bust up
the two biggest empires in Tulare Lake, the stakes couldn't be any higher.

The Salyers explained that Greer was no longer on the case and they had already lined up a top-notch lobbyist in Washington: John Harmer, a smooth actor who a couple of years earlier had been California's lieutenant governor under Ronald Reagan. Jim Boswell hated politics and politicians and, with attorney Rifkind on his side, he was more convinced than ever that he could still gain victory in the courts. But even he had to admit that another line of attack couldn't hurt. So he went along with the Salyers' plan, pledging to share the costs of an all-out lobbying effort. From that moment on, the rapport between the two companies improved markedly, years of bad blood swept aside by mutual self-interest.

By the summer of 1977, it became apparent that a hired gun like Harmer couldn't be expected to take the lead on so critical an assignment. Somebody from inside the company would have to commit to flying across the country on a regular basis, making a second home at the Hyatt on Capitol Hill.

"Well, you're going to volunteer," Boswell told Fisher.

"Hell, I'm a farmer," Fisher replied. "I don't know anything about politics."

Boswell wouldn't budge. "You're going to do it," he said. "They don't pay me enough to go to Washington."

For a greenhorn lobbyist like Jim Fisher, there couldn't have been a tougher time to descend on official Washington and ask that a giant agribusiness operation like J. G. Boswell be cut a break.

Jimmy Carter, who from his first days in office had demonstrated the greatest disdain for big water projects, occupied the White House. The Democrats, many of them reform-minded Watergate babies, controlled both houses of Congress. And the man at the helm of the Interior Department, Cecil Andrus, was as steadfast a supporter of acreage limits as anybody in the post had been since Harold Ickes himself. Though he had started out as an Idaho lumberjack—"a lunch-bucket guy," in his words—Andrus had always shown a strong conservationist bent. And as governor, he had never been afraid to stand up to the loggers, miners, ranchers and irrigation farmers that most pols kowtowed to. Besides, the way Andrus looked at things, he didn't really have any choice but to carry out the 160-acre law with the utmost vigor.

In 1976, just before Carter was elected, a California-based grassroots group called National Land for People had sued the government, arguing that the Interior Department had been illegally ignoring the reclamation law and its provisions promoting the "family farm." The group, which claimed about 1,100 members, was run by George Ballis, a hardedged former AFL-CIO newspaper editor in Fresno who had worked on Bernie Sisk's congressional campaign in the mid-1950s, only to watch in dismay as some of the area's biggest farmers contributed large sums in support. "What do these guys want?" a puzzled Ballis innocently asked in his paper, the *Valley Labor Citizen*. The answer came via a letter from Paul Taylor, whom Ballis had never heard of (though he discovered later that he was taking a photography seminar taught by Taylor's wife, Dorothea Lange). Taylor revealed that the farmers wanted what they always wanted: "to spring themselves loose from the 160-acre limitation." He sent a suggested reading list to Ballis so he could get a better grasp of the issue, and Ballis had been a crusader ever since.

The court ruled in favor of National Land for People, and Andrus felt duty bound to write up a new set of regulations that finally gave the old 1902 law some real teeth. "What disturbs me most about the current debate," Andrus told a western banker, "is the smokescreen by those who contend the proposed regulations are some new scheme of land confiscation and a threat to free enterprise and sacred property rights.

"Certainly there are tremendous problems in trying to enforce evenly a law which has had only spotty enforcement over the past seventy-five years," he added. "But a federal district court has ordered us to arrive at uniform regulations and to enforce them. Some people seem to think I should obey the letter of the court order while figuring out a way to thwart the spirit of the law. That I cannot do."

In February 1977, just weeks into the Carter presidency, things got even worse for Boswell. The Supreme Court declined to hear the Tulare Lake case, letting stand the appeals court ruling that the 160-acre limit should be placed on the lands fed by Pine Flat Dam. Simon Rifkind's promise that the farmers would win before the Supreme Court—no problem—now seemed perfectly foolish, while the land-reform gang was giddy. "This is fantastic," Ballis crowed.

Publicly, Boswell executives were full of bluster, asserting that they'd never allow their empire to be broken up by government fiat. "We have no thoughts of doing anything like that," Jim Fisher declared. Privately, though, Boswell's men fretted that a breakup of some kind might be unavoidable. When a Boswell ally walked into the Corcoran office of company attorney Jim McCain right after the Supreme Court decision came down, he found the old lawyer sitting there, grim-faced, contemplating the death of the empire. "Well," said McCain, "I guess you've come to help me bury the corpse."

The company's water chief, Stan Barnes, worked up a number of possible moves to deal with the Supreme Court setback, including "an orderly land sale." Jim Boswell termed this a "last-resort step" and seemed more inclined, should it come down to it, to forgo the benefits of the dam and let the Kings River run wild again, just as it had in the Colonel's day. Fred Salyer agreed. "There is no power in heaven that will cause me to break up that land," he said. "They can just leave the gates open and let the water run through."

Of all the options that Boswell and Salyer now had before them, the most practical was to press harder than ever for a political solution—to convince Congress to pass, and the president to sign, a law that would effectively reverse the judiciary. The onus on Jim Fisher was greater than ever.

If he was fazed by any of this, he never let on. He plunged into his assignment in Washington with the same unflinching attitude that he had displayed planting cotton in the Australian outback. The fertilizer of choice in the nation's capital, he quickly realized, was money. In the coming months, Fisher set up a political action committee—BosPAC—and began showering candidates with campaign dollars. Boswell doled out more than $150,000 over the next few years, the vast portion of that going to Republicans. Such largesse didn't buy any votes, Fisher reasoned. The system, crude as it is, didn't work like that. What money buys is access, and for Fisher that translated into black-tie fund-raisers and well-lubricated lunches with lawmakers and their aides.

The path was also greased by the energetic presence of Fisher's lobbying partner, John Penn Lee. The Salyer family lawyer (and Fred Salyer's son-in-law) was, like Fisher, a real bulldog. But Lee had a wild manner that in many respects made him the antithesis of the no-nonsense Boswell executive. A Virginia native, he was fond of folksy sayings, like when he urged others to keep their point simple: "Let's get the hay down where the goats can eat it." He traced his roots on one side of his family to Confederate General Robert E. Lee and on the other to John Penn, a signer of the Declaration of Independence from North Carolina. Such an esteemed lineage tends to make a man one of two ways: overachieving or given to utter frivolity, as if his one goal in life is to make his thoroughly accomplished kin spin in their graves. Lee picked the second course, and he spent his college years playing football, guzzling beer at the frat house and chasing skirts.

He might have been a career dilettante had he not wormed his way into law school at the University of Virginia and eventually become captivated by an issue that, by all rights, should have bored the hell out of such a freewheeling spirit: corporate taxation. It was on a tax case, in fact, that Lee had met the Salyers. Tom Greer, an old family friend, had brought him in on the matter, and Lee promptly earned his keep, fighting off a huge IRS assessment and even wangling a nice refund for Fred and Everette. Impressed, they asked Lee to handle more of their legal work, and he soon found himself diving into that most arcane of subjects, water law, and falling headlong in love with Linda Salyer.

Lee and Fisher spent a good part of 1977 jetting between California and Washington, making themselves regulars in the corridors of the Capitol. The broad-shouldered Lee, in particular, became almost a Gucci Gulch caricature, sporting a Captain Kangaroo mustache and loud ties, and devouring the biggest lobsters that the Palm served. "We were all over the town," Lee would later remember, "and we always had two or three politicians in tow."

Their argument—that the appeals court had made a grave error, misreading the intent of Congress—took more reflection than most lawmakers were inclined to engage in, even over a long, expensive dinner. So they offered up a more emotional appeal as well: Boswell and Salyer might not have been classic "family farms," but the people who worked there were all breadwinners in their own "farm families." Was one of their ranch managers any farther from the soil than somebody on an eighty-acre plot in the Midwest? Should they be penalized for prosperity? Was that the American way?

The duo could be audacious at times. In the summer of '77, John Penn Lee sat in the office of Bert Lance, the White House budget director, and offered to solve the impasse in a single stroke: Salyer and Boswell would simply buy Pine Flat Dam. "We'll just write you a goddamn check for the thing," Lee told him. After all, what were tens of millions of dollars to the biggest farmers in America? Lance was stunned. The proposal, not surprisingly, went nowhere.

As the months rolled on, Boswell and Salyer built up their lobbying team. In addition to John Harmer, now on a $10,000-a-month retainer, they hired a Democratic operative named Irv Kipnes at the same price. Also put on the payroll was Wes McAden, a slow-talkin' Virginian who, though a Republican, prided himself on maintaining good relations with both sides of the aisle—"just like the whore in Venice," he'd say, who "learned to work both sides of the canal."

By October, the group looked to be on the verge of success. Harmer had persuaded Republican Senator Malcolm Wallop of Wyoming to partake in a little subterfuge. He slipped into an unrelated bill an amendment granting the Tulare Lake area an exemption to the 160-acre rule, thereby overriding the appeals court decision. It was a sneaky way to lay to rest forty years of difficult history "very quickly and easily," as Kipnes described it.

A couple of weeks before the measure was set to be considered by the full Senate, however, the ploy was exposed. A senior staff member of the committee handling the bill complained that Harmer had been "less than candid" with him in pitching the amendment. The staffer said he'd been led to believe that California Senator Alan Cranston and the local congressman for Tulare Lake, John Krebs, supported the legislation. Actually, both of them opposed it. A month before, Harmer had even asked Cranston, who as Democratic whip was one of the most powerful men in the Senate, if he'd offer the amendment himself. Cranston had refused.

Now the senator was incensed at the double-dealing. "How can we conceivably justify reversing the court's decision with an impromptu congressional" action, he asked, especially "when Congress hasn't had even as much as one day's hearing into the matter?" Cranston would ultimately engineer the amendment's defeat. As for Krebs, he said he was keeping "an open mind" on the subject. But he, too, wasn't willing to back an outright exemption at this stage—a stance that would invite the wrath of John Penn Lee and Jim Fisher.

John Krebs wasn't exactly Fisher and Lee's kind of guy anyway. Born in Berlin in 1926, Krebs grew up in a Jewish household in Tel Aviv—and still had the accent to prove it. He had made his way to Fresno after law school and become active in local politics, eventually winning a seat on the county board of supervisors. Elected to Congress in the wake of Watergate, he was something of a political oddity. Most Central Valley Democrats are conservatives like the "yellow dogs" of the South, and they survive by adhering to a single axiom: Do whatever the big farmers say. Krebs had moved a bit to the right over the years to accommodate his district, but his style and true beliefs still seemed more like those of a San Francisco liberal. For one thing, he wasn't scared to question agricultural interests. Early on, the Salyers had tried to cozy up to him, but Krebs returned their campaign check, avoiding even the slightest intimation that he could be bought off as he continued to ponder the acreage-limit issue. "I'm sure," he said, "that's the first time an elected official ever sent them back money."

Lee had seen enough. It was high time, he was convinced, for Boswell and Salyer to get their own man in office. They would go after Krebs, as Lee put it, "hammer and tong."

Wes McAden, who had been lobbying in Washington since 1965, didn't appreciate that kind of bullying. Openly challenging a sitting congressman can be a perilous game, he knew, and he counseled a more cautious approach. Lee and Fisher "got to thinking they were pretty hot stuff," McAden said. "In the South, we call it 'showing your ass,' letting people know you've got a little power. Well, you don't show your ass like that." McAden, though, was overruled, and Lee and Fisher charged after Krebs.

The candidate they backed for the 1978 election, Charles "Chip" Pashayan, was a Rhodes scholar who had worked for the Department of Health, Education and Welfare in Washington, but he had no experience as a politician. What he did have going for him was plenty of Boswell and Salyer money—more than 15 percent of what Pashayan raised came straight out of the two companies' camps—and an opponent who was stumbling on a number of fronts. Besides angering the biggest agribusinesses in his district, Krebs also became embroiled in a controversy over a proposed ski development known as Mineral King and wound up in a scrap with the carpenters union.

It was all too much for the incumbent. Chip Pashayan, a decided underdog when the race had started, took the November election by 13,000 votes. Jim Boswell and his boys now had what they wanted in Washington: a representative who, when it came to the topic of water, was more than happy to carry theirs.

IN EARLY 1979, John Penn Lee was downing another king-size lobster at the Palm when he bumped into a friend from his Virginia law days, Donald Santarelli. He, too, would be invited to join the Kings River lobbying team.

Santarelli knew nothing about farming, and it was difficult to imagine him going anywhere near a cotton field. A chic dresser, he spoke fluent Italian, and when he ordered pasta tossed with just the right truffle oil, his slim body would writhe and he'd moan orgasmically as he talked about the food making love to his mouth. Yet Lee hired Santarelli because he knew that his flamboyance was matched by one thing: his intellect. Santarelli was a collegiate debating champion who, besides his law degree, had earned a master's in rhetoric. As a thirty-two-year-old Republican whiz kid, he had been tapped by President Nixon to run the Law Enforcement Assistance Administration, an agency that would be criticized for manhandling the Bill of Rights as it mounted a war on crime. Among the new weapons that Santarelli proposed to give cops: the power to kick down a drug suspect's door without warning, eavesdrop on doctors and clergymen, and search property not specifically mentioned in a warrant. He also cooked up the idea of automatic life sentences for criminals slapped with a third felony conviction; it was "three strikes and you're out" long before anybody had even coined the term. In June 1974, Santarelli was forced from the administration after telling a reporter, all too candidly, that he believed President Nixon should resign because of Watergate. Now, five years later, he was a Washington lawyer and lobbyist with a reputation for doing whatever it took to win. "God reinvented Machiavelli in Don Santarelli," said C. Michael Tarone, then his associate.

Santarelli's role was to work closely with Chip Pashayan, feeding the congressman the historical background and public policy pronouncements that he would need to make headway on the Hill. The fundamental point, reiterated in memo after memo: As a flood control dam built by the Army Corps of Engineers, Pine Flat was never a traditional reclamation project designed to deliver government-owned water over parched land, bringing it to bloom. Rather, the Kings River area was already rich in crops when Pine Flat was constructed, and its system of private water rights was long established.

These were reasonable arguments, persuasive even, but Santarelli knew that it would take a lot more than that to get the job done. So he also supplied strategic advice, telling the Kings River farmers that the official name of their lobbying organization—the Pine Flat Exemption

Committee—was terrible. "Exemption" was a dirty word in Washington, Santarelli explained, one that smacked of a special interest looking for favored treatment. From then on, the group would be known as the Kings River Water Users Committee.

Santarelli and Pashayan worked hard to bring together all the farmers along the Kings. They knew that Boswell and Salyer, because of their massive size, were easy targets for their political foes. But if they could convince the smaller farmers on the upper part of the river to also push for a reprieve from the acreage limit, it would blunt the perception that the Goliaths of Tulare Lake were just trying to roll over everybody. It wasn't a difficult sell. Pashayan found that even those growers with fewer than 160 acres were, by and large, a conservative bunch "viscerally opposed" to the government deciding what size a farm should be. The Kings, then, would stand united, helping Pashayan parry left-leaning lawmakers such as fellow Californians George Miller and Phil Burton. "It wasn't just Boswell and Salyer" who ought to have a pass on the 160-acre law, Pashayan would remind his colleagues at every opportunity. "The small farmers wanted" the very same thing.

The Boswell team also provoked fear where they could. If the Interior Department could apply the 160-acre limit to an Army Corps dam on the Kings, they asked, what was stopping officials from imposing it on a flood control project elsewhere, breaking up farms in Mississippi or Florida or Arkansas? It was a total red herring, and Secretary Andrus denounced as ridiculous the suggestion that reclamation law would ever be enforced outside the seventeen western states. Yet raising this fiction had the desired effect. Senators from the South, including Herman Talmadge of Georgia and Howell Heflin of Alabama, were suddenly aware of the issue and pointing out the need "for an exemption for flood-control projects throughout the 50 states." A narrow predicament for Central California had been magically transformed into a national problem.

Still, by the middle of 1979, Fisher, Lee and the rest recognized that there wasn't much chance anymore to remedy things swiftly. Clearing up the status of Army Corps dams was but one of the many items that had been lumped into the omnibus Reclamation Reform Act, a bill that promised to be the most extensive overhaul of the law since its enactment near the turn of the century. The Kings River farmers were no longer passengers on a single railroad car heading down the legislative track; they were now coupled to a whole big train that was chugging along slowly. Other cars were occupied by everybody from Wyoming cattlemen to Oregon potato growers, all with concerns about different aspects of the law: requirements for farmers to reside on their land, leasing restrictions, and the disposal of excess land by lottery, to name but a few. California's sprawling Westlands Irrigation District, where Boswell owned the 23,000-plus-acre Boston Ranch, had its own issues to deal with.

For Boswell and Salyer, the situation was now more complicated than ever. Should any one car get derailed, the whole Reform Act might grind to a halt. "Everybody who had something that needed to be fixed gradually came to understand that if each party only took care of its own interests, nobody was going to get anything," said Gordon Nelson, who ran an umbrella lobbying coalition called the Farm/Water Alliance.

As the Reform Act moved through the Senate, Santarelli zeroed in on Alan Cranston, determined to keep him from mucking things up as he had done with the Wallop amendment a couple of years before. If he couldn't win Cranston over, Santarelli figured that he could at least try "neutralizing him." Subtlety was not Santarelli's strong suit, and the tactic he came up with was, as he himself would later describe it, remarkably unrefined. He and Lee produced a campaign ad that they slipped to Cranston by way of a public relations firm. "We were like two Katzenjammer Kids poring over this mock-up," Santarelli would remember. "It wasn't very sophisticated. It didn't have a lot of facts." What it did have was a bold declaration framed by a cornucopia of fruits and vegetables: "CRANSTON AGAINST FARMERS." The warning was clear: Boswell and Salyer would use their war chests to run the ad in every newspaper in the valley come the next election, unless Cranston rethought his position on Tulare Lake.

Whether Cranston was spooked by the ad, even a little bit, is doubtful. But it did underscore the big farmers' zeal when it came to avoiding acreage restrictions. And he couldn't afford to pay them no heed; Boswell and Salyer were his constituents, after all. The tricky thing for Cranston was that the other side was also thumping him. Interior officials told Cranston's staff that letting the Boswell Company have its way would be a "blatant circumvention of the reclamation program." Paul Taylor was applying pressure as well, citing the proposed preservation of "giant landholdings in Tulare Lake" as one of many failings of the legislation now moving through the Senate.

"Senator, you have a reputation as a liberal Democrat and founder of the liberal California Democratic Council," the old Berkeley professor reminded Cranston. "Years ago at one of the Council's early Fresno meetings, I remember that we briefly discussed the 160-acre issue, and your response to my interrogation was satisfactory and reassuring. . . . Surely you recognize reclamation law's role in preserving access to land as a 'liberal' measure and, upon careful consideration, its destruction as a move to the 'right wing.'"

At the end of the day, Cranston did what any good politician would do: He tried to please everybody.

In May of 1979, Cranston announced he wouldn't "actively oppose" an exemption to the 160-acre rule for Army Corps dams—a position that prompted National Land for People, the reform group, to brand the senator "an accessory" in trying to secure yet another "loophole" for "big land and big farms."

At the same time, Cranston called for the government to reassess whether the storage of water at Pine Flat amounted to a subsidy for farmers—and, if so, to require them to pay it back before any acreage-limit exemption could kick in. The farmers, having already agreed to put up the $14.25 million that they believed fully covered the storage costs at Pine Flat, were furious. "Your proposal . . . not only does not provide relief, but also portends new and uncertain burdens on us," Alvin Quist, chairman of the Kings River Water Users Committee, told the senator. "As we understand your description of your proposal, the Bureau of Reclamation would calculate a monetary value for the so-called 'subsidy' based on 'water storage benefits.' . . . This new value, never before contemplated or calculated, in practice or in law, is so speculative, so dangerous and so potentially burdensome as to simply be unacceptable to us. We acknowledge no such 'subsidy.'"

In September, with both sides still grumbling about Cranston's posturing, the Reclamation Reform Act hit the Senate floor. The legislation, as written, would have raised the 160-acre limit to 1,280 acres. Some thought the number too high, others too low. (For Boswell, this part of the bill was irrelevant, given that even the higher figure represented but a fraction of its holdings.) Another hot-button proposal would have allowed farmers to lease as many acres as they wanted and have the federal government irrigate that land, as long as they paid "full cost" for the water—that is, without one drop being subsidized by taxpayers.

But as much as any issue, it was the Army Corps exemption that stirred passions. Gaylord Nelson of Wisconsin—the father of Earth Day and a Democrat by party affiliation but a true "progressive" in his state's venerable tradition—condemned the bill as "nothing more nor less than reclamation reform for the wealthy." As proof, he offered up "a gentleman by the name of J. G. Boswell."

The funny thing about invoking "J. G. Boswell" like this—and it would happen over and over again on the floor of the Senate and House for the next several years—was that few, if anybody, in D.C. had ever laid eyes on the man. Jim Boswell stayed as far from the place as he could, once startling Wes McAden, who had phoned back to Los Angeles with what he thought was some exciting news: He had arranged for Boswell to meet at the Capitol with Senator Cranston. "You don't know me very well, but you'll learn," Boswell responded. "You tell Alan Cranston that if he wants to see me, he can come out here. I'm not coming to Washington." Tom Foley, then chairman of the House Agriculture Committee, used to joke that there really wasn't a J. G. Boswell, referring to McAden's client as "the phantom."

Just minutes into the debate now, Senator Nelson unloaded. "Boswell's cotton farm is five times as large as the next largest cotton farm in the United States," he said, "and he is exempt under this bill—an exemption worth millions and millions and millions of dollars to one owner." Later the senator quoted straight from Judge Browning's appeals court ruling, noting that it expressly called for Boswell and the other big Tulare Lake farmers to be broken up. "Could anything in the world be more clear?" Nelson asked.

Boswell had its defenders, including Democrat Frank Church of Idaho, who suggested that if the company couldn't obtain an acreage-limit exemption, it was sure to make good on its threat to bypass Pine Flat Dam and take water out of the Kings anyway—conceivably leaving many smaller farmers high and dry. "I do know this: You will not break up the company" by trying to impose acreage limits on it, said Church. "If they insist upon the natural flow of the river, all the flood-control benefits will have been lost, the public investment in the dam will have been lost, the gates will be opened . . . and the Boswell Company will continue to irrigate its land."

If Senator Nelson's "desire is to hurt Boswell, he will have to find another way," added Malcolm Wallop, the company's old friend.

Nelson, though, was unrelenting. Boswell and the other big landowners "agreed in 1963 to go to court and abide by the decision," he said firmly. "Now . . . we are going to exempt them. Why? Well, they have been on their food stamps so long it would be a pity to take it away from them, even though it was illegal to give them their food stamps in the first place. That is about the level of the argument in support of this proposition."

The debate went on for two days, with lawmakers conjuring up the Flood Control Act of 1944 and reading the words of long dead presidents and interior secretaries into the record. Finally, with Gaylord Nelson gaining momentum, he called for a vote that would have deleted the Army Corps–Tulare Lake exemption from the bill. Boswell looked sunk. Informal head counts showed Nelson's amendment carrying by about ten votes.

Then Alan Cranston rose to speak.

Whether the land reformers wanted to admit it or not, he said, the hard truth was that Boswell and the other big growers had the upper hand. They owned the water held at Pine Flat Dam—the government didn't—and, in the last analysis, they could take it whenever they wanted. "Regardless of the merits of the Ninth Circuit Court's arguments," Cranston said, "we must face the fact that federal reclamation law will never be applied" on the Kings River. The underlying theme here—that messing with the farmers' water rights was an assault by the government on private property—was one that the Senate had to be sensitive to. For months, anger had been building throughout the West over what was perceived as increasing intrusiveness by Uncle Sam, from new strip mining regulations to Bureau of Land Management policies. In this way, the Kings River was but one more battlefield in the Sagebrush Rebellion.

The best that could be hoped for, according to Cranston, was to implement his plan for the government to calculate any storage subsidy that Boswell and the other growers were receiving and make sure they paid up. He then called for a vote on that proposal, effectively brushing aside the amendment that Gaylord Nelson had just put forward for consideration. "After sitting quietly for two days," Cranston was "all over the place, touching people, gesturing and waving to others, calling in his chips," recalled environmental lobbyist David Weiman, a former aide to Senator Nelson. "We didn't have a chance."

Later that day, with the Cranston provision in place, the Reclamation Reform Act cleared the Senate, 47–29. Senator Nelson and the land reform advocates had been fended off. But, all things considered, Boswell and Salyer had little to show for two years of lobbying. While they would keep their empires intact under Cranston's measure, the farmers faced the prospect of owing the government millions more for using Pine Flat Dam.

Meantime, the land reformers still had a chance to carve up Boswell's land in the House, where the Cranston bill would be taken up next. Lying in wait was Representative George Miller, who called the legislation's gutting of the 160-acre rule "the biggest Western stagecoach robbery of the public since Jesse James." And if that weren't worrisome enough, Jim Fisher had another mess on his hands. He had become caught up in that favorite capital city pastime: scandal.

THE WHISPERS HAD STARTED IN 1978. The Boswell Company had secretly funneled $120,000 to the head of a quasi-governmental research and marketing organization called Cotton Inc. By early 1979, with charges of inside dealing and conflicts of interest being leveled, the affair was grabbing headlines across the Farm Belt and beyond.

The man who had received Boswell's money, Cotton Inc. President J. Dukes Wooters, was not the sort who inspired neutral feelings. Many regarded him as a promotional genius who had pumped new life into a moribund industry, while some (often the same people) thought him an arrogant pitchman who wasn't worth all the trouble he brought upon the cotton trade.

A farsighted group of growers, including Jim Boswell, had hired Wooters to run Cotton Inc. in 1970, just as the public-private entity was taking shape. For years, cotton farmers had paid an assessment—$1 per bale to start—that went toward advertising the crop they grew. But the results of this early campaign, anchored by the slogan "The Fiber You Can Trust," had been uneven at best. While Boswell had little trouble making money through the 1960s, thanks to its size and sophistication, most of the country's 300,000 cotton growers were struggling. Polyester—which wouldn't shrink, wrinkle or fade—had become the fabric of choice throughout the nation. And makers of synthetic materials,

chiefly DuPont, were annually outspending their cotton industry counterparts on marketing by tens of millions of dollars. Little wonder that
cotton was considered in consumer surveys to be "cheap," "dull" and
"unfashionable."

Wooters, a fifty-three-year-old Harvard MBA who had made a mark
on Madison Avenue by finding creative ways to bring a flood of advertising into *Reader's Digest*, the magazine where he worked, was hired to
turn things around. His central challenge was to convert cotton—"that
funny little white stuff," the native New Englander called it—from a
commodity into a brand. Expectations were high. Not long into the new
job, Wooters later recounted, he visited the Mississippi Delta, where one
cotton grower invited him to meet his mother. "Young man, come into
my parlor," she beckoned. "I want you to sit under the picture of the
great one." With that, Wooters was ushered under a portrait of Robert E.
Lee, where the matriarch continued: "I would just like to hear you talk. I
understand you are going to save our farm."

And save it he did. In 1973 Wooters commissioned a San Francisco
artist to design the Seal of Cotton, which soon became an advertising
icon etched into consumers' consciousness. The following year, Cotton
Inc. began convincing textile mills to switch from blended fabrics made
mostly of synthetics to a "natural blend" consisting of at least 60 percent
cotton. By 1979, with Cotton Inc.'s marketing backed up by highly advanced technical research into textile processing, mills were churning
out 100 percent cotton bed sheets. Cotton, whose share of domestic fiber
consumption had sunk to a low of 33 percent, was now poised to reclaim
more than half the market. Dukes Wooters was arguably the most important Yankee for the American cotton industry since Eli Whitney.

Yet for all of his achievement, Wooters had a big problem: He couldn't
pay his bills, even though he had signed a contract with Cotton Inc. that
promised him about $120,000 a year. The organization in 1975 had re-
ceived a $3 million government appropriation for agricultural research.
With those federal funds came a stipulation from Congress: Wooters's
salary couldn't be higher than that of the secretary of agriculture, who
made about $63,000 a year. For Wooters, it presented a Hobson's choice.
He could quit the job he loved or take a huge mandatory cut in his
income.

His solution, blessed by the chairman and vice chairman of Cotton
Inc., was to stay on and try to make up the difference by doing some

consulting work on the side. For this, Wooters turned to Jim Fisher, who agreed to pay him $60,000 annually over two years in exchange for providing Boswell with advice about the direction of the cotton market. By undercutting Congress's attempt to regulate Wooters's salary, this side deal was especially egregious in the eyes of Paul Findley, a Republican representative from Illinois who had a history of taking on big agricultural interests. But his concerns hardly ended there. The "malfeasance and misappropriation" at Cotton Inc., Findley charged, was "like an onion"; every time he peeled back one layer, he found another that stank.

Under the government's cotton promotion program, farmers could ask that their per-bale assessment be refunded, and as it turned out Boswell had requested a refund from Cotton Inc. The amount that the company got back, $120,000, matched what it paid to Wooters. Jim Fisher insisted that this was "just a coincidence," but it sure looked awful. A chunk of money that normally would have gone into the industry's marketing and research kitty had seemingly been used to pad Wooters's salary. Evidence soon emerged, moreover, that the Boswell Company had received more than Wooters's wise counsel in return for its $120,000 payment. The Cotton Inc. chief, it was alleged, had steered as much as $8 million in business to Boswell, sending the company referrals from overseas mills that were interested in buying certain types of cotton, when he was supposed to be working on behalf of the industry as a whole. Wooters wasn't allowed to play favorites.

By early 1980, the Cotton Inc. scandal had resulted in a federal grand jury investigation, two separate inquiries by the Department of Agriculture's inspector general and a confrontational congressional hearing. But despite all that, little damage was done to Boswell or to Dukes Wooters. No one was ever charged with a crime.

Findley kept pressing for Wooters's ouster, and he was eventually joined in that effort by an even more influential lawmaker, California Democrat Tony Coelho. Wooters, though, remained defiant. When Coelho came to see him at Cotton Inc.'s tony New York headquarters, Wooters walked in carrying a silver tray with a bronze bust perched on it. "Is this what you want?" he asked the congressman, looking down at the head on the platter. Wooters managed to hang on at Cotton Inc. until 1982, when he reached age sixty-five and retired. By then, Congress had ceased subsidizing the organization, and Wooters's yearly salary was back up around $200,000.

Meanwhile, Jim Fisher scoffed at the idea that he had done anything wrong, and he sought revenge on Paul Findley, raising tens of thousands of campaign dollars to unseat the ten-term congressman. In particular, Fisher bristled at the allegation that Wooters had directed business to Boswell. This was one company, he was quick to tell people, that didn't require anybody's help. "Hell," said Fisher, "I could pick up that phone right there and in 10 minutes time sell 20,000 bales of cotton. I don't need a referral from Cotton Inc. or anybody else to do it. The accusations are ridiculous."

Whatever the case, there were certainly more crucial things to focus on. The acreage-limit fight—the fight to save the empire—was still raging.

If the Boswell-Salyer lobbying team had a glaring weakness in Washington, it was its inability to make headway with the party running the town. Sure, Wes McAden had access to Democrats, but he was the only one. Irv Kipnes had dropped out of the picture, and the rest of the bunch—Don Santarelli and John Harmer, as well as Jim Fisher and John Penn Lee—were all partisan Republicans. Cracking Cecil Andrus's Interior Department, especially, seemed hopeless.

That is, until Lee phoned home to Virginia to see if his cousin Billy might be able to lend a hand.

Bill Hopkins, a staunch Democrat who was a generation older than Lee, had served in the state Senate for two decades, rising to majority leader. Around Thanksgiving 1979, deep now into the acreage-limitation battle, it dawned on Lee that his politically savvy relative might be a good sounding board. He began ringing him up, trying to "just get a little read as to what he thought I should do from a Democratic standpoint." This went on for a month or so, Lee later recalled, until one day Hopkins volunteered, "Hey, why don't I talk to Pat."

Pat Jennings, whom Hopkins had long known through state political circles, was a former congressman from Virginia who subsequently assumed an even more powerful post: clerk of the House. "Pat knew just about everybody in Washington—especially if they were Democrats," said Hopkins. Most important for Lee and Fisher, Jennings counted as a good friend the one Democrat they truly needed to reach: Secretary Andrus.

In short order, Hopkins and Jennings were added to the Boswell-Salyer lobbying brigade, and soon after that, the two of them strode into Andrus's office on C Street, ready to argue why the farmers of Tulare Lake should be spared from reclamation law. "My God, Pat," Andrus said when Jennings told him they were representing the two cotton giants, "have they bought you, too?"

Jennings, as Hopkins would recollect, was unflustered by the reception. "No, sir, I'm not a bought man," he replied. "I've looked at this thing, and we think we're on the right side of the issue."

For the next hour, the three men talked, watched the short film *Man, Land and Water* and then talked some more. Later that day, they picked up the conversation at Duke Zeibert's, the famed power lunch spot at Connecticut and L. That Andrus would even listen to Boswell's side of the story was an enormous breakthrough. He had been, right along with Senator Nelson and Congressman Miller, an opponent of giving the Kings River area an exemption from the 160-acre rule. As he told one lawmaker, "We do not believe that principles of equity, precedent or of departmental policy support" such a move.

Now, though, Andrus's mind began to open. He asked lots of questions and agreed to follow up by getting together with Lee and Fisher at that place most conducive to mulling delicate matters of government policy: the golf course. Long afterward, Lee would still marvel at the chain of events, at the crazy way life works out sometimes. "Here I turn up on a tax case" for the Salyers, he said, "stumble into the water business, end up talking to my first cousin Billy Hopkins, who knows Pat Jennings . . . who knows Cecil Andrus. And then the next thing you know, we're standing on the first tee at the Congressional Country Club. I mean, it's just one of those things that happens."

In March 1980, at Lee and Fisher's behest, Cecil Andrus flew out to Tulare Lake to look things over for himself. It was the opportunity that Boswell and Salyer had been waiting for, a chance to persuade the secretary that, legislative history aside, there was a pragmatic reason to give the big growers what they wanted: Small farms could never endure in the unforgiving lake bottom.

Arriving in Corcoran, Andrus got the full treatment. Fred Salyer flew him around, the landscape below reinforcing the big growers' perspective. Some 15,000 acres were under water that year, a shimmering illustration of the flooding that still periodically ravaged the fields. As the narrator on *Man, Land and Water* concluded: "Despite dams, the basic geography of the Tulare Lake basin remains the same. It's still a sump . . . and man has yet to become its master." Never mind that these words were precisely the opposite of what the other Boswell film, *The Big Land*, had proclaimed more than a decade before: "It's become the lake that was; its waters controlled. . . . Once master, it's now servant." Desperate times called for a new spin.

Unlike the 1969 flood, the flood of 1980 wasn't a killer. But it was large enough to make the point. At the southern end of the lake, Andrus glimpsed another impressive scene: The farmers had diverted still more floodwater onto another 10,000 acres of unproductive land, an example of the foresight and capital required to minimize the damage to their crops. This was clearly not the work of a ma-and-pa operation.

As Andrus entered the Salyer airplane hangar, where the farmers of Corcoran planned to put on a presentation, the hostility in the room was palpable. The locals were wary of any ambassador from the Carter administration, skeptical that he'd ever be able to see things as they did. And no one was going to mince words about that. Three thousand miles outside the Beltway, in the middle of California cotton country, the obsequiousness that a cabinet member typically enjoys was in short supply.

It was up to Pat Jennings to put everyone at ease. "You know," he told the small crowd, with Andrus by his side, "this reminds me of a little story. Back home in Virginia, the county sheriff would come down to the courthouse every morning and there, sittin' in the sun, were a couple of good ol' boys who had just gotten back from fishin'. And every day, they had a string of big, fat fish that they had caught. And the sheriff says, 'Now, boys, how do you keep getting all those fish?' 'Oh,' they say, 'we're not telling you, sheriff.'

"Now, finally, the sheriff wears them down. And they say, 'OK, sheriff, we're going to take you fishin' with us.' So they all get into the boat and row out to the middle of the lake. And the sheriff says, 'Well, where are the fishing rods?' One of the guys then leans down and pulls out a stick of dynamite. He lights it, and throws it in the lake and, boom, 100 fish come belly up into the boat, stunned. The two good ol' boys start scoopin' 'em up as fast as they can. And the sheriff starts screaming, 'Wait a goddamn minute, boys! You can't do that! It's illegal.' With that, one of the good ol' boys reaches down, picks up another stick of dynamite, lights it and hands it to the sheriff. 'Sheriff,' he says, 'your turn.'"

Jennings then looked straight at Andrus. "Mr. Secretary, your turn."

The farmers howled; the tension dissolved. The metaphor was perfect: The Kings River was everybody's problem. If Andrus pushed the farmers too far—if they decided to say to hell with Pine Flat Dam and let the river flow free—the whole situation might just blow up in the government's face. The small farmers upriver, who relied on Pine Flat to help guarantee their water supply, could well find their operations disrupted, even if they complied with the acreage limit.

Andrus made no promises that day. How could he? His own staff, especially the assistant secretary for land and water resources, Guy Martin, had a decidedly different view than that of the big farmers. Years later, when he himself had become a natural resources lobbyist, Martin hung on his office wall a photograph showing him standing proudly with President Carter; but if you flipped the frame around, it revealed a photo of Martin shaking hands with President Reagan. It was his way of poking fun at the flexibility sometimes required of a Washington mercenary. As a government official, however, Martin was neither lighthearted nor willing to bend. Boswell and Salyer, he thought, were in complete error on three counts: in their interpretation of the Flood Control Act of '44, in their rationalization that they weren't squeezing subsidies out of Pine Flat, in their insistence that the government couldn't withhold benefits of the dam from the big farmers without hurting small farmers.

In late May, Andrus sent a letter to Morris Udall, chairman of the House Interior Committee, that reflected Martin's influence. "This administration believes that an exemption for the Kings River . . . is unwarranted," it read. "Congress decided against an exemption in authorizing these projects in 1944, and nothing since has shown the decision to be wrong." It was as if the secretary's trip to Tulare Lake had never happened. Boswell's men were livid. Stan Barnes, the water engineer, called Andrus's comments "invidious."

Still, Pat Jennings kept working the secretary. Fisher and Lee also never gave up on their golfing buddy. And in late November, with Jimmy Carter a lame duck, Andrus finally delivered, making one of the more astonishing reversals of his political career: He sided with Boswell and Salyer.

While Andrus continued to oppose a blanket exemption to the acreage-limit law for Army Corps projects, he was swayed that the Kings River deserved one. "This is a unique area," Andrus wrote in a new letter to Chairman Udall, "and difficult to understand unless you have actually visited. . . .

"I have concluded," he added, "that in the Tulare Lake Basin, small-scale farming is not economically feasible. . . . Only large farming operations can financially sustain the high costs of diverting the flood waters into selected large tracts of land. . . . Without large farms in Tulare Lake, large amounts of acreage would not be in agricultural production, and that would in no way further the purposes and goals of the reclamation program."

He ended by saying that if the big farmers in the lake bottom were to let the Kings River run wild, it would no doubt "impair the productivity of the small farm operations that are upstream of the Tulare Lake basin." To do anything other than grant an exemption on the Kings "would be irresponsible and would most likely jeopardize the many small operators to get at a few large operators."

Andrus's traditional allies were dumbfounded. "His arguments are now the exact arguments of the Kings River people," said Congressman Miller. "He has turned around without consulting anybody, going 180 degrees contrary to all his statements, all the evidence he submitted . . . studies he had done." David Weiman, the environmental advocate, felt personally betrayed. "Cecil Andrus asked a lot of people to stand with him and fight because it was the right thing to do, and then he jerked the rug out from under them," he said.

For Boswell and Salyer, it was difficult to overstate the importance of Andrus's turnabout. Most Republicans, including the incoming Reagan administration interior secretary, James Watt, already were on board or soon would be. And now, how many liberal Democrats were going to attack a provision that Cecil Andrus had seen the wisdom of? Simply put, said Lee, Andrus's acquiescence "changed the whole course" of the debate. It was "major, major, major," added Fisher. "From then on," explained Hopkins, "instead of pushing the rope uphill, we were pulling it downhill." The biggest trick was to make sure that the entire reclamation reform train continued to move forward, and that the Kings River exemption didn't somehow become decoupled in the process.

Not that the hard work was over for Fisher and Lee. Another political firestorm was brewing back in California, a second struggle that would prove every bit as consuming as the first.

Lobbyists, Politicians, Payola

FOR A BRIEF MOMENT, it seemed as if John Penn Lee and Jim Fisher were just one or two whiskeys away from a win in Sacramento.

The bill they so vehemently opposed—authorizing the building of the Peripheral Canal, a forty-three-mile-long channel that constituted the last major piece of plumbing in the state's gigantic water system— had already been beaten back the year before over the wishes of Governor Jerry Brown. Now the measure looked like it might be on the brink of defeat once again. The Senate Ag and Water Committee, which needed to approve the bill, was deadlocked, 5–5, on this evening in May 1979, with the crucial sixth and deciding vote in the hands of a Democrat from East Los Angeles named Alex Garcia.

Garcia was something of a joke in the California legislature. He "is prone to disappearing for days," *Sacramento Union* columnist Dan Walters once noted. "Nobody knows where he goes. Nobody inquires too deeply into his whereabouts. The work of the Senate goes on as if it had 39 instead of 40 members." Actually, that wasn't quite so. Many insiders had a pretty good idea of where Garcia went; more often than not, he was out getting drunk. The previous year, Lee and Fisher had found the senator on the town, right before an important vote on the canal was scheduled to take place, and they plied him with so much food and liquor that he never made it to the Capitol. Missing the vote, Lee recalled, "was as good as a 'no.'"

Now it looked like Garcia might vanish again, as he got up from the hearing room and the door closed behind him. Only this time, the senator had a baby-sitter.

Los Angeles Mayor Tom Bradley, who was pushing hard in favor of the canal, had personally won Garcia's assurance that he'd vote for the measure. Then the mayor assigned one of the city's lobbyists, Ray Corley, to keep tabs on the senator. A hulking man, Corley bolted across L Street, catching up to Garcia as he made his way into a dimly lit saloon called David's Brass Rail. Corley allowed the lawmaker a couple of drinks to steel his nerves. Then he coaxed him from his barstool and escorted him back to the hearing room. An hour or so later, Garcia walked out of the session a second time. Once more, Corley tracked him down and convinced him to return, just as committee members were recording their votes.

Garcia, mindful of his vow to Mayor Bradley, voiced a "yea." The place went into convulsions. Instead of voting for the bill, Garcia had mistakenly voted for an amendment to eviscerate the legislation. Ruben Ayala, chairman of the committee and the bill's sponsor, stormed over to Garcia. "You better change your damn vote or I'll break your arm," Ayala roared. After some parliamentary maneuvering, Garcia was allowed to reverse his position on the amendment. He ascribed his confusion to lack of sleep. Eventually the Peripheral Canal bill squeaked through the committee, 6–5.

Thus was born one of the most controversial measures ever to come out of the statehouse. After more bitter wrangling among lawmakers, it passed the full Senate and then the Assembly, landing on the desk of Governor Brown in July 1980. At a televised ceremony, he signed it with a flourish. But rather than end the discord, the governor's action only aroused Boswell, Salyer and the canal's other foes. Three months later, opponents announced that they had collected enough signatures to force the issue to a statewide referendum—a vote of the people. The campaign to win them over would turn into one of the most extraordinary of all time, pitting Northern California against Southern California in a kind of bloodless civil war and evoking the legacy of Governor Edmund G. "Pat" Brown Sr., Jerry's father.

Through a combination of sheer will and fiscal chicanery, Pat Brown in 1960 had brought about passage of the State Water Project, which sought to solve California's great conundrum: Most of the water is in the north, while most of the people are in the south. Not much smaller than the federal Central Valley Project, its sister conveyance system, Brown's vision called for nothing less than the most ambitious feat of engineering ever undertaken by a state. It would include the world's tallest earthen dam (Oroville); another that would rank as one of the five most massive anywhere (San Luis); the world's longest aqueduct, which would transport water 444 miles southward; and the world's highest pump lift, which would carry water 2,000 feet up and over the Tehachapi Mountains to the sprawl of Los Angeles. Best of all for the big farmers, the state project wasn't encumbered by that old bugaboo of the federal system: the 160-acre limit. After the project was up and functioning, Pat Brown boasted that he had given Southern California the resources it needed to sustain its burgeoning population while also helping to create "a Garden of Eden" in the San Joaquin Valley. But the water project was not solely for the benefit of his constituents and their children and their children's children. It was also, admittedly, a shrine to Brown himself—a grand symbol of his leadership of the Golden State. As he put it: "I wanted this to be a monument to me."

Boswell and Salyer had come to rely on the State Water Project as an essential element of their irrigation networks. Beginning in the late '60s, it provided them and the other farmers of Tulare Lake with as much water as had the Kings, Kern, Kaweah and Tule rivers. Hypothetically, a Peripheral Canal should have made things even better. The canal had long been seen as a much needed addition to the system: a means to transfer water around the eastern edge of the Sacramento–San Joaquin delta, boosting the overall flow of the project. Salyer and Boswell had no problem with that. Nor did they oppose the other facilities—new reservoirs and groundwater storage units—included in the bill that Jerry Brown had just signed. Their concerns, rather, rested with specific provisions that were supposed to make sure the environmentally sensitive delta and rivers of the north coast would be adequately protected. The way they read the legalese in the bill, they might end up with less water, not more. Taking care of wildlife, not farmers, seemed to them the legislation's primary aim.

John Lee was especially alert to this risk. A few years earlier, with California plagued by drought, the state had failed to deliver 155,000 acre-feet of water to the farmers in Tulare Lake, diverting it instead to the delta—home to 230 species of birds and more than fifty species of fish, including anadromous salmon and striped bass. Lee had sued the Brown administration on behalf of the Salyers and other state water system contractors, claiming they were owed more than $7 million in compensation. Now, Lee believed, fine print in the Peripheral Canal bill would have directly undermined his position in the pending litigation.

"My case," he said, "would have been summarily handed back to me at the courthouse."

Jim Boswell, likewise, had concerns about the state's motives under Jerry Brown. During a meeting with his top executives, he blasted the governor for using the drought "as an excuse" to "have the state decide the priority of use of water, regardless of private rights." It was "quite obvious," Boswell added, that "a large-scale farming operation such as ours" was on the "bottom of the priority list."

Salyer and Boswell weren't the only ones with misgivings about the Peripheral Canal bill. Those who were most outspoken about it, however, had fears that ran 180 degrees the other way. In the opinion of groups such as the Environmental Defense Fund and Friends of the Earth, it wasn't that the wildlife protections in the bill were too strong; they thought they weren't strong enough. "They just didn't have faith" that the safeguards would hold, said Jerry Meral, a former scientist with the Environmental Defense Fund, who had joined the Brown administration as deputy director of the Department of Water Resources. United by their contempt for the canal, the big farmers and the environmentalists—normally fierce enemies—agreed to band together to try to crush the project. Water historian Marc Reisner would term it "the oddest alliance since the Hitler-Stalin Pact." A third group, the delta farmers, also signed on to the anti-canal coalition. Like the environmentalists, they were convinced that once the canal was built, they could forget about all the pledges of improving their water quality. Odds were that L.A., the ever growing megalopolis, would suck every last drop south.

On the other side of the issue stood the elite of Southern California—civic leaders and the CEOs of energy companies, construction firms, developers and financial institutions. Each one felt that without the extra water the canal promised to bring, a shortage loomed and their world-class economy was bound to falter. Many of them took their cues from the Metropolitan Water District, a public agency that until now had formed an axis with Central Valley farmers to encourage water development projects. "The Mighty Met," as it was known, served a huge stretch of city and suburb, from Los Angeles to Orange County to San Diego. It had played a pivotal role in forging the region and was a formidable political force. As a matter of fact, the Met's clout went a long way toward explaining why its officials didn't worry about the environmental restrictions in the canal bill the way Boswell and Salyer did. Once the canal was in place, the Met presumed that the water would inevitably surge in its direction.

Unnerved by the intensity of the recently ended drought, the governor was also cheering for the canal, which he portrayed as a win-win project: good for the environment and good for the economy. The irony wasn't lost on many that Jerry Brown—the eccentric politician who had entered office declaring that "less is more" and "small is beautiful"—was now urging the multibillion-dollar completion of one of the most colossal public works projects in history. Lots of folks figured that he was simply doing it for his old man.

By EARLY 1981, a full-scale political war had broken out over the Peripheral Canal. And things didn't look so good for Boswell, Salyer and the other opponents of the project. Their mission—to overturn at the ballot box a law already signed by the governor—seemed, in many ways, quixotic. Nobody had managed such a thing since 1939. Polls showed that most Californians favored building the canal. And proponents had ample muscles to flex: the backing of the Brown administration, the promise of serious money from the business community of Los Angeles and the input of Stu Spencer, regarded as the "guru of political gurus" for having propelled Ronald Reagan and countless others into office. "They could have just buried us," said Boswell's corporate counsel, Ed Giermann.

As for the challengers' own fledgling campaign, it was in disarray. They had hired a political consulting firm in Sacramento run by two irreverent thirty-something Republicans, Sal Russo and Doug Watts, who had built their practice largely by working with agricultural interests. Now it was their job to get Boswell and Salyer, the delta farmers and the environmentalists all pulling in the same direction. The problem, said Russo, was that "nobody trusted anybody," and different factions started meeting clandestinely among themselves. "We were in so many secret meetings," recalled Russo, "that trying to remember what you could say" at the next one was almost impossible.

For each splinter group, controlling the message was imperative. "I could not tolerate being associated with a campaign that emphasizes additional water for the valley at the expense" of the delta or the north coast, Tom Graff, an Environmental Defense Fund attorney, told his green movement colleagues in a confidential memo. "I realize that . . . perhaps Boswell and Salyer are opposed" to the canal "primarily on this basis, and it is obvious we can't directly affect their positions or even their efforts in the campaign. I can and do object to them including these efforts in our campaign."

Between themselves, Russo and Watts started referring to the big farmers as "the fascists" and the enviros as "the communists." They filled their roles beautifully. Michael Storper, a bespectacled, mop-topped Friends of the Earth lobbyist, brought his flea-ridden Belgian shepherd to one of the coalition meetings. The next day, everybody was itching like mad.

In April, the coalition gathered to come up with a name. "The Unholy Alliance" was one wag's suggestion, though they settled instead for the rather lackluster appellation "Californians for a Fair Water Policy." It was about all they could agree on. By the summer of 1981, the coalition was in danger of collapse, with the environmentalists convinced that they were being excluded from critical decisions. "Our integrity is extremely important to us, and we cannot accept a situation in which we are perceived as participating in a campaign run by our historical opponents," Storper warned. "It will be necessary for us to withdraw publicly and carry out our own campaign if the committee goes this route. I say this with the understanding that our withdrawal would seriously damage our chances in the election because a split would be used by the other side. But our long-term interest . . . would be violated by playing second-fiddle to a campaign run by others."

The farmers had their own qualms, and rumors flew that they were going to defect from the coalition as well. "Big Ag was equally suspicious of us," said Watts. "Everybody was sure we were screwing them—that we were sleeping with the other guy."

Somehow, through it all, the young consultants held everyone together, while adroitly building a presence on the ground, county by county. As the months went along, it became apparent that the anti-canal position was gaining traction, especially in Northern California. There, anxiety naturally ran high about any project that would move water—their water—to the sunny south. As one Bay Area resident complained in a letter to Governor Brown, "I saw people washing their cars in Newport Beach and watering median lawns along streets" even during the drought. Fair or not, the image of Northern California river water filling Beverly Hills swimming pools and sprinkling mansion lawns in Malibu was a part of the culture that Russo and Watts knew they could exploit.

But they weren't content to stop there. Their objective was to demonize the canal from every conceivable angle.

Their clients weren't always delighted with this approach. One of the ads that Russo and Watts produced attacked the oil companies of Kern County, just over the mountains north of Los Angeles. Tenneco, Getty, Shell and others owned big tracts of farmland there, and they had thrown in their lot with the Metropolitan Water District in supporting the canal. "Who says oil and water don't mix?" went the Russo and Watts voice-over, as some of each was poured into a kitchen blender. "Major oil companies" plan to "use our tax dollars to build the Peripheral Canal. At our expense, they'll get plenty of cheap water to increase the value of their huge landholdings." As the spot ended, a wad of cash burbled from the blender. For Jim Boswell, who lunched with oil executives at the California Club, and Jim Fisher, who himself was getting beaten up on Capitol Hill for taking "cheap water" to irrigate "huge landholdings," this was not the kind of language they felt comfortable with. They ordered the ad pulled.

The heart of the Russo and Watts strategy was to tailor five different messages for five different television markets, using "local heroes"— popular politicians—to drive home these sundry points. Around the logging community of Redding, Assemblyman Stan Statham decried the "drastic effect" the Peripheral Canal would have "on our northern state water needs. . . . With it, there simply won't be enough water to go around." In the San Francisco area, the emphasis was on, in the words of county supervisor Louise Renne, "the disastrous impact" the canal would have "on our fragile Bay and delta environment." In the Central Valley, state Senator Ken Maddy came on TV screens to tell viewers that the canal "virtually guarantees that no new water development will occur in California"—a catastrophe for agriculture. In San Diego, Russo and Watts linked the Peripheral Canal to higher utility rates—an issue already rankling locals. And around Los Angeles, the refrain was as simple as it was relentless: The canal costs too much.

By early 1982, with the election to decide the canal's fate set for June, it was clear to Russo and Watts that this last message—harping on the

canal's price tag—was the key to the race. "As we know from our preliminary data, Northern California voters remain firmly opposed to the
canal," pollster Gary Lawrence told Russo in a January memo. Thus, he
advised, "Southern California must become the focus of the campaign.
Canal proponents are already freely spending to convince Southern Californians that they 'need' the water, and the effects of those efforts are
surfacing in our polls. Data suggest that if the election were held today,
the canal would carry the South with a 3-to-1 margin, enough to carry
the measure statewide, regardless of Northern California opposition."

However, Lawrence said, there was a way to reverse the tide. "When
our interviewer first explains the cost of the canal, then asks for the vote,
the opposition becomes 2 to 1, indeed a dramatic switch."

IF BOSWELL AND SALYER HAD ANY DOUBTS about the extent
of their own political power, they were dispelled one evening in March
of 1982, when John Penn Lee was home in Fresno, dressed in shorts and
a T-shirt, doing the dinner dishes. A neighbor named Jim Bouskos, a
gregarious character who had made a fortune in his family's grocery
business and was known as a generous campaign donor, knocked on
the door and told Lee that he had brought along a friend who wanted to
speak with him.

"Sure," said Lee. "Come on in."

Stepping out of the shadows was the lanky form of Jerry Brown. The
governor of California had come to make a house call.

"Would you please explain to the governor what you're doing on the
Peripheral Canal," said Bouskos. Over the next five hours, the three men
polished off a couple bottles of wine, first at Lee's and then a golf cart
ride away at Bouskos's place. Lee mostly talked, laying out the litany of
complaints he had about the canal and about state water policy in general. The governor mostly listened. Then Brown turned to what was
clearly bothering him: commercials from the anti-canal camp that were
running in the Central Valley. One of them, for example, maligned the
governor's handling of previous water matters, including a move to sat-
isfy environmentalists by holding down the level of the New Melones
reservoir. That decision "robbed valley farmers of a needed water supply," the ad said, before deriding the Peripheral Canal as another "envi-

ronmental experiment." Then the kicker: "Haven't we been burned enough by Jerry Brown?"

For the governor, who was now running for a U.S. Senate seat, the ads were a real problem, and he wanted them to go away. "Here I am giving you the time" to make a case against the canal, Brown told Lee, "and you're pounding me."

As Lee would later recount, he saw a different side of Jerry Brown that night: a charming, engaging side. To his own surprise, Lee decided that his antagonist was actually an all-right sort of guy. But he also made plain to the governor that, with all due respect, there wasn't anything he could do about the ads. The stakes were too high; there was no backing off now. Jerry Brown, said Lee, was "our best whipping boy."

In point of fact, he was no better than second best. As Russo's pollster had indicated back in January, the canal remained most vulnerable over its cost.

In April, with the election fast approaching, the anti-canal forces were given just the ammunition they needed to slam the project's price tag harder than ever: a report from a task force led by Lieutenant Governor Mike Curb, which found that the canal and other facilities would consume at least $19.3 billion. It was a staggering calculation, nearly four times as much as the Brown administration's official, inflation-adjusted estimate. And most of that, under the State Water Project's rate scheme, would be borne by the Met's customers. A "reasonable alternative" to the canal, the task force said, would be to pump Sacramento River water directly through the delta rather than around it—the very course preferred by Boswell and Salyer.

Canal advocates immediately lashed out at the task force—and for good reason. Boswell and Salyer had helped gin up the panel months before, and both Lee and Fisher belonged to the twenty-member group. They had been approached by Curb, who had started out in favor of the canal, only to swing in the big farmers' direction as he took to the stump to launch a bid for governor. His flip-flop cost him some campaign money in Southern California. But Boswell and Salyer more than made up for it, filling his treasury with tens of thousands of dollars and zipping him around on their corporate jets. "We gave him a ride in our airplane," Fisher would later say, adding in jest: "I thought it was a pretty cheap way to buy a politician myself."

The Curb task force delivered in short order. For Boswell and Salyer, the headline in the *Los Angeles Times* was all they had hoped for: "Canal Costs to Top Estimates, Curb Panel Says." The task force report also provided fodder for another round of TV ads, which played up the $19.3 billion figure and painted the canal as a boondoggle. Up north, billboards blanketing roadsides underlined the same idea: "It's Just *Too* Expensive." The ad blitz was bankrolled by none other than the Boswell and Salyer companies, which in early May put up hundreds of thousands of dollars to buy TV time.

The pro-canal side, by contrast, never quite galvanized. "Gentleman," Met Chairman Earle Blais lamented to his troops, "the tenor of this campaign is being set by the opposition!" The Met did its best to raise concerns that without the Peripheral Canal, Southern California might run out of water, bringing rationing and brown lawns. The state raised the same bogeyman, with Department of Water Resources director Ron Robie cautioning: "You can't stretch the present supplies." But the public never really seemed convinced that conservation couldn't take care of the problem. The weather also broke against the canals' backers, as the drought of the mid-1970s gave way in 1982 to a succession of El Niño–fueled storms. "We had some good fortune," acknowledged Lee. "It rained like a son of a bitch."

Jerry Brown wasn't much help to the pro-canal cause, either. The governor, with his notoriously short attention span, seemed to lose interest in the fight. Nor did the leaders of the L.A. business community ever step up as expected. "When you told them you needed a half a million bucks, they gagged," said Stu Spencer, the consultant. In the end, the canal's foes outspent their adversaries, $3.6 million to $2.7 million. Of the opposition's total, nearly $2 million came from Boswell and Salyer alone.

Some in the environmental community always felt queasy about that. "They called it 'dirty money,'" said Lorrell Long, an activist who was instrumental in the anti-canal campaign. Nobody, though, could dispute the results. The canal measure, Proposition 9, was trounced in an Election Day landslide, 63 percent to 37 percent. In Los Angeles, where proponents were counting on a 3–1 margin to push the initiative over the top, the "yes" votes outstripped the "no" votes by less than 2–1. And in Northern California—where the pro side hadn't bothered to establish any presence—the results were overwhelming. In some counties, such as Contra Costa and San Francisco, more than 95 percent of the electorate voted against the canal. "Outside of communist Bulgaria," marveled Jerry Meral, the Department of Water Resources official, "it is the widest margin I've ever seen."

The vote was a defining moment. "For the first time in modern California history," said the Environmental Defense Fund's Tom Graff, "the water developers have been beaten—and beaten badly." For the first time, "progress" hadn't been measured by building more dams and canals.

A month after the vote, Jim Fisher threw a steak-and-cocktail party at the lanai in Corcoran. "We and the Salyers think this is cause for a celebration," he said. There wasn't too long to savor the victory, however. Back in Washington, the train was now barreling down the tracks.

IN MAY 1982, the House of Representatives approved a version of the Reclamation Reform Act, which raised the number of acres that could be watered by a federal irrigation project 160 to 960. It also included the exemption to acreage limits so coveted by Boswell and Salyer. Chip Pashayan, Boswell's man in Congress, had been sure to remind his fellow lawmakers of Cecil Andrus's support for that particular provision.

The bill reached the Senate floor in July. Again, Boswell and Tulare Lake were a focus of the debate. Senator Howard Metzenbaum, a liberal Ohio Democrat, went after the company, entering into the record letters from the AFL-CIO and National Grange that damned the Army Corps exemption as a giveaway to the rich. Again Malcolm Wallop rode to Boswell's defense. "Regretfully," said Wallop, "the senator from Ohio, though he comes from a state with a good deal of farming, obviously knows nothing of farming. . . . He obviously knows nothing of reclamation projects or law. He knows nothing of the Boswell Company." Metzenbaum spent the day employing a series of delaying tactics, but he was eventually foiled, and the bill passed easily.

President Reagan signed the Reclamation Reform Act, with the Army Corps exemption intact, on October 12. It was now Public Law 97-293. The fight that had begun with the Colonel and Fred Sherrill squaring off against Harold Ickes and Franklin Roosevelt—the battle that had shaken the empire to its foundations—had at long last come to an end.

For the big farmers, triumph couldn't have been sweeter. For the hard-core land reformers, things couldn't have been more heartbreaking. "It's as if we spent twenty-five years uncovering a crime wave," said George Ballis of National Land for People, "and Congress handled it by declaring the criminal actions legal."

In Washington and Sacramento, Boswell and Salyer were now juggernauts, and they topped off the year by supporting Republican George Deukmejian for governor. "In a remarkable string of political successes," wrote William Kahrl, a close observer of water issues, "they have stopped the Peripheral Canal, . . . gutted the provisions of federal reclamation law that throughout this century have threatened to break up their landholdings, and contributed substantially to the election of the new governor, who is likely to be far more sensitive to their concerns than" anyone before him. "In consequence, agriculture is probably stronger politically today than it has been at any time since World War II."

Boswell gave out medallions to those who had helped win passage of the Reclamation Reform Act. They included the apt line from Ecclesiastes: "The race is not to the swift." Jim Fisher got his own special medal—two of them, actually. One had a triangle in the middle outlined by the words "Leadership," "Teamwork," "Persistence." The other, presented to "Big Bucks" Fisher, contained three different words: "Lobbyists," "Politicians," "Payola."

The View from the
Forty-sixth Floor

JIM BOSWELL FILED into the small basement courtroom in downtown Bakersfield, the place full of people eager to see the biggest farmer in America get licked. With the possible exception of catching Buck Owens and the Buckaroos, it was easily the most entertaining show in town.

The past few years hadn't been easy ones for Boswell. As jubilant as they had been in the wake of the Reclamation Reform Act's passage and the Peripheral Canal's defeat, Boswell and his boys found themselves by the mid-1980s beset by internal strife, a faltering farm economy, and a spate of negative publicity over federal subsidies. And now there was this—a court case whose sole purpose seemed to be to embarrass the company, to make an issue of the one thing that J. G. Boswell always felt a little defensive about: the size of his empire.

The matter had grown out of an advertisement that had run six years earlier during the battle over Proposition 9, the Peripheral Canal referendum. A group of small Kern County farmers in favor of the project had taken out the full-page ad in two valley newspapers, the *Bakersfield Californian* and the *Hanford Sentinel*, contending that Boswell and Salyer's opposition to the canal was driven by less than scrupulous motives. The headline was a screamer:

WHO ARE BOSWELL & SALYER? And why are they trying to cut off our water?

The answer, according to the small farmers, was greed. "Boswell and Salyer have built their empires on cheap water," the ad declared. By killing off the Peripheral Canal, these "privately held companies, accountable to no one," will be in position to "dominate California water policy" and to continue to avoid paying "full market rates" for irrigation.

The indictment went on from there, suggesting that Boswell and Salyer were conspiring to "freeze out competition."

The Boswell Company had sued for libel, but the courts found the case to be without merit. Everything asserted in the ad, they ruled, was free speech—protected by the Constitution. The dispute, however, didn't stop there. The small farmers turned around and filed their own suit, alleging that Boswell hadn't actually believed the ad to be libelous in the first place. Rather, they claimed, Boswell was maliciously using the libel suit to try to stifle open debate and chill their advocacy for the canal. This concept—known as SLAPP, or a Strategic Lawsuit Against Public Participation—had been gaining increasing notice in the legal world, and the Boswell case would come to be regarded as among the most important in the field. Ralph Nader, for one, would label Boswell's conduct a classic "abuse of the justice system."

After years of lawyers' maneuvers, the case had finally come to trial in June 1988, and Jim Boswell—a man who so hated the limelight—was thrust into a most disconcerting spot: compelled to take the witness stand. After being sworn in, J. G. did his best to play to the Bakersfield jury, determined to show them how earthy he was, what a regular Joe he was. But the local lawyer on the other side, a wily litigator named Ralph Wegis, did his best to make Boswell into something else entirely: an elitist from the big city.

"Now," Wegis asked him, "what did you first do when you joined the corporation?"

"I was a cowboy," Boswell replied.

"College-educated cowboy?"

"That's right."

"Stanford educated? You were a Stanford-educated cowboy?"

"Yes."

"With a degree in economics?"

"That's right."

"And how long did you ride the range?"

"Basically, all my life. There's my mark of it." And with that, Jim Boswell held up his mangled hand for all to see.

It had been J. G. Boswell's own intense anger that had brought things to this point. His in-house counsel, John Sterling, had thought about suing for libel when he first saw the ad six years earlier. Jim Fisher was also enraged over the allegations it contained. Particularly infuriating to them was the passage that questioned whether Boswell and Salyer were committing anti-trust violations:

> Salyer and Boswell have enough water resources to outlast the next drought and the years when water deliveries will be delayed if Proposition 9 is defeated. Smaller farmers don't have those resources and Boswell and Salyer know this. If the small farms go out of business, Boswell and Salyer will be able to totally dominate California agriculture—setting prices where they want them.

Sterling and Fisher said they considered this an accusation of a crime: price-fixing. Still, they weren't sure whether they should go ahead and sue the Kern County farmers without first talking to the boss.

By the time they left Jim Boswell's office, they had no doubt about the course that he wanted to pursue. "They understood bloody well what I was upset about," Boswell recalled. He had been so agitated by the advertisement—"mad as hell," in his words—that he had even called his mother, Mom Kate, to vent. "I've got a lot of pride in our name and our company," he said.

Sterling, according to his testimony, tried to gauge whether the Salyers were also interested in suing. Despite their litigious reputation, they didn't seem nearly as worked up about the ad as Boswell was. They were content to let it slide. Sterling said he then set out to get a second opinion about the ad, soliciting the advice of a Los Angeles lawyer by the name of William Chertok. A veteran trial attorney, Chertok agreed that the statement about "setting prices" amounted to libel per se. So he sent out a string of letters to other newspapers, warning them that they better not run the ad. Then he sued the three Kern County farmers who were believed to be behind it, demanding $2.5 million in damages from each. If successful, Boswell would have wiped them all out financially.

Now, having countersued, it was Wegis's turn to go on the offensive. He tried to show that Boswell, through his hired-gun attorney, wasn't ever truly interested in whether the legal threshold of libel had been crossed, making much of the fact that Chertok couldn't produce his notes showing that he had even researched the issue. (He maintained they had been lost when some files were moved.) Wegis also pointed out that Chertok had little experience on libel or First Amendment cases, portraying him as nothing more than an expedient hatchet man whose office happened to be in the same building as Boswell's, a short elevator ride away. In Boswell's hands, Wegis stressed, the claim of libel was simply a weapon, an instrument to intimidate those who wanted the Peripheral Canal to be built. "Their minds were made up," said Wegis, before they ever even got a second opinion about whether to sue.

During the next two weeks, Wegis further developed the idea that Boswell had tried to sue its political adversaries into silence. Legally, this was the key to his case. But in terms of winning over the jury, the other theme that Wegis pushed was perhaps even more important. The J. G. Boswell Company, he reminded them repeatedly, was "the biggest farmer in the world"—a brute used to throwing its weight around and getting its way.

It was a subtext struck from the very start, during jury selection, when Wegis invoked the name of J. R. Ewing, the famous scoundrel from the TV show *Dallas*. "You don't think it's humorous for a J. R. kind of person just to be able to be a little heavy-handed because they're big shots?" he asked one prospective juror. Shouldn't somebody like that "be treated by the same standard as the little guys?" It was a masterstroke—planting the seed of J. G. as J. R.

Now, with Jim Boswell himself on the stand, Wegis fired away at the company's immensity again and again: "Do you know anybody that grows more cotton than you do?"

"No."

At another point, he asked Boswell about the company's Australian operations. "What kind of acreage was involved over there?"

"Total approximately 30."

"You're not saying 30 acres, are you?"

"30,000. Excuse me."

Over and over, Wegis commented on how Jim Boswell and his top executives farmed not from Corcoran, but from the forty-sixth floor of a building in downtown L.A.—"a control center on Wilshire Boulevard." He talked about the company jet, the Turbo Commander. He noted how Boswell had driven John Krebs from Congress using "money and muscle." He spoke of the company's stable of lobbyists in Washington and Sacramento. And he reiterated time and again how Boswell had put more than $1 million into the campaign to kill the Peripheral Canal.

Wegis was a maestro with language. He kept referring to the Boswell *Corporation,* not *Company,* a technical misnomer that accentuated its enormity. The office tower in which the company leased the forty-sixth floor was the Security Pacific Building, but Wegis turned it into the "Boswell skyscraper."

He carefully led Jim Boswell through his own history, focusing on how he amassed more than 100,000 acres of land during his time as CEO. He asked Boswell about some of his trusted lieutenants, especially Jim Fisher, who had abruptly left the company in September 1984.

"Did Jim Fisher retire or what happened?" Wegis asked.

"Yes," said Boswell "He took early retirement."

"That was at his request?"

"Yes."

If Jim Boswell wasn't committing perjury on this one, he was pushing it pretty close.

JIM FISHER WASN'T THE KIND OF GUY TO look for omens, good or bad, but the way his morning had started would always stick with him. Just as he reached the office parking garage around 7:00 A.M. his car ran out of gas. He was getting ready to send a gofer to fill up the tank when Jim Boswell walked in.

As Fisher would remember it, what happened next was "totally out of the blue." Boswell, he recounted, turned to him and said flatly: "I know you're not happy. I'm not happy. I don't want to discuss with you why. . . . I want you to go down and talk to Wally Erickson. He's been working with a compensation firm up in San Francisco, and he'll tell you what your package is."

And that was it. There wasn't so much as a nod to Fisher's twenty-six years of service, his ascension to corporate president, his unflagging fidelity. "I was so goddamned loyal to that guy, I would have killed for Boswell," he said. There was nary a mention of Fisher's conquering of Australia or his role during the acreage-limitation struggle in Congress—a performance that saved the empire from being broken up. There certainly was no "I'm sorry" from J. G.

As he began to think about what had befallen him, Fisher realized he had overlooked several signs that this day was coming. Boswell had been badmouthing him around the company and to the board. Fisher had seen the same thing happen plenty of times before. Boswell would pick on you and pretty soon your credibility was shot. Then one day you'd be called in and told that it was over, asked for your resignation and shown the door.

Boswell didn't just rotate his crops. He rotated his executives.

"Most people have a cycle with him. Your star rises and then your star dims—and then it goes out," said Walt Bickett, the finance chief who had a falling out with Boswell after twenty-three years at the company.

Trying to get inside Boswell's head, figuring out why he cut people off this way, became a sort of parlor game for the dispossessed. Some former Boswell executives, noting how few ever retired at sixty-five or even made it to sixty, claimed that he was obsessed with youth, that he tried to stay young by surrounding himself with men a fraction of his age. Others said that Boswell treated his managers like houseguests who had overstayed their welcome. "He got tired of them," said Brooks Pierce, whose own exit came after twenty-eight years.

Gary Gamble, an executive ousted after more than thirty-three years, presumed that by dint of being around so long, you just "built up too many bad points" on Boswell's scorecard. Still others deduced that the upheaval was a management tool, a way of stirring things up, keeping them from getting stale. Whatever the reason, one thing was clear: "When he's through with you, he's through with you," said A. L. Vandergriff, who ran Boswell's ginning operation before he, too, was let go.

Boswell's penchant for severing relationships extended out of the office and into his personal life. John Grant, one of Boswell's Sierra hiking companions, got a call from J. G. during the court case in Bakersfield. "Got any time on your hands?" Boswell asked, his own appearance on the witness stand now completed. "I want your impression on how we're doing."

Grant agreed to start attending the trial and pass along any reactions. One day, a Boswell board member named Michael Morphy, the former chairman of California Portland Cement, was called to testify. "Mike was a dandy," Grant later recalled, "good looking, a tennis player, socially connected, terribly glib. Well, I can just see that the Kern County jury thinks this guy is a suede shoe"—exactly the image that Ralph Wegis, the small farmers' lawyer, was trying to paint. Concerned, Grant phoned Boswell and told him: "You better find a way to get Morphy out of here." For some reason, such candor wasn't what Boswell wanted to hear, according to Grant. "He barked back, 'Your observations aren't any help to me. You walk your side of the street, and I'll walk mine.' I didn't talk to him for years after that. Almost everybody has had that experience."

For Boswell's executives, getting axed wasn't all bad. The severance packages that J. G. furnished were so lavish, "we're all living lives we never dreamed we would live," said George Lawrence, who ran the company's sales department before he was pushed out. Boswell minted millionaire after millionaire this way. It was the oddest thing: Lose your job, get rich. No wonder Gamble called his firing "the best thing that ever happened to me."

A few devotees—still in the company's employ, of course—defended Boswell's tactics. "If he feels like something has to change, it isn't arbitrary," said Ross Hall. "I mean, it's a series of events over a period of time. And if certain people aren't wise enough to see that occurring, then I feel sorry for them."

In the case of Fisher's firing, Boswell seemed motivated by two things. The first and most obvious was that he needed to make room at the top of the company for his son, J. W. This quickly became the conventional wisdom for why Fisher was tossed aside, and there was surely a lot of truth to it. The other factor was the one that Boswell himself would cite as the real reason for Fisher's downfall. Once again, it went right to that tender spot: the size of the empire.

By the early 1980s, Boswell feared that the federal government was going to break up the company by whatever means it could muster. "All businesses are subject to government roadblocks and harassment," he reminded his employees. "Our principal business—corporate farming—is, however, both politically and socially unpopular." Even after the company had won the acreage-limitation fight in 1982, Boswell fretted that the government would still "find some way to shut us down," said Fisher.

Over time, the disappearance of the family farm had become a well-worn lament in the halls of Congress and academe. Farms were getting larger and larger—swelling from an average of 147 acres in 1900 to 426 acres in 1980—and rural America was going through some wrenching changes as a result. At more than 150,000 acres, Boswell wasn't really emblematic of this trend. It was a little like saying that people have gotten taller over the years, with American males growing from just over five feet seven on average during colonial days to more than five feet nine today. Yet how many people stand seven feet tall? Boswell was an anomaly—the equivalent of a circus freak or an NBA center—even in a world of agriculture fast becoming industrialized. Still, the Boswell Company was easy to single out—the direct opposite, in many eyes, of the small farmer that America has always so loved to lionize.

"The Boswells are mythical figures," said Dan Beard, who wrangled with the company as a congressional aide and Carter administration official and later became commissioner of the Bureau of Reclamation under President Clinton. "Somebody once described a classic as a book that everybody talks about and nobody's read. The Boswells are like a classic. They're a group of people everybody talks about and nobody's ever met. I've spent a career dealing with the Boswells in some way—and I've never met any of them.

"My side set up a dynamic, which was the 'big guys' against the 'little guys,'" Beard explained. "And they represented the big guys."

It was in this big-is-bad environment that Jim Fisher came to Jim Boswell with a proposal to get even bigger. He had his sights on buying Southlake Farms, a 27,000-acre spread in the Tulare Lake area. Boswell rejected the proposal instantly. "For a kid from Strathmore, you sure got some big ideas," he said.

Fisher thought about fighting Boswell on this, and he even drafted a memo on the subject for the board of directors to consider. "During our last board meeting," he wrote in December 1983, "there was an expression of concern about the size of the Boswell Company, particularly its land holdings and its public and political visibility. The essence of this discussion was that the size of the Boswell Company might be becoming a detriment to its continued success and possibly its very existence and, therefore, a change in corporate direction might be in order. . . .

"At the risk of boring you with old sayings, 'You can only get so pregnant.' We *are* pregnant. We are the largest farm operator in the world. To my knowledge, we are the largest producer of each of the four major crops that we grow"—cotton, wheat, safflower and alfalfa. "We could cut our agricultural holdings in half and still be the largest producer of at least three of these crops. . . . I clearly believe that we should not pass up a good investment in agricultural land for fear of public or political reprisal."

Fisher ultimately decided against sending the memo. It was one of the few times he had ever chickened out of anything. Boswell "would have been furious," he said. "It was a hill that I didn't want to die on."

Little did Fisher know then that he was about to be taken out anyway. The morning he was told his career was over, word spread swiftly through the office. His colleagues were flabbergasted. "I knew he had his differences with J. G.," said John Sterling, the Boswell lawyer, "but I didn't think it was that bad. And then all of sudden, bingo!" Fisher went over to see Ed Giermann, another Boswell attorney, who considered Fisher his mentor.

"I can't believe it," Giermann told him. "Every place I go, whether it's Sacramento or Washington, you *are* the Boswell Company."

"Have you ever thought that might be my problem?" Fisher replied.

As for Jim Boswell, he was never one to look back at personnel moves like this. His gaze was always fixed ahead. He had decided to remove certain people from his company, and from his life, and what was done was done. Still, after he sacked Jim Fisher, there were other losses that undoubtedly weighed heavier on Boswell—three deaths that came within a year and a half of each other: his brother, Billy Jr.; his mother, Kate; and his aunt, Ruth Chandler.

BILLY BOSWELL MET DEATH THE SAME WAY HE lived most of his life: not far from a bottle of booze.

That he had made it to age seventy was a miracle, really. He had lived hard, with vodka his drink of choice. "You could tell when he was drunk," said A. L. Vandergriff, the Boswell gin man. "He was wild and yelling and screaming at people."

Some said that Billy deliberately cultivated this image, turning it to his advantage. "He wanted people to think that he was a rebel," said Oleta Gordon, who made Billy's acquaintance one night at the Brunswick, the tavern that he owned. He was in his mid-sixties then, she in her mid-forties, and before long they fell in love. "Billy's reputation was almost a joke," Gordon said. "People thought he was an easy touch. But I'm going to tell you something: He never did business drunk. When it was time to go over his bills or make his payroll or negotiate a crop loan, that guy was stone cold sober. He knew exactly what he was doing."

The drinking stopped altogether a couple of years after the two had met, when Billy began undergoing dialysis. A nurse from Fresno taught

Gordon how to administer the treatments, and she moved into Billy's house to care for him. The process was frightening. Often Billy would lose consciousness, and Gordon wasn't certain whether he'd come to. "I was on pins and needles for four hours," she recalled. Billy, though, trusted her completely. When they'd go to the horse races in Southern California and check into St. Mary's Hospital in Long Beach, he insisted that Gordon be the one to hook him up to the dialysis machine. "She's the only one who doesn't hurt me," he'd say.

Eventually Billy's health deteriorated completely. He was attacked by throat cancer, and though Gordon pleaded with the doctors to keep the diagnosis from him, since it was already too far gone to treat, he figured it out anyway. Driving back from the hospital in Fresno one day, he said suddenly: "I have cancer, don't I?" Gordon acknowledged that he did and pulled the car over.

"How long do I have?" he asked.

"You've already passed that point," she answered.

Then Billy looked straight at her and said, "Would you have a drink with me?" She told him yes and bought some vodka and orange juice on the way back to Corcoran. This time, he quit after just one drink. He survived another month—"hung on to life, just so he could tell me, 'You're going to be all right,'" said Gordon.

J. G. and his brother had been through some rough patches. After the '69 flood, when J. G. had washed Billy out, the two didn't speak for a long, long time. Some of J. G.'s closest buddies barely knew he had a big brother. "He wouldn't spend time talking about some weakness in his family," said J. G.'s good friend Dick Jones. But J. G. reached out in the end. When Billy couldn't farm anymore and needed help financially, J. G. agreed to buy his ranch, the McCann—one share of Boswell Company stock for each acre. He also showed his appreciation for all that Gordon had done. After Billy succumbed in the summer of '86, J. G. arranged to let her stay in his brother's house for as long as she wanted.

Mom Kate died late the following year, at age ninety-one.

She had been alone since the fall of 1970, when husband Bill passed away. He had mellowed in his old age, in no small part because he had given up his own drinking—cold turkey—about ten years before he was laid to rest. He had gone into the hospital for a hemorrhoidectomy, and his doctor worried that Bill might get belligerent because he couldn't have any beer while he was there. But Bill surprised everyone with his good behavior. "He was absolutely the model patient," said his daughter Jody. "You've never seen anything like it." Once Bill was discharged, he went to Jody's house to recuperate for a week or so. His grandkids would come in while he was taking his sitz baths and ask him, "Big Bill, want me to get you a beer?" To everyone's amazement, he kept refusing. "No," he'd say, "I think I'll wait a while."

"He had spent a lot of time lying there in the hospital," Jody surmised, "and he probably thought, 'I don't need it.'" Regardless of why, it was no more Coors from then on. Bill Boswell was strictly a 7-Up man now.

Toward the end, driving around Corcoran in his big white car—his long face shaded by his fedora, his fingers and teeth stained from Pall Mall reds—"Mr. Bill," as he was known, carried himself like a dignitary, waving and smiling at passersby. He spent a lot of time at the horse races in Los Angeles and gathered around the radio with Kate listening to baseball games—he rooting for the Dodgers, she for the Giants. There, the language of their southern past would seep out. "Kate," Bill would say in his Georgia drawl, "my niggers just beat your niggers." Mr. Bill died in a hospital bed with a transistor radio pressed to his ear, taking in the World Series. He looked up at the attending physician and said, "There's a man on first and third." Those were his final words.

Kate carried on the best she could. "I'm cooking, I'm setting a pretty table, I'm arranging flowers," she said some six years after Bill was gone. "I love that. It keeps me busy. And since I lost Bill, I've been trying to pick up my music. I've been practicing my piano about an hour a day." She also let her thoughts drift back to Greensboro and Boswell's Crossing, to what she called "the happiest years of my life."

"Those days, they had those Negro servants that were passed down from one generation to another," Kate said. "And they're the best race of people on earth. They don't have them any more. Occasionally you find a good one, but—I don't know."

Ruth Chandler died a year later, in December 1987, about six months before the trial in Bakersfield started. She was ninety. J. G.'s Aunt Ruth had remained involved in Boswell Company affairs for years, with her nephew giving her regular updates on everything from projected earnings levels to crop reports, and she continued to head the Colonel's charitable foundation.

She married twice after the Colonel's passing, first to Sir William Charles Crocker, a British baronet. Then, late in life, she wed Karl

Godfrey von Platen, a retired California lumberman whom she had known since the 1930s. But it was her Boswell connection, and the big block of company stock the Colonel had left her, that helped keep Ruth Chandler von Platen ensconced on the *Forbes* list of the richest Americans.

As THE COURT CASE IN BAKERSFIELD PROCEEDED, Ralph Wegis not only did a masterful job of making Boswell into Goliath. He just as skillfully rendered each of his three clients—Jack Thomson, Jeff Thomson and Ken Wegis (his second cousin)—as David.

Four days after Jim Boswell took the stand, it was Jack Thomson's turn. He had farmed in Kern County since 1948, raising cotton, alfalfa, sugar beets, tomatoes, potatoes and wine grapes. If his 2,850 acres constituted a few too many for him to be considered a "small farmer" in the true Jeffersonian sense of the term, his operation—with his sons as partners and wife as bookkeeper—sure seemed modest compared with Boswell's.

Wegis took Thomson through the shock of getting sued for libel, from the moment the process server rapped on his door one evening and handed over the papers. "How did you feel when you got to the $2.5 million" that Boswell was seeking in damages? Wegis asked.

"Well," Thomson said, "maybe they could get that much if we sold all of our land and so on. But certainly, we didn't have that kind of money."

Wegis knew that the biggest thing he had to blunt was any perception that his clients somehow deserved to have been sued by Boswell, given the derogatory nature of their ad. So he used Thomson to establish some needed context and advance the view that the small farmers had merely been fighting fire with fire. Wegis did this by zeroing in on another advertisement, this one created by Boswell's side during the Peripheral Canal campaign. The ad, which highlighted that the big oil companies in Kern County were backers of the project, was plenty forceful in its own right. "STOP the $20 billion construction pork barrel before it drains us dry," it said. "The benefits of this immense windfall will be reaped by Shell Oil . . . Getty . . . Union Oil . . . Southern Pacific . . . Tenneco. . . . "

"Did you feel there was another side to that story, a Kern County side?" Wegis inquired of Thomson.

"Most of us here in Kern County, many hundreds of us, knew we were family farmers," Thomson replied. "We were upset by the constant portrayal as being nothing but large oil companies."

As assertive as Wegis was, Boswell attorney Harvey Means remained extraordinarily passive. During cross-examination, his questions were weak, his phrasing flat. He was also slow to make objections, allowing Wegis to keep pounding away, proclaiming that Boswell's camp had been "telling the lies about Kern County." Time and again Wegis used the word to great effect: "lies," "lies," "lies." In fact, the more Wegis went on, the more it started to look like the campaign literature generated by Boswell's side of the Peripheral Canal contest—not the advertisement his clients had put together—was the real smear job.

After being hit with the libel suit, Thomson said, his life became full of angst. Getting a crop loan was more difficult. His insurance rates went up. He and his family had to cope with the constant fear that "we were just about out of business." Becoming heavily involved in a political campaign again was unthinkable, even if it was something he really believed in. He couldn't afford to take the risk.

Over the next two weeks, a parade of witnesses lined up to tell the jury the same sob story. Mary Louise Thomson, Jack's wife, described how a $2.5 million judgment "would have taken away everything we've got." Jeff Thomson, Jack's son, talked about being "terrified of the amount of the suit." For years, as Boswell pursued the libel case, it remained "a gnawing thing," he said.

Ken Wegis, the third generation in his family to farm in Kern County, said he was stunned when his son tracked him down during a vacation in Oregon to relay that Boswell had sued him for millions. "I never dreamed such a thing would happen," he said. When they had fully digested the terrible news, said Ken's wife, Vivian, "we cried a little," thinking "that was the end of our life as farmers."

The list of victims, as presented by Ralph Wegis, went beyond his own clients. The way Boswell had originally filed its suit, 1,000 "John Does" were named as defendants, leaving open the possibility that many others could be dragged into the case later. Rumors flew through Kern County that anyone who had given money toward the advertisement faced the same $2.5 million threat. Peter Delis, a small grower who had given Jack Thomson $100 to put toward the ad, was among those convinced he had "something to worry about," even though he really didn't.

"I remember the time because I had just had a baby," Delis's wife, Davida, told the court. "It was a time when I needed a lot of support from him, moral support, and instead he would come home and be worried about . . . this suit by this huge company."

Over the course of the trial, Wegis continued to elicit testimony against Boswell that made the company seem cynical and ruthless. For example, when he asked Jim Fisher about the ad bashing the Kern County oil companies, Wegis got him to concede that he thought it was "kind of a cheap shot." But Fisher laid such qualms to rest when a pollster told him how effective the ad was. "It bothered me for three days," admitted Fisher, "and I got over it."

Through his probing, Wegis also brought out how Boswell had stopped doing business with a small agricultural chemicals supplier in Corcoran called PureGro because its parent, Union Oil, championed the Peripheral Canal. Boswell was such a dominant customer, PureGro soon had to shut down.

"When you cut them off . . . it closed them up, didn't it?" Wegis asked Fisher.

"I heard that it closed them up."

"Did that bother you at all?"

"No."

Fisher did his former employer no favors by exhibiting such arrogance. And Wegis would bait him to do it again, essentially asking whether he thought Boswell's executives opposed the canal because they were smarter than the small Kern County farmers who wanted to see it constructed.

"Did you think that this issue was so complex that it was beyond the ability of Kern County cotton farmers to understand it?"

"A good number of them, yes," Fisher said. "It's a very complex issue."

When he wasn't making them out to be haughty, Wegis made Boswell's men look positively sleazy. He revealed, for instance, how the company had given $17,700 to help bankroll a state Senate race being run by a woman named Christine Herzog. As it turned out, Herzog's candidacy was little more than a Trojan horse designed to carry Boswell's anti-canal message. A welfare mother who had been instructed to make no public appearances, Herzog never even met her "campaign manager" until after the election was over. All she had to do to win Boswell's money was agree to condemn the Peripheral Canal in a blizzard of mailers sent to homes in her Southern California district.

Toward the end of Jack Thomson's second day of testimony, Wegis circled back again to his central theme: Boswell was the big guy and therefore the bad guy. He did this by playing up the S-word that J. G. so hated—subsidies. Specifically, Wegis directed Thomson to comment on the cheap "subsidized water" that a giant operation like Boswell had access to.

Funny thing was, this was one area where the normally aggressive Wegis could have worked the facts a lot harder than he did.

THAT THE FEDERAL GOVERNMENT has readily extended its tit to the farmer, and that the farmer has happily latched on, is hardly a new notion.

"One might almost argue that the chief, and perhaps even only aim of legislation in These States is to succor and secure the farmer," H. L. Mencken wrote in 1924, a year after the Colonel first arrived in Corcoran. "When the going is good for him he robs the rest of us up to the extreme limit of our endurance; when the going is bad, he comes bawling for help out of the public till. . . . There has never been a time, in good seasons or bad, when his hands were not itching for more."

For all the truth in Mencken's acid pen, the situation at J. G. Boswell was more complicated than that.

On one level, Jim Boswell and his management team have genuinely wanted the government's heaping smorgasbord of crop subsidies to come to an end. The company has balked at participating in certain farm programs and even lobbied to curtail government intrusion in the cotton industry. And why not? Subsidies have principally helped generations of smaller farmers—especially the inefficient, high-cost growers of the South—stay alive. If these declining operations were left to fend for themselves in a truly free marketplace, many wouldn't last, and a well-run behemoth such as Boswell would be in position to gobble up an even bigger slice of the pie.

Yet Boswell was far from pure in its convictions. There have been at least a couple of junctures in its history, moments when the farm economy was dreadful, that the company needed a boost from the government to remain profitable—and gladly accepted the help.

On many other occasions, Boswell flat-out milked the system, collecting millions and millions of dollars in government aid in all its forms: flood control benefits, cheap irrigation water, subsidized crop insurance premiums, so-called payments-in-kind, "marketing loans" that eliminated the risk of planting, cash rewards for exporting. At times, the company has acted as if government relief was some kind of birthright. For instance, when flooding struck Tulare Lake during the Colonel's day and the government was slow to pick up the tab for levee repairs, Boswell's Louis Robinson bristled at the "raw deal" the company was getting. This was "assistance," he said, "to which we are entitled."

All things considered, the company's long-running affair with Uncle Sam is perhaps best characterized as a love-hate relationship, where bouts of desire have been tempered by sincere feelings of disgust for those who suck on the U.S. treasury. At best, Boswell has been ambivalent on the issue; at worst, hypocritical. Either way, its name has become, as one critic has noted, "synonymous with . . . subsidies."

The company's dance with the dole began at the start of the modern era of farm programs, during the New Deal. It was the fall of 1937, and Congress was debating a new version of the Agricultural Adjustment Act. As with the original AAA of 1933, the goal was to limit the amount of land that each state could plant to cotton, thereby curtailing supply and spurring prices higher. However, lawmakers were planning to base these restricted plantings on historical trends, and because the Western states were so new as cotton areas, this approach would have forced Boswell and other growers in the region to scale back their acreage to almost nothing. The threat was so grave, Fred Sherrill later remembered, it would have "meant the end of the cotton business in New Mexico, Arizona and California."

Colonel Boswell sent Sherrill to Washington to see if he could somehow fix things, though it seemed a long shot. The South enjoyed so much political clout when it came to cotton, it was difficult to get more than a handful of lawmakers to care about the fields of the West. Sherrill approached Senator Carl Hayden of Arizona, but the Democrat was skeptical that the problem could be resolved. The trick, according to Hayden, was to devise a new cotton formula that would "save the irrigated West" but not hurt any other part of the country—an equation, the senator noted, that he had thus far found elusive.

Sherrill chewed over the problem into the evening, only to get nowhere. Then, in the middle of the night at his hotel, the answer suddenly dawned on him. "At about two o'clock in the morning," he recalled, "I woke up, picked up paper and a pencil and wrote down a few notes." The next morning, Sherrill raced straight to Hayden, who was sold at once on the idea. Every state could grow 70 percent of what it had planted in 1937—a year in which no production controls were in place and California's output had soared. The beauty of Sherrill's brainstorm, which Hayden wrote into the legislation after an impassioned speech from the Senate floor, was that it increased the West's share of the harvest without throwing the entire national quota system too far out of whack. California alone saw its allotment grow by more than two-thirds.

Fred Sherrill's fashioning of the AAA helped launch a rivalry—West versus South—that would endure for the next thirty-five years.

From the 1940s through the 1960s, tensions flared between California growers such as Boswell, which wanted to plant all the premium cotton it could, and southern farmers who were perfectly pleased to idle their land and take government handouts. "The southerners were brought up to believe that everybody owed them a living because they're the lords of the manor," said Julian Hohenberg, whose Memphis, Tennessee, firm was one of the biggest cotton merchants in the country during this period. "The California guys said, 'Fuck subsidies.' They were pioneers."

But it wasn't so simple to just say no. Some years, the government built enough flexibility into its cotton quota program that California's big growers could increase their plantings to tolerable, if not exactly ideal, totals. In other years, however, production was greatly hampered, and all the frenzied lobbying to make it otherwise did no good. "The upshot," cried an editorial in the *Corcoran Journal*, "is an unrealistic system . . . which discourages expansion of Western cotton and sustains the fiction that Southern cotton is still a highly desirable product."

In the early 1960s, Boswell received a $50,000 check from the government for removing some of its land from production. True to its philosophy, the company sent the money back to Washington, just as Jim Boswell would later recall, though he remembered the check being "in the millions." And despite Boswell's further recollection that the practice of returning subsidy payments continued for several more years, the reality was that the company returned this one check and no more.

In 1966 the government began imposing a strict mandate on growers to take part in the cotton program. Boswell, having little choice now but to play along, raked in $21.5 million over the next five years. The $4.4 million the company received in 1970 ranked as the single biggest crop subsidy in the nation. The average payment to small farm operators, by contrast, was just $63.

It wasn't political clout or machinations that allowed Boswell to pocket so much government money. The fat checks it received naturally reflected the tremendous number of acres that the company had to leave fallow. Nor was Boswell to blame that a program purportedly designed to help the "family farmer" was throwing so much cash the big boys' way. In one sense, Boswell was only doing what the government had required. "The penalty if you plant in excess of your allotment is punitive," complained the company's Rice Ober.

Still, as the dollars piled up, so did the scorn. The *New York Times* slammed the cotton program for helping "widen the gap between the rich and the poor." The United Farm Workers branded Boswell "an international welfare bum" for receiving a bounty in Australia to grow cotton and subsidies in the United States not to grow. *Life* magazine, in November 1970, scoffed at Boswell's predicament, showing J. G. standing in front of a row of cotton pickers along with the caption: "Unhappy farmer who gets $4.4 million . . . He wants to get out of the subsidy business. But the government won't let him."

Alarmed at the outsize disbursements to Boswell and a handful of others, Congress soon capped crop subsidies at $55,000. Then, in 1973 the farm program changed its emphasis altogether. In an effort to respond to what agriculture secretary Earl Butz called the "ever-growing world-wide demand for food and fiber," the government encouraged more crops—rather than fewer—to be grown. To control federal expenditures, the payment limit for price supports was lowered to $20,000. With the shackles off, the company ramped up production, and the spotlight on Boswell as some sort of corporate welfare queen gradually cooled down.

"Everybody thought that Boswell would go broke when they didn't get government subsidies," recalled the company's A. L. Vandergriff. "But Jim Boswell said it would be the best thing that could happen to us.

'We'll make our money because we can grow cotton cheaper and of a better quality than anybody else.' And we did it." Through the mid-1970s, Boswell racked up record earnings, aided by a worldwide boom in commodity prices.

A decade later, however, the glare of the subsidy spotlight found Boswell again. In 1983 the company entered into a newfangled program that offered commodities instead of cash to farmers who left their land unplanted. By agreeing to take 14,000 acres of wheat out of production, for example, Boswell was supposed to receive a supply of the crop that had been stored in government warehouses. It was a nice little arrangement. The company would get a pile of wheat to sell without having to spend a nickel to cultivate or process it. But the government's gambit backfired. So many farmers signed up for the payment-in-kind (PIK) program that it ran out of wheat to give away. With no where else to turn, the Agriculture Department had to buy nearly 30,000 tons of the stuff that Boswell did grow for $3.7 million. Then it handed the crop right back to Boswell to honor its obligation under PIK. The company could now sell the wheat all over again, this time on the open market.

All in all, it was such a bonanza for Boswell that it prompted one California congressman, Pete Stark, to quip that government scientists had apparently "developed hybrid strains of wheat . . . that bloom with dollar bills, creating a rare new species for the enjoyment of rich corporate farmers."

Yet it wasn't just the magnitude of the windfall that raised troubling questions. As it happened, the wheat fields that Boswell had agreed to leave bare couldn't have been farmed anyway; most of the lake bottom had been flooded that year. In the view of many, it was an outrage—getting compensated for not growing crops on land that couldn't be sown to seed in the first place. But "let's face it," said John Smythe, a federal agriculture official based in Kings County, "they were allowed to do it under the program. They didn't do anything illegal." Adding insult to subsidy, Boswell and other area farmers then put in for as much as $15 million in federal disaster relief to help drain their sodden fields.

Inside Boswell, feelings were mixed over the federal funds that came pouring in. Executives from J. G. on down continued to talk about how they hated subsidies. Others, however, argued that the company had little choice but to snatch up the dough. The sums being dangled in front of Boswell were so substantial that it would have been crazy—if not an outright breach of fiduciary duty to their small group of shareholders—for the board to spurn the money. "There shouldn't be subsidies at all. It's one of the great boondoggles," said Jack Parker, the former vice chairman of General Electric who served as a Boswell director from the late 1970s through the early 1990s. "But if it's there, you can't very well walk away from it."

In 1985 the subsidy debate took a new turn, as the American farm belt was hit by its worst crisis since the Great Depression. Three years into a down cycle, agricultural prices continued to slide, even as drought plagued certain parts of the country. Some of the misfortune was self-inflicted. Betting they could get rich, farm families had bought up acre after acre during an earlier boom period when land and crop values seemed destined to escalate forever. Now they were paying the price, with increased foreign competition and a strong dollar shaking up world markets. At home, credit evaporated and farm foreclosures cut through the Midwest "like a scythe," in the words of one observer.

Magazine headline after magazine headline told the sorry tale: "The New Grapes of Wrath" (*Time*), "Bitter Harvest" (*Newsweek*), "America's Farmers Down the Tubes?" (*U.S. News*). Jesse Jackson challenged the Reagan administration to spend less on Pentagon weaponry and more on rural communities. "Farms not arms!" "Farms not arms!" the reverend said. Hollywood captured the gloom through such films as *Places in the Heart*, *The River* and *Country*, and sixty-four musical acts jammed their way through the first inaugural Farm Aid concert in Illinois.

For a time, California's farmers managed to escape the depths of the hardship. But by late in the year, they too were in trouble—the biggest of them all, no exception. "Unfortunately, the current farm situation forces us to . . . cut operations and costs," Jim Boswell wrote to his employees in November. "This means there will have to be significant reductions in payroll." Boswell attributed much of the pain to the subsidy programs that had put millions of dollars into the company's own coffers. The endless federal payments to growers, he said, were at the root of "the surpluses that keep commodity prices extremely low."

Boswell wrote to President Reagan and leaders on Capitol Hill, urging them not to repeat "the mistakes of the past" and to veer instead toward a market-oriented system that would immediately abolish acreage-reduction programs and rapidly reduce the level of income support to all farmers. "Times are tough, so it's high time to be tough," Boswell advised the president.

But rather than get off the farm, Uncle Sam dug in deeper. Republican lawmakers from the Midwest looked past their laissez-faire ideology and caved in to pressure from their hard-up constituents. Just before Christmas, President Reagan signed a farm bill that, admittedly, didn't go nearly as far as he wanted in embracing his free market principles. Among the legislation's deepest flaws in this regard were a couple that would, once again, enrich J. G. Boswell.

Ostensibly the Food Security Act of 1985 had put a $50,000 ceiling on the "deficiency payments" a farmer could receive. But the ceiling was full of holes. It didn't apply to a major program that allowed farmers to pocket the difference between the depressed world market price for cotton and the far higher "loan" rate established by the government. The cap also didn't affect certain "first-handler" payments that marketers of cotton, such as Boswell, received.

The bottom line was that the subsidies due the company in 1986 ballooned to $20 million—an amount the Agriculture Department's chief economist called "obscene." Although Jim Boswell had told President Reagan that he thought the limit on subsidies should be a hard-and-fast $25,000, his men made no apologies for taking 800 times that much. With the farm sector in peril, said Boswell lobbyist Wes McAden, grabbing the money is "mandatory economically. You don't know where the market's going." Walter Brown, another Boswell executive, said the $20 million was the difference between profit and loss. "This," he said, "is survival money."

In October, Congress went back and rejiggered the '85 law to put a more solid $250,000 subsidy ceiling in place. As an Agriculture Department spokesman explained it, there was no mystery as to what was behind the change. "It was designed to get one guy: Boswell," he said.

It wouldn't be the last time that America's biggest farmer engendered such ire.

WHEN HE WASN'T REMINDING the Bakersfield jury about Boswell's mammoth size, Ralph Wegis was working his witnesses to make another point: This was a company that was always looking for a rule to bend, a loophole to slither through.

With this in mind, Wegis asked questions about Boswell's Boston Ranch, a 23,000-plus-acre cotton farm northwest of Tulare Lake. Unlike the bulk of Boswell's land, which was irrigated by Pine Flat and the other Army Corps dams, Boston Ranch was watered by the Bureau of Reclamation's Central Valley Project. Here, acreage limits most certainly did apply.

Or at least they were meant to. Under the 1982 law, farmer-owned land that received water from a federal reclamation project was supposed to be whittled down to 960 acres. Farmers, meanwhile, could lease as much additional land as they wanted beyond the 960 acres that they owned directly. But now they had to pay "full cost" for the water dispensed to this leased land. No longer would they be eligible for a subsidized rate.

"Do you know," Wegis asked his client, Jack Thomson, "whether the Boswell Corporation is paying the full cost of the water on the Boston Ranch?"

"No. Best knowledge I have is that they are . . . negotiating with the Bureau of Reclamation" to continue farming "the full 24,000 acres" and still get water at a bargain price.

If this exchange flew right past most of the jurors—and it was bound to have—it wouldn't be long before the subject of Boston Ranch resurfaced in front of a much wider audience, finding its way in front of the probing cameras of *60 Minutes*. Indeed, in the arena of American land and water policy, few proposals would ever be considered as controversial as Boswell's plan for the property.

As a smiling picture of J. G. and J. W. flashed on the screen, Ed Bradley took his *60 Minutes* viewers into an agricultural realm that was "about as far from *Little House on the Prairie* as you can get." One expert, he noted, had calculated that the government in a single year picked up an $11 million tab, or 85 percent, of what it cost to deliver water to Boston Ranch, where Boswell grew "surplus cotton." Then came a classic *60 Minutes* zinger: "We wanted to ask James W. Boswell why taxpayers should subsidize his company's water bill, especially to grow more cotton than we need. But Boswell, who lives several hundred miles away in a plush Los Angeles neighborhood, declined to be interviewed."

Boston Ranch sat far enough north of Tulare Lake that it found itself in the Westlands Water District, a 600,000-acre plain that had its own cast of knaves, wonder boys and giants. Westlands was the largest agricultural water district in the nation, covering 1,000 square miles of ancient seabed, an area the size of Rhode Island popping with cotton, lettuce, tomatoes, garlic, onions, cantaloupes, broccoli and beans. It had been nothing more than jackrabbits and sagebrush, the hideout of Mexican bandit Joaquin Murieta, when Wylie Giffen planted his first cotton crop in 1923.

Giffen had arrived on the valley's west side flat broke, a paper millionaire who had lost everything but his appetite for gambling during the 1920s raisin bust. The land he chose for his second chance was practically free. Given the lack of even rain water, free might have been too much. There was no river, much less a lake, from which to irrigate. The prairie land did boast a vast supply of water 2,000 feet below the strata of clay. Problem was, a farmer needed a considerable chunk of money to punch holes in the clay and plant wells deep enough to siphon the water.

Giffen drilled a few wells and scratched out a crop before dying in 1936. He would leave it to his son, Russell, a high school dropout, to transform the west side and become California's first Cotton King.

Russell Giffen lived in the grand manner of a southern plantation owner. When it came time to build his dream house, he took a page right out of Tara, erecting a Georgian mansion on 300 acres along the majestic Kings River. One young visitor couldn't help admiring all the hard work that had gone into bending the mansion and its grounds to fit into a cranny of the river. "No, son," Giffen said. "We bent the river to fit the house."

He could make the federal government bend to his will, as well. In the 1950s Giffen and a group of west side growers persuaded the Bureau of Reclamation and the state of California to pony up more than $500 million to build a dam in the barren hills west of Los Banos. A system of canals alongside the Coast Range would shunt 1 million acre-feet of Northern California water to Westlands. Thousands cheered as President John F. Kennedy and Governor Pat Brown stood on a platform in the summer of 1962 and detonated the first blasts of dynamite on a project that guaranteed a cheap flow of water for Giffen and his hydraulic brotherhood.

But it wasn't long before the Westlands Water District became defined as much by its political battlefields as its cotton fields. Farmers began squabbling with the government—as well as with each other—over the subsidized water delivered by the Bureau of Reclamation. In the mid-1960s, the original 400,000-acre Westlands had merged with another water district called Westplains, and the marriage bred all sorts of difficulties. Certain farms that were part of the original Westlands area—especially Boswell's Boston Ranch—were adamant that they had first dibs on the water. In times of shortage, some of the Westplains farmers could conceivably wind up without a drop. The majority of the Westlands board of directors, however, favored a more balanced allocation that would let everybody get something, no matter the circumstances.

At the same time, the feds began to balk at the contract that had been signed with Westlands back in 1963, in the wake of President Kennedy's visit. Though the government had promised to deliver a steady supply of water at a cheap price, officials "kept interpreting the contract differently all the time," groused Ralph Brody, the fiery manager of the Westlands district. "There was no certainty in the thing."

The bickering between the feds and the growers—one government official would call the farmers "welfare queens"—lasted through the 1970s and into the mid-'80s. Finally, though, Boswell and other Westlands farmers prevailed. Among other things, the government gave them a $45 million refund on what they had been charged for their water since the late 1970s, when the Carter administration raised prices. The Boswell Company itself soaked up about $3 million of that. Donald Hodel, the new Interior secretary, said the government had no real choice at the end of the day but to abide by the 1963 contract, even if it wasn't a perfect arrangement for the government. "It has often been observed," he said, "that great men and great nations keep their word." Critics, convinced that the farmers of Westlands were getting subsidies far beyond what Congress had ever intended, blasted Hodel for capitulating. Boswell and the other Westlands farmers, fumed one, had been handed a "sweetheart deal."

Still, no deal lasts forever—or at least it wasn't intended to under reclamation law. The way the rules were written, a farmer with more than 160 acres (later 960) was required to sell off his "excess lands" within ten years if he wanted to keep getting water at a subsidized price. Land reformers such as George Ballis claimed that Westlands growers were getting around the law by setting up "paper farms," putting 960-acre chunks in the names of five children, fifteen grandchildren, cousins, aunts and uncles. This skullduggery allowed some growers to maintain hidden control over their vast empires.

Yet despite such tricks, a number of the old Westlands empires were carved up into smaller holdings, just as the law had intended. After some fits and starts, Boswell's own ten-year time clock at Boston Ranch was set to expire in February 1990. The question, then, wasn't whether Boswell would sell; the question was to whom.

The company's blueprint called for selling the more than 23,000 acres at Boston Ranch to something known as the Westhaven Trust, the beneficiaries of which would be 326 of Boswell's own employees. The way the '82 law had been written, land held by a trust could continue to receive subsidized water—even if it exceeded 960 acres. Boswell, in turn, would draw fees of about $2 million a year as "farm manager," working the land in a single big block. That wasn't as good as the $8 million to $10 million in profit that Boswell brought in from Boston Ranch when it owned the dirt outright, but it sure was better than nothing.

The company defended the proposal as giving hundreds of people a stake in the land—just the way Teddy Roosevelt envisioned when he put the reclamation program into place nearly ninety years earlier. "We could sell it to twenty-five of J. G.'s closest buddies," said Ed Giermann, the Boswell lawyer. But by selling it to a trust that includes 326 employees, "I view that as spreading reclamation benefits really widely." In short, he said, the Westhaven Trust met both the letter and the spirit of the reclamation law.

But Boswell's foes insisted that if it met the former—and that was debatable—it certainly didn't meet the latter. George Miller, the congressman, called Boswell's plan "nothing more than a transparent scheme" to dodge the law. Even other Westlands farmers had trouble swallowing the idea. "This is typical Boswell arrogance," said Jason Peltier, manager of the Central Valley Project Water Association.

When all was said and done, the name-calling accomplished little. After some initial reluctance, the first Bush administration approved the Westhaven Trust. The water subsidies would continue to flow to Boswell's Boston Ranch for another full decade. And the company would remain a poster child for corporate welfare long into the 1990s.

WITH THREE WEEKS OF TESTIMONY BEHIND HIM, Ralph Wegis began his closing arguments in the Bakersfield trial. Subtlety was not his style.

"Corporations are designed for making money," Wegis told the jury, "and that corporation can also be a mindless monster in pursuit of profit. . . . It is kind of surprising to think that when you're down here on the ground in Kern County that somebody from the forty-sixth floor of a Los Angeles skyscraper could have such an effect on your life. But when you have the kind of money and muscle and might that the Boswell Corporation has, you have your own vision. You have your vision from the forty-sixth floor and you have your vision . . . as you fly over Kern County . . . and you look out your window and you have big plans because you're a big shot, and you don't think of small things like putting your kids through college. You don't think about buying a duplex or a rental. You think about carving your mark on the face of the earth."

Harvey Means, the Boswell lawyer, made his closing the next morning. There was little he could say that would make any difference now. Ralph Wegis had already stolen the show.

"Mr. Taylor, has the jury reached a verdict in this case?"

"Yes, we have your honor."

The verdict was then handed off to the bailiff, who passed it to the clerk to be read aloud.

"Question No. 1: Is defendant J. G. Boswell Company liable for malicious prosecution as to Jeff Thomson? Answer: Yes.

"Question No. 2: Is defendant J. G. Boswell Company liable for malicious prosecution as to Jack Thomson? Answer: Yes.

"Question No. 3: Is defendant J. G. Boswell Company liable for malicious prosecution as to Ken Wegis? Answer: Yes."

The clerk made his way through ten more such questions, the answers all going against America's biggest farmer. Did Boswell act in good faith upon the advice of counsel? No. Did Boswell commit an abuse of process? Yes. Did Boswell commit an intentional violation of the constitutional right to free speech? Yes. Was Boswell's conduct malicious, oppressive or fraudulent? Yes.

It was a total victory for the three small farmers. The jury awarded the trio compensatory damages of $3 million, $1 million apiece. That was just the warm-up. A week later, the jury awarded punitive damages of

$10.5 million—then one of the largest verdicts ever returned in California. Wegis hailed the judgment (which the court ultimately trimmed by $2.4 million) as a triumph for freedom of expression. Executives at Boswell, who unsuccessfully appealed the case all the way to the U.S. Supreme Court, saw it differently. "We got home-towned," said John Sterling.

Or, to put it another way, Jim Boswell had been right all along. The public never cared much for big, rich corporate farmers. During the punitive damages phase of the proceedings, a reporter for the Bakersfield paper had noticed something telling in the crowded courtroom: Clustered to one side were officials of the Boswell Company, all clad in dark business suits. Wegis's three clients wore open shirts without ties. The four men on the jury didn't wear ties, either.

Death in the
Homeland Canal

EVER SINCE THEY BOTTLED UP THE FOUR RIVERS and shoved their flow into a thousand miles of ditch bank, the farmers of Tulare Lake had watched the fish die. For many growers, who thought of themselves as devoted anglers, it wasn't easy seeing the carp float sideways to the surface or smelling the flesh of stripers rotting in the 110-degree heat. The fish kills were one of the more heart-breaking consequences of their farming, but like so many other tradeoffs, they grew accustomed to it. The deathblow arrived with the sun each summer, the months of August and September the worst. In the stagnant channels, the heat and algae sucked the air right out of the water. Dead fish were such a routine part of the landscape that the news of another kill hardly registered at the state Department of Fish and Game. The cause was almost always the same: dissolved oxygen—D.O. as the scientists called it—was being consumed by the algae bloom. The fish had choked to death.

Gloria Marsh, the Fish and Game warden whose district took in part of the lake bottom, had an idea that the kill on September 7, 1997, might be different. For one thing, her bosses had set up a command center and were admonishing the troops to be wary of the press. She put on her big black boots, hopped into her pickup and drove across the fields and canals that had become so familiar in her nine years on the job. As she reached the Homeland Canal, a twenty-five-mile-long ribbon that sliced through the southeastern end of Boswell country, it became clear that an environmental catastrophe on a scale she had never seen before was taking place.

A band of dead fish stretched across two county lines, miles and miles of threadfin shad belly up and lapping against both banks of the canal. The fish not yet dead were squirming near the surface, gulping for air. "I started on the Tulare County side and it was pretty much the whole bank for as far as you could see," Marsh recalled. "I kept going and going and there was no end to the trail of dead fish. I had never seen such devastation."

In the coming weeks, as Marsh and her colleagues tried to track down the cause and the culprit, it would go down as one of the largest fish kills in California history and reveal a King's County bureaucracy—from the district attorney to the ag commissioner—that wanted no part of an investigation that delved into the Boswell Company. Even Fish and Game itself would end up pussyfooting around with the farming giant. As the estimate of dead shad climbed to 10 million, local farm boosters pointed the finger at poor people throwing gasoline and motor oil and "barrels full of something you don't know what" into the canal.

Marsh, the daughter of Mexican farm workers who had worked the high country as a park ranger before joining Fish and Game, knew better. A year before, not far from the Homeland, she had caught McCarthy farms red-handed as it dumped ammonia-nitrogen fertilizer into a canal. Among growers in the lake bottom, the practice had become common. The ditches were deemed a perfect place to blend fertilizer with water and then pump the mixture onto their cotton fields. For years, Fish and Game had closed their eyes to the mixing, even though the canals were arguably waterways, and dumping any pollutant into them was a violation of state and federal law. The attitude at Fish and Game was that Tulare Lake had been so altered by man that the river system had ceased being a river system long ago. A canal, even one full of fish and shorebirds, was just another tool for the farmer to employ, not unlike a tractor or a pump. A man could do pretty much what he pleased in his ditches.

The spill at McCarthy farms might have fallen outside the arm of regulators, but the McCarthy family, long considered the rogues of the lake bottom, had taken the practice to an extreme. They were mixing ammonia-nitrogen fertilizer in a large canal that connected to other canals—the equivalent of a river's fork. Thousands of bass, catfish and carp had died as a result. An eyewitness to the dumping had contacted Fish and Game, and Marsh found the so-called hot spot, a fertilizer tank on the canal bank. It had all the makings of a slam dunk case, one that could have ended with the McCarthys paying a large fine and having to set aside a swath of habitat as mitigation. Instead, the Kings County DA cut a deal with the powerful farming family to pay a small fine. "They operate with impunity," Marsh said. "It's hard to get anybody to care about the lake bottom. It's the forgotten land."

Now seventeen months later, the Homeland Canal spill, to the dismay of regulators, was catching the eye of the media. George Nokes, the Fish and Game veteran who oversaw the central San Joaquin Valley, was known among staff as a friend of farming and a half-hearted defender of the environment. So it came as no great surprise that the agency, in the initial stages of the investigation, was doing its best to keep the public focus off Boswell. Whenever a reporter asked about the company's possible role, Fish and Game managers begged off and praised Boswell for cooperating fully in the probe. Yes, the first tests were finding high amounts of ammonia in the water, they confirmed, but that didn't necessarily mean death by fertilizer. Ammonia was a natural byproduct of fish dying, they said. A drop in water level due to irrigation and evaporation and a rise in temperature could have robbed enough oxygen from the canal to set off a deadly chain reaction. "There have been some traces of pesticide, but nothing significant yet," said Pete Bontadelli, chief deputy director of Fish and Game's oil spill team. "We very well may have a combination of factors here."

What the public didn't know was that a week earlier, the incident commander for Fish and Game had interviewed Walter Bricker, a water manager for Boswell. Bricker had confirmed that the company had sprayed the fields near the Homeland Canal with a cotton defoliant on September 4. This was the same day, according to local fishermen, the shad and other species began to die. Bricker's interview, as it turned out, was the only bit of assistance the Boswell company was giving. Contrary to the public kudos from Fish and Game, the company was refusing to open its pesticide records to regulators. Without such documentation, there was no easy way to determine what else Boswell had sprayed on its fields or dumped into its canals in the days before the kill.

"Boswell did its best to stymie the investigation," said Fish and Game's incident commander, Dennis Davenport. "I tried to contact Boswell workers, but they were closed mouth. We learned later that the reason we couldn't get any information out of them was because they were told not to talk. It got to the point that the company refused to hand over its pesticide records, and we had to go to the federal grand jury and seek a subpoena."

Davenport contacted Kings County prosecutors to alert them to the extent of the kill and possible impacts to human health. He knew that the web of canals, ditches and holding ponds was a popular spot for fishermen, many of whom were poor and depended on the stocked varieties to put food on the table. Davenport said he also wanted to discuss the possibility that a crime may have been committed by a farmer dumping a pollutant into the canal. Nine days into the probe, a county prosecutor finally called back to say that her office had no knowledge of the kill and generally knew nothing about environmental cases involving farmers and Fish and Game.

Davenport then sought the assistance of the Kings County agricultural commissioner, whose job included monitoring the pesticide applications of farmers. The commissioner's office seemed nice enough and offered a few ideas on what chemicals Boswell may have used. But the commissioner's deputies didn't know anything for sure and they weren't exactly inclined to find out. For years, the ag commissioner had let Boswell and other big growers pretty much do as they pleased. One former deputy commissioner recalled how the farm giants routinely flouted the county's order to plow down their cotton by a specified date. The date was important to safeguard against pests burrowing in for the winter. There were two dates, as it turned out. Small farmers had to stick to the date set by the county. Boswell and Salyer could wait as long as they wished. One season, the ag commissioner fined a struggling black cotton farmer in Lemoore for disobeying the plow-down date on his nineteen acres. Meanwhile, the big boys farming 150,000 acres in the lake bottom got to ignore the date without worry.

During the first weeks of the Homeland Canal investigation, as Fish and Game awaited the results of more water and soil samples, a biologist on the payroll of the farm giants spoke to a sentiment shared by many in the farm belt. He underscored that it was threadfin shad—a junk fish—that was dying in droves. The species was delicate, only two to four inches long, and especially vulnerable to a loss in oxygen levels. This alone, he argued, pointed to a cause that had nothing to do with Boswell. "The threadfin are like the canary in the coal mines," said Rob Hansen, a biologist who worked as a consultant for Tulare Lake growers. "It seems reasonable to consider a lack of oxygen as a cause. If some compound or chemical had been put into the canal, it would have hammered a lot of fish, not just the threadfin."

His theory might have made fine sense except for one small oversight: More than 2,000 western grebes, a shorebird that fed on shad and other fish, were also dying along the banks of the Homeland Canal. The black-and-white birds were preening so incessantly—as if to remove some foreign substance—that their bellies were filling up with feathers.

Carmen Thomas, a biologist with the U.S. Fish and Wildlife Service, which had been called in to help with the probe, had never seen grebes that sick. They were groggy, eyes half closed, and they hardly moved when she approached. The grebe, a ducklike bird with flaps of skin around its toes rather than webbed feet, was one of the more reclusive shore dwellers around. Normally, if a human came within twenty yards the grebe would dive into the water, its lobed toes collapsing and expanding with each thrust. On closer inspection, Thomas noticed that the coat of oil that repelled water and helped keep the grebe insulated was gone. They were emerging from the canal like wet rats. "The grebes didn't even stand up or try to peck at us," Thomas recalled. "Something had wiped away all the oil in their system. Something was terribly wrong."

Pinning down a precise cause in a place where the soil and water absorbed more than 2.7 million pounds of farm chemicals each year was the equivalent of a scientific shot in the dark. State and federal biologists were not only contending with any number of defoliants, insecticides, herbicides and fertilizers applied in the days and weeks preceding the kill—an average of 6,500 pounds every square mile. They were also coming across the tracks of a few of the more infamous chemicals developed since World War II, compounds that had been banned years earlier. Several burrowing owls found dead showed high levels of DDT. Shad, carp and clams taken from ditches that fed into the Homeland revealed considerable amounts of Toxaphene, a cotton pesticide taken off the market in the early 1980s because of its lethal effects on fish and wildlife. Were these chemicals merely showing up in their nefarious half-lives? Or were they obtained on the black market and still in use?

Investigators did find forty pounds of an ammonia compound lying on one of the banks of the Homeland Canal. The discovery on Boswell land certainly pointed toward a worker dumping enough ammonia-nitrogen fertilizer into the water to kill the fish. A fertilizer, if concentrated enough, was also capable of destroying the oil glands of the grebes. And without the ability to repel water, the birds may have been dying from exposure. But further complicating the search for a smoking gun was the massive wheeling of water that took place in the lake bottom. Investigators couldn't be certain if a pollutant had migrated into the basin from a farmer upstream through one of the ditches connected to the Homeland Canal. Because Marsh and her colleagues had been called to the scene a full week after the fields had been dusted with defoliant and the fish began to die, any chemical footprint on Boswell land was by now gone.

"The lab reports came back real high on fertilizer, which makes us believe that's what is was," Marsh said. "But the way water moves in and out of the lake bottom, you just can't say with certainty that, 'Yes, it's agriculture. Yes, it's this farmer.'"

Carmen Thomas believed a combination of triggers—the defoliant spraying, the dumping of fertilizer, the strong presence of Toxaphene and other residual chemicals—came together to cause the kill. "It's a toxic soup out there, and a lot of this stuff accumulates in the system of birds and fish," Thomas said. "You add one more thing to the mix, like a fertilizer spill, and all of a sudden you've got a big kill on your hands."

Mark Grewal, Boswell's farm manager, knew enough about the history of the lake to argue that dead fish had been a part of its ebb and flow since the Yokuts. As the heavy snowmelts gave way to searing summers and periods of drought, the lake shrank and so many fish died that the stench could be smelled for miles around. "Whether you're talking about the old lake or the canals we use to deliver water, the evaporation is the same," he said. "Sooner or later, those fish are going to die. When we water, some of the fish are killed by the pumps themselves. The ones that survive are swimming in a canal where the amount of water is decreasing by the day and the amount of nitrogen is increasing. Yes, that nitrogen comes from the fertilizers that we mix into the water. But it also comes from their own shit. There's no outlet for those fish. This is the bottom of the lake. When it dries, they die. They died back in Indian time and they die today."

Grewal had set up a practice of testing the water every day at nineteen different locations in the lake bottom. It wasn't likely, he said, that his men would have been applying a fertilizer so late in the growing season. It's also not likely that a defoliant sprayed a few miles away would have drifted into the Homeland Canal, much less caused a kill of that magnitude. "The fish died from a lack of oxygen. It's no more complicated than that." As for the grebes, he said, their death puzzled him. "What happened to the shorebirds was an anomaly."

As the investigation wound down, some staffers inside Fish and Game blamed their own agency for failing to solve the mystery. By long ago ceding the lake bottom to farming, Fish and Game had failed to set up a system of game wardens and biologists to protect the still considerable natural resources of the basin. Marsh, whose office was located in the adjacent county, had little time to police the lake bottom. The position of game warden for Kings County—the person with direct oversight of the lake bottom—had long gone unfilled. In addition, Fish and Game biologists never measured the water's temperature or pH in the first hours of the probe, according to insiders. Only by establishing such a baseline could they later draw conclusions to take to court. Likewise, Fish and Game commanders showed little interest in playing tough with Boswell in the first important days of the inquiry, waiting weeks to request records or even attempt to question key employees. By that time, all mouths were shut.

"This case should have demanded the toughest approach. If a guilty party was found, you're talking about the potential for millions of dollars in damages and the creation of a large wetlands as mitigation for damages," one former staffer said. "Unfortunately, the investigation was buried by scientific incompetence and politics. And the farmers were once again untouchable."

LIKE SO MANY OTHER BIG GROWERS, Jim Boswell cared about nature in the finest tradition of an outdoorsman, but he was far too pragmatic to fret about what his farming had done to a treasured wetlands. Old files were sprinkled with letters of praise from the Nature Conservancy and others for his generous donations to preserve pristine habitat in the lake bottom and beyond. One check for fifty grand helped snatch 4,500 acres of spectacular California coastline from the paws of developers. But Boswell didn't like to talk about his environmental philanthropy any more than he liked to discuss his environmental plunder.

"Fish kills? I don't know anything about fish kills," he said two years after the Homeland Canal incident. "Somebody might have brought it up, but frankly it went in one ear and out another. Remember, I was out of this company for thirteen years."

No people ever went through a landscape faster than Californians went through their swath of the West. At the time of the Gold Rush, the state boasted 5 million acres of wetlands. More than 60 million water-fowl, it was estimated, stopped to feed or spend the winter in a string of lakes from Kern County to Sacramento. Now there were just 300,000 acres of marshland left and the birds numbered fewer than 2.5 million. It was the same story with the valley's native needle and blue grass, ex-terminated by the cattle and sheep. There was an entire ethos behind the decision to remake the land, to divert the nation's deepest snow pack to irrigation and settle two-thirds of the state's population on desert land and move the rain that fell hundreds of miles away. Blaming one farmer—even the nation's biggest farmer—for quieting rivers and denuding riparian habitat was the sort of demonizing that allowed Boswell to shake off his critics. Yet it was also true that so much destruc-tion had taken place under his direct watch that environmentalists seemed only a slight bit hyperbolic when they used words such as "rape" and "moonscape" to describe what had transpired in Tulare Lake. The way Boswell dismissed the lake's metamorphosis—from the most dominant wetlands on the western map to a dead zone where nothing grew where he didn't want it to grow—struck one as too facile. Birds didn't die in anomalous episodes but with a kind of numbing routine that made the kills seem acceptable. By the thousands, ducks and geese landing in the lake bottom on their migrations from Canada and Alaska died from botulism that thrived in the stagnant waters. Within hours of feeding on contaminated insects, the birds lost their ability to fly, walk or swim. *Newsweek*, as far back as the 1969 flood, had documented the botulism kills beneath the headline "Lethal Haven."

Then came the flood of 1983 and Boswell's attempts to drain 80,000 acres of submerged land. Moving the water meant pumping it up the Kings and San Joaquin Rivers and out to the Bay delta, a plan that came under immediate fire from environmentalists and fishermen. The resur-gent lake happened to be filled with white bass, a game fish so voracious that its release into the rivers would threaten the state's population of Chinook salmon and striped bass. Boswell spent more than $10 million erecting a small dam and a series of 350-horsepower pumps that made the Kings River flow backward. The Dairy Avenue shop run by Paul Athorp had worked day and night for three months to build the dozen pumps—each one weighing 9,000 pounds—and set them across the mouth of the Kings. To keep the white bass from escaping along with the water, John Baker, Boswell's master engineer, designed a metal fish screen 100 feet long. It was quite an invention, with thousands of round holes big enough to let the water out but small enough to hold back the spawning minnows. A series of electric-powered brushes acted as a kind of windshield wiper, scouring off any debris that might clog the holes.

Fish and Game signed off on the pumping plan without holding a public hearing or studying Boswell's contraption in any detail. Environmentalists accused the state agency of once again abdicating its role as a protector of natural resources and doing the bidding of a powerful grower. It wasn't just the enviros who were upset with Boswell, though. Farmers along the San Joaquin River, who had spent weeks battling their own floods, opposed any more lake water coming their way. Boswell shored up a few levees and did tractor work for gratis to buy the acquiescence of a few neighboring farmers. Some of the rowdy ones felt so strongly, however, that they took up shotguns in the middle of the night and began pelting the Boswell pumps from the opposite bank. One sharpshooter managed to take out the cooling system on a pump and shut it down for a while. But most of the bullets bounced right off the mammoth contraptions, as each one pulled 100 cubic feet of water per second out of the lake. In no time, Boswell drained 7,000 acres dry.

As lawyers for the environmentalists argued in federal court to stop the pumping, some of the white bass apparently found their way past Boswell's metal barrier and headed upstream. As soon as the fish were gill-netted—proof of a breach—Fish and Game had no choice but to halt the pumping. Jim Boswell himself wondered if the environmentalists were so intent on making a ruckus that they had planted the white bass on the other side of his pumps. It was anyone's guess how many white bass had migrated out to mingle with the catfish and carp. The state was forced to pour a thousand gallons of a deadly plant substance called rotenone into the river's north fork. A week later, a second mass poisoning was performed, wiping out every fish in a six-mile stretch of river around the pumps. It would take two more years for Boswell to completely dry out, moving water from low ground to high ground within the basin and diverting water through the California Aqueduct all the way to L.A.

The white bass controversy set the stage for an even more contentious battle that would cost the company millions and leave it with another environmental black eye. In the summer of 1987, government biologists were back on Boswell land, this time to study the damage to wildlife from the giant evaporation ponds that dotted the old lake bottom. The ponds were filled not with fresh water but with salty water that came off the fields after irrigation. The salts were a vestige of the valley's days as ocean bottom. Draining a land that could not always drain itself was an intractable problem as ancient as Mesopotamia. As much as Boswell ripped into the earth and reconditioned the soil, parts of the clay strata below the cotton fields remained impenetrable. The irrigation water percolated only so far before it began bubbling up to the surface again—this time as gray water laced with toxic salts. The cotton plant, which could survive almost anything, became sickly when its roots sat in a briny bog.

Rather than retire the sections of land turning white, Boswell and the other farmers in Tulare Lake turned to engineering. In the winter of 1966, they formed a 213,000-acre drainage district and began constructing a system of underground pipes to siphon off the salty water. They dug huge pits into the ground to act as cesspools. "Some ancient civilizations were forced to give up irrigated agriculture because they could not solve the salt problem," the *Hanford Sentinel* opined in a January 1966 editorial that applauded the formation of the Tulare Lake Drainage District. "Our farmers could have let nature take its course as people all too often do. Or they could have relied on federal and state governments to come in and do the job for them. That they were able to organize and tackle the problem themselves with their own resources speaks well of our farmers."

These weren't ponds like Walden's. They stretched for five square miles, and the drain water that evaporated in the summer sun had such a high concentration of salts that one federal biologist fell in and floated—just as people do in Israel's Dead Sea. "It felt like a bunch of fingers were underneath me holding me up," he recalled. Salt, though, wasn't what brought the U.S. Fish and Wildlife Service to the Tulare Lake ponds. It was the presence of selenium, a well-known killer of ducks and black-necked stilts and American avocets. Like salt, selenium was a trace element native to the soils of the valley. In small doses, it brought better health to both wildlife and humans. In higher doses, it did just the opposite and ranked as a poison more deadly than arsenic.

As far back as 1941, scientists for the U.S. Department of Agriculture had warned that the valley's soil type carried the taint of selenium. By 1981, the devastation it could deliver had become nightmarishly clear 100 miles up the road from Tulare Lake. In a federal wildlife refuge known as Kesterson, government biologists had discovered widespread deformities in the eggs of shorebirds. Curled up inside the shells were embryonic monstrosities—birds with no eyes, beaks or wings. The entire 1,280-acre marsh had been poisoned by selenium that came from the irrigation drain water of cotton and vegetable farms to the south. In the mid-1980s, after repeated newspaper stories by local reporter Lloyd Carter and a segment by Ed Bradley on *60 Minutes*, the federal government sealed off the drainage pipes and shut down Kesterson.

The toxic gumbo now filling the evaporation ponds at Tulare Lake was more or less the same brew. Boswell's water engineers and the managers of the drainage district assured the public that, as one put it, "there are no Kestersons here." But by the summer of 1989, Joe Skorupa, a biologist at the U.S. Fish and Wildlife Service, already had found the damning evidence. A tall, slender man who wore dark sunglasses and a broad-brimmed hat, Skorupa was hit with a riddle: The evaporation ponds at Tulare Lake contained nowhere near the same concentrations of selenium found at Kesterson. And yet the eggs of ducks collected along the ponds were showing the same deformities as their counterparts in the now notorious wildlife preserve.

Skorupa studied the differing landscapes and offered up an explanation. In the north part of the valley, a clean mosaic of wetlands surrounded Kesterson and gave the ducks plenty of food sources in addition to the selenium-laced crustaceans. Many of the ducks at Kesterson were feeding on the bad stuff only part of the time, thus sparing themselves a toxic buildup. To the south in the Tulare Lake basin, the ponds may have been less polluted but they happened to be the only game in town. Charmed by the deceiving blue waters that broke up the miles upon miles of farmland, the ducks were feeding full-time from the evaporation ponds. Even though the selenium concentrations in the ponds were ten times lower than at Kesterson (30 parts per billion versus 300 parts per billion), the cumulative effect of eating a little poison day and night was producing the same deformities.

"The duck eggs we were gathering in the lake basin showed almost as much selenium as the ducks eggs at Kesterson," Skorupa said. "The more data we collected, the more it backed up this landscape effect. Boswell and the other farmers didn't want to believe it. They kept protesting that something had to be wrong with our research. The work we were doing turned out to be very controversial. Not scientifically controversial because we had the numbers to back it up. But politically explosive."

In the eyes of the farmers, Skorupa was just another environmental wacko like Felix Smith, the Fish and Wildlife biologist who had blown the whistle on Kesterson. For Skorupa, there was no higher compliment. Like Smith, he relished the heat that came with staring down the big boys of California agriculture, and he wasn't afraid to upset his bosses by speaking his mind to reporters. In a front-page article in the *Los Angeles Times,* Skorupa didn't hesitate to make the case that the evaporation ponds of Tulare Lake were every bit as scandalous as Kesterson. "Kesterson wasn't a worse case," he said. "We thought it was worst case. Now we know it wasn't."

As more reporters lined up to record Skorupa's latest provocation, the farmers and their drainage district struck back. They took their complaints about Skorupa to the California Department of Water Resources, the agency helping fund the $50 million study of farm drainage in the valley. When the state agency refused to yank its financial backing, the farmers tried to gag the biologist. They made an exhibit of his newspaper quotations and presented it to his superiors at Fish and Wildlife. Skorupa wasn't acting as a dispassionate biologist, they argued, but as a zealot on a crusade. Skorupa assured his bosses that his science was solid. Of the twenty kinds of birds nesting in and around the evaporation ponds, he had found corkscrewed beaks, missing eyes, shriveled limbs and other deformities in nine species: the mallard, gadwall, pintail and redhead ducks, American avocets, eared grebes, black-necked stilts, killdeer and snowy plovers.

Good science or not, Skorupa was labeled a "dissident" by one boss and reprimanded. His research papers suddenly had a bull's-eye on them. The layers of internal review grew interminable; the pencil edits by higher-ups became so heavy that he hardly recognized the sanitized end product. What the reader got was plenty of mind-numbing technical data without any context to say what it all meant. "He was publishing his research in the newspapers instead of scientific journals,"' said David Trauger, a biologist who oversaw Skorupa's work. "And we were very concerned about that." The final muzzle came when Skorupa's superiors at the Patuxent Wildlife Research Center in Maryland banned him from making any more trips to the evaporation ponds. It was now left to a young research assistant to collect the eggs and other data—and she had to give the farmers forty-eight hours notice before each visit. "It was a classic case of trying to kill the messenger," Skorupa said. "They weren't interested in intellectual honesty. They wanted someone they could control."

By the early 1990s, more than a dozen operators, led by the Boswell-dominated Tulare Lake Drainage District, wanted to expand their ponds. The Central Valley Regional Water Quality Control Board, lacking a single environmental voice to offset the boosters of agriculture, seemed poised to approve whatever the farmers needed. A handful of local environmental activists led by Lloyd Carter, the newspaperman turned lawyer, argued that the drainage system violated the Migratory Bird Treaty Act. They pressed the Department of Interior, which oversaw Fish and Wildlife, to halt the practice and demanded that the regional water quality board deny the permits. This was the same water board that had come under stinging criticism in the 1973 book *Politics of Land*, a Ralph Nader–led study on land use practices in California. The book quoted state staffers as saying that the board made no attempt to measure the pesticides and other farm chemicals that were fouling the waters of the Central Valley. Instead, the board had essentially granted blanket immunity to the region's biggest industrial polluter. "Agriculture is the subject we're least up on," said the board's senior engineer, George Schmidt.

True to form, the regional water board ended up giving the Tulare Lake farmers so much leeway in completing their environmental impact studies that it took four years for the final report to come in. All the while, the farmers were allowed to ship more poisonous water to the ponds. When the water board finally completed its review of the drainage district's report in the summer of 1993, it announced that the public would have no more than four weeks to review tens of thousands of pages of documents before the formal hearing. To no one's surprise, the hearing ended with the regional water board granting nearly every wish of the pond operators. They would have to raise the sides of the ponds to make them less attractive to shorebirds, and they would have to adopt a hazing program—complete with shotgun blasts—to shoo away the birds from nesting. They also would have to provide enough land and fresh water to create one acre of clean habitat for every ten acres of poisoned ponds. Other than that, though, the farmers could discharge as they pleased.

The environmentalists argued that to fully make up for the impacts to wildlife, the habitat being set aside was nowhere near enough. They appealed the regional water board's decision to the State Water Resources Control Board, an agency not exactly sympathetic to environmental concerns, either. For a time, it appeared as if the state water board might demand more mitigation and the U.S. Justice Department might even prosecute under the Migratory Bird Treaty Act. In the end, as the farmers undertook a new round of environmental studies, the state board backed off and the top federal prosecutor in Sacramento declined to take the matter to court.

In the summer of 1997, after a decade of debate over the extent of their killing ponds, Boswell and the other cotton giants invited scientists and reporters on a tour of the lake bottom. What they had engineered in the way of a new habitat bore no resemblance to a wetlands. The compensation habitat, as they called it, was a 307-acre bird farm where chicks hatched in a zone free from selenium. It was a strange place in a strange land. Built in the manner of a bowling alley, the habitat boasted thirty-four lanes of spare brush and soil interspersed with thirty-four lanes of clean water. Boswell was growing birds as it grew cotton. "When people see the design, they say, 'Wait a minute, this doesn't look like a wetlands habitat,'" Tom Hurlbutt, Boswell's manager for water resources, acknowledged. "We tell them, 'We didn't want to build a wetlands habitat. We're trying to produce birds.'"

After a four-hour bus tour across cotton, safflower and wheat fields in various stages of dress and a pork loin and tri-tip barbecue at Boswell's Homeland Ranch, some of the federal biologists still weren't convinced. "I was impressed by the number of birds nesting on the new habitat. They seem to love it," said Doug Barnum, a biologist with the U.S. Geological Survey. "But there are questions that remain to be answered about the quality of water they're using and the selenium levels over the long haul. And I'm not sure that one habitat is going to be enough to compensate for all the damage from the contaminated ponds."

Six years later those questions would remain unanswered. The biologists at Fish and Wildlife had moved on to other hot spots, and the regional water quality board still hadn't made up its mind on a final environmental impact report. The farmers had used the same government red tape they so often bemoaned to tie up and exhaust their critics. The fight over selenium now won, Boswell would soon be engaged in a new battle on the environmental front, this one over its plans to milk 55,196 cows in the lake bottom.

It DIDN'T TAKE THE MOST IMAGINATIVE conspiracy buff to see something sinister in Boswell's ability to emerge, time and again, as the victor in conflicts involving fish and birds. While the company was certainly capable of pushing around its political heft and squeezing out deals in back rooms, it chose, for the most part, to make its environmental case in full public view. This was no more true than during the flood of 1998 when Jim Boswell himself flew into Corcoran on the corporate jet and helped push an idea that once again raised the bristles of the U.S. Fish and Wildlife Service. As the snowpack reached its near-record melt that spring, Boswell could see that the only thing standing between the floodwaters of the Tule River and 27,000 acres of prime farmland was the dam on Lake Success. It was a tiny bulwark by California standards, a slab of concrete 652.5 feet high and 200 feet long, and Boswell wanted to top it with a six-foot-high wall of sandbags. This extra height would add 10,000 acre-feet of storage, just enough to take the starch out of the river and save the land.

The idea might have flown through like most everything else the Boswell Company put in front of the federal agency that managed the dam, the U.S. Army Corps of Engineers. This time, though, there was a bush and a beetle and the beetle's guardian—the Fish and Wildlife Service—standing in the way. Erecting sandbags to pack 12 percent more water behind the dam was certain to drown out a clump of ninety giant elderberry bushes near the shores of Lake Success. The gnarly bush was the only known habitat of the endangered elderberry longhorn beetle, though Boswell's attorney insisted that the bush was merely the "suspected" home of the beetle—no one could say for sure. As officials from Fish and Wildlife and the Army Corps went back and forth on a plan to make both sides happy, some 20,000 acres of two-bit cotton land near Corcoran fell under water. The next to go would be the rich ground of District 749. At Boswell's urging, Senator Barbara Boxer and a local state legislator named Jim Costa jumped into the fray and began applying their own pressure. In a letter to Army Corps General J. Richard Capka in San Francisco, Senator Boxer pleaded that the sandbagging be allowed to go forward, lest an economic disaster level Kings County.

Then Boswell's water men did what they so often do in times of flood. They offered to take the bureaucrats on an aerial tour of the lake basin, hoping it would work the same magic it always worked on those who had never glimpsed the land and its rivers. For Roy Proffitt, the local Army Corps chief who had grown up in South Carolina and spent much of his career in the Midwest, the tour was all he needed to see. "It was unbelievable to me, to fathom how much land was flooded and how much more could come," he said. "I couldn't imagine this scale of agriculture, and I had seen agriculture all my life. The big farmers in the Midwest were pikers compared to the farmers in Tulare Lake."

The tour, the letters, the phone calls—not to mention Boswell's vice-president flying on the company jet to San Francisco to lobby General Capka in person—paid off. Boswell got the green light to bring in his men and machines to sandbag the spillway and the elderberry bushes, too. Boswell and wife Roz drove up to the lake to see the work firsthand, a job that included a biologist counting every stem of the elderberry bushes that measured more than one inch wide—the beetle's potential perch. As it turned out, the bushes were clumped in an area that no amount of sandbags could keep dry. So the Army Corps struck a new deal with Boswell: The cotton king need not worry about the elderberry bushes at Lake Success as long as his men would someday plant an equal number of bushes at a wildlife preserve downstream. The dam topped with sandbags measured 658.5 feet in elevation. The extra six feet had saved the 27,000 acres of cotton land. All but one of the elderberry bushes was dead—swamped as much by Boswell's ceaseless wave of lobbying as by the rising water behind the dam.

In the meantime, a far more painful scene was playing out at the end of the river's run. The leftover waters from the flood sat bottled up on the backside of the old lake, a reservoir for future irrigation. Unlike the evaporation ponds, the vast sheet of fresh water constituted some of the best wetlands in California. It was the same place that Jim Boswell liked to show off to visitors who doubted whether a farmer could moonlight as a guardian of ducks. One of the first times Michael Morse glimpsed the reservoir as a field man for Fish and Wildlife in the mid-1990s, the sun was setting against the Kettlemen Hills and the ducks were flying in by the thousands. "I've been on wetlands before, but never like those. It was just an amazing place. I got a sense, even with all the changes, of how the lake must have looked back in historic times."

Morse spent three years trudging across the basin, surveying the nests of shorebirds and counting eggs as he dodged rattlesnakes lurking in the scorched-out carcasses of jackrabbits. With each trip down, he became more convinced that this wetlands, too, was a killer. The fresh waters enticed colonies of eared grebes, 500 and 600 strong, to lay their eggs in spring only to see them float away in a wet season and shrivel up in a dry one. The pumping of water in and out of the reservoir—the way a farmer answered the call of his crops—set up a cruel cycle for the birds. "We were losing eggs as the water came up and as the water came down," Morse said. "I'd walk onto the beaches and see hundreds of eggs from different species just floating. Then in dry years, the water would be drawn down so much that the nests were left high and dry."

The farmers and drainage district managers all had the same answer: What Morse was seeing was merely part of the historic curve of mortality. As far back as the Yokuts, snowmelt and evaporation carried the same one-two blow. Pumping water into the basin was no different than the rush of snowmelt. Pumping water out was no different than evaporation. "They told me they were doing the same thing that nature did," Morse said. "But what was happening out there was nowhere near normal conditions. They wanted to control the snowpack. They were doing it by altering the timing and duration of the water. It's in a bird's nature to find the edge of the water. That's where they set up shop. And on Boswell land, there was no edge."

Fresh water on its way to the fields or dirty water on its way out, it made no difference. Birds were dying on both turns of the water's wheel.

LIKE SO MANY DEALS that had reshaped the land over the past century, this one began and ended with water. After the last paper had been signed and the 30,000 acres handed over and the two cotton kings parted ways to contemplate the end of seventy-five years of feuding and making up, the deal came down to a single temptation: the chance to vanquish the lake for good.

Fred Salyer had too much pride to pick up the phone himself. The day Jim Boswell got a call from Salyer's right-hand man saying Fred was worn out and aching to sell, Boswell's first impulse was to tell him he wasn't interested. The last thing he needed was to get bigger. The Salyer land was tired in parts, and whole stretches were crusted with salt. Boswell already farmed more marginal land than he knew what to do with, and the price of cotton didn't exactly leave him feeling sanguine about the future. Yet ever since his uncle and father had bought their first piece of the lake bottom, the Boswells had lived by the axiom that the land was dirt but the water was gold. It was hard to ignore all those water rights that came with the Salyer land, as well as that big levee that had caused so much grief between their two clans. Controlling that levee—along with the canals and water-storage cells on the Salyer side—was no small advantage in staving off floods. If Boswell was ever going to engineer the lake out of existence, Salyer's land just might be his last chance.

Besides, there were worse things than riding to the rescue of an old childhood friend. Fred was looking even more beat than his land. Salyer American had fallen into deep financial trouble with the Bank of America. And Fred and his only son, Scott, in grand Salyer tradition, were about to take their bickering to court. Fred accused Scott of racking up $833,000 in personal credit card expenses and charging them to the company. One expense, a $52,000 bill for Scott's tenth wedding anniversary bash, irked Fred to no end. Scott accused his dad of paying himself a $25,000 fee for attending the company board meeting, while denying Scott his $15,000 fee by deliberately scheduling the meeting on a day he couldn't attend.

Boswell flew into Corcoran in January 1995 and picked up Fred in his battered Chevy truck and off they drove to the lake bottom. Kicking up dust along the Schwartz levee, they talked about all the history they had shared. Boswell wanted to make sure that Salyer wasn't having any second thoughts. "I needed his farm like a hole in the head," Boswell would say later. "My attitude is that every man is in love with his land and if he wanted to back out that was fine with me. But Fred was tired and wanted to sell. We consummated the deal that day."

Salyer signed over his 30,000 acres of lake bottom land, his draw to irrigation water from the Last Chance canal, his prized airplane hangars and a new seed alfalfa plant. Boswell handed him $6 million in cash, plus whatever Fred could glean from the sale of a stunning array of farm equipment. "I didn't want to be picking over his bones," Boswell said.

Old man Clarence wasn't there to see his rival swallow up everything he had built, even the cabana where he used to throw his lavish shindigs. He had died in 1974 in classic Cockeye style, owing half a million dollars to the attorney who brokered the armistice that held together a fragile family peace. The attorney had to sue Cockeye's estate to collect. "My services as a peacemaker in connection with this bitter family feud constituted the most onerous and oppressive (and at times truly nerve-racking) professional task I have ever experienced during a half century of the practice of the law," Walter Gleason wrote in court papers. With the old man gone, his two boys started to eye each other with mistrust. Everette, the older son who ran the company's finances, fell over dead while attending a national cotton meeting in Dallas in 1982. Fred had scarcely buried him when he discovered that his brother had stolen millions of dollars from company accounts to trade in the commodity futures market for himself and his children. The armistice soon gave way to war. Everette's daughter, who operated a private school in Rancho Cucamonga with her female lover, sued to take over Salyer American. Fred and son Scott managed to keep control of the company, only to begin their own ugly feud.

"All the little wounds are cut open again," said Linda Salyer, who joined her father in the lawsuit against her brother. "It's sad. We've been fighting for three generations. Now we're going at it over the remains of something that the Boswells have already picked clean."

Boswell and Salyer tried to keep the sale a secret, but word leaked out after Fred broke the news to his 136 full-time employees. A bad mood gripped the entire company as reporters moved from the Brunswick barber shop to the Salyer cabana looking for comment. The Salyer men in their green hats and mud-caked white Chevy pickups warned journalists not to step foot on the boss's far-ranging lands. "I wouldn't be snooping around here if I was you," snapped one grim-faced foreman. Fred Salyer, a tall, thin man with a web of veins etched into his nose and his initials sewn into his white shirt, waved off one reporter in front of his tree-shaded ranch house. "I've got no comment," he said.

It would take seven more years and a twist of the arm from Jim Boswell to get Salyer to sit down in his office next to the airport runway (he had sold that, too) and talk about a lake bottom where only one cotton king was left standing. For the first fifteen minutes, Salyer looked down at the floor and barely spoke. When he did, it was mostly a mutter. Boswell, who had taken a seat next to him, then got things rolling with a story about their days training together at an army base in Texas and driving back to California in his '37 Buick, a gift from his Aunt Ruth Chandler.

"I picked up Fred and another soldier and we came around a corner and there was a carload of black boys."

"Say 'colored gentlemen,'" Fred interrupted, smiling for the first time.

"No," Boswell said. "I was going to say 'niggers.' Well, they nicked our fender and they skidded to a stop. I told Fred to reach into the glove compartment and get out that .45."

"When you pull back that slide and turn it loose, it speaks a universal language," Fred said. "It don't make any difference if you're Japanese, Hindu or German. They decided that three honky soldiers weren't what they wanted to mess with."

With Fred now loosened up, Boswell directed the conversation to the day they shook hands on the historic deal. "I didn't want to see it parceled out or turned over to incompetents," Salyer explained. "The Boswell family and our family had a good relationship and mutual respect for a very long time."

The Boswell Company, envisioning a new day with the acquisition of Salyer, had begun to raise the levees and reengineer the water delivery system with an eye toward planting permanent crops. Already, Mark Grewal had managed to put in a vineyard or two without the boss knowing.

Salyer thought the whole notion was folly. He recalled driving around the lake basin with his father in the 1930s, after a long drought had given way to flooding on the Kings River. His father pointed to the washed-out bridges built by the state highway boys and then to the still standing bridges built by the railroad men. Those bridges had been built with a respect for the river; they'd been built to last. Nobody was going to lick that lake, he said, not even if they raised the dams another twenty feet. "It's never going to happen."

He was asked what quality, above all, was needed to farm the lake bottom. "You have to have balls," Salyer said. "That's a given."

"Either that," said Boswell, and then he laughed, "or be dumb as hell."

PART FOUR

AUTUMN

"Me? Pick cotton?" cried Scarlett aghast,

as if Grandma had been suggesting

some repulsive crime.

"Like a field hand?"

—**Margaret Mitchell,** *Gone with the Wind*

Autumn

SUMMER HAD PICKED CLEAN THE VINEYARDS and orchards on both sides of Highway 99, and now it was autumn's turn to take the cotton from Tulare Lake. The valley—land, man and machine—looked beat up and dog tired. It was as if one great flogging had ended and another was set to begin, and no one—except maybe the boys at Boswell—was quite ready. The nights had turned chilly, but autumn didn't arrive in purple or gold, colors of surrender. Rather, a leathery brown that started in the hills and worked down to the fields had scorched itself into the land. Bled of its green, the vine had nothing else to give. On the road from Fresno to Corcoran, we saw the last of the Thompsons still cooking into raisins under a sky that hadn't seen rain in seven months. October was the cruelest month. The air was thick and foul, and the breeze out of the northwest had lost its gumption. As each row of raisins finished baking, the farmers gathered up their paper drying trays—6 million tons in all—and set them ablaze. Dark gray ashes smudged the powdery loam, and the smoke from the vineyards rose to meet the dust from the almond harvest. The cloud lifted skyward and mingled with the rain of chemicals that the airplanes were dumping on the cotton fields each night. Around the valley, the fits of coughing and wheezing that accompanied autumn were known as the "defoliant cold." The cotton farmers insisted that the timing didn't mean anything. What ailed the children wasn't the Ginstar or Starfire that singed the cotton leaves in advance of the pick, but a virus that happened to swoop down the same week as the crop dusters. Sheer coincidence was all.

Corcoran was too busy setting up for its annual Cotton Festival parade to let a little asthma get in the way. The town, with the help of the Boswell Company, had erected a new park and train station and converted the old Corcoran Hotel into a senior citizens lodge—all within a few blocks of each other on Whitley Avenue. The parade was a chance to show off the new additions and watch the main street, if only for a day, find its old bustle. Ever since Tolbert's had closed a few years earlier, the town hardly had a homegrown place to eat. On the far end of Whitley, across from the Boswell lanai, McDonald's and Taco Bell were cashing in on the void, serving a mix of locals and out-of-towners who came to visit loved ones behind the prison's walls.

Lining the parade route were a snow cone stand and a tri-tip barbecue pit and the booth of a huckster barking out the letters MSM, a miracle elixir to "heal America." As both sides of the street filled with townsfolk—a scattering of white and black faces amid hundreds of Mexicans—a local state assemblyman named Dean Florez handed out cardboard fans to wave off the 95-degree heat. Duane Dye, who had run the Brunswick barbershop for twenty-two years on this same block, these days took care of the local cemetery. It was a perch that gave him a lot of time to consider what the town had been and what it amounted to now. "We used to be the 'Farming Capital of America.' Now we're the capital of bad people. It used to be when somebody heard the name Corcoran, they'd say, 'That's J. G. Boswell. That's where they grow the best cotton in the world.' But now they say, 'Charles Manson lives in your town. Juan Corona lives in your town.'"

For forty years and running, the Cotton Festival had celebrated the county's past as the Little Kingdom of Kings. This year's version, a tad threadbare, began with the lights and siren of the town's shiniest police car. The children from Fremont Elementary marched past the hardware store in their crowns of cotton, chanting, "Two bits, four bits, six bits, a dollar. All for cotton, stand up and holler." They rushed ahead of the Chrysler convertible carrying Cotton Queen Jana Halverstadt and fell in behind the Corcoran High School marching band—Panthers dressed all in blue and gold. Someone had gone to the trouble of building a plastic miniature train with fifteen cars now filled with fifteen kids. What made it such a marvel was that each car had been recycled from the farm—cut out of a used pesticide container, no less. The tractors thundered by in all sizes—little tiny Caterpillars and Massey Fergusons as big as houses and one John Deere that was pulling what seemed like pro wrestling tag teams but turned out to be the county board of supervisors. There were more Shriners than politicians, and they seemed not the least bit self-aware as they bounded past the Mexican trinket store wearing their tasseled fezzes and carrying their samurai swords, singing "When the

Saints Go Marching In." There were Shriners on scooters and Shriners in Tin Lizzies and Flying Shriners putt-putting in little wheeled airplanes. Then came the green berets of the 4-H Club, young agriculturalists pledging their heads to clear thinking, their hearts to great loyalty, their hands to larger service and their health to better living for the community. The prison sent its honor guard, and the sheriff's posse sent its maroon-shirted cowboys astride beautiful horses. Bringing up the rear, just behind the tractor driver with no teeth and a dragon tattooed on his arm, were the mariachis in their blue jackets and turquoise scarves. Beneath a Cotton Festival banner that had been shot full of holes, Corcoran was strutting.

Cliff Hill, who came out from Oklahoma in 1948 and owned The Club tavern, hadn't seen this many potential customers since the prior year's festival. "Back in the old days, there were so many people you couldn't even walk down the street here," he grumbled. "It just got slow. Real slow. Everybody wants to live out of town now. Nobody wants to live here. Corcoran's gone to the dogs."

Not even the most optimistic booster could see the picture changing anytime soon, not with Boswell swallowing up Salyer and downtown littered with so many vacant storefronts. Corcoran was now a Latino town, its fate hitched to a peasant class from Mexico that had come north to do the valley's hardest labor. The immigrants from Sinaloa and Michoacan and Zacatecas who weren't working in cotton were taking the highway at 5:30 in the morning to pick and prune the vineyards and orchards. Their children and grandchildren, born to this side, could hardly be blamed for turning their backs on the fields. Yet they were rejecting not only the life of a farmhand but the work ethic that came with it. Instead of embracing college or jobs that demanded higher skills, the children plunged into a world of out-of-wedlock births, welfare dependency, drugs and gangs. Their parents had found a low but firm rung on the economic ladder only to see the next generation give it away, taking a collective step downward.

The numbers told the story of a despair that ran as deep, if not deeper, than in any urban barrio. One out of ten Latino girls between the ages of fifteen and nineteen had given birth in Tulare and Kings counties. This was double the rate for the United States as a whole and higher than in Texas and Mississippi. Even the Third World countries of Namibia, Haiti and Cambodia had a lower rate of teen births. The cen-

tral San Joaquin Valley had become the number one source of methamphetamine in the country, farms turned into illicit labs by drug gangs
from Mexico. One young dealer, half Latino and half black, had been
found lying dead at the edge of a cotton field, naked and hog tied, his
back riddled with nine bullet holes and a squeegee handle protruding
from his rectum. His fingers bore the marks of electrical shock. "There's
no word to describe his end," said his mother, Hallie Jones, whose oldest child, also a drug dealer, had been murdered two years earlier. "'Tor-
ture' doesn't do it. Neither does 'desecration.' What they did to my son is
unlisted."

Nearly one out of four residents in Kings County lived in poverty, and
the jobless rate in Corcoran had been stuck at 16 percent for more than a
decade. As a measure of the desperation, 800 people gathered outside
John Muir Junior High on a winter morning to sign up for two low-wage
clerical jobs at the prison. If not for the welfare checks sent out each
month from Washington—to the mailboxes of the poor and the post office boxes of the rich—Corcoran might have sunk right into the clay bot-
toms. Seven families with roots three and four generations deep—the
Boswells, Boyetts, Gilkeys, Newtons, Hansens, Howes and McCarthys—
had collected $23 million in federal crop subsidy payments since 1996.
Not a penny of the aid went to the fruit, vegetable and nut farms of Kings
County. They weren't eligible. Instead, the money went to prop up cotton, wheat and barley—crops that existed in surplus. After two years of
debate, Congress decided to not only perpetuate the program conceived
in the Great Depression as a way to help struggling family farmers, but it
had beefed up the payments to a handful of the biggest growers.

The new list of beneficiaries for California no longer found Boswell at
the top. The honor went instead to his neighbor to the south, the Mc-
Carthy family, which had collected $7.2 million in federal crop payments
for cotton land now knee-deep in human waste. The McCarthys had
found an unusual ticket back to farming after years of bankruptcy and
federal investigations for fraud. They had become a dumping ground for
L.A.'s sewage, more than 200,000 wet tons of it each year. For the trouble
of cooking the sludge into compost and spreading it on their land, Los
Angeles County was paying the McCarthys $5.6 million a year. The
money was so good that Ceal Howe Jr., who farmed on the other side of
the lake basin, had decided to fallow 15,000 acres of cotton land—nearly
a third of his holdings—and turn it into a shit farm for seventy-eight

cities on the other side of the mountain. "We can't justify planting any cotton this year," Howe explained, citing high labor and water costs and abysmal crop prices. "The future of agriculture looks bleak."

Naturally the farmers were a little touchy on the subject of subsidies. The crop payments, they argued, didn't line their pockets but were plowed right back into the soil. The money went to the local tractor dealer and hardware store and to the men and women earning $9 and $10 an hour to drive a cotton picker.

But the town, at least those not in the employ of the farming giants, wasn't buying it. For seventy-five years, the rich and poor of Corcoran had coexisted mostly on the growers' terms. Now a fundamental shift was taking place, and it played out on an autumn night in the halls of the Corcoran City Council. On the surface, the issue was nothing more than a feud between a white city manager loyal to the growers and a white police chief beloved by the Latino community. Underneath, though, the town was roiling with the question of wealth and why it hadn't trickled down from the growers to the community at large.

Walking into city hall, the first thing you noticed was the display of miniature cotton bales, no bigger than a pint of milk, embossed with the names of each big grower. Every crop raised and ginned and pressed in Corcoran—from garbanzo beans to safflower seeds—had been given a place of honor on the shelf. As the growers and their top managers filed past the display into the main chamber, an old farmhand in the back muttered that the bad blood about to become public was a half century brewing. "They brought the plantation to Corcoran," said Santiago Delgado, who worked thirty years for Salyer before a knee injury forced him to retired. "Except they never called it the plantation. They called it 'The Company.'"

The whites of Corcoran had managed to control the spoils long after their share of the population had dwindled to 15 percent. They did this mostly by backing candidates—Latino and otherwise—who either worked for the growers or had relatives who did. Now, in the wake of two Latino activists winning seats on the council and a white councilwoman resigning, the growers had lost their majority. On matters dear to their hearts, they could count on only two votes—with an uncertain swing vote hanging in the balance.

The matter before the council this night seemed straightforward enough: Who would replace city manager Don Pauley? In his nine years on the job, Pauley had done little to alter the perception that he was the growers' lapdog. When he wasn't tending to their needs directly, he was hard at work securing government disaster relief for flooded-out farmland. Pauley's reports were a testament to the wonders of the multiplier effect—calculations that transformed a one-third loss to Boswell's cotton crop into a countywide calamity. The city manager was less adept at keeping Latino power brokers happy. Instead of mending fences, Pauley tried to muscle out the new police chief, Jim Harbottle. The move so angered the Latino community that Pauley had no choice but to resign.

"The Latinos thought I was doing the bidding of the big growers, and a few of the big growers thought I was doing the bidding of the Latinos. What I was trying to do was both—to the benefit of both," Pauley said. "It was a fine balancing act, and I pissed off everybody."

As to the question of a successor, Harbottle himself wanted to take the job. His wife, a flashy attorney from Visalia, opened the meeting by proclaiming her husband a fit enough administrator that he could juggle both the duties of police chief and city manager, at least on an interim basis. No prospect could have irked the growers more. They were convinced that Harbottle carried a particular contempt for Big Ag, but rather than say so straight out, they picked at the obvious conflict of interest. A police chief moonlighting as city manager was an invitation to unchecked power, they said. If chosen, Harbottle would answer to no one but himself.

Both sides appealed to the one man who seemed to hold all the cards, a blind black preacher named Sir Lee Shoals who had been appointed the fifth member of the city council, the swing vote. The growers wasted no time trying to butter him up. They talked about his willingness to be an honest broker and the historic divide he had just crossed as the city's first black councilman. Shoals, who wore big round sunglasses and a Nehru jacket, nodded his head and smiled, but he was a tough act to read. For most of his life, he could see well enough to watch his mother fall for one fellow after another, unions that produced sixteen children in all. The Latinos in the crowd figured that any man of color who grew up fighting for crumbs like Shoals had was a man on their side. As the two camps weighed their next words, Don Gilkey, the former college football lineman who took over the family farming business years before, rose up to speak.

"Something's been brought up several times that I need to address, and that is fear. People are accusing us of trying to use scare tactics. That's ridiculous. We're not trying to scare anybody, " he said. "You folks gotta show a little leadership and start working with the whole community. We got people who have been living here for years, people who have contributed to this community and just done everything they can for it. And you don't even want to consider them."

Joan Garcia-Munoz, one of the new council members, let the hulking Gilkey finish. Then she launched into a lecture about changing demographics and ballot-box power and the dawn of a new Latino day that the growers and their boosters were simply too blind to see. She pointed to the full-page letter that they had taken out in the *Corcoran Journal* a few days before. It was addressed to the good citizens of Corcoran and criticized a new "extremist" element for booting out the city manager. Written in English and Spanish, the letter was signed by a group of growers calling themselves "The Silent Majority." Garcia-Munoz suspected, quite correctly, that the whole thing had been orchestrated by Boswell's Mark Grewal.

"See, I was elected by the people to serve the people," she said. "I wasn't elected to serve the self-imposed affluent people in the community. I'm not going to answer to a group of twenty or thirty people because we have almost 10,000 people here in Corcoran. And you know, Don, only because you brought it up, but there has been an individual businessman here in town who has tried to intimidate me. This individual has sent word to me to back off of Boswell. I've had other prominent businessmen in this town make the allegations that I've had affairs. You should know that I've contacted the FBI because trying to intimidate an elected official is against the law. It's a felony."

The moment was ripe for Sir Lee Shoals to step in and act as peacemaker, but he said little, perhaps calculating that the divide was too wide. The group of fifteen farmers led by Gilkey walked out during a break and didn't come back to see Harbottle named interim city manager. Corcoran, it seemed, had staggered into a war that had nothing to do with floodwaters or levees and everything to do with race and the injustices of a company town. If Garcia-Munoz had the cheek to stand up to the growers, it didn't hurt that she worked at the one place in town where Boswell held no sway: Corcoran State Prison. The high-tech, maximum-security lockup had risen behind the Boswell bale yard in the late 1980s and was now infamous. This is where guards set up fist-fights between rival inmates in tiny yards and then used the brawling as an excuse to shoot them dead. More inmates were killed at Corcoran in the 1990s than at any other prison in America.

The prison's entrance, a river rock fence with a silhouette of the state of California and a coat of arms that featured a cotton field, was the only concession to its location. Otherwise, the prison sat as a separate world where thousands were herded into two distinct facilities known as Corcoran I and Corcoran II. The first Corcoran was where a Who's Who of killers shared space in a dorm known as the Protective Housing Unit, or PHU. The state Department of Corrections didn't like opening up its thirty-three prisons to visitors, especially the media, but if you were on tour with the director or, strangely, a contestant in the Miss California pageant in Fresno, you got a peek inside the yard where a senile Juan Corona grew chili peppers and a sullen Sirhan Sirhan played chess and a sentimental Charles Manson strummed his guitar.

As a general rule, no inmate moved unless a guard in a remote booth pushed a series of buttons. The men inside the PHU, on the other hand, could roam here and there without handcuffs or leg chains. Each one's fame had put a bounty on his head that kept him apart from the general population in the other yards. They read newspapers and drank coffee and spread files with epic legal maneuvers across big round tables in a wide-open setting. During one director's tour, they gathered around a *Los Angeles Times* photographer and mugged for the camera like school kids. A former shot caller for the Aryan Brotherhood, a lifer with five murders to his jacket, wanted the photographer, a black man, to take his picture holding up a Bible. Manson spotted the photographer from across the room and, with a balled up fist, gave him the old Black Power salute. America's most notorious killer from the '60s was wearing three layers of beads around his neck, shorts over sweatpants and sandals that did a poor job of hiding his toenails. His teeth were rotten and his beard was mostly gray. His eyes had softened, but the swastika between them had hardly faded. "You're with the media, huh?" he said with a snicker. "Man, you guys aren't even on base anymore. You're all tagged out."

Corcoran had looked to the prison as a savior, a tenant that would bring jobs and new houses and something more than a Taco Bell and a McDonald's along the commuter route. Because the town's official population had doubled on account of the inmates, Corcoran received more than $800,000 a year in extra state revenues. For citizens who tended to blame the town's economic woes not on The Farm but on The Prison, the state's largess was little more than blood money. "We were told that a lot of guards were going to be living in town, that the tule fog was going to be the big thing to keep them living here, that no one was going to want to commute," said Victor Castillo, a police officer and life-long resident. "Instead the guards and most of the staff just blow through town. I don't think more than a handful of them actually live here." Even Marion Salyer, Fred's wife and Corcoran's staunch matriarch, had moved away because of the prison. She lived half the year in Palm Springs and the other half in Pebble Beach. "My mom could literally see the lights of the prison from our house," daughter Linda Salyer said. "It feels like it's next door, and she hears this rustling and the dog barks and she thinks some prisoner has escaped."

If the prison had become a shame, the focus of newspaper exposes and a *60 Minutes* investigation, the town itself bore no blame. The same, sadly, could not be said for Corcoran's other disgrace: the remnants of a black community that had progressed in only the barest of ways since the 1940s. Gertha Toney, the matriarch of Sunny Acres, still lived in the shotgun shack her husband, Howard, had built on old man Matheny's land. She was ninety-five years old, and the Lord had preserved her like good apricot jam, the preacher said, cooked down so nice in sugar. Toney had watched her youngest son, Joe Boy, die from a heroin addiction and one of her grandsons take a summer job with Boswell only to be killed two weeks later when his truck rolled over on a levee. He was sixteen, a running back at Corcoran High, and didn't have a license. She had seen numerous daughters and granddaughters fall victim to the cycle of out-of-wedlock births and welfare that cut short the dreams of other families. Yet the gallery of photos that took up two walls in her living room was filled with bright and smiling children who knew nothing of her toils in the camps and cotton fields. "You know," she said, tracing their Asian eyes and Latino skin and Caucasian hair, "this all started from a simple black family. Just me and Howard."

She never questioned the wisdom of settling outside a town that refused for so long to extend even the basics of a water system. If her husband ever entertained the notion of moving elsewhere, say Fresno or Bakersfield, it never got much beyond pretend talk. They owned the land free and clear and their church was just a walk across the field. One year faded into the next and pretty soon she looked up and Howard was gone and she was still here. Her daughter Dorothy, the rebellious one, had come home after working fifteen years in convalescent hospitals in Southern California. Like so many other children of the black Okies, she had left the valley for a decent job and returned for the simple reason that her mother needed caring and her Social Security checks stretched a lot further here.

Dorothy had visited enough times to follow the fate of Sunny Acres and the other black enclaves in and around the old lake. Her family and their neighbors from Texas and Arkansas, Oklahoma and Louisiana, had come west because they saw an ethic in the land. Their desire to be rural and liberated at the same time was what made them different from millions of other Southern blacks. Their isolation had preserved a freshness in them that wasn't present in their city cousins. They were straight-up country, stripped of everything but the essential. At the same time, the decision to stay rural, at least rural as defined by the San Joaquin Valley, had magnified their misery. They had watched their dreams turn putrid in the middle of a stale alkali desert, and the cotton, in a cruel twist, was still all around them. It was farmed by giants and picked by machines, and those machines were driven by men and women who weren't black but brown. Field hands fresh from Mexico were working their way up from the bottom of the American dream with a life of promise, however illusory, in front of them. Old black field hands like Gertha Toney were still stuck at the same bottom with their lives, America's history, behind them. The poverty they lived in was the poverty they would die in.

Dorothy, now a grandmother with gray hair, had a difficult time reconciling what mechanization had done to the land and its people. The machine had freed blacks from the fields and driven their children to find a new life of hope and peril in the city. But for towns like Corcoran, where the older generation and the stubbornly rural tended to remain, it was hard to deny that the machine had brought a slow, choking death. The workers who used to stream by the thousands into town, clutching cotton tickets as good as cash, were long gone. With them had vanished the stores and shops and bars that made Corcoran a boomtown. "The cotton picker came and the people folded up their sacks and moved on," Dorothy said. "All the little stores I grew up with have shut down. The J.C. Penney's and Safeway Market and Sprouse & Reitz and Maroot's

Jewelry. You walk down Whitely today and Corcoran's nothing but a ghost town."

THE PLANES LINED UP one behind the other on the concrete strip, a fat hose stuck into each belly, filling the tank with another load of chemicals to knock the leaves off 180,000 acres of cotton plants. The sun was coming down on the Coast Range, and there wasn't much wind to delay things. Blackbirds by the thousands were flying into the sunset to feast on the army worms. The worms no longer concerned the growers—for this was picking time. Kerosene pots lined the field next to a levee to light up the takeoffs and landings. On a big night of spraying, the planes guzzled 2,000 gallons of fuel, enough for a trip across the country. From the middle of September to the end of October, the pilots dropped some 90,000 gallons of paraquat and other defoliants on the fields below. The whole basin smelled like rotten flesh.

As the days stayed hot and the bolls began to pop, Mark Grewal calculated that the first pick was seven to ten days away. Jim Boswell, still the boss, took one look and wanted to start right away—on a Friday. Grewal broke the news to his district managers, who didn't take it well. They were all convinced that a big crop was in the offing—three bales to the acre and maybe more. Now with the early start, some bolls would never ripen in time for the first or second pick, and the district managers would lose their bragging rights at the coffee shop in Lemoore. "None of them want to go in early. It means leaving behind too much cotton that hasn't matured," Grewal said. "I told them, if this was the only field you had to worry about, we'd wait a few more weeks and pick the cotton clean. But when you're as big as we are, some fields have to come off early or you risk catching rain and fog on the back end."

After November 15, the chance of rain shot up 50 percent. The moisture reduced yields and messed with the precision of the gins. It was one thing to hit rain and tule fog in the days of the hand pickers. They'd wait a few hours for the plants to dry and then start filling their gunnysacks. In an age of mechanization, however, the fog complicated the movement of the machines from equipment yard to field. The fog was said to be an old Yokut curse on the white man. *Take our lake and you will be haunted by its vestige, the creeping mists, for generations.* The Boswell

Company, refusing to concede the upper hand to nature or Indian curse, poured miles of lime in the dirt and guided its pickers through the fog and into the fields.

Grewal had speeded up the cotton's maturation by cutting off the water early and spraying a double dose of paraquat that burned holes in the seam of each boll. The holes let in enough air to push out the fiber from its shell. The plant, thinking it was dying, released a gas that put every row into an even deeper sleep. All the green had been bleached from the fields and the stems were adorned only in white. A huge machine that looked like a prehistoric animal was now eating its way across the land, six rows at a time. The driver didn't need to navigate—a computer chip did that for him—so it was possible to catch a moment of shuteye, although one driver a few years back took it too far and landed in an irrigation ditch. Inside each of the six snouts, a whirling spindle snatched the lint and stripped it from the bolls. Then a series of spinning pads grabbed the cotton and shoved it into an air stream that blew into a big basket overhead—6,000 pounds of fiber and seed by row's end. A good hand picker might have bagged a quarter of an acre in a day. Each Boswell machine, working twelve-hour shifts, turned forty acres of white-flecked field into a swath of brown. At that pace, Grewal predicted, the pick would be finished in forty days.

"Go fly over the valley and see how many guys are picking cotton," he said, smiling at the thought that no one but Boswell had begun. "We've got seventy-two machines running right now. Look at that cotton picker right there. He went up the field, a mile run, and he's busting at the seams. I mean the cotton's coming out from all ends. We're slamming it."

Gone were the metal trailers where the cotton was dumped and tramped down and hauled to the gin. The trailers had been replaced by four-sided, bottomless containers the size of railroad cars. They were known as module builders, and each one came equipped with a giant hydraulic arm that packed the cotton into a forty-foot-long, ten-foot-wide loaf of bread. As each driver reached the end of the row, he turned off the navigator and maneuvered alongside the module builder. He idled for a second and then let the RPMs rip, raising the basket high in the air and spilling forth the cotton. Any fiber that came tumbling out was raked up by a crew of workers and dumped back in. As a precaution against rain, each finished module was then covered in a Boswell blue tarp until a special truck could come by and haul it to the gin. By harvest's end, 10,000 loaves of bread had been punched out under the autumn sun.

"Look how beautiful those modules are," Grewal said, pointing to a 640-acre section lined with cotton under tarps on both ends. "Look at this. That is money."

For five years, we had kept a grand ledger book in our minds to try to reckon the Boswell empire—and by extension Big Ag—in way that broke free of the dogmatic screeds of the 1930s and 1940s. What were the pluses? What were the minuses? We had studied the works of Carey McWilliams and Paul Taylor and read one of the masters of historical scholarship, the Frenchman Fernand Braudel, who had looked at the development of the Mediterranean world and concluded that valleys almost always led to a plantation-like system that produced slaves and oligarchies. To conquer the plains had been a dream of man since the dawn of history, but the dream required more than man himself. The reclamation of the flatlands, draining swamps and controlling rivers, relied on large-scale government investment. And that investment rarely, if ever, worked its way down to the working class. Land that was flat and endless became the easy domain of the machine. In such a place, the rich became very rich and the poor became very poor.

The same conclusion, more or less, had been reached in a seminal 1946 study of two towns in the San Joaquin Valley: the company town of Arvin, dominated by the 12,000-acre grape operation of the DiGiorgo family, and the town of Dinuba, surrounded by seventy-five-acre vineyards and fruit orchards owned by different family farmers. The two cities were the same size, and the farms that skirted them brought in the same aggregate income, and yet the way they developed couldn't have been more different. Social scientist Walter Goldschmidt found Dinuba to be a pleasant town with four grammar schools and a modern high school. College-educated residents lived in tree-lined neighborhoods, belonged to community clubs and patronized a thriving downtown. Arvin, by contrast, was squalid and dilapidated with a single school, unpaved streets, no sidewalks and residents who went to church in tents and socialized in a knot of bars that lined the main street. DiGiorgo and the other big farmers lived elsewhere and cared little about the community. One visiting minister called it "the worst town I ever saw," and business owners freely admitted that they lived there to make a "killing" and were going to flee the first chance.

Goldschmidt's study, commissioned by the U.S. Department of Agriculture, vilified the kind of corporate farming that defined Tulare Lake and the valley's west side. The same Associated Farmers that had beaten back the labor strikes of the 1930s reacted with such vehemence to the study that government bureaucrats prevented Goldschmidt from digging deeper and tried to bury his report. Congressman Alfred Elliott, whose district took in both towns, feared the study would be seized on by those in favor of acreage limits for farms receiving federal water. "Some silly professor, some dreamer, some man who knows nothing about agriculture, knows nothing about the living conditions of the farmers and the people who work there, is doing this," he told a subcommittee hearing. Goldschmidt soon found himself investigated by the FBI.

No one could argue, at least not safely, that Corcoran matched Arvin or that Jim Boswell resembled DiGiorgio. As far back as the 1960s, in an effort to rebuild the town, Boswell threw a big dinner at the Cotton Club and invited all the growers to kick off the Corcoran Community Foundation. He was looking for each family to set aside a chunk of its fortune for Corcoran's future. They dressed up in coats and ties and ate big steaks and failed to come through with a single big donation. So Boswell tried to fill the gap all by himself. His company built the park, the hospital, the YMCA, the ball fields, the high school track. Every year, he paid the college tuition for a new pair of Boswell scholars, kids from the local high school. He encouraged his top men to live in town, and they coached Little League and sat on the planning commission, the school board and chamber of commerce. He set down a policy that all goods and services had to be purchased from local vendors, even if they charged as much as 10 percent more than outside competitors. Tractor drivers with a sixth-grade education were making $30,000 a year at Boswell. "If you look at Corcoran and the cotton towns on the west side, there's no comparison," Grewal said. "Those aren't towns but labor camps with maybe a liquor store or two. To say the Boswell Company hasn't put wealth into this town isn't true. The trickle down is here through this company."

A drive through valley farm towns such as Kerman and Kingsburg showed another truth: main drags more vibrant and neighborhoods more healthy than Corcoran's, even though their populations were nearly the same. Kingsburg shared the same Kings River but it differed from Corcoran in one essential way: It didn't sit near the bottom of an on-again, off-again lake. Farmers had gone about planting all manner of fruit and nuts and other permanent crops. The soil—loam rather than clay—produced some of the highest yields in the nation. Likewise, the machine hadn't reshaped Kingsburg or Kerman to the degree that it had transformed Corcoran. Men were still needed to prune and pick the vineyards and orchards, and their wages came back through town. Mechanization, of course, was the cotton grower's prerogative. At some point in the 1950s, the economics of hand labor made no sense. But mechanization also turned cotton into an economic engine of far less horsepower for a town. The math really hadn't changed since Gold-schmidt. Corcoran may have generated more tax revenue than compa-rable towns, but the sales flowed to only a few hands: the farm equipment dealer, the crop dusting service, the petroleum fuel distribu-tor. Kerman and Kingsburg, by comparison, boasted dozens of small shops and restaurants that kept turning over cash. In the end, a village of small farmers created more community wealth than one giant farmer, however philanthropic.

Boswell and the other big growers made perfect sense when they ar-gued that Corcoran could never have been farmed small. The Jefferson-ian ideal of 160 acres was pure folly at the bottom of a lake. If nothing else, large land holdings were needed to wheel around the floodwaters. But rather than being a rationale for big farms, the flood risks seemed a stronger argument for the lake never being farmed at all, or farmed as it was in the 1920s and 1930s—without the dams and other government subsidies. In other words, if a farmer in the lake bottom wanted to gam-ble on three years of crops for every one year of flood, it should be his gamble, not the taxpayers'.

Of course, such a view needed to account for the rather substantial role that government bureaucrats played in encouraging such depend-ency. The notion of the Kings River as a utility in which every drop be-longed to man and to waste a drop was akin to a crime had been formalized as far back as 1918, in a joint study by the U.S. Department of Agriculture and the California Department of Engineering. "With an in-creasing population and consequent demand for foodstuffs, public pol-icy requires and will eventually demand that the water be so handled that every possible acre will be bought into cultivation," the report stated. "It is quite obvious that the Kings River is not doing its fullest duty in this respect." For every drop to be utilized and every acre to bloom, the report concluded, the government had to intervene.

And intervene it did, forging a hydraulic society that paved the way for more government subsidies and land concentrated among a privileged few. From the very beginning, the building of Pine Flat dam had been a manipulation of the word "flood." In the days of the big wheat farms, a flood was welcomed because there was no better way to water the crops. Farmers didn't fret over the land lost for a season or two. Indeed, they set aside certain pieces of marginal ground to serve as water storage areas. The force of nature they feared most wasn't flood but drought. Then the notion of every drop utilized and every acre farmed began to take root with a considerable nudge from the reclamation boys in Sacramento and Washington. A handful of big farmers began buying up more land and erecting more levees, a frenzy that diminished the capacity to endure a flood. At some point in the 1930s, too much of the lake bottom had been claimed by farming and too little had been left to store the floodwaters.

Farmers who had always lived side by side with a flood suddenly were no longer fine with nature being nature. They wanted nothing less than to control the Kings, Tule, Kaweah and Kern rivers. Because they had purchased water rights up the river to safeguard against drought, they now had the luxury of worrying about the floods to come. Thus began the levee wars and the decades-long fight for the dams. To bring even more marginal land into production, the farmers ended up trading water storage systems. They went from holding water in the lake bottom to holding water in the hilltop. Pine Flat Dam and the dams on the other rivers came down to one spoil above all: the precise delivery of snowmelt. Farmers could now keep the water up on the mountain when they didn't want it and summon it down when they did. But America's investment in Tulare Lake—hundreds of millions of dollars that funded the dams and rebuilt the damaged levees and covered the subsidy payments for cotton and grain—ended up going to a handful of big growers. It allowed the Boswells and Salyers and a couple of neighbors to overengineer and overcrop the land, bringing more flooding, more excess fiber and more subsidy. All this didn't include the other costs—to the fish and wildlife and riparian habitat that had to be destroyed to control the rivers, or to the health of the people who lived beneath the cloud of dust and defoliants.

What might the valley have looked like if government officials had insisted on a balancing of interests at the outset of the West's development? What if the pioneers had been required to set aside 30 or 40 percent of the river's flow for fish and game and let the farmers have the rest? Such a division might have guaranteed that only the best land was tilled, and it would have gone a long way toward avoiding the periodic glut of raisins, grapes, plums, peaches, cotton and grain that bedeviled agriculture. Even as Grewal was talking about planting grapes and other permanent crops in the lake bottom, raisin growers across the valley were yanking out their Thompson vineyards and holding communal bonfires, the price of the dried grape too low to pay off the bank. By subsidizing the grab of rivers and land, America had created a surplus hell in the valley.

Some farmers growing high-dollar crops such as almonds were now making the case that the value of the water dictated an end to California's experiment with cotton. To grow one T-shirt, they pointed out, took 257 gallons of water. One acre of cotton swallowed two and a half acre-feet of irrigation. The Pima that came off that acre sold for $1,100 in an average market. Had the water been shipped to the houses of Southern California instead, it would have generated $750 in income for the farmer—with no real costs—plus income for the water distributor and housing tract developer. From the perspective of California's fruit and nut growers and urban dwellers, water was simply too precious a resource to waste on a surplus crop grown in sixteen other states. Texas alone accounted for 5 million of the nation's 17 million bales of cotton, and now China was growing a river of cotton too. If government was going to subsidize farming in California, why not subsidize crops that only the valley could grow, crops that matched the uniqueness of its soil and weather and the value of its water? Why not underwrite almonds and pistachios and stone fruit and table grapes that were in demand around the world?

After all we had been through with Jim Boswell—the two years chasing him down, the long rides across the lake basin, our barbed questions and his heated answers over Jack Daniels, the extraordinary interview with Fred Salyer in which we prodded him to talk about his father, Cockeye, and the murder in Pixley—we owed him one last shot.

As we drove out to meet him, a big storm was coming in from the west and the crews were racing to make the pick. The front had reached Yosemite, just a mountain away, and the whole basin was aswirl. Cotton and dust were blowing everywhere, the blue tarps of the modules flapping like geese.

Boswell was standing inside one of his gleaming gins, the floor with the inlaid diamond B polished clean like a ship's deck. The modules were being fed into a giant oven that blew 400 degrees of hot air, cooking moisture out of the fiber. The gin was nothing more than a series of stations a block long that removed the leaves and sticks, separated fiber from seed and spit out 500-pound bales squeezed tight with steel wire. This gin, a Pima gin, produced 600 bales a day.

Boswell greeted us with a warm handshake; there was no offer of a bribe this time. He seemed more than a year older, as if twelve months had settled in on him like five years on a younger man. He was about to get the toughest news of his life, that his wife, Roz, was dying of pancreatic cancer, and he was still quarreling with his son over the reins of the company. Only after they would bury her together, father and son, and he had accomplished his list of nine goals and turned the company around would he hand back control to J. W. The two of them would smooth over their rift and become, in the words of J. W., "tighter than ever."

But for now, their relationship remained tense. During one interview at the corporate office in Pasadena, J. W. conceded that he took some glee in his father and Grewal bragging about record yields when, in fact, the gin was showing something quite mundane—2.2 bales to an acre. J. W. considered what would happen if his father died in the midst of their bickering. How would he remember him?

"If I had to eulogize the man, what would I say? I don't know. I don't know. I've pictured the funeral at the senior citizens center in Corcoran, people overflowing out onto the street. Most of these people think the guy is an icon—the King of Kings County and all that. Well, I can say all the right positive things. And I'll do that. It'll be classy. That's all I can tell you. I mean, the people who are closest to me know the truth—my mom, the other top managers, the big investors.

"Ten or fifteen years ago we had a meeting in my office—it used to be his office—and at the end, J. G. looked at everybody and said: 'God damn it, stand up and lie like gentlemen.' So I guess that's what I'll do." Then the son paused. "Of course, the guy might be eulogizing me for all I know."

As the bales were sheathed in plastic and trucked out to the yard that day—the end of another season—there was something in the old man's voice that felt like an opening. He talked about his best friend, Harvey Ruth, the old cowboy at the Yokohl Ranch along the eastern Sierra, and wondered who was more content. Sure, he had access to the Bohemian Club, a jet airplane and the symphony, but Ruth lived a beautifully simple life. "I know where he is right now. He's sitting in this place we call the Cateron Cabin and there's an old red-tinged bear up in the tree and he sits there and talks to the bear. So you know, who's the happiest? Who knows?"

We asked him if he was afraid of dying, and his answer was classic Boswell. "Hell no. I could have been dead fifty years ago. I never was going to be captured alive by a Jap." Then we eased into the biggest question of all—the one that went to Boswell's legacy. Had the astonishing success of his company penciled out for the town of Corcoran? We guessed that his defense of what he had built—and big agriculture in general—would be at least as spirited as Grewal's. His answer, like so much else about him, caught us leaning the wrong way.

"I can't deny we're a company town. Once we acquired Salyer, you win on that point. But we have a degree of stability here that company towns in the South don't have. And I think we've made huge inroads into alleviating future floods, which will help with the economy. But you're right. Our type of farming doesn't create the kind of community that a bunch of small vineyards or peach orchards might. But we're cotton. That's the hand that was dealt to us. And we can do cotton better than anybody."

A part of him, bluster aside, had already closed the books on a cotton culture that had shaped the lives of Boswells for five generations. He had lost the fight to the environmentalists to build a super-dairy on the fringe of the lake bottom, but he was already envisioning other ventures—beyond farming and the lake basin. "My grandson will probably read this book and wonder, well, 'Why didn't that dumb bastard move out of the lake and go someplace better.' There probably won't be any cotton growing in California ten years from now. What the hell you doing growing cotton? I mean, let the Chinese grow cotton."

Some of his old pals could already see the day when his son would choose to fallow the fields and sell the water to Los Angeles. To them, the company had always been about water. Boswell controlled enough of it to feed the Metropolitan Water District all the way past the Mojave Desert. The company's 15 percent draw of the Kings River was worth billions of dollars all by itself. The Boswells could be filthy rich for generations. And the family foundation would be free to give in a way that J. G. never imagined.

Boswell pretended not to care what his son did with his water. "I'm going to be scattered someplace," he said. "I'm delighted I lived when I lived. . . . I think it's ridiculous to try to set the future for somebody else. I'm not going to tell you how to run your life except to get it right."

We took one last drive across the lake bottom and watched the tractor discs shred the cotton stalks into dust. By week's end, the field would be clean and planted with winter wheat. Skidding along old Cockeye's levee, he repeated his contention that he hadn't gone hunting for Salyer any more than he had set his sights on Crocket & Gambogy or the tens of thousands of other acres he had gathered. He wanted it understood that what he had built in his corner of the West, even as it extended across an ocean to Australia, was more by serendipity than design. The Accidental Empire.

"This talk of empires and kings is just bullshit," he said. "I don't covet money. I'm not out to accumulate it. I'm an old dog. I did the battle. I never backed away from anything in my life." Then he repeated the old fish story.

"It was a Sunday afternoon right over there. The El Rico levee. I was standing out there with a shovel. The only guy left. That's when ol' Wes Hansen come out and said, 'What they hell you doing out here?' And I said, 'If I can just stop these waves.' . . . They were literally starting to come over the top. The water's coming in and the wind's up a little, and I said, 'Wes, I ain't gonna quit.' He said, 'What can I do?' And I said, 'I need something to help me stop these waves.' And he said, 'You know that old colored guy on the other side of town? He's got forty or fifty junk cars.' So he had a bobtail truck, and him and I pushed those old cars off and stuffed them into the levee. And it worked, but the water wouldn't stop. It was still coming up. So I put the word out that we'll pay $15 a wrecked car, no questions asked. Cash. And we captured the junk car market between here and Los Angeles. By the time we finished, we had six miles of cars stashed in and we saved some of the crop. I wasn't going to quit because I had been taught that you never quit on a flood. You don't quit."

Notes

ABBREVIATIONS

AC	The papers of Senator Alan Cranston, Bancroft Library, University of California at Berkeley
BC	*Bakersfield Californian*
BCA	Boswell company archives, Pasadena, California
CCJ	*California Cotton Journal*
CJ	*Corcoran* (California) *Journal*
CN	*Corcoran News*
CP	The papers of Representative Charles Pashayan, Central Valley Political Archive, California State University at Fresno
CR	*Congressional Record*
CTA	Archives of the California Institute of Technology, Pasadena
CMcW	Files of Carey McWilliams, Department of Special Collections, UCLA Research Library
DM	Cecil B. DeMille Papers, Special Collections, Harold B. Lee Library, Brigham Young University
DV	Interview conducted by David Vowell; transcripts are located at BCA
FB	*Fresno* (California) *Bee*
FEC	Federal Election Commission records
FS	Boswell executive Fred Sherrill
FSal	Fred Salyer; interview with Salyer
FSH	An undated internal company history at BCA, written by Sherrill
FSP	The papers of Fred Sherrill, a collection maintained by his family in Pasadena
GB	Letters of George C.S. Benson, Archives of Claremont McKenna University
GDAH	Georgia Department of Archives and History, Atlanta
GHJ	*Greensboro* (Georgia) *Herald Journal*
HJW	The papers of Hobart J. Whitley, Department of Special Collections, UCLA Research Library
HMJ	*Hanford* (California) *Morning Journal*
HS	*Hanford Sentinel*

JF   Jim Fisher, former Boswell Company president; interview with Fisher
JGB   Lieutenant Colonel J. G. Boswell
JGB2   J. G. Boswell II; interview with J. G. Boswell II
JPL   John Penn Lee, onetime Salyer lawyer and FSal's former son-in-law;
            interview with Lee
JWB   James Walter Boswell, JGB2's son; interview with James Walter Boswell
LAT   *Los Angeles Times*
LC   Library of Congress, Washington, D.C.
LF   La Follette Committee Report/Hearings
LSal   Interview with Linda Salyer, FSal's daughter
NA   National Archives, Washington, D.C.
PST   The papers of Paul Schuster Taylor, Bancroft Library, University of
            California at Berkeley
PTOH   Paul Taylor oral history, Regional Oral History Office, Bancroft Library,
            University of California at Berkeley
RWA   Files of political strategy firm Russo, Watts & Associates, Sacramento,
            California
SB   Former Boswell water chief Stan Barnes; interview with Barnes
SD   The papers of Sam Darcy, Tamiment Library, New York University
STH   Sidney T. Harding Papers, Water Resources Center Archives,
            University of California at Berkeley
TDAR   *Tulare* (California) *Daily Advance-Register*
WBC   The papers of Wofford B. Camp, Shafter (California) Historical Society
WCOH   Wofford Camp oral history, Regional Oral History Office, Bancroft
            Library, University of California at Berkeley
WW   Interview with Warren "Spud" Williamson, Ruth Chandler's son and
            JGB's stepson

Winter

1   **"Never seen no cotton":** Steinbeck, *The Grapes of Wrath*, p. 521.

Chapter 1: Winter

Many of the subjects touched on in this chapter, such as the history of Tulare
Lake and Boswell's receiving of subsidy payments, are explored at much greater
depth later in the book, and the relevant sources are noted then.

3   **"Never harpooned":** Interview with Rosalind Seysses, the granddaughter
       of JGB's brother Walter.
4   **Nine straight beers:** Interview with longtime Corcoran resident Verdo Gre-
       gory. **"Misery of hemorrhoids":** DV with Josephine (Jody) Boswell Larsen.

6   **The biggest farmer in America:** JWB confirmed that through 1999, Boswell owned about 200,000 acres in California, farming about 160,000 of that; the company has downsized a bit since then. In addition, the company owns another 60,000 acres in Australia, of which it farms about 35,000. Though there are certainly farming operations with higher revenue, nobody plants more acres than Boswell does, according to the company itself; a rare article written about Boswell, which appeared in *Forbes,* April 17, 1989; and interviews with other industry executives. A 1997 *Worth* magazine article listed Boswell as the forty-fourth biggest landowner in the United States; all of those above him on the list are into cattle, timber or oil and don't actually farm on the scale that Boswell does. **More irrigated wheat, safflower and seed alfalfa:** Interviews with company executives; *Forbes* article.

6   **Liked to tease his wife:** JGB2.

7   **Boxer shorts:** Extrapolated from "What Can You Make with a Bale of Cotton?" by the National Cotton Council. **His water rights alone:** Interviews with Boswell investors, including Dick Jones, and water industry executives; *Forbes* article.

7   **A cowboy:** *Wegis, Thomson & Thomson v. J. G. Boswell,* case no. 179027, filed 1988 in state Superior Court for Kern County.

7   **"Play your game":** Interview with JGB2's friend Dick Jones, who also played golf with Palmer that day. In a separate interview, Palmer noted that he met JGB2 in Australia and the two became "good friends." As for JGB2's golf game, Palmer remarked: He's "a nice player, not the national amateur champion, but a nice player."

8   **"A shooter":** Interview with Jones. **"Maverick sort":** Interview with Welch.

9   **Largest body of freshwater:** Brown, *The Story of Kings County California,* p. 94.

10   **David Vowell:** Biographical detail from CJ, September 30, 1976.

11   **"Geologic freak":** From a Boswell-commissioned film called *Man, Land and Water.*

11   **Flood of 1969:** Details on Boswell's efforts are from numerous stories in CJ; JGB2; SB.

13   **5,200 jalopies:** JF.

14   **Bobbing like corks:** Interview with Doan.

14   **Salyer clan:** Details on the Salyers are from "Salyer Clan: Feud Among the Wealthy" by Charles P. Wallace, LAT, March 27, 1983; FSal; LSal; interview with Corcoran resident Verdo Gregory.

14   **88,000 acres:** Interview with former Salyer employee Earl Bankhead.

15   **"Never miss a payday":** Interview with Bankhead. **"Anybody else here":** Quoted in LAT, March 27, 1983.

15   **Brown's bagman:** LSal.

16   **$216,000 check:** CJ, October 16, 1969.

18   **List of products:** "Cotton: Foremost Fiber of the World," *National Geographic,* February 1941; "Cottonseed and Its Products," 9th ed., National Cottonseed Products Association Inc., pp. 12–13; "What Can You Make with a Bale of Cotton?" National Cotton Council.

21   **"Ribboned Dukes":** Cited in McWilliams, *Factories in the Field*, p. 25. He is quoting Henry George.

22   **"His boots on":** JWB.

CHAPTER 2: THE COTTON KINGDOM

23   **This dwelling:** Details on the history of the house are from an interview with Penfield resident Mary Ben Boswell Brown; E. F. Beckemeyer's notes for a "Front Porch Conversation about the Village of Penfield," September 27, 1998.

24   **Brunswick stew:** Armor's recipe for a 10-gallon pot of Brunswick stew starts with grinding 1 pound each of onions and green peppers. Take a quarter of the vegetable mixture and set it aside for the sauce. Put the rest into a gallon of boiling water and leave it until tender. Then add 10 pounds of chicken, preferably thighs with the legs on, and 10 pounds of Boston butt. Boil the meat until it falls off the bone, and "grind it with a coarse blade." Then add 3 gallons of ground tomatoes and 3 large, ground-up Irish potatoes. Keep the pot boiling. Skim off the fat if it is too thick. "Stir like hell all this time." Add salt and pepper to taste. Next, take the ground-up onions and peppers that have been set aside and simmer them with 1 quart of vinegar, 1 quart of tomato juice and 1 quart of water. After an hour, add a half cup of hot sauce. Take 1 ½ "dipperfuls" of this and add it to the big pot. Serve.

24   **"Some graciousness":** Warren, *Segregation*, p. 98.

24   **Buried beside them:** Armor, *The Cemeteries of Greene County Georgia*, p. 32. **Trace their lineage:** From the Boswell family Bible, compiled by Leila Boswell McCommons in the 1940s. **Began with George Boswell:** This and all other details on the Boswell ancestors, except where noted, are from the family Bible. For a history of Wilkes County, see "An Overview of Local History" by Robert Willingham, part of the USGenWeb Project.

25   **Boswell grew corn:** Though it's possible that George also planted cotton, a surviving business record, included with the family Bible, shows that he sold eight barrels of corn in 1818. A ledger, also included with the Bible, shows his account with a local merchant from whom he bought his whiskey and rum in 1819. **After George and Sarah died:** He died in 1820, she in 1833.

25   **King Cotton:** The details on cotton's global and national impact are from a

wide variety of sources, including Cohn, *The Life and Times of King Cotton*, pp. 18–25; Dodge, *Cotton: The Plant That Would Be King*, pp. 4–13; "Cotton: Foremost Fiber of the World," *National Geographic*, February 1941; "Cotton, King of Fibers," *National Geographic*, June 1994; Woodman, *King Cotton and His Retainers*, pp. 81–82.

25   **Sousa:** He composed the "King Cotton" march in 1895 for the Cotton States and International Exposition in Atlanta. **Degas:** The painting is *Portraits in an Office: The New Orleans Cotton Exchange*, 1873. **Dow Jones:** The company in the original Industrial Average of 1896 was American Cotton Oil. **"Fine-fibered and smooth":** From the Tennessee Williams play *27 Wagons Full of Cotton*.

26   **"Furs, cattle, oil, gold":** Cohn, p. vii.

26   **Spanish had grown cotton:** Turner, *White Gold Comes to California*, p. 2. **Leaders of the new nation:** Green, *Eli Whitney and the Birth of American Technology*, pp. 7–12. **English textile industry:** Dodge, pp. 18–36; Green, pp. 13–14; Jacobson and Smith, *Cotton's Renaissance*, p. 43. For a particularly colorful account, see Hobhouse, *The Seeds of Change*, pp. 146–148. The innovations included John Kay's flying shuttle, John Wyatt's machine for "spinning by rollers," the jennies of James Hargreaves and Richard Arkwright and Samuel Crompton's "spinning mule." Jacobson and Smith note that this industry emerging "around the cotton plant . . . formed the very core of the Industrial Revolution. Around cotton coalesced new 'high technologies' in steam power and automated machinery as well as coordinated marketing and managerial systems that linked the continuous stages of transformation from harvest to woven textile, from the field to the final customer" (p. 41).

26   **Eli Whitney:** Whitney's story—including how his patent got stolen—is recounted in Green, pp. 19–62; Cohn, pp. 3–16; Hobhouse, pp. 150–153; Britton, *Bale o' Cotton*, pp. 12–20; Bruchey, *Cotton and the Growth of the American Economy: 1790–1860*, pp. 53–62. **Plantation cat:** Jacobson and Smith, p. 46.

27   **First three cotton millionaires:** This and the other details on Greene's blossoming cotton culture are from Bryant, *How Curious a Land*, pp. 16–17.

27   **As well-to-do:** Walter, *Oconee River: Tales to Tell*, p. 218; Bryant, pp. 187, 189. **Owned thirty-three slaves:** Population and slave schedules, federal census of Georgia for 1860, GDAH.

27   **Mercer University:** The history of the school is from Rice and Williams, *History of Green County, Georgia*, pp. 72–73; Beckemeyer, "Front Porch Conversation."

27   **Johnson Boswell:** Details are from the Boswell family Bible. For a history of the Penfield Presbyterian church, see GHJ of August 20, 1926. Regarding

the Good Templars Lodge, see the *Herald* of June 3, 1875. **Each other's services:** Interview with Mary Ben Boswell Brown.

27   **Greene County thrived:** All details on activity and attitudes in the county from the 1840s through the 1860s are from Bryant, pp. 21–25, 61–62.

28   **"Suspicious white men":** Raper, *Tenants of the Almighty,* p. 57.

28   **Calhoun:** Ellison, *A Self-governing Dominion,* pp. 87–88. See also Worster, *Rivers of Empire,* p. 218, in which he notes that Jefferson Davis "insisted that slavery was better suited than free labor 'to an agriculture which depends upon irrigation.'"

28   **By 1859:** Cohn, p. 17. **Slave labor:** For a compelling discussion of the nexus between cotton and slavery, see Hobhouse, pp. 160–169. **"No power on earth":** Quoted in Dodge, p. 2.

29   **Grays was mustered:** Bryant, pp.64–67. The Grays were formally known as Company C of the Third Georgia Volunteer Infantry. **Johnson Boswell's son:** Company muster roll, GDAH.

29   **"Heard today":** The diary of Mercer professor Shelton P. Sanford was made available by E. F. Beckemeyer, Penfield.

29   **The Grays fought:** Folsom, *Heroes and Martyrs of Georgia.* **Another unit:** Company muster roll, GDAH. His new unit, the Stocks Volunteers, was formally known as Company B of the 55th Regiment of the Georgia Infantry. **Cumberland Gap:** Details of the Confederate surrender are from Kincaid, *The Wilderness Road,* pp. 262–273; interview with Richard Beeler, a ranger at Cumberland Gap National Historical Park.

30   **Remainder of the war:** Roll of Prisoners of War, GDAH.

30   **"Blessing in disguise":** Cohn, p. 121. Foote, in *The Civil War: A Narrative,* p. 113, notes that "cotton, the raw material of Great Britain's second leading industry, as well as the answer to France's feverish quest for prosperity, was the white gold key that would unlock and swing ajar the door through which foreign intervention would come marching."

30   **Textile workers:** Cohn, p. 138; Hobhouse, p. 175. In *Battle Cry of Freedom,* pp. 550–551, McPherson notes that "a number of historians have discovered cracks in the apparent pro-Union unity of working men" in England. But, he concludes, this "revisionist interpretation overcorrects the traditional view. . . . A good deal of truth still clings to the old notion of democratic principle transcending economic self-interest" among the textile workers. **"Could extort":** Cited by McPherson, p. 384.

30   **North's supremacy:** From AmericanCivilWar.com.

30   **"Swallowing the yellow dog":** W. Emerson Wilson, "Fort Delaware in the Civil War," Fort Delaware Society.

31   **Crusade of terror:** Bryant's book paints a vivid picture of the horror that befell blacks in the county after the Civil War. The black leader quoted is Abram Colby, p. 131.

31   **In that direction:** For a thorough look at cotton's move west, see Bruchey,

pp. 79–162. See also Jacobson and Smith, pp. 55–56. **Gone to Texas:** Cohn, p. 153. Bruchey, p. 117, also cites this "Texas fever."

31   **Boswell & McCommons:** Numerous editions of the GHJ from that era. **Minnie Griffin:** Details on Minnie and her father are from the Griffin family Bible, compiled by Emily Griffin Langford in 1997.

31   **Joseph farmed:** Details on Joseph's life are from his will and estate records, GDHA; his obituary in the GHJ, November 2, 1906; and other editions of the paper over the years that covered his business activities. **Cotton mill:** It was the Mary-Leila Cotton Mill.

Chapter 3: The Little Sahara

This chapter is built around the kind of road trip that JGB would have likely taken in the fall of 1922. Though no records exist showing exactly when Boswell made these drives or what he encountered along the way, interviews and corporate records confirm that he worked the Yuma Valley as a cotton merchant during this period. A variety of sources were used to reconstruct what it would have been like to travel from Pasadena to Yuma at this time, including maps and other material from the Automobile Club of Southern California Archives, Los Angeles; information from the Arizona Historical Society in Yuma and the Palm Springs Historical Association; and interviews with Coachella Valley historian Pat Laflin and Gadsden, Arizona, resident Louis Gradias.

32   **Best cotton lands:** Exam results taken from JGB's military personnel file at NA.

32   **Opened an office:** *Los Angeles City Directory*, 1922. The description of the office comes from a visit to the site. **"Office in my hat":** FSH. **Stored his ledgers:** JGB2.

33   **Hobbled:** WW; interview with Barbara Hamilton, the niece of JGB's first wife. Nephew J. O. Boswell, in an interview with DV, remembered JGB as walking with a limp. **Officer's uniform:** WW; DV with J. B. Long, who owned a gin in Arizona at the time. Details on the uniform itself are from the Army Quartermaster Museum and the Army Institute of Heraldry.

33   **Ranch-style house:** Details on the house at 300 S. Arroyo Blvd. in Pasadena are from property records and a visit to the site. Though there is no record of the kind of car that JGB drove in these early days, WW remembered that later in life he would only drive a Buick. "He was absolutely married to that particular type of automobile," he said. It's reasonable to assume, therefore, that JGB may have always driven one.

33   **Crop had plummeted:** Musoke and Olmstead, "A History of Cotton in California." Turner, p. 35, reports an even more dramatic falloff in 1920, from forty-nine cents a pound to six cents. **On the rebound:** The *Yuma Morning*

Sun of October 24, 1922, reported that Pima cotton was fetching thirty-four cents a pound in Boston, and noted that the market was functioning "smoothly and well." **A single maxim:** FSH.

33   **Uttering a swearword:** JGB2. **Indulging in a drink:** JGB2. WW remembered JGB telling him that when he was a boy, someone had forced alcohol down his throat, and he stayed far away from the stuff ever after.

34   **Scour the region:** JGB2. JGB's job illustrates the decline of cotton middlemen known as "factors," who in the pre–Civil War South instructed planters when and where to sell at the best prices, and the rise of "merchants," whose specialized knowledge ensured that their spinning mill customers received the quality and quantity of cotton they needed. The factor provided many other services as well, including extending credit and purchasing goods for his principal. But technological changes—the telegraph, the telephone, the transatlantic cable, fast rail transportation, high-capacity gins and compresses—and new sources of credit rendered most of the factor's work obsolete. JGB's relationship with Goodyear, in particular, provides a great example of how the merchant would use his expertise to deliver specific types of cotton to particular customers. The definitive book on cotton financing and marketing from 1800 to 1925, which covers this transition in great detail, is Woodman's *King Cotton and His Retainers.*

34   **Coachella Valley:** Turner, p. 49.

34   **Salton Sea:** The sea had been created only a short while before, in 1905, when an attempt to divert the Colorado River for irrigation went horribly awry. The mighty waterway surged out of its channel and rushed toward the Salton sink, the low point in the dry alkaline basin of the Imperial Valley. The water flowed for two years before the flooding could be stopped. Suddenly the sink was a sea. The disaster, which ruined fields and swept away homes and businesses, might well have served as a reminder of man's limited ability to control his environment. For many, though, especially the powerful capitalists who saw a fortune in bringing water to the desert, the sea came to symbolize something else entirely. It was "an assertion of water against aridity," as historian Kevin Starr put it (*Material Dreams*, p. 41), an almost biblical sign that "irrigation offered human beings the opportunity to co-create with divinity itself." For details on the Salton Sea, see deBuys and Myers, *Salt Dreams*, pp. 99–121; Reisner, *Cadillac Desert*, pp. 122–123; Worster, *Rivers of Empire*, p. 197.

34   **California Development Company:** The development of the Imperial Valley is one of the most fascinating—and important—chapters in California history, full of backstabbing and corporate greed. For a particularly lively account, see Starr, *Material Dreams*, pp. 20–44. As for the details of the crops that were growing in the area by the early 1920s, see *Southern California Business*, "Prosperity Reigns in Fertile Valley," May 1925.

35 **Algodones dunes:** Details on the dunes and the Plank Road are from "The Plank Road of Imperial County," Bureau of Land Management, 1989; "A Brief History of the Plank Road of Imperial County," *Branding Iron,* Los Angeles Corral, Spring 1999; Pierson, "The Growth of a Western Town."

36 **Next great cotton state:** Early California cotton history is from Turner, pp. 9–25. **"Labor is abundant":** Strong's comments are from an 1871 newspaper clipping in the files of the U.S. Department of Agriculture's cotton research station in Shafter, California. This quote is also cited in Johns, "Field Workers in California Cotton," p. 13.

36 **Plan to recruit:** Morgan, *History of Kern County California;* notes from an interview with William Henry Pinkney, one of the blacks who was recruited to came to Kern County in 1884 (Kern County Museum Library, Bakersfield, Calif). Haggin's Kern County agent, who was heavily involved in wooing the black pickers, was W. B. Carr. Haggin would become famously embroiled in a landmark California water law case against fellow land baron Henry Miller, an episode covered in chapter 5.

36 **All but died out:** This was blamed on a variety of factors, including a shortage of labor, high transportation costs, a lack of irrigation and competition from cereals and fruit, which brought higher rewards. See Johns, pp. 16–18. **Began to experiment:** This era of Imperial Valley cotton is captured in Turner, pp. 29–35; Stemen, "Forty Acres and a Ditch." Working out of an office known as the "monkey farm," because of the thicket of palm trees and rubber plants that shaded it from the desert heat, the researchers eventually urged the growing of a medium-staple variety called Durango.

36 **Wofford B. Camp:** Details on Camp are from Briggs and Cauthen, *The Cotton Man,* pp. 39–43; WCOH.

37 **Had wrestled away:** JGB2. **Anderson, Clayton:** Details on Anderson, Clayton are from Fleming, *Growth of the Business of Anderson, Clayton & Company.* The background on McFadden is from U.S. Senate Hearing to Investigate the Causes of the Decline of Cotton Prices, Committee on Agriculture and Forestry, March 10, 1936, pp. 488–489.

38 **Tire plant:** Details on the factory are from Rodengen, *The Legend of Goodyear,* p. 56; *Southern California Business,* April 1922, p. 6; June 1922, p. 42; September 1922, p. 62. See also Sitton and Deverell, *Metropolis in the Making,* p. 103.

38 **Talking his way:** JGB2

38 **Unmistakable imprint:** Details are from advertisements in the *Yuma Morning Sun* from 1917 through 1922. **At least eight gins:** Ginning report in the *Somerston Star,* October 6, 1922.

38 **Long had helped teach:** DV with Long.

39 **Classer's hands:** Garside, *Cotton Goes to Market,* p. 77. Garside gives a

great amount of technical information on cotton classing. Also see Jacobson and Smith, pp. 202–205.

39   **Loved this part:** FS once noted in a letter that JGB's fondness for haggling "is a part of his insatiable desire to figure out what the other fellow has in mind and try to beat him." **Gin man proffered:** In DV, Long recalled dispensing such advice at one point to JGB, though it's not clear exactly when that happened.

39   **Well-rehearsed riff:** Though it's not clear how JGB reacted to Long's suggestion, JGB2 says this is precisely the kind of story his uncle would spin out to make a point.

40   **Military remained:** WW.

40   **Walter Osgood Boswell:** Details on his academic performance and later military career are from information provided by the U.S. Military Academy archives; the *Official Register of the Officers and Cadets of the U.S. Military Academy*, 1899; *Register of Graduates and Former Cadets, U.S. Military Academy*, 1990. **Get kicked out:** The official explanation was that Walter flunked mathematics. Family lore had it that he hoisted a cannon onto the roof of the cadet barracks as a prank, and it took the Army Post Engineers a week to get it down. (From the unpublished memoir of Walter's son, James O. Boswell.) **Anne Decker Orr:** Vandiver, *Black Jack: The Life and Times of John J. Pershing*, p. 476.

40   **Pershing's aide-de-camp:** Vandiver, pp. 531, 537. **Fire ripped:** Vandiver, pp. 595–596. **"My heart bleeds":** Undated letter, probably from September 1915, from Walter Boswell to Pershing, Pershing Papers, LC. **Crossed the Rhine:** Back home, the newspaper reported: "Greene County Solider One of First to Enter Germany." **Upper ranks:** JGB2.

40   **Hospital bed:** GHJ, January 3, 1919. **Dreamed of flying:** JGB2; WW. In a note written in April 1918 to his future wife, Alaine Buck, JGB noted that "I am on duty with the Air Service . . . and hope to go over soon."

41   **Signed up in 1903:** All details on JGB's military career are from his personnel file at NA and from a 1921 volume of the *Army Register*.

41   **Showed a toughness:** While serving as inspector-instructor of the South Carolina militia, JGB got into a scrape with Governor Cole Blease over whether state troops met federal standards. JGB held his ground so firmly, the governor—known as "the stormy petrel of South Carolina politics"—relieved him of duty. Army higher-ups expressed dismay over JGB's sacking and, praising him as a "conscientious and efficient officer," reassigned him to the militia of New Jersey. For details, see *The State* newspaper, Columbia, South Carolina, for June 25 and 26 and July 6, 9 and 11, 1913.

41   **Weevil had stormed:** *The South in the Building of the Nation*, p. 102. **"Counterpart of the elephant":** Hearing to Investigate the Causes of the Decline of Cotton Prices, p. 348. **"Traditions of tragedy":** Sandburg, *The*

American Songbag, p. 8. **"Weevil said to the farmer":** This version of the boll weevil song is cited in Litwack, *Trouble in Mind*, p. 176. For a lively account of the boll weevil's march through the South, see Jacobson and Smith, *Cotton's Renaissance*, pp. 70–76.

42   **First spotted:** GHJ, June 15, 1917. **"Boll weevils and politicians":** GHJ, December 31, 1920. **Child who could catch:** GHJ of June 24, 1921; Raper, *Tenants of the Almighty*, p. 163. **Mass exodus:** Rice and Williams, pp. 364–365.

42   **Settled in the fast-growing city:** *Thurston's Directory of Pasadena*, 1921.
      They had met: Though no one knows for sure how JGB and Buck met, the only time that their paths seemed to have crossed was when he was stationed at Fort Logan and she was living in Denver. When Buck's first husband died in 1918, JGB sent her a condolence card that was signed, "As always, your friend."

43   **Military retirement check:** JGB would have earned between $133 and $234 a month, according to historical tables provided by the Department of Defense Office of the Actuary. **Denver insurance man:** Federal Census of Colorado for 1900 (available through the Colorado Historical Society, Denver). Information on Buck's father, who worked for the Liverpool and London and Globe Insurance Company, is from a "Historical and Descriptive Review of Denver," circa 1893. **Raymond McPhee:** Details on him and his family are from "The Pinnacled Glory of the West," Cathedral of the Immaculate Conception, 1912; Report of the Colorado Bar Association, twenty-first annual meeting at Colorado Springs, July 12–13, 1918. **Alaine inherited $188,000:** Records of the Denver probate court.

43   **Boswell's scheme:** JGB2; FSH.

43   **Presbyterian college:** It was Davidson College; Bill Boswell attended during the 1910–11 and 1911–12 school years, according to the college archives.
      "When Bill walked home": All quotes by Kate Hall Boswell are from DV.

44   **Bill sold the farm:** GHJ, August 15, 1919. As for his other dealings, the details are from interviews with E. H. Armor and articles in the GHJ of November 21, 1919 and January 7, 1921.

44   **Hotel Miramar:** The fact that JGB learned of Tulare Lake from Harold S. Doulton is noted in FSH. Details on what the Miramar was like in 1923 are from Harold K. Doulton, son of Harold S., who saved the old hotel registries.

45   **A newsreel:** CJ, October 10, 1919.

Chapter 4: The Lake of the Tules

46   **30,000 fresh souls:** Estimates of the Yokut population in the San Joaquin Valley and Tulare Lake basin vary widely, with numbers ranging from 17,500 to 70,000. The most knowledgeable historian of the Yokut culture was Frank

F. Latta, a Kern County museum director and ethnographer who conducted scores of interviews with surviving Yokuts from the 1920s to the 1960s and wrote several books, including *Tailholt Tales*, in which he estimates that the Yokut population was somewhere between 25,000 and 35,000.

46   **No better spot:** Bancroft, *The Works of Hubert Howe Bancroft*, 2:204; Mitchell, *The Way It Was*, p. 16.

46   **Some 800 square miles:** PG&E series, *The Lakes of California;* 1892 description by Thomas H. Thompson, historian of Tulare. **Four distinct tribes:** Brown, pp. 10–13.

47   **"Republic of hell":** Mitchell, *The Way It Was*, p. 21. **Mystery and malaria:** Austin, *The Land of Little Rain*, pp. 240–241. **Wild mustangs:** Castro, *James H. Carson's California*, p. 63; Smith, *Garden of the Sun*, p. 165. **"Huge broiler":** Smith, p. 22. **Beastly condors:** Smith, p. 22; *Memorial and Biographical History of the Counties of Fresno, Tulare and Kern*, p. 143.

47   **Save the souls:** Like the estimates of the Yokut population, the estimates of California's total native population vary widely, from 150,000 to 300,000. See Rolle, *California: A History*, p. 23.

47   **Rather innovative way:** Mitchell, *King of the Tulares*, p. 12.

47   **"Soul and body":** Rolle, p. 61. **First aqueduct:** Adams, *California Missions*, p. 26. **Inveterate gamblers:** Rolle, p. 19. **Clerical mortification:** Rolle, p. 61.

48   **Smallpox and syphilis:** Mitchell, *King of the Tulares*, p. 20; Preston, *Vanishing Landscapes*, pp. 56–58; Smith, p. 19.

48   **Father Juan Martin:** Mitchell, *The Way It Was*, p. 16.

48   **Tule rafts:** Latta, *Tailholt Tales*, p. 77. **Chinese fisherman:** Brown, p. 104.

49   **Yokuts found language:** Latta, *Tailholt Tales*, pp. 37–40.

49   **"Their disposition":** Smith, p. 32.

50   **"Taken troops along":** Smith, p. 47.

50   **Deepest glacial canyons:** *Sierra Magazine*, January–February 1987, p. 40. **Christened the river:** Mitchell, *The Way It Was*, p. 17.

51   **American explorers:** Smith, pp. 55–75. **Tamed mission mustangs:** Mitchell, *The Way It Was*, p. 23.

51   **Baptizing two dozen:** Smith, p. 41.

52   **Barely eighteen years old:** This and other details from Castro, p. 215.

52   **Journey around the tip:** Castro, p. 218.

53   **"Proof positive":** Castro, p. 2.

53   **Dutch John:** Castro, p. 5. **"The parson":** Castro, pp. 8–9.

54   **"Most miserable country":** Quoted in Preston, p. 70.

55   **"Climate of this valley":** Quoted in Preston, p. 68.

55   **"Honey-bloom":** Muir, *The Mountains of California*, p. 339.

56   **"Crop of barley":** Castro, p. 50.

56   **"Treaties stand?":** Castro, p. 46.

57   **"To thee, California":** Castro, p. 135.

57   **Murder of John Wood:** Recounted in Latta, *Tailholt Tales*; Mitchell, *King of the Tulares.*

58   **Target practice:** *Memorial and Biographical History of the Counties of Fresno, Tulare and Kern*, p. 180.

58   **Patch of skin:** Latta, *Tailholt Tales*, pp. 153-155.

59   **Known as Pikes:** Nordhoff, *California for Travellers and Settlers*, pp. 192-196.

59   **El Rey Tulares:** Mitchell, *King of the Tulares*, p. 94.

60   **On a moonbeam:** Mitchell, *King of the Tulares*, p. 99.

60   **"Savage cruelty":** Mitchell, *King of the Tulares*, p. 108.

60   **First known whites:** Smith, p. 339. **"No dog":** Mitchell, *King of the Tulares*, p. 116.

61   **900,000 acres:** "James D. Savage," by Annie R. Mitchell, *Los Tulares Quarterly* no. 82, Tulare County Historical Society, p. 2.

61   **"Good horse, bridle, spurs":** Mitchell, *King of the Tulares*, p. 219.

62   **"Our Indian shoot, bang!":** Latta, *Handbook of Yokuts Indians*, p. 664.

62   **"Profound manifestations":** Mitchell, *King of the Tulares*, p. 222.

63   **Poland china hogs:** CJ, October 5, 1923.

63   **Yoimut:** Her story is drawn from Latta, *Handbook of Yokuts Indians*, pp. 667-730.

64   **Cargo of valley honey:** Rider, *Rider's California*, p. 403.

CHAPTER 5: THE LITTLE KINGDOM OF KINGS

66   **Hopping-mad jackrabbits:** "The Abattoir of the Prairie," by William Deverell and David Igler, *Rethinking History*, 1999, pp. 321-323. Norris paints a vivid picture of a rabbit drive in *The Octopus*, pp. 334-336. A photo of a jackrabbit kill is found in Johnson, Haslam and Dawson, *The Great Central Valley*, p. 42.

66   **Pikes pitted:** Nordhoff, pp. 192-196.

67   **Visalia's southern loyalties:** See Mitchell, *Visalia: Her First Fifty Years.*

67   **"Tired, aching limbs":** Quoted in McFarland, *Water for a Thirsty Land*, p. 17.

67   **Botanist and schoolteacher:** Biographical details on Sanders come from an interview with McFarland, who is the historian of the Kings River Water Association.

68   **Church had come:** Maass and Anderson, *And the Desert Shall Rejoice*, p. 159.

68   **"Weeks passed away":** McFarland, p. 17.

69   **"Wildest, maddest":** Cited in McFarland, p. 31.

69   **The Nile and the Indus:** Worster, *Rivers of Empire*, p. 234. **Breaking with the rich tradition:** See Elwood Mead, "Irrigation Investigations in California," U.S. Department of Agriculture, 1901, pp. 259-260.

70   **Cattle had died:** *Memorial and Biographical History of the Counties of Fresno, Tulare and Kern*, p. 165.

70   **Independence Day:** McFarland, p. 35.

70   **Taken lightly:** Shafer became something of a legend after townsfolk watched him gun down the former police chief of San Diego in the summer of 1903. The lawman, J. E. Harris, had bought a twenty-acre fruit farm just north of Selma a few weeks earlier. On a nearby ranch, he and Shafer came to blows while arguing over possession of a pile of wood. The lawman, his breath thick with whiskey, threatened to kill Shafer. The pioneer irrigationist changed his clothes, rode into town and swore out a criminal complaint. He was chatting with friends in front of Price and Dusy's drugstore when Harris showed up toting a shotgun. In broad daylight, Harris took aim but hesitated for a moment, not sure that the man in the crisp suit was the same man he had just tussled with; the delay provided Shafer with just enough time to pull out his revolver and fire one fatal bullet into Harris's brain. Shafer then dropped his gun, threw up his hands and demanded to be arrested. A coroner's jury, finding sufficient cause for self-defense, wasted no time exonerating him.

71   **"Large and varied treasure":** Twain, *The Autobiography of Mark Twain*, p. 266.

71   **William Randolph Hearst:** Moffett was swept up in the fervor of the antiriparian movement of 1883 and wrote a series of impassioned newspaper pieces advocating the rights of farmers to appropriate river water. One of those essays caught the eye of Hearst, who immediately telegraphed the postmaster at Kingsburg inquiring, "Who is this Moffett fellow?" (Smith, p. 459). At least once, Twain gave his nephew a little lesson in writing: "You should write from the standpoint that the reader knows nothing at all, and as a general rule he doesn't know anything. This can be done without his suspecting—and without pedantry or immodesty, by simply saying 'as the reader will remember'—and then go on and tell him something which he ought to have known but which he never heard before" (From a July 2, 1885, letter found in the Mark Twain Papers, Bancroft Library, University of California at Berkeley).

72   **"Doubt your courage":** This and the other Shafer quotes are from McFarland, p. 36.

72   **Biggest land grabs:** Pisani, *Water, Land and Law in the West*, p. 87. See also McWilliams, *Factories in the Field*, pp. 11–27; Daniel, *Bitter Harvest*, pp. 18–39. For more on the impact of the gold rush on large-scale California agriculture, see Holliday, *Rush for Riches*, pp. 194–201.

73   **Twenty-four of the thirty grants:** Maass and Anderson, p. 157. **More wild speculation:** This entailed selling huge tracts of land to a group of San Francisco capitalists, men who had built their fortunes servicing the gold rush. President James Buchanan then dumped 11 million acres of Califor-

nia land onto the open market in 1857, a move to raise money for a nation mired in depression. At the same time, California offered more than 3 million acres of so-called swampland to homesteaders willing to drain the muck and plant crops. All of it came at incredibly cheap prices: 1 to $2.50 an acre, much of it rebated if the settler actually built a levee or two and reclaimed the mud (Pisani, *Water, Land and Law in the West*, p. 89).

73   **$1.25 government scrip:** Pisani, *Water, Land and Law in the West*, p. 90; Preston, pp. 101–103. **William Chapman and Isaac Freidlander:** The prospect of cheap land lured Chapman from Minnesota to San Francisco in the 1860s. Within a decade, using cash and scrip and other dubious means, he had acquired a million acres of government land. His partner, the German immigrant Freidlander, had migrated to California in 1849. Gold fever soon gave way to speculating on grain. By the age of twenty-nine, he had cornered the flour market and became known as the grain king, a moniker brought home by his six-foot-seven-inch, 300-pound frame (Maass and Anderson, p. 157). For more on Chapman, see Opie, *The Law of the Land*, pp. 136–138. **Amassed 170,000 acres:** Preston, p. 104; Maass and Anderson, p. 158. **516 men:** McWilliams, *Factories in the Field*, p. 20.

74   **Controlled two rivers:** This and all other biographical details on Miller are from Treadwell's hagiography, *The Cattle King;* Igler, *Industrial Cowboys;* interview with Igler.

75   **"Spit on":** Nordhoff, p. 192.

76   **"Wheat, wheat, wheat":** Nordhoff, p. 102.

76   **Largest grain farm:** Preston, p. 135. **Wooden shoes:** Recollections of early Corcoran pioneer George Smith, found in BCA. **Led the nation:** Pisani, *From the Family Farm to Agribusiness*, p. 9.

76   **"Gee dock":** Gist, *The Years Between*, p. 47.

76   **Tule River Reservation:** *Memorial and Biographical History of the Counties of Fresno, Tulare and Kern*, p. 177. **Rollicking hamlet:** See the *Traver Business Directory*, 1887. **World record:** Smith, p. 246.

77   **Mussel Slough:** For a particularly engaging account of this much-covered episode, see Brown's Kings County history.

78   **Spreckels:** Smith, p. 214.

78   **Nothing like it:** See *Report of the Board of Commissioners on the Irrigation of the San Joaquin, Tulare and Sacramento Valleys*, U.S. Army Corps of Engineers, 1873, pp. 10–11.

79   **Fresno scrapers:** McFarland, p. 29.

79   **"Monopolized the land":** *Sacramento Bee*, February 14, 1873.

80   **Shameless manipulation:** Worster, *Rivers of Empire*, p. 103. **Finest thoroughbreds:** Smith, p. 190.

80   **Miller & Lux sued:** For details of this important case and its aftermath, see Hundley, *The Great Thirst*, pp. 91–97; Worster, *Rivers of Empire*, pp. 104–107; Irvine, *A History of the New California*, chap. 12.

80   **State found a way:** Worster, *Rivers of Empire*, pp. 106–110. The constitu-
      tional amendment, adopted in 1928, is codified at Article X, Section 2.

81   **Santa Fe passenger car:** This rendering of early Corcoran is from HJW and
      numerous articles in the CJ, including a February 15, 1952, piece based on
      the recollections of the late J. W. Guiberson, who had been Whitley's agent
      in Kings County.

81   **Worms in their eyes:** Unnamed newspaper story dated February 1914, Tu-
      lare County (California) Historical Library. **"Tulare Lake is gone":** *San
      Francisco Chronicle,* undated editorial, Tulare County Historical Library.

81   **60 men and 300 horses:** *Visalia Times,* April 28, 1905.

82   **As pitchman:** Mulholland, *The Owensmouth Baby*, p. 12. **Domain of
      celebrities:** *Los Angeles*, September 2000, p. 94. **"One more million":** Mul-
      holland, p. 4. Whitley's life is also covered in Davis, *Rivers in the Desert*.

82   **Traver had vanished:** Mitchell, *The Way It Was*, p. 43.

82   **Briggs:** *Memorial and Biographical History of the Counties of Fresno, Tu-
      lare and Kern*, p. 169.

83   **Security Land and Loan:** Company records, including plans for Corcoran,
      are found in HJW.

83   **Otis had fought:** Biographical details on Otis are from Halberstam, *The
      Powers That Be*, pp. 95–112; Gottlieb and Wolt, *Thinking Big*, pp. 17–117.

84   **"Indefatigable and fruitful":** Board meeting minutes, HJW.

84   **"Daddy" Lewis:** The biographical sketch is taken from the "Saga of Sandy
      Crocket," an unpublished memoir written by the former Corcoran farmer.
      Crocket wrote it in 1994, when he was ninety-two years old. **Fence posts:**
      Recollections of George Smith, BCA.

85   **Went on for ten years:** Thompson and Dutra, *The Tule Breakers*, p. 246.

85   **The "Square Lake":** *Ripley's Believe It or Not,* July 20, 1940.

86   **Top agricultural commodities:** 1922 Kings County Chamber of Commerce
      promotional booklet, found in BCA.

Chapter 6: Spring

90   **"Moved the rain":** *The Central Valley Project,* the Northern California Writ-
      ers' Program of the Work Projects Administration, 1942.

91   **Zip code 93711:** FB, March 18, 1995, citing a study by the Washington-based
      Environmental Working Group. It found that from 1985 to 1995, $21 million
      in subsidy checks had been sent to the wealthiest part of Fresno.

91   **Doubling of the population:** "Putting the Brakes on Growth," LAT, Octo-
      ber 6, 1999.

92   **Prunes, when you're feelin' blue:** Arax and Pickford, *California Light*,
      p. 268.

92   **$14 billion a year:** See "Economic Forecast for California's Central Valley,"

Great Valley Center, January 2001; "Rescuing the San Joaquin," FB special report, 1999. **More than Texas, Iowa:** 1999 data from the California Farm Bureau Federation.

92   **Truck driver's Shangri-La:** See James J. Parsons's 1987 Carl Sauer Memorial Lecture at the University of California at Berkeley, "A Geographer Looks at the San Joaquin Valley."

102   **The diary of one:** This was Frank Latta's account.

102   **Stunned the conservancy:** Interview with Steve McCormick, president of the Nature Conservancy; Conservation Associates Records collection, Bancroft Library, University of California at Berkeley.

103   **Audubon:** "Death in a Black Desert," by Ted Williams, January-February 1994.

103   **Master-planned community:** This was Boswell's EastLake development, just outside Chula Vista, which is set to have about 8,900 housing units.

106   **"If you beat 'em up":** JWB.

106   **Taxonomy tables:** Smith and Cothren, *Cotton: Origin, History, Technology and Production*, pp. 36–42. **Gossypol:** This and other technical details on the cotton plant are from Smith and Cothren; Hake, Kerby and Hake, *Cotton Production Manual*; material from the cottonsjourney.com website; "Cotton: Foremost Fiber of the World," *National Geographic*, February 1941.

107   **Stuffing inside dolls:** Smith and Cothren, p. vii.

CHAPTER 7: WHITE GOLD, BLACK FACES

113   The opening of this chapter is set at a minstrel show, which was held in Corcoran on October 2, 1929, according to a CJ advertisement. Although the exact performance given that night can't be recreated, it is possible to accurately depict what a typical minstrel show from the era would have been like. That was done with the scholarly assistance of William Mahar, a professor of humanities and music at Penn State Harrisburg-Capital College and the author of *Behind the Burnt Cork Mask: Early Blackface Minstrelsy and Antebellum American Popular Culture*. His research found that the production that came to Corcoran, Richard and Pringle's Georgia Minstrels, was an all-black ensemble. The material used in the chapter was taken from *New Minstrel Jokes: Containing a Select Collection of the Latest Conundrums, Jokes, Monologues, End Men Gags, Cross Fires, Stories and Etc. Used by all the Leading Minstrels of the Day* (Philadelphia: Royal, 1913).

114   **The Dixie Jubilee Quartet:** CJ, January 25, 1929. **"A southern Cinderella":** CJ, November 1, 1929.

114   **Corcoran fire department:** The program for the show was reprinted in the CJ of May 5, 1922.

114   **"Coolidge prosperity":** The state of America just before the Depression is
from Watkins, *The Hungry Years*, pp. 8–24. **"Unparalleled plenty":** Cited in
Watkins, *The Hungry Years*, pp. 15–16. Allen's 1931 book, *Only Yesterday*, is
considered a classic.

115   **$70 tailored suits:** This and the details on what other merchants were
peddling are gleaned from advertisements in the CJ beginning in early
1929. **J.C. Penney:** CJ, October 25, 1929.

115   **Hotel Corcoran:** The guest list was regularly published in the CJ. William
Cromlie's bootleg liquor was described in the CJ of September 30, 1982.
Vices beyond: Charlie Quong's gambling parlor is from the CJ of December 6, 1929, and the reference to Mrs. Richmond's whorehouse is from an
interview with longtime Corcoran resident Lino Cardoso, whose father
owned various properties in that part of town. **The Harvester:** CJ, September 20, 1929. **Newly paved sidewalks:** The CJ of May 13, 1928, reported that
some 500,000 square feet in town had just been paved, including more
than 7,000 square feet of sidewalk. **A dozen lights:** CJ, August 16, 1929.
"Line averaged 130": Gentry, *The Way the Ball Bounces*, p. 16. **Surge in
homebuilding:** The CJ of March 1, 1929, noted that "homes are no longer
temporary," as "the old lake bed is better known than it was a few years
ago and the permanency of farming opportunities has been firmly established." During 1928, the paper reported, a hefty $200,000-plus was spent
on new buildings in Corcoran. The first half of that year was said to be the
greatest building period in the history of the town. **"It cannot pay":** CJ,
October 18, 1929.

115   **Incomes on the nation's farms:** Cohn, pp. 242–243; Watkins, *The Hungry
Years*, p. 345. Watkins writes that the average per-acre value of American
farms fell to $29.68 in 1932 from $69.31 in 1920—the biggest decline ever.
Hamilton's *From New Day to New Deal*, pp. 10–11, presents a more complex
picture, however. Though farmers were "facing rapid changes and a more
competitive marketplace that left them in a less secure and less stable
condition," Hamilton says, after 1923, "the farm economy began a spotty
and sluggish recovery that continued until 1929." One critical issue,
Hamilton says, was heightened regional competition. The more efficient
cotton farms of California and the West were "putting the screws to cotton
growers in Georgia and regions farther east" (correspondence with the
authors). **"Concealed the fact":** Quoted in Culver and Hyde, *American
Dreamer*, p. 72. **European demand:** Bush, *An American Harvest*, p. 146. He
notes that 1926 was a "last glorious spasm" for U.S. cotton producers and
merchants, who exported "more than 11 million bales, an unprecedented,
and since unequalled record." But in 1927, "the export business in the
Eastern Belt began a steady, gradual decline, which lasted until October
1929, when the stock market crash and the ensuing Depression knocked
the last stilts out from under the wobbly world economy."

116   **Mood was the opposite:** This was true throughout rural California. Daniel notes that "consistent with its industrial character, large-scale commercial farming in the state shared in the relative prosperity that the nation's urban economy experienced in the 1920s rather than in the decline and demoralization that plagued the country's agricultural sector throughout the decade" (p. 68). Smith, p. 249, notes that the grain fields of Kings County continued to provide harvests so prodigious, they conjured images straight out of *The Octopus.* "The bonanza wheat farms which Norris described with a master's hand can still be seen in all their pristine glory." He goes on to tell of a 12,000-acre barley and wheat ranch near Corcoran, where the work crew's camp was so big during the 1929 season, it resembled "a small town in itself." **Weeds and willows:** From local writer Mae Weis's extensive history of Tulare Lake, which appeared in the CJ of June 3, 1938. **"Vitally interested":** CJ, August 16, 1928. **"Hatfield the rainmaker":** His presence in Tulare Lake is recounted in a local history, which ran in the CJ of January 24, 1941. Biographical details and his quote are from McWilliams, *Southern California Country,* pp. 196–199. McWilliams notes that "for eight consecutive years, the farmers of the San Joaquin Valley contracted with Hatfield to 'make rain.'"

116   **187,868 acres:** CJ, February 27, 1931. **"Foolhardy enough":** From an October 11, 1930, article by L. H. Storgaard in the *Pacific Rural Press.*

116   **Oil fever:** See Latta, *Black Gold in the Joaquin,* pp. 339–344. **"A second Tulsa":** *Delano Record* of July 12, 1929, headlined "Kings County Center of Oil and Gas Boom." The Kettleman North Dome field had by 1998 yielded 458 million barrels of oil and nearly 3 trillion cubic feet of natural gas. But the field never was the producer that some thought it would be; the story in the *Record,* for instance, suggested that the field would be the "greatest . . . in the world if present predictions of its expanse are accurate." Richard Curtin, an official with the state Department of Conservation, says that the field could have produced twice as much oil as it has over the years had it been managed more wisely from the start. He says that in their reckless haste to get at the oil, the original operators burned off massive amounts of gas coming out of the field, which greatly reduced the pressure in the reservoir. Eventually, that made it too difficult and expensive to tap the field to its full potential.

117   **Sitting pretty:** Such was the analysis of one cotton-industry expert, A. H. Fowler of Cutter Bros., which appeared in the CJ of May 4, 1928.

117   **Little pay:** The standard wage of $1.25 for every 100 pounds of cotton picked was adopted by the San Joaquin Valley Agricultural Labor Bureau (CJ of September 6, 1929). This was actually better than the U.S. average of $1.06. **Jasper Palacio:** The fates of Palacio and the unidentified African American murder victim were reported in the CJ of December 6, 1929. By 1926, Mexicans constituted 80 percent of the cotton workforce in the val-

ley, according to Weber, *Dark Sweat, White Gold*, p. 35. There was clearly, though, a black contingent in the valley at the time, as the CJ regularly reported on the town's "colored" residents, and state enforcement records from 1929 and 1930 show numerous cases brought against small black growers or labor contractors for failure to pay for cotton work.

117   **Two more gins:** CJ, August 2, 1929. By the mid-1920s, California's fields were yielding more cotton on average than any other state's, according to the *Cotton Year Book of the New York Cotton Exchange* (1935). In 1929, for instance, California farms produced 400 pounds of cotton per harvested acre compared with 190 for Georgia, 112 for Texas and 164 for the United States overall. Within California, state agriculture data shows that the cotton farms of Kings County produced a higher yield per harvested acre than those of any other county in the valley, save for Kern. The average from 1928 through 1932 for Kings County was 506 pounds, compared with 467 for the entire San Joaquin Valley and 431 for the state; Kern County's mark was 563 pounds (From CMcW). **A premium:** The CJ of October 18, 1929, reported that J. W. Guiberson sold 750 bales to the California Cotton Mill Company in Oakland at "65 points on New York December quotation." According to the New York Cotton Exchange yearbook, the quotation for December delivery for the 1929–30 season ranged from a low of 16.55 cents per pound to a high of 19.85 cents per pound. One key reason that Corcoran growers were able to fetch a premium was that in September 1929 the state set up a cotton-classing office in town, giving farmers "definite knowledge concerning the quality of their commodity," according to the March 1930 *Western Cotton Journal and Farm Review.* **"Center of the cotton industry":** Though there was certainly a good deal of truth in Corcoran's claim, this title was probably best bestowed on Kern County, which by 1930 was home to the U.S. Department of Agriculture's Shafter Research Station, the Farm Bureau Planting Cotton Seed Distributors and the California Cotton Cooperative Association.

117   **Building up his acreage:** Salyer started his empire building in 1921, according to the LAT of March 27, 1983. **"He'd haul cottonseed":** Interview with Keith.

117   **Unhook a trailer:** JGB2 recalled that he and his dad came across Cockeye right after the accident, and they rushed him to the hospital: "There was a wreck in front of us. . . . He was just a mangled mess." This account was confirmed by a CJ dispatch of November 6, 1931. **Grower swore:** Interview with Sandy Crocket. **"Into a tit!":** LSal.

118   **Lost a fortune:** Morgan's *Merchants of Grain*, p. 75, notes that the commodity markets through the 1920s were marked by "instability. . . . The entire decade had been a speculator's hayride." The 1921–22 period, in particular, was one of crisis for American farms. The story of von Glahn's

own lost fortune is from an interview with his friend Sam Crookshanks; DV with von Glahn's wife, Frances. Other details on von Glahn's life are from those sources, plus JGB2 and an interview with Corcoran trucker Don Keith. **Toy airplane:** CJ, January 18 and July 25, 1924. **Newsweek:** July 14, 1947.

118   **Jimmy Grier:** Famed for his steady gig at the Biltmore Hotel in Los Angeles, Grier played at the dedication of von Glahn's warehouse on May 4, 1940, according to articles in the CN.

119   **All of $26:** From Crocket's obituary, as carried by the Associated Press, July 3, 2000. **"Make a lot:"** From Crocket's unpublished memoir. **First sections:** CJ, September 6, 1929. **80,000 acres:** FB, July 1, 1948. Crocket's partner was Albert (Casey) Gambogy.

119   **1,000 or so:** CJ, April 20, 1923. **Better than a bale:** CJ, October 5, 1923.

119   **Two train-car loads:** CJ, October 26, 1923. **Having bought up:** CJ, December 7, 1923. **"The richest land":** DV with Kate Boswell.

120   **Imperial and Yuma areas:** For the decline in cotton culture there, see Stemen's work; *Valley Grower*, November–December 1997, p. 33. In 1926 Bill Camp noted to a Bank of Italy executive that in the Imperial Valley, "most of the good land on the American side has been more recently devoted to the growing of melons, lettuce and other crops, cotton being planted only on the inferior types of soil" (WBC). **Cocopah Indian:** Interview with Gadsden resident John Haynes.

120   **The local farm bureau:** CJ, January 10, 1919. Much of the early planting was also urged by California Products Company of Fresno, a major purchaser of cotton in the San Joaquin Valley. **"Of the South":** CJ, December 5, 1919. **A growers association:** The first cotton ever ginned in Corcoran was a batch of Durango turned out by the Corcoran Cotton Growers Association's newly built facility on December 22, 1920. See CJ, December 24.

120   **"You'll make a killing":** DV with Kate Boswell. **800 acres:** CJ, November 30, 1923. **$36,000 profit:** JGB2.

120   **Remained a fixture:** Various issues of the GHJ for 1922 and 1923.

121   **"Blue blood":** This and all other quotes and recollections from Kate Boswell are drawn from DV.

124   **"Mom Kate's no-nos":** LSal.

125   **San Francisco earthquake:** Recounted in the CJ of September 30, 1976, in an article about Nis Hansen.

127   **Entourage of minstrels:** CJ, September 26, 1924. **Firmly enmeshed:** A March 1924 Bank of Italy memo noted that there were six "principal concerns engaged in the cotton brokerage business" in the valley, including Boswell and his two old rivals, McFadden and Anderson Clayton. (A copy was provided by historian Devra Weber, University of California at Riverside.) **Leased the Corcoran gin:** This and other details are from FSH. **Two**

twelve-hour shifts: CCJ, December 1925. "Largest industrial unit": Quoted in Weber, p. 24.

127   Partnership dissolved: JGB2. The J. G. Boswell Company was incorporated in October 1925. Three shares of stock, each in the amount of $100, were issued—one to JGB, one to his wife Alaine and one to R. L. Curtis, his first employee (Records of the Secretary of State of California).

127   The Colonel financed: Figures are from a June 12, 1931, memo to JGB, found in FSP. Displayed a bale: CCJ, September 1926. "Take its place": CCJ, December 1925.

128   Tough to spin: Weber, p. 29. "Part of my job": WCOH, p. 74. One-Variety Cotton Law: For background, see Weber, p. 30; Briggs and Cauthen, pp. 90–100; Turner, pp. 57–59; and numerous letters and other documents in WBC. Interestingly, Camp first endorsed growing Pima cotton, not Acala, in the San Joaquin Valley. By 1924, though, he was wholeheartedly behind Acala, and saw broad merit in promoting it alone. "The need of protecting the community against the introduction of new varieties," he wrote, "brings the growers in closer relationship with each other, promotes their interest in other phases of production improvement, and tends to establish a more sympathetic contact with the local agricultural agencies" (Cited in Briggs and Cauthen, p. 100). Interestingly, growers in Kings County—Boswell foremost among them—were leaders in trying to get around the one-variety law over the years. In WCOH, Camp recalls an attempt by Boswell in the late 1930s to wangle an exemption to the law in the state legislature. "A special meeting of the Senate was called," Camp recounted, "and it went into the afternoon and into the evening, and it was after midnight before we got through the arguments, the most impassioned pleas that some of us ever made." At last, after 2:00 A.M., the Senate voted, and the one-variety law was kept intact. In 1935 Boswell looked to have the legislature amend the law to permit the planting of Pima cotton (CN, March 15, 1935). In 1941 another such effort was made (CJ, March 14, 1941). And in 1943, a group called the Corcoran Farmers Committee for Egyptian Cotton again sought such an amendment (FSP; FB, February 5, 1943). Clearly those in Kings County weren't the only ones interested in such a change. In December 1927, for instance, the Kern County horticultural commissioner wrote with great alarm that many cotton buyers "were desirous of introducing varieties other than Acala" (WBC). Later on, beginning in the mid-1950s, growers in Tulare, Madera and Merced Counties complained of low cotton yields and urged the seed breeders at the U.S. Department of Agriculture's Shafter station to develop a new variety that was better suited to their land. These requests were ignored. And in 1975, responding to complaints that the one-variety system amounted to a "seed monopoly," the U.S. Justice Department initiated an antitrust inves-

tigation. In 1978, concerned about possible action by Justice, the state legislature finally amended the one-variety law to permit private companies
to develop new Acala types, though they had to meet certain "quality standards" (Constantine, Alston and Smith, "Economic Impacts of the California One-Variety Cotton Law"). The law was again revised in 1991, this time
allowing Pima to be grown under the oversight of a governing body called
the San Joaquin Valley Cotton Board. Then in 1999, a new law was passed
that further relaxes Camp's original vision. Now cottons not blessed by the
cotton board can be planted, though "strict regulations" were imple-
mented to protect the approved Acalas and Pimas "from contamination
by the nonapproved varieties" (Information from the San Joaquin Valley
Cotton Board, California Department of Food & Agriculture). **Ready to
pounce:** In January 1927, for example, the horticultural commissioner of
Kern County warned of possible attempts "to bring illegitimate seed into
California this season," and all inspectors were ordered to "keep a close
watch on all material arriving from out of the state and act accordingly."
Later in the year, he added: "Our leaders here look with horror upon the
smuggling of this seed or bringing it in through clandestine channels from
the standpoint of quarantine and protection to the industry" (WBC).

128   **Wild fluctuations:** All pricing figures in the book, unless otherwise noted,
are from Turner, pp. 51, 94. **Took on debt:** Over the years, Boswell carried
debt on its books, but that merely reflected its own financing operation; it
would borrow money from an institution such as the Federal Intermedi-
ate Credit Bank to, in turn, lend to farmers. What JGB shunned was buying
land or equipment with a long-term loan; any such purchases were routinely financed with cash (JGB2; interview with Wally Erickson, who was
Boswell's outside auditor at Arthur Young & Company starting in the mid-
1940s). Nor did JGB expand just for the sake of expanding. "Believe me, J.
G. Boswell never fooled around," Sherrill said. "'Are we in competition
with ourselves?' 'Will the thing make money?' If the answer was yes on the
former or no on the latter, we changed the slide awfully fast" (FSH). **Speculated hog wild:** JGB had the savvy to use the futures market as a hedge—
an insurance policy—against swings in the market, yet he never got
caught speculating, as did some of his less disciplined contemporaries.
For a comprehensive lesson in cotton "hedging" and "straddling," see
Garside, *Cotton Goes to Market,* pp. 206–315. The turmoil that could be
wrought in the cotton markets by speculators became a big issue before
Congress in the 1920s and '30s. Senator Ellison DuRant Smith held a series
of hearings in which he explored whether there had been "any manipulation, direct or indirect, of the cotton markets" (U.S. Senate Hearing to Investigate the Causes of the Decline of Cotton Prices, Committee on
Agriculture and Forestry, February 17, 1936, p. 2). Smith's principal target

was Anderson Clayton, whose chief executive, Will Clayton, successfully parried most of the senator's attacks. "I was determined to give them more than they bargained for" Clayton recalled (William Lockhart Clayton oral history, Columbia University, New York, p. 91). In fact, Clayton used the first round of hearings to help promote a pet cause. He was fighting for a new delivery system whereby cotton could be routed through southern ports, not strictly through New York—an anachronistic and costly mandate that benefited a few New York traders. He ultimately prevailed in establishing "southern delivery." (See Fossedal, *Our Finest Hour,* pp. 47–55.)

A steady hand: In the January 1927 CCJ, JGB wrote: "Cotton growers should consider their probable price and return, just as an insurance company or any other line of business, from the standpoint of averages. In other words, what average price should a cotton grower reasonably expect for his product over a 10-year period?" If this logic is followed, he said, "the conservative cotton farmer . . . will find it is not as gloomy as it appears" for there are more than enough chances to reap "a handsome return." If only the grower "would insist upon playing it safe and not holding his cotton for the highest possible price," JGB added, he could consistently post strong profits.

128   **A pure intellectual:** FS background from FSP; interviews with son Steve Sherrill and daughter Sue Sherrill. **Tape recording:** Steve Sherrill provided tapes of his father, including a speech he made to a gathering of young Republicans in 1961. **"Course of study":** May 19, 1960, letter from FS to General Leslie Groves, Groves Papers, NA. **"Doctor, lawyer, merchant":** January 21, 1964, letter from FS to Colonel H. B. Wharfield, FSP.

129   **"Brother of yours":** November 1, 1926, letter from FS to Walter Boswell, FSP.

129   **R. L. Curtis:** Background from his niece, Lorene Brooks. **"Used to growing":** This and all other quotes by Louis Robinson are from DV. **Hammond brothers:** From Tommy Hammond's retirement album, provided by his widow, Ruth.

129   **Spread his reach:** FSH; issues of the CCJ; records found in DM. **"Cotton that grew taller":** Litchfield, *Industrial Voyage,* p. 161. Goodyear's Arizona farming operation was called Southwest Cotton Company. **Goodyear invited:** Reflected in numerous Southwest Cotton Company records, Architecture & Environmental Design Library, Arizona State University; Smith, "Litchfield Park and Vicinity," p. 32. **As many as ten gins:** December 17, 1929, letter from JGB to Arthur King of Cecil B. DeMille Productions, DM.

130   **Anderson Clayton foremost:** See Weber, pp. 31–32. **Verged on usury:** See, for instance, the comments from Elmer Lobre, a justice of the peace, in the *Delano Record,* October 27, 1933. **"Biggest bouquet":** DV with Anderson. **"Man, land and water":** JGB2.

130   **Suit and garters:** JGB2. **"Sharp pencil":** Interview with Robinson's daughter Anne Woolley.

130   **Fifteen cotton dealers:** *Los Angeles City Directory*, 1930. **Cotton Exchange Building:** The Pacific Cotton Exchange was established in L.A. in 1920 to handle futures contracts. "Without exaggeration at all," the secretary of the group assured members in July of that year, "your stock in this Exchange will ultimately be one of your most valued possessions. Its possibilities are unlimited" (Letter in WBC). Because of cotton's growing prominence, the Los Angeles Chamber of Commerce had high hopes that a textile industry would develop in the city, and it commissioned reports studying the idea in 1922, 1924 and 1926. (Copies available at the Regional History Center, University of Southern California, Department of Special Collections.) However, the great spinning center that was envisioned never materialized. **More and more land:** Property records, Los Angeles County archives. **Flair for the dramatic:** This and other details on Alaine Boswell are from interviews with her niece Barbara Hamilton and with Dorothy Northcote, whose husband was Alaine's first cousin.

131   **Hips protruding:** From "Dinner Dance," p. 110 of Boswell's *Stories & Poems*.

131   **$25,000 a year:** From a 1932 audit of the Boswell Company by Floid S. Day's Income Tax Accounting Service, found in DM. **Hobnobbed:** JGB2; WW. **Cecil B. DeMille:** Dealings between JGB and DeMille over the Lone Butte Farms Company of Arizona are detailed extensively in DM. **"Crop failure":** From a memo found in FSP recounting a meeting between DeMille and JGB. **"Honesty and integrity":** From the same memo.

131   **An unscheduled stop:** JGB2. **"Shiny faces":** Interview with Durand.

132   **"Shine up everything":** Interview with Medina.

132   **"Used to get frustrated":** Interview with Ober.

132   **"Didn't want to wait":** This and all other quotes by Josephine (Jody) Boswell Larsen are from DV.

132   **Brief stop:** DV with Kate Boswell. **"Run her children":** This and all other quotes by James O. Boswell are from DV. Before long, Walter's wife packed her bags and returned to Washington, taking her children with her. Though they never divorced, she and Walter remained estranged for the rest of their lives (DV interview with James O.; JGB2; interview with Walter's granddaughter Rosalind Seysses).

133   **"Seen Alpaugh yet":** JGB2. In truth, Bill did travel some—to Mexico and to Texas, for instance, on cattle business. But he never went overseas.

133   **Feedlot:** See CJ, July 3, 1931. **Kate was bitter:** DV with Kate Boswell; JGB2.

133   **"Miraculous escape":** The article from the Porterville paper was reprinted in the GHJ of July 26, 1929. Details on the accident are from this account, plus DV with Josephine (Jody) Boswell Larsen; JGB2; interview with Skyler Dunlop, the daughter of the doctor who treated Bill, John Dunlop.

133   **Escape the summer heat:** For a fictionalized account of Doyle Springs, see Allen Drury's *That Summer*.

134   **Filled the bathtub:** Interview with Don Keith. Bill's incredibly heavy beer drinking was recounted by Keith, Verdo Gregory, DV with Louis Robinson and many others. Kate herself told DV: "I would laugh and say (to Bill), 'If they ever have a contest for the man who could drink the most beer, I'm going to enter you.'"

134   **"Agriculture is sick":** *Western Cotton Journal and Farm Review,* October 1930. Scholars are careful to point out that the stock market crash of 1929 didn't directly bring on the depression that followed. However, the crash did herald an American recession, which in turn contributed to economic weakness that quickly snowballed around the globe.

134   **"Bottom would drop":** FSH. He said the contracts were liquidated in December 1930.

134   **Nearly $114,000:** From the Floid S. Day audit, DM. This was for the year ending June 30, 1931. Boswell suffered another blow in July 1931 when a fire destroyed the oil mill, refinery and two seed houses. By late September, a new plant was in full swing. **Gossip swirled:** "It is rumored that Anderson, Clayton & Company have bought Boswell out or have control of his interests in some way," one manager of Goodyear Tire's cotton farm wrote to a colleague. "I have heard that Boswell owes $900,000 to the Citizens Bank—they have been compelled to sell cotton for 8–1/2 cents that they had held at 23 cents. . . . If this is true, it looks as though it were but a matter of time until Anderson Clayton will control all the cottonseed business" in the West (April 28, 1931, letter to George Sherry from K. B. McMicken, Southwest Cotton Company records, Architecture & Environmental Design Library, Arizona State University). **Cut his own salary:** Noted in the Floid S. Day audit, DM. **Six men:** A list is provided in the audit. John McWilliams Jr., whose family investment firm controlled mineral rights in Kern County and 4,000 acres of rice land in Arkansas, was a Pasadena neighbor and JGB's closest friend (detail from p. 17 of Fitch's *Appetite for Life*, the biography of McWilliams's daughter, chef Julia Child). Another of the Boswell investors, Frank Bacon, was the chairman of electrical equipment maker Cutler-Hammer Inc. Though Cutler-Hammer (now part of Eaton Corp.) was based in Milwaukee, Bacon wintered in Pasadena (interview with his granddaughter, Elizabeth Cutler). Stanton Forsman ran a paper company and had married into the Pillsbury flour family (interview with his son, Fred Forsman). Two of the others who bought shares from JGB had southern ties of their own. J. D. Heiskell, who operated a feed and grain company in Tulare, had been born in California, but his family took pride in their Virginia ancestry; the J. D. was for Jefferson Davis (interviews with his daughters, Eleanor Heiskell and Pat Hill-

man). And Alexander Macbeth, the president of Southern California Gas Company, was a South Carolina native whose grandfather was secretary of the Confederate treasury (Obituary in LAT, March 21, 1945). The final investor was Arthur King, a distant relative of and factotum for Cecil B. DeMille (interview with Richard DeMille, Cecil's adopted son; DM).

134   **End his partnership:** In a memo found in FSP, FS reconstructs this encounter between JGB and DeMille in great detail.

135   **Closed down:** Interviews with Ruth Hammond and Gordon Felder, whose father was among those laid off; JGB2.

135   **Clear that the blame:** In a June 9, 1934, letter to Will Clayton, JGB wrote: "I agree with you that the high protective tariff and the War Debts are the principle causes of our present day troubles" (BCA). **Clogged overseas markets:** The company tried to change things by taking its own plan straight to the Oval Office. Specifically, Fred Sherrill's proposal to aid the cotton farmers called for the banking system to tighten up on credit and thereby force growers to hold back production. He sent the idea off to C. B. Hodges, a senior aide to President Hoover whom Sherrill knew from his days in the military (as noted in correspondence between FS and Hodges, FSP). But a few days later, Hodges scrawled in his diary, "The president told me he didn't think so much of the suggestions." (See Hodges's diary entry of September 15, 1931, Herbert Hoover Presidential Library, West Branch, Iowa.) Hamilton, p. 127, notes that around the same time, the American Cotton Shippers Association (of which Boswell was a prominent western member) was similarly urging "big people at the top" to enforce credit control as a means of cutting production. But, writes Hamilton, one major weakness of the plan was that "the vast proportion of rural banks, the principal suppliers of production credit, were not members of the Federal Reserve System and were only loosely controlled by state regulatory agencies. There was no machinery for disciplining noncompliers."

135   **Hoover's farm program:** Hamilton discusses the program in copious detail; also see "History of Agricultural Price-Support and Adjustment Programs, 1933–84," U.S. Department of Agriculture, Economic Research Service, Bulletin no. 485. **"Immediate and drastic action":** Hamilton, p. 125. **"Every third member":** Hamilton, p. 125.

135   **Wasn't until:** Hamilton, p. 128. **About a quarter:** Culver and Hyde, p. 123. **"Honest, intelligent":** June 9, 1934, letter from JGB to Will Clayton, BCA.

135   **Pushing sharecroppers:** McWilliams, *Ill Fares the Land*, p. 218; Gregory, *American Exodus*, p. 12; Stein, *California and the Dust Bowl Migration*, pp. 7–8, 76. **Lot of many farmers:** "The farmer," *Fortune* magazine wrote in 1934, "has had . . . a great deal done for him" (cited in Watkins, *The Hungry Years*, p. 361). In the AAA's very first year, farmers' actual cash income shot

up 30 percent, according to Culver and Hyde, p. 125. And after AAA checks began to appear in rural mailboxes, a spasm of violence that had erupted in the Midwest to protest farm foreclosures started to subside. The most famous incident—recounted in Watkins, Hamilton, and Culver and Hyde—occurred in April 1933 in Le Mars, Iowa, when a mob of farmers burst into the courtroom during a farm mortgage hearing. They drove the judge out of town, where they roughed him up and threatened to lynch him. **Bigger operations:** Weber, pp. 115–119 ; Stein, p. 145. **Tens of thousands of dollars:** In 1938 alone Boswell received $53,822 in AAA payments (LF Report, pt. 3, p. 279). A complete accounting of the amount of AAA money that the company took in over the years isn't available, however. According to NA officials, records of AAA payments to individual growers have been destroyed. **Loans that Boswell made:** Weber, p. 114.

136   **Many smaller farmers:** Weber, p. 117. **Boswell bought thousands:** 1976 memo from Boswell executive Len Evers to DV laying out the history of the company's land purchases in the lake bottom (BCA). **Thirteen different ranches:** LF Report, pt. 3, p. 279. **Two-thirds of the cotton:** LF Report, pt. 3, pp. 278–279; also cited in Weber, p. 122. By 1937, Boswell was the second biggest ginner in California, trailing only Anderson Clayton. **Key water rights:** Among its most important acquisitions through this period was the 6,500-acre Chamberlain ranch, just north of the lake. The land wasn't all that good, but on it sat a bunch of water wells, and, best of all, it carried with it a stake in the old People's Ditch Company—holder of one of the most valuable water rights on the Kings River (HMJ, October 1, 1933; SB). **Wheedling Portuguese farmers:** DV with Louis Robinson.

136   **Lost their fondness:** FS, for one, had voted for FDR in 1932 but went on to attack him as the years passed, FSP. **"This New Deal":** Walter Boswell to General Pershing, Pershing Papers, LC. **"Every man is entitled":** WW. **"Socialism, communism":** April 12, 1944, letter from JGB to FS, FSP.

136   **"Our Friends":** Poem courtesy of Fred Forsman.

Chapter 8: La Mordida

137   **First time he glimpsed:** Castro's story is based on numerous interviews with him.

137   **Rapid industrial expansion:** Weber, pp. 49–53. She notes that many of California's cotton workers "had come originally from areas of Mexico that had been disrupted by capitalist expansion since the 1880s: the central plateau area . . . which encompassed Michoacan, Jalisco, Aguascalientes, Zacatecas and Guanajuato; and the western Mexican states, near the U.S. border, of Baja California, Sinaloa, Sonora and, to the north, Chihuahua, Coahuila, Durango and Nuevo Leon." **150,000 Mexican nationals:** Stein, p.

36. **Procession of exiles:** The history of these groups is discussed at length in McWilliams, *Factories in the Field*, pp. 103–133; Daniel, pp. 23–70. **"Class of labor":** Cited in LF Report, pt. 3, p. 499; also quoted in Weber, p. 39.

139   **Tents were later upgraded:** The CJ of October 18, 1929, noted that the lake area was having an easier time finding cotton pickers than were other parts of the valley, which faced a labor shortage. The paper attributed this, in part, to the "up-to-date cabins and sanitary living conditions prevailing in the Corcoran district." It added, with its usual rah-rah touch, that Boswell and others built cabins and provided "running water and modern conveniences that the cotton pickers appreciate." In 1927, a surveyor had counted 180 families at the Boswell camp (Weber, p. 43).

140   **Patriarch of the Medina family:** Interviews with Frank Medina.

140   **Couldn't believe his eyes:** Interview with Dominguez.

141   **On the dance floor:** Interview with Felicitas Tamez Marin.

142   **Cat-and-mouse game:** Interview with Ray Magana.

144   **"It's life":** Interview with Rivas.

147   **Fanned out:** This description of the union's office is found in Winter, *And Not to Yield*, pp. 193–194. The Cannery and Agricultural Workers was an offshoot of the Communist Party's Trade Union Unity League.

147   **A toehold:** There had been efforts to organize California farm workers in the early 1900s by the AFL and the IWW, in the late 1920s by the Confederacion de Uniones de Obreros Mexicanos, and in the early 1930s by the Communist Party. But other crops, as opposed to cotton, had been the focus of these mostly ineffective attempts. **"Heavy cotton ginning season":** CN, September 22, 1933; HMJ, September 26, 1933.

147   **As late as October 1:** HMJ, October 1, 1933, under the headline: "No Strike Expected in Corcoran Cotton Area; Pickers Arrive."

148   **200 pickers:** HMJ, October 5, 1933.

148   **Struck by a car:** Interview with Rudy Castro.

148   **Boswells managed:** The HMJ of October 7, 1933, reported "some picking on the J. G. Boswell ranches." **She laid out:** CN, October 13, 1933. **Some 12,000:** This is the estimate of the number of strikers by October 9, which is noted in Daniel, p. 194; Weber, p. 100. Weber, p. 79, puts the final number of strike participants at 18,000, citing a U.S. Labor Department report. Paul Taylor and Clark Kerr, in their "Documentary History of the Strike of the Cotton Pickers in California," are more conservative, saying that the walkout "directly involved over 10,000." Their report is appended to the La Follette Committee hearings transcript and also included in Taylor's *On the Ground in the Thirties*. All endnotes citing the "Documentary History" refer to page numbers in that book.

148   **Bill Boswell howled:** Interview with Rudy Castro. **Stood steadfast:** Weber, p. 93.

148   **45 percent:** LF Hearings, pt. 51, p. 18792; also cited in Johns, p. 40. **Agricultural Labor Bureau:** The activities of the ALB, a sister group to the Associated Farmers, are covered widely in LF Hearings, Part 51; see also Weber, pp. 37–42. Besides setting wages, the ALB was responsible for recruiting workers, often Mexicans from Los Angeles, to come to the valley to pick crops.

149   **Stormed their labor camps:** Taylor and Kerr, pp. 41–43. **"Did not know":** Tape-recorded interview with Roberto Castro by Devra Weber. (A copy was provided by Roberto's daughter Anna.) All quotes by Roberto are from this source.

149   **"Going to rid":** Taylor and Kerr, p. 45.

149   **Four-acre lot:** Taylor and Kerr, p. 61. **"Little Mexico City":** Daniel, p. 187. Union leaders arranged to use the land by talking with J. E. Morgan, who leased the lot from the Santa Fe Railroad. That led the growers to call it "Morgan's strike." **3,000 workers:** This is the number reported in Daniel, p. 187. Others put the figure as high as 6,000. But Taylor and Kerr, p. 62, caution that even the 3,000 may be overstated given that the "growers wished to enhance the menace, law officers the importance of their problem, and strikers the success of the strike." **Barbed wire fence:** This and the other details on Camp Corcoran are primarily drawn from Taylor and Kerr, but Weber, pp. 90–94, also paints a particularly vivid picture of the place.

149   **Wages of the pickers:** Daniel, p. 179. **Also superior:** The average wage for cotton pickers across the U.S. was fifty-three cents (From the "Crops and Markets" table in CMcW).

150   **"To mow down":** Interview with Rudy Castro.

150   **Roberto's job:** From Devra Weber's interview.

150   **"We stumbled along":** Winter, p. 196. The union-related activities of Winter and Steffens are discussed thoroughly in Loftis, *Witnesses to the Struggle.*

151   **"Complete discipline":** Interview with Chambers by George Ewart, who tape-recorded the recollections of a number of the players in the cotton strike. These oral histories are found at the Bancroft Library, University of California at Berkeley.

151   **Previous strikes:** These are discussed in detail in Daniel, pp. 105–166; Jamieson, *Labor Unionism in American Agriculture*, pp. 80–100. Perhaps the biggest failure was the El Monte berry picker's strike in which noncommunist Mexicans took control of the action and negotiated their own deal with growers, completely cutting out the Cannery and Agricultural Workers. Among the most important union victories was at the Tagus ranch, where Pat Chambers pushed fruit pickers' wages to twenty-five cents from fifteen cents an hour. **Part of the process:** From "The Storm Must Be Ridden," an unpublished autobiographical typescript by Sam Darcy in SD; Daniel, pp. 185–186.

151   **Not to flaunt:** Daniel, p. 216, quotes an AFL official who said admiringly:

"The Communists talked strike and not revolution." **"Prepared to give my life":** This quote and the biographical detail on Decker are from Ewart's interview with her. **"Exact two pounds":** Taylor and Kerr, p. 105.

151   **Deputies seized:** Darcy autobiography, SD.

151   **"Frowsy woman":** Winter, p. 197.

152   **To buy off:** Taylor and Kerr, p. 57. **More than 100:** Taylor and Kerr, p. 91. The HMJ reported that 113 strikers were arrested during the course of the conflict. By comparison, only eight growers were ever arrested. **Brought in strikebreakers:** Taylor and Kerr, p. 50; Weber, p. 102. **Turn white worker:** Darcy autobiography, SD; Taylor and Kerr, p. 58. **Threatened the Mexicans:** Weber, pp. 107–108.

152   **"Hell of a time":** Interview with Buckner found in PST. **Tossed a noose:** JGB2.

152   **Boswell fired Castro:** Debra Weber interview with Roberto Castro; authors' interview with Rudy Castro.

152   **"Have the guts":** Interview with Guiberson found in PST.

153   **"This is no time":** Taylor and Kerr, p. 84.

153   **W. D. Hamett:** Hamett's account of what happened at Pixley is reported in Taylor and Kerr, p. 86. Biographical details on Hamett are from an interview with his grandson Roy Hamett. Hamett's role in the strike is also discussed in Loftis and Daniel. Daniel, p. 185, notes that Hamett earned his tough-guy reputation during a brawl with a group of growers at the town of Woodville in which he and his sons more than stood their ground.

153   **"Let 'em have it":** Starr, *Endangered Dreams*, p. 78. The strikers were also violent at times, attempting to burn cotton at the J. B. Long gin near Corcoran and firing shots into the home of one of the farmers indicted in the Pixley affair, among other things (Taylor and Kerr, p. 80). But the bulk of the violence and intimidation was clearly perpetrated by the growers.

153   **Town of Arvin:** Taylor and Kerr, pp. 88–89. **Twenty workers:** HMJ, October 11, 1933.

153   **The Colt .38 Special:** This whole scene is drawn from FSal. **Slain Dolores Hernandez:** It's unclear which person Salyer thought he had shot and killed—Hernandez or D'Avila—but more farmers had apparently fired at the former, making it a reasonable assumption that Salyer was in this group.

155   **Initial foot-dragging:** Taylor and Kerr, p. 84. Daniel, p. 201, notes that "no attempt was ever made by authorities to determine which of the growers had fired the shot that killed Pedro Subia," the one picket who was cut down in Arvin. Absurdly, nine strikers were arrested for Subia's murder— perhaps, writes Starr, "to keep in equipoise . . . the eight cotton growers" indicted in the Pixley slayings (*Endangered Dreams*, p. 78).

155   **"Comrades died":** Weber, p. 102.

155   **James Cagney:** Loftis, p. 20; Mitchell, *The Campaign of the Century*, pp. 142–143. **To tap dance:** Loftis, p. 46. Steffens appointed himself chairman, secretary and treasurer of the "Caroline Decker Typewriter Fund," soliciting a small contribution from none other than James Rolph, the Republican governor of California. "I am going to ask you," he wrote, "to contribute personally a dollar or so toward a fund to buy a typewriter for Caroline Decker, the tiny little labor agitator who is doing what no big AFL leader had ever dared undertake: to organize the migratory workers . . . of California." Rolph's passivity throughout the cotton strike had been seen as an endorsement of the growers' bloodthirsty behavior, and Steffens acknowledged that his request of the governor was little more than an "experimental probing for humor in high places." (See *The Letters of Lincoln Steffens*, vol. 2, pp. 974–975.)

155   **Blood on the Cotton:** A copy of the work, at one point known as *Blood on the Fields*, is included in the Langston Hughes Papers, Beinecke Rare Book and Manuscript Library, Yale University. **Too unwieldy:** Loftis, p. 46, says that the play, written in collaboration with Ella Winter and a Carmel writer named Ann Hawkins, "was a long, artificial propaganda vehicle too complicated and too cumbersome to be performed." Another artist who found inspiration in the cotton strike was poet Marie de L. Welch. She composed an ode called "Camp Corcoran" after sensing the "pride . . . and dignity" among those holed up in the strikers' outpost (Welch, *This Is Our Own*, p. 57. Welch's poetry is discussed in Loftis).

155   **Many details:** Benson, *The True Adventures of John Steinbeck, Writer*, pp. 296–309; Loftis, p. 55. One of Steinbeck's sources was Cicil McKiddy, a cousin of rank-and-file leader Big Bill Hamett. **London:** Daniel, p. 321. **Jim Casy:** Benson, p. 342.

155   **Eight other:** Reisler, *By the Sweat of Their Brow*, p. 241. **"About as sanitary":** Taylor and Kerr, p. 64.

156   **Roosevelt didn't like:** Daniel, pp. 168–176. If left unchecked, the president believed, both big business and big labor tended to act selfishly and undermine the national interest. **Government compelled:** Daniel, p. 212. **"No water":** Testimony of Lilly Dunn, cited in LF Hearings, pt. 54, p. 19934. **Afford shoes:** Testimony of Pauline Dominguez, cited in LF Hearings, pt. 54, p. 19932. Dominguez's story is also recounted in Daniel, p. 214.

156   **"60-cent picking rate":** HMJ, October 20, 1933. **Extortionate terms:** As one small farmer told the panel, "We will be glad to pay more to the poor slave who has to do the work. If the finance companies . . . were put in their right place," the pickers would get paid enough "so they could eat" (LF Hearings, pt. 54, p. 19936).

156   **By the feds:** Daniel, p. 215. **Lead federal negotiator:** George Creel was the central figure in bringing about a strike settlement. See his autobiography,

Rebel at Large. **Source of sustenance:** Though Governor Rolph did little to defend the strikers, taking the position that valley authorities were fit to keep the peace, he ordered the state administrator for the Federal Emergency Relief Administration to get food to the workers (Daniel, p. 204; Watkins, *The Hungry Years*, p. 413).

156   **Precursor to Cesar Chavez:** See Ferriss and Sandoval, *The Fight in the Fields*, pp. 23–25. **Unprecedented:** Daniel, p. 219. **"Faulting":** Taylor and Kerr, p. 17. The strike produced a few political ripples as well. A master propagandist who had worked in that capacity for Woodrow Wilson and was a friend of FDR, George Creel would cite his role in the strike's resolution in his bid to be California's next governor. Astonishingly, Creel was upset in the 1934 Democrat primary by Upton Sinclair, the erstwhile socialist and iconoclastic author, whose plan to end poverty in California incited a full-blown populist insurgency across the state (Starr, *Endangered Dreams*, p. 121. Also see Creel, pp. 276–288).

156   **"One square meal":** Darcy autobiography, S. D. **Withered away:** Daniel, pp. 254–257, explains that union leaders had a hard time holding on to the rank and file after the Communist Party ordered them to stop focusing on "immediate economic goals" and instead start preaching "Marxist tenets." **Eight union leaders:** Watkins, *The Hungry Years*, p. 419.

157   **Found very odd:** Ewart interview with Tapson.

157   **Asked Boswell:** Interview with Lawrence Galvan, who played second base for the Bears. Weber, p. 242, suggests that Boswell bought the Bears uniforms in a paternalistic effort to stave off labor problems, but Galvan remembers it differently, saying that he and some other baseball-loving workers simply approached the company and asked to be outfitted. **Cotton pickers league:** CN, October 6, 1939. **The Bears won:** CN, November 10, 1939.

CHAPTER 9: "GOON-SQUAD TACTICS"

158   **Alvin Odle said good-bye:** The Odles' story is based on numerous interviews with Fred Odle.

158   **Dust Bowl:** Gregory, p. 11, notes that, although the term "dust bowl" is generally applied to the mass migration of southwesterners, the actual area of "severe wind erosion" consisted of a relatively small section of the wheat belt near the intersection of Kansas, Colorado and the Oklahoma and Texas panhandles. "Despite the popular perceptions, less than 16,000 people from the Dust Bowl proper ended up in California, barely six percent of the total from the Southwestern states."

160   **The American heartland:** A caveat: Gregory, p. 15, cautions against stereotyping the migrants. One in six, he shows, was a professional, a proprietor

or a white-collar employee. "Dispossessed and sometimes desperately poor farm families . . . were indeed an important element of the migration, but there were also many other participants who defied the popular image of the rural Dust Bowl migrant." **"Gypsies by force":** Steinbeck, "Their Blood Is Strong," p. 3. The Simon J. Lubin Society published this booklet in April 1938. It was based on a series of stories that Steinbeck had written for the *San Francisco News* under the title "The Harvest Gypsies." **More than 900,000:** See Gregory, p. 6.

160   **Repatriation program:** Stein, p. 37; McWilliams, *Factories in the Field*, p. 129. McWilliams writes that "in excess of 75,000 Mexicans had been shipped out of Los Angeles alone." And Stein says that by 1937, an estimated 150,000 had returned home. **No longer necessary:** Cross and Cross, *Newcomers and Nomads in California*, p. 53; Gregory, p. 62. This situation wouldn't change, Gregory notes, until the 1950s with the "large-scale reentry of Mexicans" to the valley.

160   **String Bean:** Interview with Roy Hamett.

160   **As much as $1:** See Weber, p. 215.

161   **Black labor camps:** Gregory, p. 174, says more than 125,000 southwestern blacks moved west in the 1940s.

163   **"What's the idea?":** This quote, and much of the rest of the chapter, is taken from a 3,000-page transcript of a National Labor Relations Board hearing held in Corcoran in the summer of 1939: NLRB case no. XXI-C–1025, the matter of *J. G. Boswell Company, the Associated Farmers of Kings County, the Corcoran Telephone Exchange and Cotton Products and Grain Mill Workers' Union, Local 21798.* The testimony captured in the document, unearthed at NA, paints a detailed picture of the unrest at the Boswell gins during the prior fall and winter. Not surprisingly, different witnesses gave conflicting accounts of what happened when the AFL and its foes clashed. But many times, the union's opponents simply stonewalled when they were questioned, giving us confidence that the union's version of events was probably more correct. In cases where we couldn't resolve what likely occurred, we have pointed out the discrepancy or simply avoided recounting that part of the story. All of the dialogue in the chapter comes from the testimony of at least one witness, and often multiple witnesses. All other quotes and facts to the end of the chapter are also gleaned from the testimony, except where noted.

163   **Plant foreman:** During the NLRB hearing, Boswell's attorney disputed that a group of men identified by the union as supervisors actually had such a role. But the testimony leaves no doubt that workers in the plant routinely took orders from these men, regardless of whether they held a formal title. Bill Robinson was one of these de facto foremen.

163   **Broad-shouldered:** The physical description of and biographical material on Farr is from an interview with his brother Virgil.

164   **Fourteen of them:** TDAR, May 20, 1939; CN, May 12, 1939.

164   **Hidebound federation:** Zieger, *American Workers, American Unions*, pp. 51–58.

165   **Average hourly wage:** *Historical Statistics of the United States: Colonial Times to 1957*, U.S. Government Printing Office, p. 92.

165   **"Slave labor":** Interview with Farr's brother Virgil. Information on Farr's union activism is from Virgil, plus information provided by his widow Esther and daughter Juanita Bettencourt.

165   **J. B. Boyett:** He was president of the Associated Farmers unit in Kings County. **Left a stain:** Daniel, pp. 274–275, points out that the AFL had historically given only tepid support to agricultural workers, and through the mid–1930s AFL leaders were quite friendly toward the Associated Farmers. But over time, that relationship changed. Writes Stein: "Claiming at first that they opposed only 'Communist unions,' the Associated Farmers, when confronted with non-Communist AFL unions, opposed them too. . . . So useful was the Communist label throughout the period that the Associated Farmers applied it everywhere" (pp. 234–235).

165   **Didn't see unionization:** Stein, pp. 260–273; Weber, pp. 147–151; Gregory, pp. 154–164. Gregory cautions against overgeneralizing about the Okies' antiunion feelings, noting that "though definitely in the minority, the number of union supporters was by no means negligible." Still, he, Weber and Stein all point to the range of factors that made it difficult for the unions to sign up Okie workers. Though all three authors write primarily of the CIO's unsuccessful attempts to organize the cotton fields in 1938 and '39, many of these same phenomena were clearly present during the Boswell gin strike. **"Rugged individualism":** Stein, p. 264.

166   **Cult of drudgery:** Gordon Hammond's legendary work ethic is recounted in a letter from Boswell worker Mel Carr to Tommy Hammond on the occasion of the latter's retirement in 1971 (Courtesy of Ruth Hammond).

166   **Crop surpluses:** By 1939, even with the AAA, the United States had a surplus of 13 million bales—twice as much as the country would normally consume in a busy year (*Time*, April 10, 1939). **Falling prices:** The average price per pound of California cotton was 12.6 cents in 1936, but it fell to under 10 cents for the next three years. It then began a steady recovery in 1940.

166   **Nearly 40 percent:** U.S. Census data shows that the total number of Kings County residents increased to 35,168 in 1940 from 25,385 in 1930. That increase, of 38.5 percent, compares with a 15 percent rise from 1920 to 1930 and a 33 percent jump from 1940 to 1950. **"Oklahoma City":** CN, November 25, 1938. **Relief rolls:** The California State Relief Administration caseload in Kings County grew steadily from just four in November 1936 to 245 in November 1939 (CMcW). **Tuberculosis:** In 1939 and the first half of 1940, Kings County reported sixty-seven cases of the disease (Hearings before

the Select Committee to Investigate the Interstate Migration of Destitute Citizens, U.S. House of Representatives, 76th Congress, September 24–25, 1940, p. 2959). **"Ham and Eggs":** A thorough account of the initiative's history is found in Starr, *Endangered Dreams*, pp. 197–222. **"Glory, glory":** Starr, p. 206.

166   **Ballot box:** The vote, Starr reports, was 1,398,999 to 1,143,670. In Kings County, the measure was defeated 5,465 to 4,650 (CN, August 18, 1939). In 1939, a second Ham and Eggs referendum also failed, this one by a much bigger margin. But Gregory, pp. 153–154, notes that many Okie areas enthusiastically supported the proposal. Ham and Eggs, he says, seemed "especially dear to the hearts of much of the migrant population" as it played perfectly to the group's "suspicions of bankers, corporations and pretentious wealth." **"Ramblin' cotton picker":** The song, by Lloyd Stalcup, is from "Voices from the Dust Bowl: The Charles L. Todd and Robert Sonkin Migrant Worker Collection," American Folklife Center, LC.

167   **Sharp reduction:** The total cotton acreage harvested in California slipped to 327,390 in 1938 from 587,600 in 1937. In Kings County, the decline in acreage was even steeper—to just 24,260 from 66,300 (Weber, p. 211). **Compounded the glut:** Stein, p. 86; Gregory, p. 88. **Inundated 100,000 acres:** Board of Army Engineers Report, June 26, 1939. It noted that this "high-grade agricultural land" would be "out of production for at least two years" (Joseph B. Lippincott Papers, Water Resources Center Archives, University of California at Berkeley).

167   **Only twenty-nine:** CN, January 27, 1939.

167   **Every time a truck:** Undated letter from Associated Farmers field secretary S. H. Strathman to the group's executive secretary, Harold Pomeroy (LF Hearings, pt. 72, p. 26433). **Cotton by rail:** HMJ, February 10, 1939.

168   **Since the 1933 strike:** The cotton strike had been the climax of a great many union actions in '33, which also affected the pea, berry, sugar-beet, apricot, pear, peach, lettuce and grape industries. All told, an estimated 47,575 laborers were involved. Twenty-five of the strikes, or four-fifths of the total, had been under the direction of the Agricultural and Cannery Workers (Jamieson, p. 87). **Local chapter:** The Kings County unit was formed in September 1938. **Dedicated to trouncing:** The Associated Farmers agenda is covered extensively in LF and summarized nicely in Johns, p. 79.

168   **Who's Who:** Among the contributors were Bank of America, Times Mirror and Standard Oil (Johns, p. 78). **Labor spies:** To cite but one example, the Associated Farmers of San Joaquin County used "an undercover man" during the grape harvests in 1937 and '38 to penetrate union activities (LF Hearings, pt. 48, p. 18232). **Secret files:** Johns, p. 80. **Baseball bats:** Again, to provide but one example, during the Madera cotton strike of 1939, "300 Associated Farmers' deputies under command of the local sheriff

attacked a strikers' rally with bludgeons, fan belts, tire chains and pick handles" (Stein, p. 257).

169   **"Embattled farmers":** TDAR, January 31, 1939. **Better American Federation:** Starr, *Endangered Dreams*, p. 156.

169   **"They want you":** FB, January 31, 1939. **Afforded the same:** Farm workers were excluded from the benefits of the National Industrial Recovery Act in 1933 and the National Labor Relations (or Wagner) Act in 1935. These developments, along with "explicit denials of equal protection to them under both the Social Security Act and the Fair Labor Standards Act," meant that "New Deal lawmakers effectively codified the traditional powerlessness of farm laborers," according to Daniel, p. 261.

170   **Union sympathies:** Zieger, p. 40. **Extremely chummy:** Auerbach, *Labor and Liberty*, p. 179. He writes, "Olson had waged a vigorous and successful campaign against Proposition 1," an antipicketing initiative, and was "elected with overwhelming labor support." Olson then "further endeared himself" to the union movement during his first week in office by pardoning jailed labor radical Tom Mooney.

170   **"Entitled to choose":** FB, February 2, 1939. In fact, most of the Boswell employees joined together in their own "company union" as an alternative to the AFL.

170   **"You're not dealing":** FB, February 2, 1939; HMJ, February 2, 1939.

170   **"Grow, harvest and market":** TDAR, February 8, 1939.

170   **"Not a yokel":** TDAR, February 9, 1939.

171   **Two had flirted:** In an interview, Dorothy confirmed that she and Sprecher wound up dating.

171   **"Last thing":** Bill Boswell denied saying this.

172   **Marked the first time:** *Propaganda Analysis*, August 1, 1939, p. 10 (CMcW). The case was also reported on in the September 1939 issue of *Survey Graphic*, a leading social research journal.

172   **Just weeks earlier:** *The Grapes of Wrath* was published in April 1939. **Organizations attacked:** The Kings County chapter of the Associated Farmers, for one, sponsored a special train trip so that locals could attend a convention in Stockton at which the book was to be disparaged (CN, November 17, 1939). **Offended:** Gregory, p. 111, writes, "Steinbeck may have invested his characters with an underlying human dignity which won the hearts of millions of readers, but to many Southwesterners who read the novel what stood out was the graphic sexuality, the crude living habits, the illiterate-sounding speech and the general lack of modern sophistication." **Bill Camp:** Briggs and Cauthen, p. 169. A photo of Camp and cotton picker Homer Hester watching the book being dropped into a flaming trash can was published in *Look*; the photo is also available at the Kern County Museum Library, Bakersfield, California. Camp always contended that he was

mostly upset at the way Steinbeck portrayed the migrants (Interview with his son Don Camp). In addition, supervisors in Kern County banned the book (BC, August 21, 1939). The novel was also banned in Buffalo, New York, and East St. Louis, Illinois (Benson, p. 418).

173 **"Some 200,000":** McWilliams, *Factories in the Field*, p. 9. **Mutinous solution:** pp. 324–325.

173 **A nuisance:** The biographical material on McWilliams is drawn from Loftis, pp. 62–64 ; McWilliams, *The Education of Carey McWilliams.* **"Fair" wage:** In May 1939, McWilliams led a hearing in Madera to decide on a fair wage for chopping cotton. This was followed up in September by a Fresno hearing to determine a fair wage for picking cotton. Under the policy McWilliams helped to develop, laborers who refused to work for less than the amount recommended by the state wouldn't be taken off the relief rolls. The wage proceedings—and the growers' strong reaction to them— are discussed extensively in Johns, pp. 112–124. **"Pest No. 1":** McWilliams, *The Education of Carey McWilliams*, p. 77. **"Regret to report":** "Steinbeck and the 1930s in California" (Typescript in CMcW). The growers struck back at *Factories in the Field*, however feebly. Associated Farmers leader Roy Pike released a tract called "Facts from the Fields" as an answer to McWilliams's book. And farmers applauded *Of Human Kindness*, a mawkish counterpoint to *The Grapes of Wrath* in which author Ruth Comfort Mitchell, the wife of a union-busting state senator, depicted a whole different side of the growers through heroine Mary Banner and her farming family. (The Mitchell book is discussed in Loftis, pp. 169–170; Benson, pp. 419–420; Stein, pp. 206–208.)

173 **American Exodus:** The book and its importance are discussed in Loftis, pp. 178–181; Starr, *Endangered Dreams*, pp. 264–265. Unlike *The Grapes of Wrath* (top of the best-seller list) and *Factories in the Field* (five printings in three months), *American Exodus* didn't fare well commercially. **Very own footsteps:** Taylor, who later in life became a harsh critic of the Boswell Company in connection with water policy, recalled hearing that the Boswells were chased west by the boll weevil when he and Lange went to Georgia (PTOH, p. 45). Taylor's approach for spotting trends—traveling the country, listening to ordinary folks tell of their daily struggles to make ends meet—was anathema to much of his profession. "Paul was always out there finding out what was really happening while others played around with their theoretical models and ran their regression analyses," Clark Kerr, who as a graduate student in 1933 had assisted Taylor and later became UC president, noted in *On the Ground in the Thirties*. For additional biographical details on Taylor, see Loftis, pp. 115–140; PTOH; Clark Kerr's essay "Paul and Dorothea" in *Dorothea Lange: A Visual Life*; "The Power of A Tenacious Man," *The Nation*, October 12, 1974.

174   **Largest financial contributor**: In 1938, the first year the Kings County
chapter existed, Boswell contributed $287.09 to the Associated Farmers.
This was based on a 1-½-cent payment for each bale of cotton ginned (LF
Hearings, pt. 48, Exhibit 8203). Boswell also made a 1-½-cent per bale payment to the Agricultural Labor Bureau. **"Vigilantism"**: McWilliams, *Ill
Fares the Land*, pp. 26–27. He writes, "The alliance formed between big
business in agriculture and these allied industrial interests is really what is
behind vigilantism in California agriculture," citing, as an example,
Boswell's participation in the Associated Farmers.

175   **NLRB ruled**: CN, January 26, 1940; CJ, October 3, 1941; information pro-
vided by Dorothy Dunn; proceedings of the 41st annual convention of the
California State Federation of Labor, September 23–28, 1940 (Labor
Archives & Research Center, San Francisco State University). **"Fascist
Farmers"**: *Rural Observer*, July 31, 1940. The *Observer* was published by the
Simon J. Lubin Society. For details on the group, see Stein, p. 244. Lubin
had been the first chief of the California Division of Immigration and
Housing and a tireless opponent of the state's large growers. **Subpoenaed
Boswell's**: LF Hearings, pt. 72.

175   **Just faded away**: In 1942, through the Cottonseed and Vegetable Oil Workers, the AFL made one more serious run at organizing Boswell's Corcoran
plant, but the employees voted down the union by the slimmest of margins, 41–40 (NLRB records, NA). **CIO tried**: The first attempt was in 1938
(Johns, pp. 99–111; Jamieson, pp. 171–172; Stein, pp. 256–257; Weber pp.
180–184). But labor leaders had very little success. Johns notes that the CIO
union—the United Cannery, Agricultural, Packing and Allied Workers of
America—claimed 4,000 pickers participated in the strike, but the growers put the figure at only 500. Whatever the number, the unrest was confined to Kern County. And so many workers were willing to defy the union
that the fields there were picked without a hitch; in fact, the growers questioned whether what was taking place could even be called "a strike." The
AFL and CIO were, of course, bitter rivals at this point. For a discussion of
the interunion conflict, see Daniel, pp. 277–281; Jamieson, pp. 149–192.
"Two Men Killed": Flier found in BCA.

175   **Sang to a packed house**: From Klein, *Woody Guthrie: A Life*, pp. 136–145.

175   **A hundred field laborers**: CN, October 13, 1939. The '39 strike, which was
centered in Madera and turned violent there, is discussed in great detail in
Johns, pp. 125–145. **Even Bill Hamett**: Interview with his grandson. **"Deal
with the devil"**: Bill Hamett himself told an interviewer in 1936: "I ain't no
Communist. I hold the American flag just as good here and now as when
Betsy Ross finished her stitchin' and handed it over to George Washington" ("Migratory Labor in California," State Relief Administration of California, Division of Special Surveys and Studies, p. 140). The man quoted is

given the pseudonym Clayton Bennett in the report, but Loftis has identified him as Hamett.

176   **Fourteen-room castle:** Noted in the CJ, November 30, 1967.

CHAPTER 10: RIVER OF EMPIRE

177   **Battelle stood:** The account of the boating accident on Tulare Lake is based on an article in the CN of April 16, 1937, as well as numerous other articles from area newspapers. In addition, several interviews were conducted with two of the survivors: Claire (Battelle) Plannette and Doris (Butcher) Verboon.

177   **Thirteen years:** From Mae Weis's Tulare Lake history in the CJ of June 3, 1938. Also see Brown, p. 95. Brown puts the dry spell at fourteen years, Weis at thirteen. C. L. Kaupke, the Kings River water master, was even more specific. The lake, he noted, dried up in 1919 and remained dry until 1922, when about 40,000 acres went under water. Then no water flowed into the lake until 1927, when about twelve sections were flooded. Then, once again, no water entered the lake until 1936. (See his testimony before the House Committee on Flood Control, February 1–23, 1944.) **Incredible pace:** This and other images throughout this part of the chapter are taken largely from a home movie of the late-1930s flooding that is found in BCA.

178   **"Barracuda":** Interview with Corcoran old-timer Stuart Bartlett.

179   **"Jumping up and down":** Crocket recounts this dialogue with his dad in his unpublished memoir.

181   **Word around town:** Interviews with long-time Corcoran residents Sam Crookshanks, Stu Bartlett and Gordon Felder. Felder had been invited by his good friend Betty Battelle to go out on her dad's speedboat but, as fate would have it, couldn't make it that day.

181   **Blanketed more than:** Weis account in CJ; article in May 21, 1937, CN headlined: "Tulare Lake now covers close to 100 sections." **"Farmers felt":** Weis account in CJ.

181   **Huge winter storm:** Brown, p. 96. Also see Kaupke, *Forty Years on Kings River*, pp. 59–60.

181   **For the Okies:** Stein, pp. 75–80. **Eighty-seven people:** From the Boswell-produced film *Man, Land and Water*, BCA. **Grant-in-aid:** Stein, p. 85.

182   **Over 40,000 acres:** CN, March 18, 1938. **Four-foot waves:** CN, April 29, 1938. **The Boswell Company saw:** September 6, 1938, letter from FS to his friend Colonel E. R. W. McCabe at the War Department, in which he notes: "On June 5 we lost by flood several thousand acres of our California farm land" (FSP).

182   **Hysterectomy:** Interview with Barbara Hamilton, Alaine's niece. **"If I were told":** Boswell's *Stories & Poems*, p. 36.

182   **Lace evening dress:** Details from Alaine Boswell's estate, Los Angeles Superior Court. **"Go mad":** June 15, 1938, letter from Frederick Buck to FS, FSP.

182   **Red Heads:** CN, March 4, 1938. **"Tallest man":** CN, September 9, 1938. **Lake Theatre:** CN ads, September 1938. **Frenetic race:** "The Lake That Comes and Goes" by Ron Taylor, *Westways*, October 1967.

183   **"Making of Elmer":** Interview with Crookshanks.

183   **Bought the Cousins:** Boswell didn't actually buy the ranch until 1946, but it was still under water from the 1938 flood. See testimony of real estate expert and former Boswell employee Bryce Sherman in U.S. Court of Appeals for the Ninth Circuit, transcript of proceedings, *J. G. Boswell Company v. Commissioner of Internal Revenue*, October 16, 1959, pp. 215–217. **"Water in Tulare Lake":** CN, August 26, 1938.

183   **Twenty districts in all:** November 1, 1946, report from the sate Division of Water Resources. **More than 140,000 acres:** May 1942 engineering report, "Flood Control in Tulare Lake Basin," by Roy May and S. T. Harding, STH; An April 1947 report by the Bureau of Reclamation put the number as high as 170,000 acres. **Waters crept:** Interviews with Claire Plannette and Jack Fossett. **Airplane pilots:** Interview with Henry Coultrap. **$7.5 million:** The figure was given by R. F. Schmeiser, president of the Southern San Joaquin Flood Control Association, at a public meeting held in Fresno, July 30, 1946. A transcript is in PST. **Looked to shift:** HMJ, November 20, 1938.

184   **Ed Bradley:** CN, July 15, 1938. **Robert Scutt:** CN, August 6, 1938. **Two duck hunters:** CN, December 1, 1939.

184   **Too little water:** Maas and Anderson, p. 237. They note that "the very dry years of 1913 and 1914 prompted mass meetings of farmers to discuss the need for river storage, and several committees of leaders representing the major users of irrigation water were formed to examine the possibilities." See also the *Fresno Morning Republican,* December 17, 1916, which lays out the report of Louis C. Hill, a Bureau of Reclamation engineer charged with investigating the feasibility of constructing a large dam at Pine Flat (Article found in the Kings River Collection, Water Resources Center Archives, University of California at Berkeley). **Name of flood control:** The Army Corps of Engineers, in its June 1939 report on Pine Flat Dam, notes that "during the 20-year period prior to 1936, runoff from the Kings River was generally below normal and few flood-protection measures were undertaken by local interests. However, as a result of the extensive flooding in the last few years an active demand has developed for adequate flood protection." Also see the testimony of Brigadier General Thomas M. Robins of the Corps of Engineers before the House Committee on Flood Control, March 18 to April 9, 1940.

184   **Sherrill brimmed:** All details on the Denver meetings are from various

papers found in FSP; S. T. Harding's notes on the sessions, dated August 1, 1941, in STH; Harding's "History of Work on Kings River," pp. 34–40, STH; and Harding's deposition in *U.S. v. Tulare Lake Canal Company*, a copy of which is in STH. **Majestic sculptures:** Description of the site is from the Historic American Buildings Survey Inventory. **Compelled to intervene:** In *Muddy Waters*, pp. 212–214, Maass writes: Roosevelt was so concerned about the "duplication of water development functions in the Central Valley," he called for a conference in his office to discuss the issue on July 19, 1939. "As for the Kings River project, the president agreed to an arrangement whereby the two agencies would cooperate in preparing independent reports, but these reports should contain agreement on both design and economic features of the project." He also instructed "the departments of War, Interior and Agriculture, in cooperation with the National Resources Planning Board to draw up a memorandum of agreement, which, by insuring consultation in the early stages of project planning, would preclude the possibility of similar conflicts" in the future. By early 1940, however, both the bureau and the corps had submitted to Congress reports on Pine Flat that differed in crucial aspects. "The Planning Board scheme of insuring integration prior to the crystallization of findings and recommendations had failed."

185   **Like most growers:** Two farmers groups, the Kings River Water Association and the Kings River Pine Flat Association, met in Fresno on March 9, 1940, and adopted a resolution backing the corps plan (Reflected in a June 1941 statement by the two associations). **Since 1936:** The corps calls the 1936 Flood Control Act "one of the most important events" in its history because "for the first time, Congress declared that flood control was a proper activity of the federal government" (From "U.S. Army Corps of Engineers: Brief History" by Martin Reuss and Charles Hendricks). **Studies indicated:** The Bureau put the annual irrigation benefit from Pine Flat at $1.255 million and flood control at $1.185 million. The corps, meantime, found flood control was worth $1.185 million a year and irrigation only $995,000. (From the agencies' original Pine Flat reports, which are found in the Joseph B. Lippincott Papers, Water Resources Center Archives, University of California at Berkeley. The reports were also reprinted in February 1940 as U.S. House of Representatives Documents 630 and 631.)

185   **$19.5 million:** This amount doesn't include the cost of power development. Under the bureau's proposal, the farmers would repay the entire amount allocated to irrigation ($9.75 million) over a period of forty years. As with all reclamation projects, no interest would be charged—itself a huge subsidy. Under the army's plan, the farmers would contribute a lump sum of $5.2 million to cover irrigation costs; if they borrowed this amount and the debt was retired in forty years (with 3.5 percent interest),

the cost to the growers would be equal to that in the bureau plan. (See the agencies' original reports; Maass, pp. 215–216.) **Half of that:** As prescribed by law, the flood control portion of the project was free to the farmers under both agencies' plans. **Crucial difference:** CN, March 15, 1940. In its original Pine Flat report, the bureau said that the "complexities" of the project "make it highly advisable" for the government to operate the dam. The Army Corps proposal suggested, however, that because "water rights in the area under consideration are complex," local interests should maintain and operate the project. **Pet interest:** Watkins, *Righteous Pilgrim*, p. 776. **Ickes himself:** Expressed in an April 17, 1940, letter to state legislator S. L. Heisinger, Kings River files, National Archives, Rocky Mountain Region, Denver. **"We are afraid":** Testimony before the House Committee on Flood Control, March 18 to April 9, 1940.

185   **"It shakes down":** "The West Against Itself," *Harper's*, January 1947.

186   **Generally accepted:** See especially *Muddy Waters*, as well as an August 1949 *Harper's* article by Maass and Robert de Roos, "The Lobby That Can't Be Licked." Also see Reisner, pp. 175–179; Worster, *Rivers of Empire*, pp. 252–253. For an opposing view, see Cooper, *Aqueduct Empire*, pp. 158–159. **"Spoilsmen in spirit":** From the foreword to *Muddy Waters*. **"Grab off":** May 28, 1940, letter from Ickes to Roosevelt, Ickes Papers, LC.

186   **Already had convinced:** As early as May 1940, Roosevelt told Ickes that he considered Pine Flat "dominantly an irrigation undertaking" that "should be constructed by the Bureau of Reclamation" (Letter included in the Kings River Project supplemental report, House Document 631). **"In a coma":** FS letter to Leslie Groves, February 5, 1941, FSP.

186   **Served as chairman:** It was Senator Alva Adams, a Colorado Democrat. **"Shut the army out":** August 18, 1941, letter from FS to W. C. Benton, a West Point classmate from Colorado and close friend who helped arrange the meeting with Senator Adams (FSP).

187   **Exuded passion:** Drawn from the definitive biography on Ickes, Watkins's *Righteous Pilgrim*. **"Sexually alert":** In his unpublished memoirs, Ickes recalls how "a spark passed from her to me which made me aware that she was attractive physically, as well as alert sexually and that I would like to possess her" (Watkins, *Righteous Pilgrim*, p. 355). **Page was unemotional:** Details on Page are from an interview with his daughter, Milly Oppenheimer. **Enough concrete:** From "The Story of Hoover Dam," U.S. Bureau of Reclamation.

187   **"Further in our favor":** Harding's "History of Work on Kings River," p. 35, STH. **"Practically everything":** FS, in his August 20, 1941, letter to his friend E. L. Thurston of the Federal Reserve Board, describing his trip to Denver, FSP.

187   **Page had signed off:** The Page memo is included in the "Report on Kings

River Project," February 15, 1943, by the Kings River Pine Flat Association. The report, known as the "Green Book" because of the color of its cover, was prepared with the bureau's assistance.

188 **World's largest:** In an April 1942 memo, FS noted that Boswell owned 37,809 acres, mostly in California with some land in Arizona. "This firm," he added, "conducts the largest cotton farming operation in the world, having produced in one year in excess of 35,000 bales" (FSP).

188 **Formally approved:** Ickes's approval is included in the Green Book.

188 **Engulfed:** The June 20, 1941, CJ reported that 80,000 acres in the lake area were flooded. **"Dry-land fishing":** CJ, May 23, 1941. **Rattlesnakes:** CJ, May 23, 1941. **Walter Winchell:** CJ, June 20, 1941.

189 **"Hector's pup":** August 31, 1942, letter from JGB to FS, FSP.

189 **Serious doubts:** A history of the Sacramento District of the corps from 1929 to 1973 notes that "with the outbreak of war, most domestic programs were shelved—Pine Flat among them." Bureau of Reclamation Commissioner Harry Bashore also addresses the issue in his testimony before the House Committee on Flood Control, February 1–23, 1944. **"I can get":** August 19, 1942, letter from Gearhart to Tulare Lake engineer Roy May, FSP.

189 **Big a worry:** Upriver engineer H. L. Haehl said in an August 19, 1942, letter to FS (FSP): "I think we all appreciate that the people in the lake are concerned chiefly with flood control and will receive a large free flood control benefit with relatively minor contribution toward the irrigation feature. To them, of course, the financial burden will be relatively light compared with the benefits received. The opposite conditions affect the remainder of the Kings River area concerned principally with irrigation." Harding went so far as to tell the bureau's Bashore that those in the lake basin have "90 percent of the flood damage and flood control interest on the Kings River" (July 20, 1943, letter, STH).

189 **"The attitude":** August 19, 1942, letter from Haehl to FS, FSP.

190 **"In the face of":** September 4, 1942, letter from FS to Haehl, FSP.

190 **"You can't fight":** November 15, 1944, letter from JGB to FS, FSP.

190 **Ease the crunch:** CJ of October 30, 1942; November 6, 1942; October 29, 1943. **"Prettied herself up":** CJ, October 30, 1942.

190 **Mexican laborers:** CJ of May 14, 1943; June 18, 1943; September 22, 1944. **Prisoners of war:** Regarding the German POWs, see the CJ of November 17, 1944; December 8, 1944; January 5, 1945; January 12, 1945; April 13, 1945; November 30, 1945. Regarding the Japanese POWs, see the CJ of September 28, 1945. **"Feel the disgrace":** CJ, December 7, 1945. **He hanged himself:** Interview with Corcoran resident Bonnie Liggett.

191 **4,000 jitterbugged:** CJ, May 8, 1941. **Pilot training:** CJ, October 17, 1941. **Jimmy Boswell:** JGB2's letter was reprinted in the CJ of January 28, 1944.

191   **112,000 acres:** Cited by Army Corps Colonel George R. Goethals in his tes-
timony before the House Committee on Flood Control, June 1–11, 1943.
"Make it imperative": June 3, 1943, letter from Harding to Bashore, FSP.
See also "Tulare Lake Water Storage District Wants Dam Built, Doesn't
Care Who Does It," CJ of July 2, 1943.

192   **Cut their own deal:** In the summer of 1943, Harding and fellow Tulare Lake
Basin Water Storage District engineer Roy May told the district directors
that they were at a crossroads. "We have arrived at the point in Kings River
and Pine Flat matters where we will have to decide between continuing to
drift under the leadership of other Kings River interests or proceed under
our own initiative" (STH). Louis Robinson, who served as president of the
storage district, noted that a report by Harding and May proposing to link
Pine Flat with the bureau's Central Valley Project "would probably be correctly interpreted as an open break between the Lake and the River." Then
he asked FS: "Do you think we should now spring this report or drift further along?" (June 29, 1943, letter from Robinson to FS, FSP). In the end, the
report was not released (Harding's "History of Work on Kings River," p. 98,
STH). **Go so far:** Harding, in his June 3, 1943, letter to Bashore said: "It is also
my opinion that the operation of the (Pine Flat) project should and eventually will be assigned to the Bureau because of the physical advantages of its
coordination with the Central Valley Project." **Leviathan system:** Details on
the CVP are from the "Layperson's Guide to the Central Valley Project," prepared by the Water Education Foundation, Sacramento. See also Kahrl, *The
California Water Atlas*, pp. 46–50. **"Mind-boggling":** Reisner, p. 9.

192   **"Never seemed reasonable":** June 29, 1943, letter from Robinson to FSP.

192   **Unilateral power:** Bashore discusses the mechanism by which a reclamation project is authorized in his testimony before the House Committee
on Flood Control, February 1–23, 1944.

192   **Start of the hearing:** All details and quotes come straight from the official
transcript of the House Committee on Flood Control, February 1–23, 1944.
Gotten sick: Interview with Page's daughter. **Totally by surprise:** Harding
said that Bashore's testimony marked "a departure and practically a rever-
sal of the previous basis of understanding with the Bureau" ("Memorandum for Tulare Lake Basin Water Storage District" February 17, 1944, STH).
Interestingly, two of the fiercest opponents of acreage limits—Representative Alfred Elliott and Senator Sheridan Downey—never seemed to blame
Bashore for the hardening of the bureau's position; rather, they heaped all
of their enmity on Secretary Ickes. Downey even assured Bashore in a December 17, 1947, letter of the "high opinion" he had of his "integrity and
ability" (Harry W. Bashore Collection, Western History Research Center,
University of Wyoming).

193   **"As solemn":** An analysis by FS, dated January 1949, responding to Maass's criticism of how the corps pursued Pine Flat Dam (FSP). Maass's work on Pine Flat first appeared in a report for a Hoover Commission task force on natural resources. Later he included the material in his book *Muddy Waters.*

193   **"Don't want any dam":** April 12, 1944, letter from JGB to FS, FSP.

194   **"Woe to you":** Isaiah 5:8.

194   **"Small landholders":** *Landownership Survey on Federal Reclamation Projects,* U.S. Bureau of Reclamation, 1946, p. 64; also quoted in Opie, p. 10; Griswold, *Farming and Democracy*, p. 45. The irony is that Jefferson himself had 11,000 acres and was of the planter class (Dies, *Titans of the Soil*, p. 24). Jefferson's vision for agriculture is also discussed in Griswold, pp. 18–46; Daniel, p. 15. Opie, pp. 28–33, notes how Jefferson's ideas grew out of the natural-rights philosophies of James Harrington and John Locke. He also discusses how Jefferson's notions of land policy clashed with those of Alexander Hamilton. **Rebuke of the landed aristocracy:** Opie, p. 24.

194   **"I am in favor":** Quoted in *Landownership Survey*, p. 73; Griswold, p. 142.

194   **Antebellum plantation:** Griswold, p. 25, notes: "Lack of capital and a wilderness that yielded only to hard, slow manual labor made small-scale farming the rule long before Jefferson became its advocate." However, "the tobacco, rice and cotton plantations of his southern compatriots were exceptions to the rule." See also Jacobson and Smith, p. 56, for how the cotton culture of the South was at odds with the "yeoman tradition." **Founded in California:** By 1890, as Hubert Bancroft noted in his seminal history of the state, California already was characterized by "her speculative spirit, which delights in operations of a large scale" (quoted in Daniel, p. 19). Writing of the period 1850–1860, Opie adds: "The national agrarian myth, still attached to the 160-acre quarter for the family farm . . . was not California's primary mode" (p. 133). In the eyes of critics, the land hogs of the Central Valley—the Henry Millers, the William Chapmans, the Isaac Friedlanders—produced an economy marked by wild swings and insidious class divisions.

194   **"Most important":** Worster, *Rivers of Empire*, p. 130. **"Irrigation crusaders":** The life of the leading crusader, William Smythe, is discussed in Worster, *Rivers of Empire*, pp. 118–125; Hundley, pp. 114–115. See, too, Smythe's *The Conquest of Arid America.*

194   **John Wesley Powell:** Biographical details from Stegner, *Beyond the Hundredth Meridian*; Worster, *A River Running West*; Worster, *Rivers of Empire*, pp. 132–143.

195   **Doubled Powell's:** Worster, *A River Running West*, p. 569. Originally, the 1902 Act was to remain true to Powell's vision of 80-acre farms. But the railroads objected, "fearful that 80 acres would sound dreadfully small to

prospective settlers," so it was raised to 160 (Worster, *Rivers of Empire*, p. 161). **Somewhat arbitrary:** See "The 160-Acre Limitation Policy," by L. T. Wallace, University of California at Berkeley, June 1967, PST. **Land Survey Ordinance of 1785:** Opie, p. xvii. **Archetypal "family farm":** The 160-acre mark was incorporated in the Preemption Act of 1841, which let western settlers lay claim to unsurveyed plots and later buy them from the government (*Landownership Survey*, p. 69; Griswold, p. 140; Opie, p. 56). Next was the Homestead Act, which gave people 160 government acres free and clear if they lived on the land for five years and improved it by putting up a dwelling and growing crops (*Landownership Survey*, pp. 73–74; Griswold, p. 140; Gates, *The Jeffersonian Dream*, pp. 40–55). On pp. 65–67 and p. 147, Opie discusses shortcomings with the Homestead Act. He notes that from 1862 to 1956, "a disappointing" 147 million acres had gone to homesteaders; by comparison, more than 450 million acres had been sold to private interests. In addition, the Carey Act of 1894, which encouraged irrigation by the states, likewise specified that land shouldn't be sold off in hunks of more than 160 acres (*Landownership Survey*, p. 75). A history of government land programs is also found in Clawson, *Uncle Sam's Acres*.

195   **"So bothersome":** Quoted in Rowley, *Reclaiming the Arid West*, p. 104. Worster, *Rivers of Empire*, pp. 167–169, argues that Newlands and many others behind the 1902 act were hardly promoters of a great agricultural democracy. They were "overwhelmingly an elite group promoting an elite program," he writes. "Their overriding aim was to enlarge, for their own ends, the country's wealth and influence. To secure peace and stability at home. To earn profits at home and abroad. And to pursue power, always power."

195   **Newlands expected:** Worster, *Rivers of Empire*, p. 160. **But private property:** For a rundown of the rules, see the *Landownership Survey*, pp. 29–57.

195   **"Verges on fraud":** Worster, *Rivers of Empire*, p. 174.

195   **Part of the problem:** Koppes, "Public Water, Private Land," *Pacific Historical Review*, November 1978, p. 611.

196   **To crack down:** Opie, p. 118. **"A dead letter":** Koppes, p. 613. **Abetted Congress:** Koppes, p. 613. Ickes, who had never even been told by his subordinates about the trio of exemptions in Colorado and Nevada, was enraged to learn of them belatedly (Koppes, p. 613; Watkins, *Righteous Pilgrim*, p. 777).

196   **Walter Packard:** For details on the man, see Worster, *Rivers of Empire*, pp. 249–251. **"Basically unsound":** Packard, *The Economic Implications of the Central Valley Project*, pp. 86. **"I asked him":** PTOH, p. 145.

196   **Obsessed:** See "The Power of a Tenacious Man," *The Nation*, October 12, 1974. Taylor spent much of the rest of his long career studying the history of the 1902 act, testifying before Congress about it, quarreling over its fine

points with big farmers, and writing about it in the most hallowed terms. Among his many essays on the topic was one that went right to the question that Boswell was now apoplectic over: "Whose Dam is Pine Flat?" That article originally appeared in *Pacific Spectator* 8 (1954). It is included in Taylor's collection, *Essays on Land, Water and the Law in California*. **Access to Ickes:** PTOH, p. 181. **"Gapes of Wrath families":** Koppes, p. 617; also cited by Hundley, p. 260.

196   **"Slipped it through":** PTOH, p. 170.

196   **Elliott was:** Biographical details are from interviews with Elliott's son, I. J. Elliott, and Jack Kelly, who grew up in Tulare and included Elliott's picture in a museum he founded to honor heroes of the valley's African American community.

197   **"Trying to socialize":** From Koppes, p. 617.

197   **Downey had taken:** A wonderful description of Downey is provided in Starr, *Endangered Dreams*, pp. 200–201. **"I abhor":** Downey, *Onward America*, p. 83. **"More liberal":** PTOH, p. 119.

197   **Having stumped:** Starr, *Endangered Dreams*, p. 211. **He pushed:** Auerbach, p. 179. **"Grotesquely inappropriate":** Downey, *They Would Rule the Valley*, p. 5. Also see Cooper, pp. 156–157.

198   **Generating headlines:** See, for instance, *Business Week*, May 13, 1944. **Two-pronged:** Letter from FS to JGB, April 12, 1944, FSP.

198   **May 10:** JGB's itinerary is included in an April 12, 1944, letter from JGB to FS, FSP.

198   **"Rumors are circulating":** March 29, 1944, letter from JGB to FS, FSP. JGB reiterated this scenario in his April 12 letter.

199   **"Down our throats":** JGB to FS, March 29, 1944. **"A tough fight":** JGB to FS, April 12, 1944.

199   **"Present extremity":** Letter from FS to Lieutenant Colonel P. M. Robinett of the Army General Staff, December 15, 1941, FSP.

199   **Will Clayton:** Details from Fossedal, *Our Finest Hour*. **"Acts of statecraft":** Fossedal, p. 3. Clayton also played a key role in the Bretton Woods agreement on international monetary policy and trade, as well as the formation of the Truman Doctrine, which brought economic and military aid to Greece and Turkey when they faced threats from the Soviets in the late 1940s.

199   **"Cruel and heartless":** April 1, 1942, letter from FS to Leslie Groves, FSP. FS's closeness with Groves is underscored in a series of letters over the years between the two men that can be found in Groves's personal papers, NA.

200   **"Having friends":** FS to Groves, October 13, 1941, FSP; Groves Papers, NA. **Manhattan Project:** Lawren, *The General and the Bomb*, p. 22.

200   **Twenty-three billion:** From the official recommendation that FS be

awarded the Distinguished Service Medal, FSP. **Tax Question:** Transcript of phone conversation between FS and JGB, June 26, 1942, FSP. **"Bottleneck":** October 1, 1942, telegram to FS, FSP.

200   **"A real service":** Robinson to FS, June 2, 1943.

201   **"National proportions":** *Business Week,* May 13, 1944. **"Hot June days":** PTOH, p. 172. **The "welfare":** *Landownership Survey,* p. 87.

201   **House-Senate conference:** The Senate's version of the rivers and harbors bill did not include any exemption from the acreage limit, despite Sheridan Downey's efforts. However, the Elliott amendment was restored in conference. **College classmate:** PTOH, p. 174. The basic story line of the Elliott rider is covered by many sources, including De Roos, *The Thirsty Land,* pp. 94–95; Koppes, pp. 617–619; Worster, *Rivers of Empire,* p. 252; Hundley, p. 261.

201   **"Unchecked force":** Auerbach, p. 190. **"If you press":** PTOH, p. 176. **"Gravy train":** PTOH, p. 174.

201   **Should be "built by":** Roosevelt to the secretary of war, May 16, 1944.

202   **"Insubordination":** *Harper's,* August 1949. **"Into the bung":** Stegner, p. 359.

202   **Itself acknowledged:** The bureau's acting chief engineer, in a March 20, 1940, telegram to agency officials in California said that the 160-acre limit wouldn't apply on the Kings River. The "present trend," he explained, "is to eliminate restriction on projects where water is predominantly . . . storage and servicing of water previously appropriated as distinguished from projects where entire water supply is furnished to new lands" (Kings River files, National Archives, Rocky Mountain Region, Denver). **"No need whatever":** From Kaupke's testimony before the Senate Committee on Commerce, May 29 to June 15, 1944.

202   **"Not irrigation":** From his testimony before the Senate Committee on Commerce, May 29 to June 15, 1944.

202   **Request nixed:** Maass, pp. 227–228.

203   **The secret was:** PTOH, pp. 182–183.

203   **More discussion:** Trying to make sure that there would be no question about Congress's intent, Interior officials arranged for a colloquy on the Senate floor between the acting majority leader, Joseph Lister Hill, and John Overton, the chairman of the flood control subcommittee (PTOH, p. 183; Maass, p. 236). "There still seems to be some confusion on the part of some senators with reference to the application of reclamation laws in regard to some of these projects," Hill said, his lines fed to him by Ickes's staff. "No project in this bill which may include irrigation features," replied Overton, following the same script, "is exempted from the reclamation laws." **"Sweeping defeat":** Maass, p. 237.

203   **Even Senator John Overton:** July 16, 1946, letter from Overton to Reclamation Commissioner Michael Straus, FSP. **Attorney general:** December 15,

1958, letter from Attorney General William P. Rogers to Interior Secretary Fred A. Seaton. Rogers did find that section 8 "makes the reclamation laws applicable . . . even if no additional works are required to be constructed" in order to make irrigation benefits available.

204   **Within the scope:** Quoted in a July 12, 1950, letter from Commissioner Straus to Senator Paul Douglas, PST.

204   **"Reluctantly":** Statement by the president, April 2, 1945.

204   **"Terrific pain":** Goodwin, *No Ordinary Time*, p. 602.

204   **"2:50 Thursday":** An undated note from Rothery to FS, FSP.

205   **"Leans definitely":** Letter from FS to JGB, April 24, 1945.

205   **"Clearly established":** Truman to the secretary of war, June 2, 1945.

205   **Senate appropriated:** For details on this and Truman's impounding of the money, see Maass, pp. 243–246.

206   **Distrust abounded:** There was "an undercurrent of feeling in the Bureau that we were trying to reduce the amount" the farmers should pay "and thereby favor special interests," the corps noted in its summary of the discussions (July 16, 1946, memo entitled "Report on Conference re Allocation of Cost of Kings River Project, California" to the chief of Engineers from underlings George Beard and Carter Page, FSP).

206   **Fifteen conditions:** These were laid out at a July 30, 1946, public meeting in Fresno. A transcript of that meeting is in PST. The list of conditions is also included in an October 28, 1946, report by the engineers representing the Kings River water users.

206   **"Catechism":** July 19, 1946, letter from Straus to Senator Overton, FSP. **"May not apply":** From "Bureau of Reclamation Response to Statement of Terms Submitted by Kings River Water Association Covering Participation in Kings River Project, California," October 1946, PST.

207   **"Difficulty to delineate":** September 12, 1947, bureau memo from the chief of the Economics and Statistics Division to the regional director (Bureau of Reclamation regional library, Sacramento, California).

207   **Notoriously loose:** Reisner, pp. 137–139, offers a great description of Straus. **"Mike Straus pose":** Quoted in Reisner, p. 138.

207   **"Nothing on earth":** Reisner, p. 137.

207   **"Historic policy":** From a Straus essay entitled "Family Farming Becomes a Made-in-America Commodity." A typescript copy is in PST. In the late 1940s, Straus even got to play martyr. Sheridan Downey, as part of another ill-fated attempt to exempt the Central Valley Project from the acreage law, tried to pressure Straus and then rammed through a bill terminating his salary as commissioner. It was later restored with back pay (PTOH, pp. 203–211). The bureau's regional director, Richard Boke, also had his salary temporarily cut off.

207   **"Technical compliance":** See Straus's testimony before the Senate Committee on Public Lands, May 5–June 2, 1947; Koppes, pp. 624–636. Also see Worster, *Rivers of Empire*, pp. 254–255; Hundley, pp. 265–266.

208   **"Grave mistake":** December 20, 1948, memo, from Taylor to Boke (Bureau of Reclamation regional library, Sacramento, California). In PTOH, p. 207, Taylor notes that he resented Straus's "readiness to compromise."

208   **"Getting dried out":** Cited in an August 19, 1947, letter from Sid Harding to Louis Robinson, STH. **But to Taylor:** See undated memo, marked "Confidential," commenting on positions taken by representatives of the Tulare Lake Basin Water Storage District, PST. **Taylor persuaded:** PTOH, p. 207. Taylor, while criticizing Straus for his willingness to compromise on the 160-acre limit, praised regional director Boke as someone who "was staunchly in support of the law under all the pressure."

208   **"Proper place":** Harding to FS, September 19, 1946, FSP.

209   **"No project":** FS to Harding, September 17, 1946, FSP. **Break the logjam:** As usual, Sherrill was operating with the benefit of inside information. A week before writing to Harding, he had received a confidential correspondence at his home from R. A. "Speck" Wheeler, the corps chief of engineers in Washington. In it, Wheeler said that the army's $16.75 million estimate had been "frankly an effort to approach an agreement with the Bureau." But "after considerable further study, we now feel that the Bureau's contention is entirely too high and that a repayment of from $12 million to $14 million would be reasonable from both federal and local standpoints."

209   **Elliott telephoned:** Memo from Robinson to FS, February 6, 1947, FSP.

209   **Truman released:** U.S. Bureau of Reclamation Region II "Press Digest" item, PST.

209   **Tortured sounds:** The description of the groundbreaking, including all quotes, is from the FB of May 28, 1947. **Charles Kaupke:** Background on him is from *Forty Years on Kings River*; interview with his son, John Kaupke.

209   **Gauging and divvying:** Kaupke had been instrumental in putting together historic agreements in 1927 and 1949 that brought together the heretofore feuding farmers up and down the Kings, setting up a schedule by which water from the river was to be diverted. (See Kaupke's paper, "The Kings River Problem," Kings River Collection, Water Resources Center Archives, University of California at Berkeley.)

210   **"Flooding the country":** June 12, 1947, letter, from FS to Representative Clair Engle, FSP.

210   **Such recriminations:** Details on the acrimonious contract talks can be gleaned from numerous pieces of correspondence between the bureau and the Kings River Water Association in PST; Harding's "History of Work on Kings River," pp. 70–75, STH; Maass and Anderson, pp. 256–263.

"**Knocked him flat**": Quoted by Maass and Anderson, p. 258. The bill to integrate Pine Flat into the CVP was pushed by White in 1950. For more details on that battle, see PST; "Water Supply of the South Fork of Kings River and Tulare Lake Under Past Conditions and Under the 1949 Kings River Agreement," by S. T. Harding, September 1950, STH; Harding's "History of Work on Kings River," pp. 102–103, STH. For further details on White, see the papers of Representative Bud Gearhart, Fresno Historical Society.

211   **Completed in 1954:** This, Pine Flat's final price tag and information on the other three dams is found in *Water Resources Development in California*, Army Corps of Engineers, 1995.

211   **Go-between:** See *The Memoir of Bernie Sisk*, pp. 90–93.

CHAPTER 11: BLUE BLOOD

212   "**Come on in here**": This anecdote and all direct quotes in the chapter by Spud Williamson are from WW.

212   **Tudor mansion:** All details on the house are from visits to the site; interviews with the current owner, Geneva Thornton; a January 30 article on the mansion in *California Arts & Architecture*; and material from "Myron Hunt, 1868–1952: The Search for A Regional Architecture," a 1984 exhibition sponsored by the California Institute of Technology's Baxter Art Galley. The latter two items are found in the Architectural Drawing Collection, University of California at Santa Barbara. "**More responsible**": *Time*, October 2, 1944.

212   "**Sage of Penfield**": GHJ obituary, January 21, 1916.

213   "**Not a rebel**": December 31, 1952, note from Munro to Robert Millikan, Millikan Papers, CTA.

213   **Harry Chandler:** Background on Chandler is from "The Ancestry of Harry Chandler," which was put together by family member Gwendolyn Garland Babcock (Pasadena Historical Museum Library); Gottlieb and Wolt, pp. 121–237; Halberstam, pp. 94–122; Starr, *Material Dreams*, pp. 102–103; Kahrl, *Water and Power*, pp. 186–188; Reisner, pp. 56–57; *Time*, October 2, 1944; *Newsweek*, April 30, 1956; *Who's Who in California*, 1939–40. For a particularly critical view of Chandler, see Bonelli's *Billion Dollar Blackjack*.

214   "**Good for real estate**": Halberstam, p. 115.

214   **HOLLYWOOD sign:** Williams, *The Story of Hollywoodland*, p. 15. The sign, constructed in 1924, originally spelled out HOLLYWOODLAND to advertise a real estate development. Williams (p. 36) notes that maintenance of the sign was discontinued in 1939. Ten years later, the Hollywood Chamber of Commerce restored it to just read HOLLYWOOD. "**To count them**": Gottlieb and Wolt, p. 125.

215   "**Marriage was perfect**": McWilliams is quoted in an undated note from

JGB's secretary, Alice Rothery, to FS, FSP. **"Widow's table":** WW. **First husband:** Williamson's biography is in the *Pasadena Community Book,* 1943.

215   **Physical education:** LAT, April 26, 1943, in an article noting Ruth's election to the Pomona College board of trustees. **"Sit up":** Interview with Sue (Williamson) Dulin.

215   **Stand her ground:** Halberstam, p. 280; McDougal, *Privileged Son*, pp. 191–192. **"Snow her":** Interview with Casey.

215   **"Homework done?":** Interview with Norman Williamson.

215   **"Getting old":** Undated note from Rothery to FS, FSP. **"Loved Alaine":** JGB2.

216   **Imperious reputation:** McDougal, in his biography of Ruth's nephew Otis Chandler, colors Ruth as "pompous" and "often unprincipled." A similar portrait emerges in "The Chandlers: L.A.'s First Family," a series of articles in the September-October 1999 *Santa Monica Mirror* by Mark Dowie. **Down-to-earth:** This is in fact how many Boswell managers and folks back in Penfield remember her. **Rusted pickup:** Interview with Greensboro resident and Boswell relative Miles Walker Lewis. The truck belonged to the wife of Kyle Smith, a close friend of JGB's. **Expressed concern:** JGB2.

216   **Chandler eschewed:** Halberstam, p. 96.

216   **Twelve richest men:** Gottlieb and Wolt, p. 125. **"No Easterner":** Halberstam, p. 94.

217   **"Kept man":** Interview with Norman Williamson. **Real fetish:** JGB2.

218   **"Georgia cotton patch":** As quoted in Ruth Chandler's obituary, LAT, December 11, 1987.

218   **"In the army":** Interview with Norman Williamson.

218   **"A good feeder":** Interview with Walter Boswell's daughter-in-law, Winthrop P. (Pat) Boswell.

218   **Smithsonian:** The slide, which is part of the Garden Club of America Collection, is housed at the Archives of American Gardens in the Smithsonian's Arts and Industries Building. The garden was copied from Chateau La Napoule in Monte Carlo. **Florence Yoch:** See the brochure text from "Personal Edens: The Gardens and Film Sets of Florence Yoch," a 1997 show at the Huntington Library, which is available through the California Historical Society. **Plant magnolias:** This and other details on how JGB oversaw the grounds at Oak Grove are from WW.

219   **"Buckingham Palace":** Interview with Winthrop P. Boswell. **"Very pleased":** February 8, 1985, letter from Walter's son James O. Boswell to his cousin Virginia (Boswell) Mason. **"That contrast":** DV with Josephine (Jody) Boswell Larsen.

219   **"I'll go anywhere":** DV with Kate Boswell.

220   **"She's my ward":** DV with Josephine (Jody) Boswell Larsen.

220   **Wed a sailor:** Interviews with Virginia (Boswell) Mason and Louise

Boswell. JGB also sent their father Herbert—the family black sheep who had been trained as a lawyer but had wound up a car dealer in Memphis—an extra $300 every month.

220   **Lots of friends:** WW.

221   **Biggest stars:** See Goodstein, *Millikan's School*; Starr, *The Dream Endures*, pp. 61–89. **"Next to Einstein":** Goodstein, p. 105. **Luminaries:** Starr, pp. 86–87; Millikan, *The Autobiography of Robert A. Millikan*, pp. 226–227, pp. 238–250. **The Colonel joined:** California Institute of Technology Catalogue, 1951–52, CTA.

221   **"Contaminated":** February 7, 1949, letter from JGB to George Benson, GB. **"Menace to":** July 22, 1949, letter from Millikan to JGB, Millikan Papers, CTA. Also see Millikan's speech at the dedication of the Earhart Foundation Plant Research Laboratory in which he discusses "tendencies toward totalitarianism" in the United States, CTA.

221   **George C. S. Benson:** Details on his life are gleaned from an obituary put out by Claremont McKenna, March 23, 1999. **"The pinkos":** February 7, 1949, letter from JGB to Benson, GB; see also January 26, 1949, letter from Benson to JGB, GB, in which he cites his speech "How Pink Are Our Colleges?" and notes "our frequent conversations on this subject." **Boswell Professorship:** August 28, 1948, agreement establishing the professorship and other documents in GB; Starr, *Commerce and Civilization*, p. 60. **"See that the graduates":** April 18, 1950, letter from Benson to JGB, GB.

222   **"Goose feathers":** Ruth Chandler obituary, LAT, December 11, 1987. **"Your desires":** December 6, 1948, letter from Benson to JGB, GB.

222   **Charitable foundation:** The foundation's articles of incorporation and a series of tax returns beginning in 1950 are located in the Foundation Center Historical Foundation Collection, Indiana University-Purdue University, Indianapolis. **By 1999:** Data from the Foundation Center, New York. **Loan funds:** JGB's obituary, GHJ, September 19, 1952. **Bestowed $50,000:** JGB's support is reflected in a wide range of documents found in the papers of Spright Dowell, the former president of Mercer University, which has kept up the Penfield cemetery (University special collections, Macon, Georgia). **Rural hospital:** Information provided by the Minnie G. Boswell Memorial Hospital; various articles in the GHJ; interview with Arthur Stewart.

223   **"Outstanding commitments":** February 11, 1939, letter from JGB to Millikan, Millikan Papers, CTA. **"Come a time":** Memo in BCA.

223   **"Cotton Rush":** *Fortune*, May 1949.

223   **More than tripled:** Production data supplied by Kent Lanclos, economist for the National Cotton Council, Memphis. **Surpassed Alabama:** FS speech to unidentified industry group, October 11, 1951, FSP; *Collier's* article, "Away Out West in the Land of Cotton," June 23, 1951. **Topped citrus:** *Fortune* article; *Life*, January 1, 1951. **Motion picture production:** FS speech. **"10 greatest":** *Collier's* article.

224 **"Identified with gold":** FS speech. **Suzanne Howell:** *Fortune* article; *The Nation,* February 19, 1949. **Annual convention:** *Fortune* article; *Time,* March 21, 1949. **Noting the ascent:** See the aforementioned articles, plus *Newsweek,* March 22, 1948.

224 **"Gold strike":** *The Nation,* February 19, 1949.

224 **"Smart operator":** *The Nation,* February 10, 1951.

224 **"Little panicky":** June 29, 1945, note from Rothery to FS, FSP. **First choice:** DV with Kate Boswell; interview with Oleta Gordon, who lived with Billy and took care of him at the end of his life. **Empress of Britain:** The trip is noted in CN, September 22, 1939. Don Keith, a good friend of Billy's, said in an interview: "You see, J. G. tried to get Billy to be president of the company. Took him across to England, France and all that. Big trip with him."

225 **"Rat's ass":** Interview with Gregory. **"On his own":** DV with Kate Boswell. **Told the story:** Interview with Oleta Gordon.

225 **Nacho Gomez:** JGB2. **"My scoutmaster":** Oleta Gordon interview. **Others swore by:** Interviews with Corcoran residents Verdo Gregory, Duane Dye and Lino Cardoso. **"Too numerous":** Oleta Gordon interview.

225 **Grown disenchanted:** JGB2. **Start a rodeo:** Interview with Verdo Gregory.

226 **"Even think":** JGB2.

226 **"Charting a course":** May 29, 1945, letter from FS to JGB, FSP. **"Conservative position":** June 2, 1945, letter from JGB to FS, FSP.

227 **Peek into the unsaid:** Rothery's secret notes to FS, which run from June through August 1945, are in FSP.

228 **"The nation needs":** June 21, 1945, letter from FS to Rothery, FSP.

228 **Henry Crown:** Background from Crown's obituary in the *Chicago Sun-Times,* August 15, 1990. **Close friendship:** Interviews with Steve Sherrill and Lester Crown, Henry's son. **Billionaires:** The *Chicago Sun-Times* put the Crown family net worth at $2.8 billion in October 2000. **"Was interested":** September 20, 1945, letter from Crown to attorney Robert F. Gaylord Sr. of San Francisco, FSP.

228 **Crown later told:** Interview with FS's son, Steve Sherrill. **Own version:** From Crocket's unpublished memoir. **Merrill Lynch:** JGB2; interview with the Boswell accountant from that time, Wally Erickson.

229 **"Rest of the day":** Interview with Erickson. **"Popping off":** McWilliams is quoted in a June 28, 1945, letter to FS from W. B. Coberly of the California Cotton Oil Corporation, FSP.

229 **Shingles:** JGB and Ruth canceled a trip to Europe in August 1949 because he was hospitalized (GHJ, August 12, 1949; August 30, 1949, letter from FS to Mercer University's Spright Dowell). **Assist in the drafting:** Interview with Steve Sherrill. **A third would go:** Steve Sherrill interview; JGB2; DV with Walter's son James O. Boswell.

229 **$225,000 donation:** Reflected in numerous pieces of correspondence between JGB and G. W. Beadle, chairman of the Division of Biology at Cal-

tech, CTA; minutes from the March 3, 1952, meeting of the Caltech board of trustees (made available by the president's office).

230   **Old-fashioned:** WW; Norman Williamson interview.

230   **Dr. Williams asked:** This whole scene is from JGB2.

231   **"Worth a damn":** JGB2.

231   **The night nurse:** Interview with Walter's daughter-in-law, Winthrop P. Boswell. When JGB's estate paid off his last bills, it did in fact write checks to four nurses, including one Mary O. Sherman.

231   **Well sodded:** As reflected in an August 17, 1942, letter from FS to JGB, FSP. **Board proclaimed:** Its statement is found in the Millikan Papers, CTA. **"Analytical mind":** GHJ, September 19, 1952. **"Philanthropic acts":** CJ, September 12, 1952. **All stood:** Board of trustees meeting minutes for January 5, 1953 (made available by the president's office).

231   **Wasn't the one:** Records of the Los Angeles Superior Court; JGB2; DV with Walter's son James O. Boswell. **Vowed to contest:** JGB2.

232   **"The great exploiter":** Interview with Steve Sherrill. **"Sucker":** Interview with Martha Sherrill.

SUMMER

233   **"Me? Pick cotton?":** Mitchell, *Gone with the Wind*, p. 448.

CHAPTER 12: SUMMER

235   **From the orchards came:** All figures are from California Agricultural Statistics Service, 2001; County agricultural commissioners' data, 2001.

236   **August 9, 1999:** Details on the accident and its aftermath, including quotes from the families, are from FB of August 10 and August 22, 1999; LAT of August 10 and 11, 1999.

237   **California Senate passed:** LAT, August 20, 1999.

238   **Sensors inside the snouts:** All details on Boswell's equipment and growing processes are from numerous interviews with Grewal and other Boswell executives.

243   **$30 million a year:** Interview with Grewal.

243   **Ciba-Geigy:** Noted in *Forbes*, April 17, 1989. **Dow Chemical:** Phytogen Seed Company is a joint venture of Boswell and Dow Chemical's Dow Agro-Sciences unit.

245   **Budding soil scientist:** Biographical details on Doan are from interviews with him; JGB2.

245   **Teamed up with Bell:** Details on Doan and Bell's growing practices are from interviews with the two of them, as well as with Mark Grewal. **"Cousinitis":** Interviews with Doan, Bell and Grewal.

246   **"Esprit de corps":** Interview with John Grant.

247   **Idaho sheep farm:** Interview with Bickett. **Good Neighbor Policy:** JGB2; interview with Alta Ober, Rice's wife. **Len Evers:** JF; JGB2; DV with Evers.

247   **Paul Athorp:** Interview with Athorp; JGB2.

247   **Twenty-eight patents:** Listed in Vandergriff, *Ginning Cotton: An Entrepreneur's Story*, pp. 285–286. **Land's End catalog:** April 1997. **Stanley Barnes:** All material on him is from SB.

248   **Data processing equipment:** Noted in "Historical Sketch," found in BCA. **Plastic swaddle:** Interviews with Boswell executive Ross Hall, A. L. Vandergriff; **Killed off a whole industry:** See Vandergriff, p. 56. **Marry the virtues:** Interview with Ross Hall; visit to the Boswell gins in Corcoran. Starting in the late '90s, Boswell began using equipment made by a third company, Consolidated Cotton Gin Company.

249   **Biotech in a blender:** Interview with Bell.

249   **Using a Landsat:** Material on the project is found in BCA. From June of 1973 until late the following year, GE personnel came to Corcoran to help manage Boswell's crops through this space-age imagery. For GE's big-city technicians, it was a chance to gather some knowledge about cotton, while Boswell's men certainly gained some new insights, as well. "More precisely," noted project manager Earle Schaller, "the GE people learned about nodes and bolls, and the farm people learned about spectra and pixels." (From "Agricultural Applications of Remote Sensing: A True Life Adventure," a paper delivered by Schaller to a 1975 NASA symposium, BCA.) **Satellite receiver:** See *California Vegetable Journal*, June 2000.

249   **"To check it":** Quoted in *Forbes*, April 17, 1989.

251   **"Two or three times a day":** Interview with Tom Lewis, former minister of lands for New South Wales.

252   **List of nine goals:** JGB2 provided the authors with the list.

252   **Building a dairy:** Details on the proposed dairy project are from FB special report, *King of Kings*, October 31, 1999.

253   **"Environmentally sensitive":** FB, October 31, 1999.

253   **Attorney general's office:** FB, October 31, 1999.

253   **Carelessly run afoul:** JGB2. **Robert Rowling:** Biographical detail from "Rowling Balances Mix of 'Moral Spirit' and Business Savvy," *San Antonio Express-News*, February 9, 1997; "Can an Oil Man Burnish Image of Hotel Chain?" *Wall Street Journal*, January 29, 1996. **"Print them like him":** Interview with Rowling.

254   **Boswell invited Rowling:** Interview with Rowling; JGB2. **"Really, God owns":** *Forbes*, October 12, 1998. **$20 million:** Interview with Rowling.

254   **"Threw him the keys":** These and the other quotes are from JWB. **Tears to her eyes:** JWB.

CHAPTER 13: BOOT HILL

256   **Western Cotton Journal:** March 1929 issue.

257   **Mexican guest workers:** Stein, p. 37. **Japanese cotton pickers:** For a history of Japanese farmers in the valley, see Masumoto, *Country Voices: The Oral History of a Japanese American Family Farm Community*; McWilliams, *Factories in the Field*, pp. 104–116. **Okie pickers had moved up:** Gregory, pp. 182–190. The economic advance of one Okie family in Central California, the Tathams, is told in Morgan, *Rising in the West*.

257   **Forty-one of the machines:** CJ. For background on the development of the mechanical cotton picker and its impact on society, see Lemann, *The Promised Land*, pp. 4–7; Street, *The New Revolution in the Cotton Economy*; numerous articles, letters and reports found in PST. **Product from American Cyanamid:** CJ, August 18, 1944.

258   **Robert Parker:** Details on Boots Parker and his exploits are from interviews with Parker and Nancy Irons, Gus's widow.

258   **First black settlers:** The history of the black and Native American communities in Oklahoma is from Burton, *Black, Red and Deadly*; Johnson, *Black Wall Street*; interviews with Idabel historian Lewis Coleman and author Art Burton.

259   **Tulsa Tribune trumpeted:** Cited in Johnson, p. 50.

260   **That lynching:** Details from research shared by historians Lewis Coleman and Charles Clark.

261   **Gus and Walter Irons:** Their story is from interviews with Gus's widow, Nancy; his daughter, Marcia Fulkerson; and his sister, Emma Wilds.

261   **Sheriff Irons was beloved:** Interviews with Robert Parker and Luke Etta Hill.

262   **"Took me everywhere":** Interview with Fulkerson.

262   **"Going to California":** Interview with Hill.

263   **"Oh my goodness":** Patterson's story is from an interview with her.

263   **"Guam":** Interview with former FB and LAT reporter Ron Taylor. **Las Moscas:** Interview with Vince Banales.

264   **Between 30,000 and 40,000:** From U.S. Census data. **7,000 of them settling:** From U.S. Census data. **Number of Mexicans:** *Mexican Farm Labor Program, Consultant's Report*, President's Committee on Migratory Labor, October 1959.

265   **"Until you met the plow":** Interview with Grigsby.

265   **"Game of tonk":** Interview with Noble.

265   **"The good Lord":** Interview with Handsbur.

266   **Body parts pressed:** Interview with Beulah Handsbur.

266   **"Dad was tough":** Interview with Allen.

267   **"Oh, Lord, if you just save me":** Interview with Wilds.

267   **Howard and Gertha Toney:** Their story is from interviews with Gertha Toney; daughter Dorothy Toney; and son Bobby Toney.

270   **Edwin Matheny:** Details on his land-sale business are from an interview with his daughter, Barbara Girton.

272   **Not Forrest Riley:** Details on his life and farming operation are from interviews with Fred Odle; Riley's daughter, Frances Nanasy.

273   **Tournament of Roses parades:** CJ, January 9, 1953. **Quickie in the fields:** From interviews with Fred Odle and Lino Cardoso.

273   **"Weren't equals":** Interview with Nanasy.

273   **Allensworth:** Information from Allensworth State Historic Park.

273   **Honky-tonks and whorehouses:** Background on this part of town and its colorful characters is from interviews with Fred Odle, Willie Johnson, Evelyn Grigsby and the Toney family.

274   **Snake killed three men:** Interview with Odle.

275   **Began an affair:** Details on this incident are from CJ, November 24, 1960.

275   **Two grad students:** Their story is from interviews with Anne Chomet Petrovich and Chuck Chomet; the files of Chuck Chomet.

276   **More members in Corcoran:** February 26, 1963, letter from Cesar Chavez to Sue Carhart (Copy courtesy of Sue Carhart Darweesh).

277   **"People of Sunny Acres":** August 22, 1964, letter from Howard Toney to Paul O'Rourke (Copy courtesy of Chuck Chomet).

277   **More than 60 percent:** CJ, November 19, 1954.

278   **Jim Crow, California-style:** For a history of blacks in California's Central Valley, see Mark Arax's LAT series on the "Black Okies," August 25–27, 2002. **"Klan Will Hold Picnic":** CJ, July 3, 1931. **"Proud to be associated":** CJ, May 12, 1922.

278   **Ruby Hill:** His story is from an interview with Hill.

279   **Did it slick:** Interview with Evelyn Grigsby.

280   **"Hayrides and cookouts":** Interview with Fortson.

280   **"Those darkies":** Interview with Janie Toney.

281   **Election in 1962:** Details are from interview with Sue Carhart Darweesh; numerous issues of the CJ, March-April 1962.

282   **Percy Whiteside:** Details on him are from an interview with former CJ part owner Henry Coultrap.

283   **"Socialist poppycock":** CJ, April 5, 1962.

283   **"Reach into his pocket":** Interview with Hatley. **"Grandmother's mouth":** LSal.

284   **Lorenzo G. Luna:** His story is from interviews with the Luna family.

284   **"God damn geese":** Interview with Ron Taylor. **Employing birds:** CJ, June 5, 1953.

285   **Shut their doors:** Noted in the CJ. When J. C. Penney closed in the summer of 1961, it cited, among other factors, declining sales amid farm mechanization (CJ, July 22, 1961). **$46,000 federal loan grant:** CJ, July 13, 1967.

CHAPTER 14: THE STUD

286   **Rain poured:** This episode is drawn from interviews with John Grant and Chris Reynolds.

286   **Some variation:** Interviews with Jim Maddox and Dick Jones, among others. **"The Cheetah":** Interview with John James. **Quit smoking:** JGB2. **Dean River:** Drawn from interviews with several of those on the trip, including JF, SB and Jon Rachford.

287   **Had worked for the company:** For his part, Reynolds had worked for Boswell in California, Australia and Arizona. His dad, Bob, was also pals with JGB2. The two formed a venture, dubbed Reybos, that drilled for natural gas in western Colorado. A former Stanford football star (where he played alongside his best friend, Pop Grant), Bob Reynolds founded Golden West Broadcasters with Gene Autry in 1952 and then served as president of the California Angels baseball team. (From his obituary, LAT, February 10, 1994; interview with Chris Reynolds.)

287   **"Put me to work":** Interview with Grant.

288   **"Gutsy trekker":** Interview with Bonnington. **"One of the fittest":** Bonnington, *Kongur: China's Elusive Summit*, p. 104.

288   **John Muir Trail:** JWB. **"Got sick":** JGB2.

288   **Even as a boy:** All recollections of Boswell's early life, unless otherwise noted, are from JGB2.

291   **Depleted squad or not:** Interviews with Stanford athletic and public affairs officials.

291   **"As a baseball player":** Interview with Bobby Brown.

293   **Gill brothers:** The Gill Cattle Company, based in Exeter, California, owned or leased more than 6 million acres of land at one point (Information from the Exter, California, Chamber of Commerce).

294   **"Madam chairman":** This scene is rendered entirely from JGB2.

294   **"In the palm":** Interview with McInerny.

295   **"Knee-jerk reaction":** JGB2. **"Matter of economics":** DV with Evers.

295   **45,000 or so:** JGB2; interview with former Boswell executive Wally Erickson. **Manifest Destiny:** Interview with Bruce Roberts, University of California Extension Service farm adviser for cotton for Kings County.

295   **Flood in 1952:** Details on the flood are found *in J. G. Boswell v. Commissioner of Internal Revenue*, filed 1956 in the Tax Court of the U.S., docket no. 61846. **Rigging prices:** CJ, April 3, 1953; correspondence and other documents related to the matter are found in Harlen Hagen's papers, Califor-

nia State University at Bakersfield. Boswell and six other San Joaquin Valley cotton companies faced an antitrust suit in 1956, but that case was dismissed (CJ, August 30, 1957).

296   **"Interference from any":** Quoted in CJ, December 25, 1953. **Contract was signed:** CJ, June 25, 1954.

296   **Union struck:** HS, October 12, 1954.

296   **Those who crossed:** Details from an interview with James Edward Hatley, who was among the laborers who stayed loyal to the company and went to work. **Set ablaze:** CJ, October 15, 1954; HS, October 25, 1954. **Sugar and honey:** Noted in a handwritten list—"Fire, Matches, Tacks-Nails, Stolen Car, Trailer Fire, Acid, Dynamite, Sugar-Honey"—found in BCA. **Hear a boom:** Interview with Alta Ober. See also CJ; October 29, 1954. HS, October 25, 1954. **"Big firecrackers":** Interview with Jones.

297   **"She'd cry":** Interview with Rudy Castro.

297   **$1.3 million:** Net income for year ended June 30, 1952. It earned $1.6 million during the following twelve-month period (Consolidated financial statements, BCA).

297   **"Screw yourselves":** JGB2; interview with Charles Jones, former Machinists official. See also "Conferences Fail to Solve Strike Here," CJ, October 29, 1954.

297   **Decertified:** JGB2.

297   **"Too demanding":** September 25, 1957, letter from JGB to Ruth Chandler, BCA.

297   **"Gadd wasn't working":** September 25, 1957, letter from JGB to Ruth Chandler, BCA.

298   **"Fucking me":** This scene, including all the quotes, is according to JGB2.

298   **100,000 acres:** FB, September 24, 1956; also noted in CJ, January 4, 1957.

298   **Siphoning:** Crocket, in an interview, denied doing anything improper. **Kidney stones:** From Crocket's unpublished memoir.

299   **"Grasp more information":** Interview with Bickett. **"Jim was floating":** DV with Evers.

299   **Boston Land Company:** Details on the deal are from JGB2; HS, April 14, 1955.

300   **Team of CPAs:** JGB2; interview with Erickson. Crocket put down a completely different version of events in his unpublished memoir, making it seem as if he simply got tired of the business and it was time to sell. **"Shenanigans":** Interview with Erickson.

300   **"Pussyfooting around":** JGB2. **$10 million:** FB, September 24, 1956.

300   **Nasty legal scrap:** The merger was opposed by certain Tulare Lake Land shareholders and ultimately had to be settled before the state Corporation Commission (JGB2; interview with John Pollock, the lawyer who handled the case for Boswell).

301   **"Covet thy neighbor's":** Transcript of remarks by JGB2 at company meeting, January 9–12, 1980, Scottsdale, Arizona, BCA.

301   **Fundamental principles:** JGB2; interviews with former Boswell financial men Wally Erickson and Walt Bickett. **"Conservative person":** DV with Evers.

301   **Key canal systems:** SB.

301   **Being hijacked:** Interviews with SB and Don Mills, manager of the Kings County Water District. See also "Kings County Water Dispute is Settled," CJ, February 19, 1954. **"Level of arrogance":** SB. **Straight at them:** Interviews with smaller farmers around the lake basin.

302   **Political reach:** See *King of Kings*, FB special report, October 31, 1999. **Personal lawman:** Ely's memories of working for Boswell are recounted in the 1979 issue of "Ten-Four," the annual publication of the Kings County Peace Officers' Association. See also CJ, July 11, 1958.

303   **"Like Disneyland":** DV with Jorgenson. **Ginning operation:** In the late 1960s, Boswell developed a "superplant" in Corcoran that churned out forty bales of cotton an hour—more than double that of most top-caliber facilities of the era (Vandergriff, p. 48).

303   **A tough sell:** JGB2; interview with Mike Quinn, a North Carolina cotton broker and the son of Paul Quinn, who was Boswell's long-time East Coast representative. **"Getting along good":** Interview with Kimbrell. **Cracked them all:** JGB2; Quinn interview.

303   **Aircraft:** Boswell acquired Ted Smith Aircraft Company in 1960 and exchanged it for the stock of American Cement Company in 1969; it later disposed of that stock. **Oil and gas:** Reybos, which drilled for natural gas in western Colorado, was acquired in 1971 and disposed of in 1975. Boswell also acquired oil fields in the Tulare Lake area. **Cattle:** The Oregon ranch, Prineville, was acquired in 1967 and traded for development property in Colorado in 1982. Yokohl Valley was acquired in 1965. The Corcoran feed pens were closed in 1975 and the Arizona pens in 1976. **Pig business:** Boswell's swine operation was established in 1971. The company later formed a 50 percent partnership with Clougherty Packing Company before exiting the business.

303   **Hopelessly expensive:** Interview with former Boswell executive Brent Bowen, who became the company's liaison with Del Webb.

303   **Del Webb:** Biographical details from Finnerty, Blanc and McCann, *Del Webb: A Man. A Company.*

304   **Alerted to the project:** Interview with Bob Johnson.

304   **"Interested in buying":** As recounted by JGB2. Roughly the same dialogue is captured in Freeman and Sanberg, *Silver Anniversary Jubilee: A History of Sun City, Arizona*, p. 21.

304   **Each side put up:** Freeman and Sanberg, pp. 22–23. **"Pick it apart":** Inter-

view with Childress. In numerous interviews, others who worked with Boswell made the same point: "He cuts through the wrapping, down to the essence," said real estate baron Peter Munk, who in the 1970s developed property with Boswell in the South Pacific. At General Electric's board meetings, Boswell "provided counsel when asked for it—and sometimes when not," said Jack Parker, former GE vice chairman. "His questioning was always astute." Likewise, at Safeway Stores, where Boswell also served on the board, he consistently "asked all the best questions," said fellow director Merrill Magowan, the grandson of investment tycoon Charles Merrill, who in 1926 had purchased the grocery chain.

305   **On this scale:** Youngtown, started in the mid-1950s, was the first full-fledged retirement community in the nation. Relatively small, it served as part of the inspiration for Sun City. (Interview with Bob Johnson, former Del Webb president; Freeman and Sanberg, p. 15.) **"Cannibalism":** Freeman and Sanberg, pp. 19–20. **"30-year mortgage":** Freeman and Sanberg, p. 29.

305   **Line of cars:** Opening-day details are from Freeman and Sanberg, p. 30. **Sociological phenomenon:** Freeman and Sanberg, pp. 13, 32. **"Webb's speeches":** JGB2. He also made similar comments in an interview with the Sun City Historical Society.

305   **Many millions:** Through the 1970s, Boswell consistently earned more than $1 million a year—sometimes substantially more—from its development activities in Sun City, according to financial records found in BCA. In 1983 Del Webb purchased Boswell's 49 percent interest in the venture for $42 million. **"Up to our ass":** DV with James O. Boswell.

306   **Ran things Down Under:** JF; interview with Jim Blasdell.

306   **"Broke every law":** Interview with Blasdell. **Bent and twisted:** Those he didn't bend, he helped to write. Blasdell knew that the Australian government, eager to improve its balance of trade, was hoping to turn cotton into an export crop. So he pushed for the Bounty Act, under which farmers would be paid a premium on top of the market price for every pound of fiber produced. He met with officials in Canberra, lobbying and shaping the legislation, which passed in November 1963. It seemed a little silly: The U.S. government was subsidizing farmers not to grow cotton, and the Australians were now paying them to grow it. But money was money (Interview with Charles Wannan, Blasdell's lawyer in Australia).

306   **Weeds stretched:** DV with JF.

306   **"Don't have an option":** Interview with Blasdell.

307   **"Red ink":** JF. Also see CJ, November 14, 1963.

307   **Parliamentary approval:** JGB2; JF; interviews with Charles Wannan and Tom Lewis. **"Sydney Harbor Bridge":** Interview with Lewis. **Afternoon tea:** JF.

307   **"Broke my pick":** JF. **Biggest cotton grower:** The May 8, 1986, *Australian Financial Review* noted that Boswell grew 10 percent of the Australian cotton crop and processed a third of it. Boswell sold its Australian operation in 1987 to mogul Kerry Packer and then bought it back from him three years later. See also *Sun Herald,* August 26, 1990. Boswell was eventually overtaken as Australia's largest cotton grower by Colly Cotton.

307   **Magnates went under:** In remarks to the Boswell board of directors in June 1973, Boswell said: "We are the only company of 12 that come to mind that has survived our particular line of business over the years" (Remarks found in BCA). **Foray into Mexico:** Boswell exited that country in 1963. **"Turned to money":** Interview with Vandergriff. **"Boswell luck":** JF.

308   **"Surround himself":** Interview with Rachford.

308   **See the fear:** Observed during lunch with JGB2, Mark Grewal and the authors in July 2002.

308   **"Red ass":** JF. **"The bubble":** Interview with Erickson. **"A boot":** Interview with McMicken. Outside of the company, JGB2 didn't mince words, either. At Security Pacific Corporation, where Boswell would play a crucial role in the firm's eventual buyout by Bank of America, Chairman Robert Smith viewed him as "undeniably the strongest voice" on the board. "When he spoke, you were always sure where he stood" (Smith, *Dead Bank Walking,* p. 116).

308   **"Minute before ten":** JF. **"On Fridays":** JGB2.

309   **Make decisions:** Interview with, among others, Boswell executive Bill Ostrem.

309   **"Marine Corps":** Interview with Pierce.

309   **Critics perceived:** Former Salyer lawyer Tom Greer, for one, leveled this charge. **"A delegator":** JF. **"All times":** SB.

310   **"I'll be goddamned":** JF. In a 1969 interview with DV, found in BCA, JGB2 acknowledged that JF was the one who came up with the car scheme. But as the years went on, he apparently became determined to put himself in the middle of the story.

310   **"The hourly level":** Interview with Hall.

310   **By name:** Interviews with Rudy Castro and Karl Schneider. **"One of them":** Interview with Ruth.

310   **"Very folksy":** Interview with Munk. Munk controlled the Sears Tower and Watergate complex through his real estate investment trust, Trizec Properties.

311   **Watts riots:** JWB. **Spotty record:** Interviews with many members of Corcoran's black community. **"Out of our way":** JGB2.

311   **Up With People:** Founder Blanton Belk said that "no board member advised or counseled me more" than Boswell. He served on the executive committee, chaired the compensation and management committee, and

funded Up With People's historic trip to China in 1978. In 1993 the Boswell Foundation provided the money for Up With People to buy the land on which it built its new headquarters near Denver. **"Pretty bold":** Interview with Belk.

311   **"Grapes of Wrath days":** JGB2. A September 11, 1964, letter from CJ publisher Percy Whiteside to JGB2 thanks him for his help in "erasing the stigma stamped upon" Corcoran "during the era of migrants" (BCA). **Home loans:** SB. **"Pissed off":** SB. **"Good conscience":** May 15, 1970, letter from JGB2 to William O. Boswell, BCA.

312   **"Be a hero":** The encounter with JGB2 was recounted by Jones in an interview. **500-acre:** That made Boswell's operation, known as Cactus Lane, the largest table-grape ranch in Arizona, according to material produced at the time by the United Farm Workers Organizing Committee and the union's office in Tolleson, Arizona (Walter P. Reuther Library, Wayne State University, Detroit).

312   **Early-morning pushups:** Interview with Hurtado.

312   **Exploiting the poor:** Flier found in BCA.

313   **Only about fifty:** The union claimed that as many as 350 of Boswell's 500 workers walked out. But Boswell's men—and contemporaneous newspaper accounts in the *Phoenix Gazette* and elsewhere—put the number at fifty or so.

313   **"Genuinely felt":** Interview with Jones. **"His own terms":** Interview with Rankin.

313   **Signed a bill:** Chavez's showdown with Governor Williams is captured in Ferriss and Sandoval, pp. 197–198. **"In thanking you":** Recounted in a July 10, 1972, note to JGB2 from Raymond, BCA.

313   **"My regards":** June 30, 1972, letter from Governor Williams to Raymond, BCA.

313   **200,000 acres:** JF. **"That simple":** JF.

314   **Eight terrorists:** The incident occurred March 2, 1973. **"Ass shot off":** As recounted by JF.

314   **Fifty-three lawyers later:** Noted in Birmingham, *California Rich*, p. 48. **Full potential:** Details on the Bowleses's position—and Boswell's counterarguments—are gleaned from a July 12, 1972, memo from JGB2 to Les Doan and three other top executives, BCA.

315   **Stiff resistance:** As reflected in numerous pieces of correspondence found in BCA. **"Made a mistake":** Interview with Nickel.

315   **128,000 acres:** From his obituary in BC, January 17, 1987. **"Horse trader":** From JGB2's testimony in *Wegis, Thomson & Thomson v. J. G. Boswell.*

315   **"On their knees":** Interview with Williams. **C. Arnholt Smith:** Roberts's relationship with Smith is discussed in the *New York Times*, October 29, 1973, and July 1, 1974. Also see Smith's obituary in the *San Diego Union-Tribune*,

June 10, 1996. **Bankruptcy:** BC, September 12, 1977. Boswell helped Roberts work through some of his financial troubles with bank regulators, leading the old boy to commission a painting showing a squad of cavalry kicking up dust. "That's the J. G. Boswell Company riding over the hill to save Hollis Roberts," he said (Interview with Walt Bickett).

CHAPTER 15: TRUCE

316   **Much dithering:** Anderson, *Water Rights*, p. 79. **Leave it to the courts:** Interior Department news release, October 4, 1963. Copies of the "test case" contracts between the government and the farmers using Pine Flat can be found in numerous archives including PST. **Nine long years:** JGB2 noted that the growers' strategy had been "one of delay" with the idea that "dragging our feet" would allow for "a change in the makeup of the Supreme Court" (December 10, 1969, memo to SB). **Crocker had decreed:** For his ruling, as well as the opposing sides' arguments in the case, see *U.S. v. Tulare Lake Canal Company*, case no. CV 2483, filed 1963 in the U.S. District Court for the Southern District of California, Northern Division. Crocker issued his findings of fact and conclusions of law on March 15, 1972.

317   **Flitting back and forth:** In October 1947, an associate solicitor in the Interior Department named Felix Cohen issued a five-page opinion stating that a lump-sum payout would lift acreage limits for big farmers. (A copy of the document can be found in, among other archives, CP.) Through the 1950s, several interior secretaries confirmed the payout principle, as Cohen had articulated it. Indeed, Worster, *Rivers of Empire* (p. 288), describes how during the Eisenhower administration's first term, Interior Secretary Douglas McKay offered in the case of Pine Flat Dam to accept the $14 million lump-sum payment. But McKay's successor, Fred Seaton, "felt compelled to take that offer back." Actually, what Seaton did was refer the matter for additional legal review on the eve of his departure from office in January 1961. (Seaton's position is reflected in numerous papers in his files, which are housed at the Dwight D. Eisenhower Library, Abilene, Kansas. See also the library's "Kings River, Pine Flat Project" files. Also relevant are the papers of Senator Clair Engle [D., California], California State University at Chico.) Then in December 1961, President Kennedy's new solicitor of the Interior Department, Frank Barry, issued an opinion stating strongly that the lump-sum payment shouldn't negate the acreage-limit law. Cohen's 1947 opinion, said Barry, had been "in error." To accept it, he added, "would be to brand" the lawmakers who passed the original 1902 Reclamation Act "as either hypocritical knaves or fools. They were, of course, neither." (Details are spelled out in a December 31, 1961,

Interior Department news release, CP.) Attorney General Robert Kennedy concurred with Barry (December 29, 1961 letter from Kennedy to Interior Secretary Stewart Udall, CP). In turn, Secretary Udall rejected the proposed repayment contracts on the Kings River, as well as on the Kern, pushing the case into court. **Extensive paper trail:** Copies of the correspondence referred to, as well as a trove of other key background documents, are found in CP. In addition, see "Statement of Position of Members of Kings River Water Association with Memorandum of Supporting Authorities Re Execution of Pine Flat Repayment Contracts," November 16, 1961, Engle Papers, California State University at Chico.

317 **Stunning news:** 535 F.2nd 1093 (1976), ruling of the U.S. Court of Appeals, Ninth Circuit, April 5, 1976. A rehearing *en banc* was denied June 7, 1976.

317 **More than 88,000 acres:** Noted in a July 23, 1979, letter from Interior Secretary Cecil Andrus to Senator Gaylord Nelson, PST; CR, May 5, 1982, p. H 1827. According to the table, Boswell also owned 21,036 acres watered by the Kern River that were affected by the Army Corps acreage-limit issue.

317 **"Out of trouble":** JGB2's remarks and the anniversary dinner menu are found in BCA. **Simon H. Rifkind:** Details on Rifkind are from his November 16, 1995, obituary in LAT. **Hand-delivered to him:** From a list titled "Items to be delivered by J. G. Boswell II to Simon H. Rifkind," found in BCA.

318 **"Lives and livelihoods":** June 16, 1976, letter from JGB2 to Rifkind.

318 **"Hate to take":** JGB2.

318 **30,000 "excess acres":** Noted in July 23, 1979, letter from interior secretary Cecil Andrus to Senator Gaylord Nelson, PST; CR, May 5, 1982, p. H 1827.

319 **"Worst enemy":** FSal.

319 **"Little money, no land":** CJ Golden Jubilee Edition, October 1, 1964.

319 **Feud that pitted:** Details can be found in *Walter Gleason v. Salyer Land Company*, case no. 24734, filed August 4, 1975 in Kings County Superior Court; "Salyer Clan: Feud Among the Wealthy" by Charles P. Wallace, LAT, March 27, 1983.

319 **"Grievous mistake":** July 2, 1963, letter found in the Gleason lawsuit.

320 **"Blue blood?":** LSal.

320 **"One of the most brilliant":** Interview with Greer.

320 **"Think I'm pretty?":** CJ, August 17, 1967. **Dirty dirt:** FSal. **"John Birch":** JGB2.

321 **"Violent and dangerous":** This and all other recollections about Cockeye, except where noted, are from FSal.

322 **"His false teeth":** LSal.

322 **7,001-foot runway:** FSal. **Paying Sergeant Ely:** LAT, March 27, 1983.

322 **Three flamingoes:** CJ, September 21, 1961. **Mississippi riverboat:** CJ, September 23, 1965.

323   **"The biggest taxpayer":** Quoted in the arrest report, which was obtained by the authors.

323   **Collecting cash money:** LSal. **"And his pushy wife":** FSal.

323   **State Board of Agriculture:** CJ, March 24, 1960. **Brown fondly recalled:** LAT, May 16, 1982.

324   **"Forgot-my-checkbook":** Interview with Cherry.

324   **Sometimes slip:** Interview with Verdo Gregory.

324   **Grabbed a shotgun:** LAT, March 27, 1983. **Clarence knocked down:** Referenced in the Gleason lawsuit.

325   **"The guts":** From the Gleason lawsuit.

325   **Bickered over:** Alvin Rockwell—Boswell's choice—was an able enough lawyer, the Salyers acknowledged. He was a partner in a prominent San Francisco firm. And in 1959 he won a modicum of fame as president of the World Affairs Council of Northern California for helping defuse the tension around a highly charged visit to the United States by Soviet Premier Nikita Khrushchev. But U.S.-Soviet relations proved a lot warmer than Boswell-Salyer relations, and Fred and Everette alleged that Rockwell had a conflict of interest they just couldn't abide. The Salyers objected that Rockwell also represented a large grower around Bakersfield; now that Boswell was farming the Buena Vista ranch in that same area, they feared their own interests might somehow be compromised. (The Salyers' concerns are spelled out in numerous documents in BCA, including a February 20, 1964, letter from Everette Salyer to the Boswell board; a March 2, 1964, letter from Rockwell to Boswell executive Len Evers; and a March 3, 1964, memo from JGB2 to his files. Biographical details on Rockwell are from his May, 26, 1999, obituary in the *San Francisco Chronicle.*) **$150 million:** The amount is noted in the minutes of a special March 3, 1964, meeting of the Boswell board of directors, BCA. The document adds that the Salyers later lowered their demand for damages to $31 million. **"No interest in bluff":** February 20, 1964, letter from Everette Salyer to the Boswell board. At one point, Fred and Everette accused Boswell of trying to subvert them by plotting to acquire their father Cockeye's interest in the Salyer operation (February 1, 1965, letter from attorney Webster Clark to Boswell attorney Robert Newell).

326   **Two families fought:** With the Salyers threatening litigation at every turn, Louis Robinson advised Boswell that it was time to get tough with Fred and Everette. "Appeasement has been a dismal failure," he said, "and the more we appease the deeper we sink in the mire" (November 14, 1966, letter from Robinson to JGB2, BCA). Boswell, for his part, agreed that the Salyers were being "obstructionist (to put it mildly)," and he proposed moving his water operations from Corcoran to L.A. so that "the local people won't be harassed to the extent they presently are" (May 31, 1966, letter

from JGB2 to Ruth Chandler, BCA). A year later, Len Evers was still complaining that the Boswell employees on the water district were being "vilified, harassed and condemned" by the Salyers (April 26, 1967, memo from Evers to JGB2, BCA).

326   **Transcontinental commuter:** Details on Greer are from interviews with him; JPL. **"Never in doubt":** SB. **"Stirring the pot":** JF.

327   **Plan was tabled:** Details are found in *Salyer Land Company v. Tulare Lake Basin Water Storage District*, 410 U.S. 719 (1973). For a good synopsis, see Goodall, Sullivan and De Young, *California Water: A New Political Economy*, pp. 98–99. **One man, one vote:** Ballard, in her thesis "And Conflict Shall Prevail," points out the irony of Salyer opposing a property-weighted voting system. In 1976, she notes, Salyer led a fight to change the Kings River Conservation District to just such a system.

327   **"Robber barons":** Douglas, *The Court Years, 1939–1975*, p. 165.

327   **Fresh in everyone's minds:** JF, for one, had taken to calling the Salyer brothers the "Suer brothers."

328   **No longer on the case:** JF. **John Harmer:** Harmer himself declined to be interviewed, but his role was recalled by many others involved in the lobbying battle.

328   **"Going to volunteer":** As recounted by JF.

328   **The greatest disdain:** Carter's infamous "hit list," in which he put Congress on notice that he wanted to cut off funding for nineteen major water projects, was widely seen as a political disaster. The episode is described in many places, including Reisner (pp. 314–322) and Frederick and Hanson, *Water for Western Agriculture* (pp. 122–123). **"Lunch-bucket guy":** Andrus, *Politics Western Style*, p. 9.

328   **Sued the government:** *National Land for People v. Bureau of Reclamation*, case no. CV 76-0928, filed May 1976 in the U.S. District Court for the District of Columbia. **1,100 members:** Testimony of George Ballis to the House Committee on Interior and Insular Affairs, March 18, 1980, p. 279. **George Ballis:** Details are from an interview with him. **"These guys want?":** As recounted by Ballis. **"Spring themselves":** Ballis interview.

329   **"Disturbs me most":** December 5, 1977, letter from Andrus to Ralph Comstock Jr., chairman of the First Security Bank of Idaho (Found in the Andrus papers, Special Collections, Albertsons Library, Boise State University).

329   **Declined to hear:** Nonetheless, two Boswell lawyers, Ernest Clark and Robert Newell, continued to pursue the matter in court on certain constitutional grounds. **"Fantastic":** FB, February 23, 1977.

329   **"No thoughts":** Quoted in LAT, February 23, 1977. **"Bury the corpse":** JPL.

330   **"Orderly land sale":** SB memo dated March 25, 1977. **"Last-resort step":** April 20, 1977, memo from JGB2 to SB. **"No power in heaven":** FB, August 21, 1977.

330   **BosPAC:** A statement of organization was filed November 22, 1977, according to FEC. **More than $150,000:** FEC. See also "Agriculture Giant Sows the Political Field," LAT, November 18, 1979. **Black-tie fund-raisers:** JF; JPL.

330   **John Penn Lee:** Biographical details are from JPL. **"Get the hay down":** Interview with former California assemblyman and U.S. Representative Rick Lehman.

331   **"All over the town":** JPL.

331   **More emotional appeal:** JF.

331   **"Goddamn check":** JPL. A Boswell lobbyist confirmed the Lance meeting. Boswell and Salyer also floated the idea of buying Pine Flat Dam on Capitol Hill. (See FB, August 21, 1977.)

332   **$10,000-a-month retainer:** Federal lobbying reports, NA; FB, September 20, 1978. **"Whore in Venice":** Interview with McAden.

332   **Harmer had persuaded:** JF; JPL; FB, October 21, 1977. **"Quickly and easily":** Kipnes interview.

332   **"Less than candid":** FB, October 21, 1977. **Cranston had refused:** Noted in an October 20, 1977, staff memo found in AC. Cranston was hardly pure on the acreage-limit issue, however; he pushed hard for an exemption in California's Imperial Valley.

332   **"Conceivably justify":** October 27, 1977, letter from Cranston to Senator Jennings Randolph, AC. **"Open mind":** FB, October 21, 1977. Krebs finally supported an exemption for the Kings River area from federal reclamation law, but his move in January 1978 was too little, too late in the eyes of Boswell and Salyer.

332   **John Krebs:** Details on Krebs's life and his political career are from interviews with Krebs; "The Anti-Krebs Elixir," an article by Tom Kennedy in the January 1979 issue of *California Journal*; FB, November 9, 1978. **"Back money":** Krebs interview.

333   **"Hammer and tong":** JPL.

333   **"Pretty hot stuff":** McAden interview.

333   **"Chip" Pashayan:** Details are from an interview with Pashayan; *California Journal*; and the *Biographical Directory of the United States Congress*. **More than 15 percent:** FEC.

334   **Donald Santarelli:** Details are from an interview with Santarelli; Baum, *Smoke and Mirrors*, pp. 3–7, 14–17; "My Truth, Their Consequences," a piece in the August 15, 1993, *Washington Post* by Robert Williams, the journalist who brought about Santarelli's downfall in the Nixon administration; and the *New York Times*, June 5, 1974.

334   **Santarelli's role:** Interview with Santarelli. This is also borne out in numerous documents found in CP.

335   **A dirty word:** Noted in a February 23, 1979, letter to the Clark's Fork Reclamation District from attorney Lawrence Clawson, who had attended a presentation by Santarelli, CP.

335   **"Viscerally opposed":** Pashayan interview. **"Small farmers wanted":** Pashayan interview.

335   **Denounced as ridiculous:** July 23, 1979, letter from Andrus to Senator Gaylord Nelson, PST; testimony before the House Committee on Interior and Insular Affairs, November 13, 1979; May 21, 1980, letter from Andrus to Representative Morris Udall, CP. **"Throughout the 50":** June 8, 1979, letter to Senator Henry Jackson from six of his colleagues, AC.

335   **Big train:** The wide range of reclamation-related issues is reflected in numerous congressional hearings held throughout the late 1970s and early '80s, as well as documents found throughout CP; AC; PST.

336   **"Going to get anything":** Interview with Nelson.

336   **"Neutralizing him":** This and the other quotes about the incident are from an interview with Santarelli.

336   **Is doubtful:** Interview with long-time Cranston chief of staff Roy Greenaway. **"Blatant circumvention":** April 9, 1979, staff memo to Cranston recapping the Carter administration's arguments, AC.

337   **"Have a reputation":** July 20, 1979, letter from Taylor to Cranston, PST.

337   **"Actively oppose":** FB, May 18, 1979; May 21, 1979, letter from Cranston to Alvin Quist, chairman of the Kings River Water Users Committee, AC. **"An accessory":** From a National Land for People newsletter (undated) found in AC.

337   **"Uncertain burdens":** April 2, 1979, letter from Alvin Quist to Cranston, AC; CP.

338   **"More nor less":** All quotes and details from the Senate floor debate are taken from CR, September 13–14, 1979.

338   **"You'll learn":** Interview with McAden. **"The phantom":** Interview with McAden.

339   **Informal head counts:** "Cranston's Choice," an article in *New West* by Ehud Yonay, July 14, 1980.

339   **"Touching people":** Quoted in *New West*, July 14, 1980.

339   **Sagebrush Rebellion:** See *Newsweek* cover story, "The Angry West," September 17, 1979.

340   **"Stagecoach robbery":** Quoted in the *Washington Post*, July 14, 1979; *Congressional Quarterly*, September 29, 1979, p. 2121.

340   **In 1978:** In an August 22, 1979, news release, Representative Paul Findley noted that in "January 1978, I learned from a confidential source that Wooters . . . had entered into a secret agreement with the J. G. Boswell Company."

340   **Many regarded him:** Interview with Neal Gillen, executive vice president and general counsel of the American Cotton Shippers Association.

340   **Including Jim Boswell:** JGB2 wasn't part of the formal hiring committee. But, in an interview, Wooters noted that JGB2 played a critical role in establishing the criteria that led to his selection. He also provided Wooters

with invaluable backing once he was in the job. "Jim was a great pillar of support," said Wooters. "We spoke the same language." **Paid an assessment:** The history of Cotton Inc. is drawn primarily from Jacobson and Smith. They discuss the industry's early marketing efforts on pp. 109–135. **Annually outspending:** In 1979, the synthetics industry outspent Cotton Inc. on advertising $70 million to $5 million. (See "Major Accomplishments of the Cotton Research and Promotion Program," submitted as part of the hearing record before the House Committee on Agriculture, February 26–27, 1980, p. 341.) **"Cheap," "dull":** Hearings before the House Committee on Agriculture, February 26–27, 1980, p. 340.

341　**Harvard MBA:** Wooters's background is from Jacobson and Smith, pp. 142–150; interview with Wooters. **"Little white stuff":** Jacobson and Smith, p. 156. **"The great one":** The anecdote is recounted in Jacobson and Smith, p. 150.

341　**Seal of Cotton:** Jacobson and Smith, pp. 162–165. **"Natural blend":** Jacobson and Smith, pp. 177–178. **Technical research:** Jacobson and Smith, pp. 197–232. **Mills were churning:** In late 1979, J.P. Stevens Inc. and Dan River Inc. introduced 100 percent cotton Durable Press bedsheets. (Hearings before the House Committee on Agriculture, February 26–27, 1980, p. 340.); Jacobson and Smith, p. 223. **Half the market:** Market share for cotton—defined as its portion of U.S. mill fiber consumption—bottomed out in 1973 at 33 percent. It began to improve during the early 1980s and accelerated after 1985. By the end of the 1990s, it had reached 60 percent (Jacobson and Smith, pp. 165, 171, 315).

341　**Big problem:** The situation regarding Wooters's salary is laid out in the hearings before the House Committee on Agriculture, February 26–27, 1980; statement by the Board of Directors of Cotton Inc., February 1979.

342　**"Like an onion":** August 22, 1979, news release from Findley.

342　**Requested a refund:** Cotton was unique among agricultural promotion programs in having a refund mechanism (Jacobson and Smith, p. 136). **"Just a coincidence":** Hearings before the House Committee on Agriculture, February 26–27, 1980, p. 264. **$8 million:** FB, October 30, 1979; *Daily News Record,* November 1, 1979. Wooters was alleged to have passed on referrals to a number of other companies as well, including Salyer.

342　**Grand jury:** FB, February 25, 1979; *Women's Wear Daily,* March 9 and March 12, 1979. **Inspector general:** Noted in August 22, 1979, news release from Representative Findley; FB, March 8, 1979. **Ever charged:** FB, June 12, 1979.

342　**Bronze bust:** Interview with Coelho. Wooters confirmed the story in an interview, though he remembered the details a bit differently. **To hang on:** *Wall Street Journal,* July 3, 1980; FB, July 21, 1981. **Around $200,000:** Interview with Wooters.

343　**Scoffed at the idea:** JF. **Sought revenge:** JF; interview with Findley; FEC. **"In 10 minutes":** Quoted in the HS, February 8, 1980.

343   **Lee phoned home:** JPL.

343   **"Pat knew":** Interview with Hopkins.

344   **Lobbying brigade:** Technically Hopkins represented an entity known as the Tulare Lake Water Users Committee (Federal lobbying reports, NA). **"My God, Pat":** Interview with Hopkins; LAT, March 27, 1983. Andrus declined to be interviewed.

344   **"Principles of equity":** June 7, 1978, letter from Andrus to Representative Morris Udall, CP.

344   **Golf course:** JF; JPL. **"One of those things":** JPL.

344   **March 1980:** The March 31 date is noted in LAT, December 12, 1980.

345   **Salyer flew him:** FSal. **15,000 acres:** Interview with Tom Hurlbutt of the Boswell water department. **"Still a sump":** As quoted in CJ, April 20, 1978. **The Big Land:** Videotape in BCA.

345   **Farmers had diverted:** Hurlbutt interview.

345   **Hostility in the room:** Interview with Don Santarelli, who was there that day.

345   **"A little story":** Interviews with Don Santarelli; Billy Hopkins; JPL.

346   **Operations disrupted:** The farmers pressed this point extremely hard and were outraged when, in March 1979, the Bureau of Reclamation found in a report that growers who didn't comply with acreage limits could be denied storage at Pine Flat Dam without hurting any other users of the facility. SB, Boswell's water chief, was particularly incensed, telling Senator Henry Jackson: "The Bureau's report is, in all candor, totally shocking. . . . " (August 13, 1979, letter, CP.)

346   **Martin hung:** Visit to Martin's Washington office; interview with him. **Complete error:** Reflected, for example, in a July 24, 1979, letter from Martin to Senator Henry Jackson, CP.

346   **"Is unwarranted":** From "Interim Report: Acreage Limitation Executive Summary," sent to Representative Morris Udall by Secretary Andrus, May 19, 1980, CP. **"Invidious":** May 27, 1980, letter from SB to Pat Jennings, CP.

347   **"Unique area":** November 19, 1980, letter from Andrus to Representative Morris Udall, CP.

347   **"Going 180 degrees":** LAT, November 26, 1980. **"Jerked the rug":** LAT, November 26, 1980.

347   **James Watt:** For example, he stated his support for an Army Corps exemption in a December 7, 1981, letter to Representative James McClure, CP. **"The whole course":** JPL. **"Major, major, major":** JF. **"Pulling it downhill":** Interview with Hopkins.

CHAPTER 16: LOBBYISTS, POLITICIANS, PAYOLA

349   **One or two whiskeys:** This incident is drawn from JPL; interviews with lobbyist Ray Corley and former Assemblyman Ruben Ayala; Dan Walters's

column "End of the line for Alex Garcia?", *Sacramento Union*, December 11, 1981; Michael and Walters, *The Third House*, p. 26.

349   **"Prone to disappearing":** Walters's column, December 11, 1981. **"As a no":** JPL.

350   **"Break your arm":** Interviews with Ayala and Corley.

350   **Bitter wrangling:** For the legislative history, see LAT of December 22, 1979; January 8, 1980; May 21, 1980; June 27, 1980; July 8, 1980. See also *New West*, June 16, 1980; and the five-part series on the canal by George Baker in the FB, March 16–20, 1980. Though Boswell has always been a tough operation to penetrate, Baker—more than any other reporter—did an exceptional job through the years of staying on top of the farming giant. **A flourish:** As described in the *Sacramento Bee*, May 20, 1982. **Enough signatures:** LAT, September 25, 1980.

350   **Great conundrum:** Rita Schmidt Sudman, executive director of the Water Education Foundation in Sacramento, points out that the equation is far more complex than most of the people being in the south; agriculture, she notes, uses more than three-quarters of the water in the state. **Most ambitious feat:** Details on the State Water Project are from the California Department of Water Resources Office of Water Education; Kahrl, *The California Water Atlas*, pp. 50–56; Reisner, pp. 347–360; *California State Water Project Atlas*. **Old bugaboo:** *New West*, June 16, 1980. **"Garden of Eden":** LAT, "Jerry Holds Pat's Dream in the Palm of His Hand," June 26, 1980. **"Monument to me":** From p. 29 of an oral history with Brown conducted in 1979 by Malca Chall of the Regional Oral History Office, Bancroft Library, University of California at Berkeley. Reisner, p. 349, also quotes from the same source.

351   **As much water:** From 1968 through 1980, the mean river runoff was 106,000 acre-feet a year while the state system supplied a mean of 108,000 acre-feet ("Report on Irrigation, Drainage and Flooding in the Tulare Lake Basin," September 1981, Tulare Lake Basin Water Storage District). Boswell saw the need for the State Water Project early on. "We cannot long maintain our economy without additional water," the company's Louis Robinson had warned in 1962. Unless farmers served by the Kings River also contracted with the State Water Project, he said, they'd face a shortage so severe that it would be impossible "to sustain existing agriculture" (CJ, March 15, 1962). **Long been seen:** See, for instance, "The Peripheral Canal of the Sacramento–San Joaquin Delta," a summary prepared by the Department of Water Resources, December 1966. **Less water, not more:** JF; JPL. LAT, July 20, 1980. Heightening their concerns was Proposition 8, which was passed by California voters in November 1980. The constitutional amendment would have mitigated the environmental impacts of building the Peripheral Canal, while also making it exceedingly difficult to

construct water projects on north coast rivers. However, for the environmental safeguards in Proposition 8 to kick in, the Peripheral Canal measure had to be approved in a statewide vote. Meantime, Boswell officials had other concerns about the canal proposal, including a giant fish screen that would have been erected on the Sacramento River. (See testimony of JF and Ed Giermann in *Wegis, Thomson & Thomson v. J. G. Boswell.*)

351 **Failed to deliver:** See *Salyer v. the State of California*, Case no. 267012, filed in 1977 in state Superior Court for Sacramento. **230 species:** Details from *Sacramento-San Joaquin Delta Atlas*, Department of Water Resources. **"Summarily handed":** JPL.

352 **"As an excuse":** Remarks prepared for Drought Operations Meeting, March 7, 1977, BCA.

352 **"Didn't have faith":** Interview with Meral. **"Hitler-Stalin Pact":** Reisner, p. 363. **Were convinced:** Interview with Dante Nomellini, a lawyer representing the delta farmers.

352 **Shortage loomed:** See, for example, the op-ed by Metropolitan Water District chairman Earle Blais in the October 11, 1981, LAT. **"Mighty Met":** Gottlieb and FitzSimmons, *Thirst for Growth*, pp. 5–6. **The Met presumed:** Interview with the Met's point man on the canal, David Kennedy. JF also says that Met officials told him as much during the battle. (See JF's testimony in *Wegis, Thomson & Thomson v. J. G. Boswell.*)

353 **Win-win:** See "Canal Law Tries to Meet Water Demand While Protecting Delta," LAT, July 20, 1980. **Irony wasn't lost:** Resiner (p. 360) makes this very point. Also see LAT, June 26, 1980.

353 **Since 1939:** Records of the California secretary of state's office. **Polls showed:** Interview with Sal Russo, cocampaign manager for the anti-canal forces. **"Guru":** So called by LAT political columnist George Skelton, December 12, 1996. **"Buried us":** Interview with Giermann.

353 **"Nobody trusted":** Interview with Russo.

353 **"Could not tolerate":** Memo dated March 31, 1981, from Graff to eight of his environmental community colleagues, RWA.

354 **"The fascists":** Interview with Russo. **Belgian shepherd:** Interviews with Russo, Watts, JPL and Storper.

354 **"Unholy Alliance":** Handwritten notes from a file labeled "California Coalition To Stop the Peripheral Canal, Executive Committee Meeting, April 11, 1981," RWA. **"Our integrity":** June 13, 1981, memo from Storper to the Campaign Executive Committee, RWA. For more on Storper's contemporaneous view of the canal, see "Water & Power" in the April 11, 1980, *Los Angeles Reader.*

354 **"Equally suspicious":** Interview with Watts.

354 **County by county:** The meticulous structuring of the campaign is reflected in numerous documents found in RWA. **In Northern California:**

Some in California's water world speculated that Boswell and Salyer tried to appeal to one Northern California lawmaker—their nemesis, Representative George Miller. The idea was that if they could all be allies in opposition to the Peripheral Canal, Boswell and Salyer might get Miller to back off on pushing for acreage limits in Washington. But JPL strenuously denied that the companies ever sought such a quid quo pro. (See FB, March 18, 1980.) **"Washing their cars":** June 5, 1979, letter to Governor Brown from Ann Hoisington of Lafayette, California, Peripheral Canal Collection, Water Resources Center Archives, University of California at Berkeley.

355   **Demonize:** Interview with Russo.

355   **"Oil and water":** A videotape of all the anti-canal ads are archived at Russo Marsh & Rogers in Sacramento, Sal Russo's new firm. **The ad pulled:** Interviews with Russo and Watts.

355   **"Local heroes":** Interview with Russo.

356   **"Preliminary data":** January 26, 1982, memo to Russo from Gary Lawrence, California Polling Corporation, RWA.

356   **A house call:** The scene of Governor Brown's visit to JPL is based on interviews with JPL, Bouskos and Brown. The visit was also noted in the *Sacramento Bee*, May 20, 1982.

356   **New Melones:** Brown's policy, formed after conservationist Mark Dubois chained himself to a rock to block the filling of the reservoir, is discussed in Reisner, pp. 508–510.

357   **"Whipping boy":** JPL.

357   **$19.3 billion:** The task force records are housed at the Water Resources Center Archives, University of California at Berkeley.

357   **Helped gin up:** To give the task force a patina of respectability, Fisher and Lee's Sacramento lobbyist had arranged to put in charge of its day-to-day activities Alan Post, former state legislative analyst. It was a brilliant idea. Post was seen as scrupulously nonpartisan and independent, to the point of being "a minor God around here," the lobbyist, Don "Big O" BrOwn, noted in an interview. Those on the other side couldn't believe that Post had agreed to participate. David Kennedy, then the Met's point man on the Peripheral Canal, said in an interview: "Alan Post is a very honorable guy who was naïve. I thought he let himself be used." Post, for his part, defended the task force's fact-gathering process and its conclusions, saying that he "didn't feel used" at all. But Sal Russo couldn't help but smile when he thought of what had been orchestrated. "It was a great scam," he said. For their part, pro-canal forces attacked the task force as "political" and worse. (See LAT editorial, "Curb's Carnival," Dec. 28, 1981; LAT articles on the controversy, February 16 and 18 and March 31, 1982.) **Approached by Curb:** See JF's testimony in *Wegis, Thomson & Thomson v. J. G. Boswell*. **Only to swing:** LAT, October 8, 1981. **Flip-flop cost:** Industrialist Robert

Fluor canceled his commitment to contribute $10,000 to a Curb fundraiser. However, in 1980 and 1981, Boswell and Salyer gave Curb a combined $41,330, including in-kind contributions. (Expenditure and Major Donor Committee Campaign Statements, California secretary of state's office.) **"Cheap way":** Cited during JF's trial testimony in *Wegis, Thomson & Thomson v. J. G. Boswell.*

358   **"Top Estimates":** LAT, April 17, 1982. **"Too Expensive":** Noted in Reisner, p. 365. **TV time:** From May 1–15 alone, according to records in RWA, Boswell put up $233,472 and Salyer $111,528.

358   **"By the opposition!":** Noted in a January 12, 1982, memo to Russo and Watts from Burt Wilson, who headed the anticanal campaign's Southern California office. **"Can't stretch":** Quoted in LAT, April 17, 1982. **El Niño–fueled:** Reisner, p. 366. **"It rained like":** JPL.

358   **Lose interest:** Interview with Jerry Meral, former Department of Water Resources official; "Brown Won't Step Up Efforts to Sell Canal to the Voters," *Sacramento Bee,* May 20, 1982. **"They gagged":** Interview with Spencer. **Foes outspent:** LAT, August 6, 1982. Total spending by both sides made the Peripheral Canal referendum the third most expensive ballot measure in California history.

358   **"Dirty money":** Interview with Long. **Landslide:** County-by-county results were reported in LAT, June 10, 1982. **"Communist Bulgaria":** Interview with Meral.

359   **"Beaten badly":** Graff op-ed, LAT, June 13, 1982. See also an oral history conducted in 1994 with Graff and David R. Yardas by the Regional Oral History Office, Bancroft Library, University of California at Berkeley.

359   **"A celebration":** June 25, 1982, letter from JF to Sal Russo, RWA.

359   **Approved a version:** For a recap of the bill's movement through Congress, see the *Congressional Quarterly Almanac,* 97th Congress, 2nd session, vol. 38, pp. 353–356. **Sure to remind:** Pashayan entered Andrus's comments backing a Kings River exemption into the record during the House floor debate (CR, May 5, 1982, p. H 1842).

359   **Focus of the debate:** Details are drawn from CR, July 15–16, 1982.

360   **"Crime wave":** Quoted in *Congressional Quarterly Almanac,* 97th Congress, p. 355.

360   **George Deukmejian:** He beat Mike Curb in the GOP primary. **"Remarkable string":** Kahrl op-ed, LAT, November 4, 1982.

360   **Ecclesiastes:** SB. **Special medal:** JF.

CHAPTER 17: THE VIEW FROM THE FORTY-SIXTH FLOOR

361   **Full of people:** The July 13, 1988, BC noted that the trial, in Judge Marvin Ferguson's courtroom, was the most attended in Superior Court.

361   **Full-page ad:** The ad had run May 10, 1982.

362   **Sued for libel:** *J. G. Boswell Company v. Family Farmers for Proposition 9, Ken Wegis and Does 1 through 1000*, case no. 179027, filed 1988 in State Superior Court for Kern County. **Their own suit:** This was *Wegis, Thomson & Thomson v. J. G. Boswell*. **"Justice system":** Nader and Smith, *No Contest*, p. 171.

362   **"What did you first do":** This and all other quotes and details from the case are taken directly from the trial transcript, unless otherwise noted.

363   **"Mad as hell":** These particular words are from JGB2's deposition in the case (BC, February 24, 1985).

365   **"Out of the blue":** JF. For his part, JGB2 would always deny that JF was fired. As he'd tell it, JF quit out of frustration when JGB2 wouldn't agree to buy Southlake Farms. "I'd never fire him," said JGB2. "He was too valuable a man. I hold Fisher up on a pedestal."

365   **"Goddamned loyal":** JF.

365   **Badmouthing him:** JF; interview with former Boswell board member Jack Parker, who recalled that JGB2 was complaining to the directors that JF focused too much on politics and not enough on farming. When told of this criticism, JF called it bunk, noting that his time-consuming political work (the Reclamation Reform Act, the Peripheral Canal) was already behind him.

366   **"Star rises":** Interview with Bickett.

366   **Obsessed with youth:** Interview with A. L. Vandergriff. **"Tired of them":** Interview with Pierce.

366   **"Bad points":** Interview with Gamble. **"Through with you":** Interview with Vandergriff.

366   **"Time on your hands?":** As recalled by Grant.

367   **"Living lives":** Interview with Lawrence. **"Best thing":** Interview with Gamble.

367   **"Isn't arbitrary":** Interview with Hall.

367   **"Roadblocks and harassment":** Transcript of remarks by JGB2 at company meeting, January 9–12, 1980, Scottsdale, Arizona, BCA. **"Shut us down":** JF.

367   **147 acres:** "Agricultural Backgrounder," Agriculture Council of America. **426 acres:** U.S. Department of Agriculture statistics.

368   **"Mythical figures":** Interview with Beard.

368   **27,000-acre spread:** Figure cited in CR, May 5, 1982, p. H 1827. **"Kid from Strathmore":** JF.

368   **"Expression of concern":** Rough draft dated December 29, 1983, courtesy of JF.

368   **"Been furious":** JF.

369   **"Bingo!":** Interview with Sterling.

369   **"My problem":** As recalled by JF.

369   **"Wild and yelling":** Interview with Vandergriff.

369   **"A rebel":** Interview with Gordon.

370   **"Doesn't hurt me":** Interview with Gordon.

370   Two didn't speak: Interview with Gordon. **"Some weakness":** Interview with Jones. **Buy his ranch:** JGB2.

370   **His doctor worried:** Bill Boswell's physician was his son-in-law, William Larsen. **"Model patient":** DV with Josephine (Jody) Boswell Larsen.

371   **"My niggers":** Interview with A. L. Vandergriff, who along with his wife often socialized with Bill and Kate. **"First and third":** DV with Josephine (Jody) Boswell Larsen.

371   **"I'm cooking":** DV with Kate Boswell.

371   **"Best race":** DV with Kate Boswell.

371   **Remained involved:** Reflected in correspondence between Ruth Chandler and JGB2, BCA.

371   **Married twice:** See Ruth Chandler von Platen's obituary, LAT, December 11, 1987. **Forbes list:** See, for example, the October 28, 1985, issue of the magazine.

375   **"Succor and secure":** From the essay "The Husbandman," as reprinted in Mencken, *The American Scene: A Reader*, pp. 40–41.

376   **Flood control benefits:** JGB2 never looked at Pine Flat Dam as providing any kind of subsidy for the company, likening the government's building of the project to general highway construction. But when the government picked up the tab for the flood control portion of the dam, Boswell benefited hugely. What's more, even the minor part of the dam paid for by Boswell and other farmers for irrigation was interest-free. For a discussion of flood control projects as government giveaways, see *The Distribution of Farm Subsidies: Who Gets the Benefits?* by Charles L. Schultze, Brookings Institution, 1971. **Crop insurance:** After a Freedom of Information request by the authors, the Agriculture Department calculated that Boswell's crop insurance premiums had been subsidized to the tune of $6.9 million from 1995 through 2000. **Rewards for exporting:** So-called Step-2 Payments to Boswell totaled $11.3 million from 1991 through 2000. (Information supplied by the Agriculture Department's Commodity Office in Kansas City, Missouri, pursuant to a Freedom of Information request by the authors.) **"Raw deal":** Louis Robinson to FS, August 7, 1943, FSP. By September, the first of more than $800,000 started flowing, and the levee mending in the lake bottom was on its way to completion. Specifically, the army spent $813,036 in Fiscal 1944 and $180,537 in FY '45 to reinforce and repair levees that protect Reclamation Districts 761, 2071, 812 and 1616, according to an October 15, 1948, letter to Paul Taylor from the Army Corps in Sacramento, PST.

376   **"Synonymous":** From *The Agribusiness Examiner*, published by A. V. Krebs, April 16, 2000.

376   **As with the original:** The 1938 AAA grew out of the Soil Conservation and Domestic Allotment Act of 1936, which was passed after the Supreme Court invalidated the production control provisions of the original AAA. **So grave:** FS recounted this episode in great detail in a 1965 memo to JGB2 found in FSP. The biggest threat to the West came from the House, which contemplated using a ten-year average of cotton plantings to determine state allotments. (See CN, October 1, 1937; CR, *Proceedings for the 75th Congress*, 2nd session, pp. 510–512, 1317–1318.)

376   **Political clout:** In the Congress that produced the 1938 AAA, southerners chaired the four main agricultural subcommittees. (David P. Ernstes, Joe L. Outlaw and Ronald D. Knutson, *Southern Representation in Congress and U.S. Agricultural Legislation*, Agricultural and Food Policy Center, Texas A&M University, September 1997.) **Carl Hayden:** FS met Hayden through Senator Hiram Johnson, a California Republican with whom he was especially close. The depth of FS's relationship with Johnson is made plain in numerous letters found in the senator's papers at the Bancroft Library, University of California at Berkeley.

377   **Which Hayden wrote:** FS's 1965 memo was sent to Senator Hayden, who in his reply to FS commented, "I have never forgotten that without the help that you and Hiram Johnson gave to me, my cotton amendment would not have been a part of that law."

377   **"Lords of the manor":** Interview with Hohenberg.

377   **Wasn't so simple:** Numerous issues of the CJ through the early '60s show that western cotton growers, including those in Tulare Lake, strongly opposed continuation of the government's market quota system when the issue was put to a referendum of the farmers. However, as noted in a story dated December 12, 1963, the system continued "due to the heavily favorable vote in the Southern states." See also "Cotton-Pickin' Solution," *Time*, September 14, 1962, p. 88. **"Unrealistic system":** CJ, February 2, 1967.

378   **Returned this one:** According to Boswell executive Rice Ober in an undated article, apparently from 1969, in the *Chicago Daily News* headlined "Is Time Ripe for Farm Reform?" (Found in BCA).

378   **Strict mandate:** *Chicago Daily News* article. **$21.5 million:** FB, December 11, 1977. A similar figure is noted in *The Agribusiness Examiner*, February 23, 1999. **Biggest crop subsidy:** Gates, *Land and Law in California*, p. 332. On top of the $4.4 million for taking its lake basin lands out of production, Boswell also received $677,225 in crop subsidies through its Boston Ranch subsidiary. See also Fellmeth, *Politics of Land*, Ralph Nader's study group report on Land Use in California, p. 166. **Just $63:** Gates, *Land and Law in California*, p. 332.

378 **"The penalty":** Quoted in the *Chicago Daily News* article.

378 **"Widen the gap":** Gates, *Land and Law in California*, p. 332. **"International welfare bum":** Union flier found in BCA. **"Unhappy farmer":** *Life,* November 6, 1970, "Parting Shots."

378 **Soon capped:** Noted in, among other places, Gates, *Land and Law in California*, p. 332. **"Food and fiber":** Quoted in "History of Agricultural Price-Support and Adjustment Programs, 1933–84," U.S. Department of Agriculture, Economic Research Service, bulletin no. 485.

378 **"Would go broke":** Interview with Vandergriff.

379 **14,000 acres:** All details on Boswell's involvement in the program are from the FB of October 2, November 2 and November 6, 1983.

379 **"Hybrid strains":** Quoted in FB, November 2, 1983.

379 **"Let's face it":** Quoted in FB, October 2, 1983.

380 **"Great boondoggles":** Interview with Parker.

380 **Worst crisis:** Details drawn from *Newsweek,* February 18, 1985. **"Like a scythe":** *Newsweek* quoting Iowa State University agricultural economist Neil Harl.

380 **"New Grapes":** *Time,* January 28, 1985. **"Bitter Harvest":** *Newsweek,* February 18, 1985. **"The Tubes":** *U.S. News,* February 4, 1985. **"Farms not arms":** See *Jet,* April 22, 1985.

380 **Managed to escape:** See *Business Week,* October 28, 1985, "The Farm Crisis Has Finally Migrated to California." **"Significant reductions":** November 18, 1985, letter from JGB2 to "fellow employees" offering an early retirement program, BCA. See also December 2, 1985, LAT, article by Peter H. King, "Retrenchment by Major Cotton Firm Jolts Small Town Rooted in Farming." **"Surpluses that keep":** November 15, 1985, letter from JGB2 to company shareholders, BCA.

380 **"Mistakes of the past":** This and the other quotes are from letters dated September 17, 1985, from JGB2 to President Reagan, Vice President Bush and seven lawmakers, BCA.

381 **Caved in to pressure:** See the *Congressional Quarterly Almanac,* 99th Congress, 1st Session, vol. 41, pp. 515–539. **Admittedly:** See remarks of the president from signing ceremony for H.R. 2100, the Food Security Act, December 23, 1985.

381 **$20 million:** The figure was widely reported. See for example, *Time,* August 18, 1986. The Associated Press, in July 1986, noted that Boswell would receive some $10.4 million in cotton and wheat payments through the marketing loan program. The remainder would come from the first-handler program. **"Obscene":** *Omaha World-Herald,* August 10, 1986. **"Mandatory economically":** Quoted in July 1986 AP dispatch. **"Survival money":** *Time,* August 18, 1986.

381 **"To get one guy":** *Omaha World-Herald,* October 23, 1986.

382   **Whittled down:** For a good summary of what the Reclamation Reform Act was supposed to do but in many cases did not, see "The Broken Promise of Reclamation Reform" by Hamilton Candee of the Natural Resources Defense Council, *Hastings Law Journal*, March 1989.

382   **60 Minutes:** The segment "Water, Water Everywhere" was broadcast in November 1989. **One expert:** Although not named by Bradley, this was David F. Schuy, an acreage-limitation specialist with the Bureau of Reclamation. (See Schuy declaration in *Westlands Water District v. U.S.*, Case no. CV-F–81–245-EDP, filed 1981 in the U.S. District Court for the Eastern District of California.) Others, using a different calculation, came up with a much lower subsidy estimate than Schuy's $11.1 million for 1987. For instance, a June 1990 report by the General Accounting Office put the annual subsidy to Boston Ranch at about $2 million. ("Water Subsidies: The Westhaven Trust Reinforces the Need to Change Reclamation Law," GAO/RCED–90–198.)

383   **Largest agricultural water district:** Details from the Central Valley Project Water Association.

383   **Giffen had arrived:** Details on the Giffen family and life on the west side are from *Fortune*, May 1949; interview with Carolyn Peck, Russell Giffen's daughter.

383   **"Bent the river":** Interview with Jason Peltier, longtime manager of the Central Valley Project Water Association.

383   **Persuaded the Bureau:** See "Westlands Water District: The First 25 Years, 1952–1977."

383   **Began squabbling:** The details on fighting within Westlands, as well as its battle with the government, are drawn from interviews with Boswell attorneys Bill Smiland and Ken Khachigian; court documents found in *Barcellos & Wolfsen Inc. v. Westlands Water District*, case no. CV–79–106-EDP, filed 1979 in the U.S. District Court for the Eastern District of California (which was partly consolidated with *Westlands Water District v. U.S.*); "Westlands Water District and Its Federal Water: A Case Study of Water District Politics," by Bryan J. Wilson, Stanford Environmental Law Journal, 1987–1988; Wahl, *Markets for Federal Water: Subsidies, Property Rights, and the Bureau of Reclamation*, pp. 107–124. **Westlands had merged:** Westlands itself was formed in 1952. Westplains was established in 1962. The two merged in 1965.

384   **Back in 1963:** Equally important in the eyes of Westlands farmers was a 1964 memo written by Kenneth Holum, an assistant secretary of the interior, to Interior Secretary Stewart Udall. Among other things, the Holum memorandum—which Udall formally approved—said that the Bureau of Reclamation should provide all of the water promised in the 1963 contract at a price of $7.50 per acre-foot (plus an additional fifty cents for drainage

service). In addition, Holum said that provisions should be made for the government to provide additional supplies of water at the same price to cover the acreage to be added by the Westplains merger. **"No certainty":** From p. 72 of an oral history with Brody conducted in 1980 by Malca Chall of the Regional Oral History Office, Bancroft Library, University of California at Berkeley.

384   **"Welfare queens":** Interview with Smiland. That wasn't the sum total of the name calling, either. Ronald Reagan's interior secretary, James Watt, was hardly known for taking positions that the environmental community applauded. But on the issue of water subsidies for Westlands, he was firm: He wanted them stopped, and he threatened to cut off all water to the farmers unless the district paid full cost—then $42 an acre-foot. (See "Paying for Water" by Lawrence Mosher, *National Journal*, May 31, 1986.) When Khachigian and other Westlands advocates tried to win over Watt, Khachigian recalled, the secretary growled at the group the minute he walked into the room. "Hello," Watt said. "Why am I meeting with the PLO of Westlands?" **Lasted through the 1970s:** Like Brody, executives at Boswell argued that the 1963 agreement was unassailable. Other district farmers, however, were willing to make concessions, hoping that their compromises would help secure a long-term water delivery agreement with the bureau. Finally, in 1977, the fight spilled into the open. A new Interior Department legal opinion, written by solicitor Leo Krulitz, undid the terms of the Holum memorandum. The Carter administration then announced that it was going to charge $14.50 an acre-foot for water in Westlands— nearly double the original contract rate. "Sometimes the government makes improvident bargains, and it has to change those improvident bargains" when it is "in the public interest," said a lawyer for the Interior Department. (Cited in *Amicus Curiae* of Boswell counsel Edward C. Giermann, *Natural Resources Defense Council et al. v. C. Dale Duvall, as commissioner of the Bureau of Reclamation, and Manuel Lujan Jr., as secretary of the Department of the Interior*, case no. S–88–0375-LKK-EM, filed 1988 in the U.S. District Court for the Eastern District of California.) To Boswell, this was the worst kind of doublecross. The way the company saw things, the government had to honor the water supply deal that it had made in 1963—even if the bureau's costs had climbed far more than anticipated at the outset. "Inflation caught up with them," Ken Khachigian, a speechwriter for Ronald Reagan and a prominent Republican lawyer who represented Boswell's side, said in an interview. "But a contract is a contract." Boswell urged the district board to try to enforce the 1963 accord by taking the government to court. The board, though, declined. So Boswell and five other farmers took action on their own, suing Westlands. Eventually, the government also was named a defendant. All the while, the hostil-

ity continued. "We got pounded," Bill Smiland, the attorney who brought the suit for Boswell and the other hard-nosed Westlands farmers, said in an interview. (That suit was *Barcellos & Wolfsen Inc. v. Westlands Water District.*) **"Keep their word":** From an August 5, 1986, letter from Hodel to Congress in support of the settlement. (Cited in *Amicus Curiae* of Boswell counsel Edward C. Giermann, *Natural Resources Defense Council et al. v. C. Dale Duvall, as Commissioner of the Bureau of Reclamation, and Manuel Lujan Jr., as Secretary of the Department of the Interior.*) **Beyond what Congress:** See, for example, LeVeen and King, *Turning Off the Tap on Federal Water Subsidies.* Also see *Special Task Force Report on San Luis Unit, Central Valley Project, California,* U.S. Department of the Interior, 1978. **Blasted Hodel:** See remarks by George Miller in CR, March 20, 1986, p. E921. The Natural Resources Defense Council was especially critical of the settlement, estimating that it would cost the government $266 million in lost water service revenues alone, contrary to the sanguine claims of the Interior Department (August 28, 1986, NRDC memorandum to "interested parties"). **"Sweetheart deal":** As characterized by a July 31, 1986, *Philadelphia Inquirer* editorial. Also critical of the settlement were the editorial pages of the LAT, *Washington Post* and *San Francisco Chronicle.*

385   **Empires were carved up:** Land that had been owned by Russell Giffen and the old cotton company Anderson Clayton, for example, was subdivided into seventy-seven different farms (Westlands Water District Fact Sheet, December 1976, Ralph M. Brody Papers, Water Resources Center Archives, University of California at Berkeley). Southern Pacific Railroad, which at one time held more than 100,000 acres in Westlands, also broke up its holdings (FB, August 6, 1987). **Fits and starts:** Boswell had originally signed a so-called recordable contract with the government in 1972, obligating it to sell off its excess lands at Boston Ranch within ten years. But because of the case *National Land for People v. Bureau of Reclamation,* a moratorium on all such excess land sales was put into effect in 1976. This stretched out the clock to 1990. Meantime, as the deadline neared, Boswell had an added incentive to sell off the land. In 1987 Congress tightened up on the reclamation law so that those farmers who were in the process of shedding their excess acres had to pay more for their water in the meantime. (See "Subsidies Gush from Loophole," FB, October 11, 1987; FB, December 17, 1987.) Boswell was thus facing a $2.2 million jump in its annual costs at Boston Ranch. (From declaration of Boswell counsel John C. Sterling in *Barcellos and Wolfsen Inc. et al v. Westlands Water District and U.S. Department of Interior,* 899 F.2nd 814 [9th Cir. 1990].)

385   **The company's blueprint:** Details on the Westhaven Trust are gleaned from Interior Department documents obtained through a Freedom of Information request by the authors; interviews with former Boswell execu-

tives Ed Giermann and John Sterling; interviews with Boswell lawyer Bill Smiland; interview with Natural Resources Defense Council attorney Hal Candee; and material in NRDC files, San Francisco. **326 of Boswell's own:** The original number of 258 employee-beneficiaries was expanded over time. **Draw fees:** The $2 million figure is from the Giermann and Sterling interviews. **Wasn't as good:** Profit figures are from Giermann interview.

385 **"J. G.'s closest buddies":** Interview with Giermann.

385 **"Transparent scheme":** June 2, 1989, letter from Miller to Interior Secretary Manuel Lujan, NRDC files. **"Boswell arrogance":** Quoted in FB, June 1, 1989. Jack Stone, president of the Westlands board, also had words for Boswell. He had put his own farmland into a trust for his family—the kind of prudent estate planning, he suggested, that was intended by the Reclamation Reform Act (FB, June 1, 1989). But the Boswell trust, he said, was a different animal—one that "could undermine the effectiveness of reclamation law throughout the West" (May 30, 1989, letter from Stone to C. Dale Duvall, the commissioner of the Bureau of Reclamation). On May 31, JWB fired off a threatening rebuke to Stone that began: "We are appalled and upset by your letter. . . ."

385 **Some initial reluctance:** The man who had succeeded Don Hodel as secretary of the interior, former New Mexico congressman Manuel Lujan Jr., wasn't the sharpest member of George H. W. Bush's cabinet. ("Nobody's told me the difference between a red squirrel, a black one or a brown one," he once said in discussing the Endangered Species Act. "Do we have to save every subspecies?") But on the Westhaven Trust, he seemed to have little trouble assessing the situation. Boswell is "trying to circumvent the law," he said flatly, "and I don't think that's right" (*Albuquerque Journal*, June 15, 1989). Establishment of the Westhaven Trust was thus denied. But company officials took another run at Lujan, hoping to shake him from his position. It wound up being a fairly easy task. Lujan's only real objection to the Westhaven Trust was structural. If one of the 326 employees left the company, either voluntarily or as a result of being laid off, he or she would no longer be a beneficiary of the trust. This wrinkle, the secretary decided, gave Boswell undue control over the land it was supposed to be selling. But once the company agreed to change the details slightly, so that the beneficiaries would remain beneficiaries even if they moved on from Boswell, Lujan said he had no choice but to approve the arrangement. Angry over Lujan's flip-flop, the environmental community smelled a rat. The solicitor of the Interior Department, Ralph Tarr, had worked at the very law firm in Fresno representing the trustee in the deal, Security Pacific Bank (where Jim Boswell sat on the board). When he was at the firm, Baker, Manock & Jensen, Tarr also had worked on a case that touched on some of the same acreage-limitation issues that Boswell was now con-

fronting. Senator Bill Bradley was among those who questioned Tarr's involvement in the Westhaven matter, saying that he had "serious concern" (June 14, 1989, letter from Bradley to Interior Secretary Manuel Lujan). But the government ethics police gave Tarr its blessing to work on the Interior Department's review of the trust. And, in an interview, Tarr stressed that he had no need to recuse himself and took every possible step to ensure that he handled his review of the Westhaven Trust professionally and ethically. **Full decade:** The Westhaven Trust could have continued for twenty more years, according to Boswell's Ed Giermann, but water became harder and harder to come by in the Westlands area over time, and the trustees decided to break up the property after ten years. Meanwhile, the trust remained a poster child for "corporate welfare" (see *Time*, November 16, 1998), and the Natural Resources Defense Council and others continued to fight it. Some critics would also assert that if the Bush administration did Boswell a favor by approving the Westhaven Trust, Boswell certainly showed its gratitude: JWB gave $125,000 in "soft money" to the Republican Party, becoming one of 249 wealthy contributors known as President Bush's Team 100. (1992 *Common Cause* magazine investigation.)

387   **Wegis hailed:** Interview with Wegis; articles in BC. **"Home-towned":** Interview with Sterling.

387   **Dark business suits:** July 13, 1988, BC article by Michael Trihey.

Chapter 18: Death in the Homeland Canal

388   **Kill hardly registered:** Interviews with several Fish and Game personnel, including Captain Ron Stukey, Lieutenant Dennis Davenport and Game Warden Gloria Marsh.

388   **A band of dead fish**: Interviews with Game Warden Gloria Marsh and U.S. Fish and Wildlife Service biologist Carmen Thomas.

389   **One of the largest fish kills:** October 2, 1997, FB. **Wanted no part:** Interview with Lieutenant Dennis Davenport and other Fish and Game personnel.

389   **Caught McCarthy:** Interviews with Game Warden Gloria Marsh and Captain Ron Stukey; a copy of the April 3, 1996, citation obtained by the authors.

389   **Long considered the rogues:** See June 10, 2002, LAT article on farm subsidies.

390   **Managers begged off:** FB, September 25, 1997. **"Traces of pesticide":** FB, September 23, 1997.

390   **Bricker had confirmed:** Fish and Game incident action plan from September 16, 1997.

390   **"Stymie the investigation":** Interview with Davenport.

391   **Routinely flouted:** Interview with the former deputy commissioner, who requested anonymity.

391   **"Like the canary":** FB, October 2, 1997.

392   **"Try to peck":** Interview with Thomas.

392   **2.7 million pounds:** Pesticide and other farm chemical figures are based on official state reports compiled by Susan Kegly of Pesticide Action Network.

392   **Did find forty pounds:** Fish and Game incident action plan from September 16, 1997.

393   **"Lab reports came back":** Interview with Marsh.

393   **"Toxic soup":** Interview with Thomas.

393   **"Sooner or later":** Interview with Grewal.

394   **"Once again untouchable":** Interview with the former staffer, who requested anonymity.

394   **Old files were sprinkled:** Conservation Associates Records Collection, Bancroft Library, University of California at Berkeley. **"In one ear":** JGB2.

395   **60 million waterfowl:** "The Killing Fields" by Robert H. Boyle, *Sports Illustrated*, March 22, 1993. **Now there were just 300,000 acres:** *Sports Illustrated* article. **"Lethal Haven":** *Newsweek*, September 8, 1969.

396   **Spent more than $10 million:** LAT, July 25, 1984; interviews with Paul Athorp and SB.

396   **Signed off on the pumping plan:** Op-ed piece in the LAT of October 14, 1983, by William Kahrl. **Some of the rowdy ones:** Interviews with Paul Athorp and Riverdale farmer George McKean.

396   **Deadly plant substance:** Johnson, Haslam and Dawson, p. 159.

397   **"Bunch of fingers":** Interview with Michael Morse, U.S. Fish and Wildlife official.

398   **Far back as 1941:** *Sports Illustrated*, March 22, 1993. **Kesterson:** The environmental devastation at Kesterson was widely covered. For a particularly good retelling, see Harris, *Death in the Marsh*.

398   **"No Kestersons here":** Interview with Joe Skorupa. **Joe Skorupa:** His story is drawn, in part, from Clemings, *Mirage: The False Promise of Desert Agriculture*, which is an excellent account of Skorupa and the evaporation ponds of Tulare Lake.

398   **Ten times lower:** Interview with Skorupa.

399   **"Politically explosive":** Interview with Skorupa.

400   **"He was publishing":** *Sports Illustrated*, March 22, 1993. **"Kill the messenger":** Interview with Skorupa.

400   **Violated the Migratory Bird Treaty Act:** Interviews with environmental advocates Lloyd Carter and Patrick Porgans, who filed a petition with the California State Water Resources Control Board, August 25, 1995. **"Least up on":** Fellmeth, p. 131.

400   **No more than four weeks:** Interviews with Carter and Porgans; their petition to the state.

401   **State board backed off:** Interviews with Skorupa and Carter.

401   **Invited scientists and reporters:** All facts and quotes are from Mark Arax's
June 16, 1997, LAT story on this event.

402   **Jim Boswell himself flew:** JGB2.

402   **Elderberry longhorn beetle:** Details on this episode are drawn from JGB2;
interviews with Army Corps chief Roy Proffitt; Boxer aide Tom Bohigian;
and staffers at Fish and Wildlife and the Army Corps. **Senator Boxer
pleaded:** May 18, 1998, letter from Boxer to Capka.

403   **"Unbelievable to me":** Interview with Proffitt.

404   **"Never like those":** Interview with Morse.

404   **Like so many deals:** All details on Boswell's purchase of Salyer are from
JGB2 and FSal.

405   **Fred accused Scott:** *Salyer American v. F. Scott Salyer*, case no. 115999, filed
1998 in Monterey County Superior Court; *F. Scott Salyer v. Salyer American*,
case no. 115481, filed 1998 in Monterey County Superior Court.

406   **Owing half a million dollars:** *Walter Gleason v. Salyer Land Co.* **"Services
as a peacemaker":** The letter is part of the court file.

406   **"The little wounds":** LSal.

406   **Word leaked out:** See "Two Farm Giants End Decades of Rivalry with Land
Deal," LAT, February 10, 1995.

CHAPTER 19: AUTUMN

411   **Paper drying trays:** See "A Bumper Crop of Bad Air in San Joaquin Valley,"
by Mark Arax and Gary Polakovic, LAT, December 8, 2002.

412   **"Capital of bad people":** Interview with Dye.

413   **"Real slow":** Interview with Hill.

413   **One out of ten Latino girls:** See "Maternity Before Maturity," a report by
Hans Johnson of the Public Policy Institute of California, February 2003.
Methamphetamine: LAT, March 13, 1995. **One young drug dealer:** "A Myth
of Hope in a Land of Tragedy," by Mark Arax, LAT, August 26, 2002.

414   **Lived in poverty:** BC, November 9, 1999. **Jobless rate:** FB, January 9, 2000.
Seven families: See LAT, June 10, 2002. Data is from the Environmental
Working Group's farm subsidy database.

414   **The McCarthys had found:** LAT, June 10, 2002. **Ceal Howe Jr.:** See *Capital
Press*, February 23, 2001.

415   **A fundamental shift:** All details and quotes are from the author's attendance at the city council session.

416   **"Pissed off everybody":** Interview with Pauley.

418   **More inmates were killed:** See a series of 1998 articles in the LAT by Mark Arax and Mark Gladstone.

418   **A senile Juan Corona:** This scene is drawn from the author's visit to the prison.

419   **"We were told":** Interview with Castillo. **"Could literally see":** LSal.

419   **"Just me and Howard":** Interview with Gertha Toney.

420   **"Folded up their sacks":** Interview with Dorothy Toney.

421   **The pilots dropped:** The amount of defoliants sprayed in the lake basin came from a field supervisor for Lakeland Dusters-Aviation, based in Corcoran, and from Mark Grewal of Boswell. The Lakeland supervisor, Daniel Saenz, said an average of two pints per acre is used on 180,000 acres. There are two such applications in autumn, meaning that 720,000 pints are used in an average year. This amounts to 90,000 gallons of defoliating agents, which include sodium chlorate, paraquat and other herbicides.

421   **An old Yokut curse:** From Miller, *First the Blade*, p. 369.

422   **A good hand picker:** Interview with Grewal. **Each Boswell machine:** Interview with Grewal.

423   **10,000 loaves of bread:** Unlike many areas of business, where bigger operations invariably enjoy certain economies of scale, huge farming operations don't typically see such benefits. "Efficient farmers are not usually large operators," Kramer notes in *Three Farms* (p. 192). "Size inevitably breeds clumsiness—and the modern management structures that are designed to work flexibility back into stiffening systems themselves add layers of costly management. . . ." One economist (Kramer, p. 267) puts the "optimal-sized" farm at no bigger than 2,000 acres—a fraction of Boswell's holdings. But Boswell was clearly an exception to this idea; its bottom line, year in and year out, defied the notion that smaller is better.

423   **Fernand Braudel:** See *The Mediterranean and the Mediterranean World in the Age of Philip II.*

423   **Walter Goldschmidt:** The Arvin-Dinuba study later appeared in book form under the title *As You Sow.* His saga is drawn from an interview with Goldschmidt; "Walter Goldschmidt's Baptism by Fire: Central Valley Water Politics," by Paul Taylor, *Anthropology UCLA,* 1976; *New Republic,* February 3, 1947.

424   **Failed to come through:** JGB2.

425   **Corcoran may have generated:** From interviews with city staff and a review of Corcoran city budgets from the 1940s, '50s, '60s and '70s.

425   **Kings River as a utility:** "Use of Water from Kings River, California," by Harry Barnes, Bulletin no. 7, State of California, Department of Engineering, 1918.

427   **To grow one T-shirt:** Figures from the California Cotton Ginners and

Growers Associations. **$1,100 in an average market:** Based on 2002 market price. **Had the water been shipped:** Interviews with several San Joaquin Valley farmers engaged in water marketing talks with the Metropolitan Water District of Southern California.

428   **Produced 600 bales a day:** Interview with Boswell executive Ross Hall.

428   **"Tighter than ever":** JWB.

430   **15 percent draw:** Interviews with Mark Grewal, SB.

Bibliography

Books

The books listed are those used by the authors, not necessarily the original editions, so that the page numbers cited in the endnotes will correspond.

Adams, Kenneth C. *California Missions*. Los Angeles: California Mission Trails Association, 1947.

Agee, James and Evans, Walker. *Let Us Now Praise Famous Men*. Boston: Houghton Mifflin Company, 1960.

Allen, Hugh. *The House of Goodyear: Fifty Years of Men and Industry*. Cleveland: Corday & Gross, 1949.

Anderson, Terry L. (editor). *Water Rights: Scarce Resource Allocation, Bureaucracy, and the Environment*. San Francisco: Pacific Institute for Public Policy Research, 1983.

Andrus, Cecil. *Politics Western Style*. Seattle: Sasquatch Books, 1998.

Anon., *Memorial and Biographical History of the Counties of Fresno, Tulare and Kern California*. Chicago: Lewis Publishing Company, 1892.

Arax, Mark and Pickford, Joel. *California Light: The Watercolors of Rollin Pickford*. Fresno, California: CSUF Press, 1998.

Armor, E.H. *The Cemeteries of Green County Georgia*. Agee Publishers Inc., 1987.

Auerbach, Jerold S. *Labor and Liberty: The La Follette Committee and the New Deal*. Indianapolis, Indiana: Bobbs-Merrill Company, 1966.

Bancroft, Hubert Howe. *The Works of Hubert Howe Bancroft*. San Francisco: The History Company, 1890.

Barry, Raymond P. (ed.). *A Documentary History of Migratory Labor in California: Labor in California Cotton Fields*. Oakland, California: Federal Writers Project, 1938.

Baum, Dan: *Smoke and Mirrors: The War on Drugs and The Politics of Failure*. Boston: Little, Brown and Company, 1996.

Benson, Jackson J. *The True Adventures of John Steinbeck, Writer*. New York: Viking Press, 1984.

Birmingham, Stephen. *California Rich*. New York: Simon & Schuster, 1980.

Bonelli, William G. *Billion Dollar Blackjack*. Beverly Hills, California: Civic Research Press, 1954.

Bonington, Chris. *Kongur: China's Elusive Summit*. London: Hodder and Stoughton, 1982.

Boswell, Alaine Buck. *Stories & Poems And a Play, Still Waters*. 1940.

Braudel, Fernand. *The Mediterranean and the Mediterranean World in the Age of Philip II*. Berkeley, California: University of California Press, 1996.

Breimyer, Harold F. *Farm Policy: 13 Essays*. Ames, Iowa: Iowa State University Press, 1977.

Briggs, William J. and Cauthen, Henry. *The Cotton Man: Notes on the Life and Times of Wofford B. ("Bill") Camp*. Columbia, South Carolina: University of South Carolina Press, 1983.

Britton, Karen Gerhardt. *Bale o' Cotton: The Mechanical Art of Cotton Ginning*. College Station, Texas: Texas A&M University Press, 1992.

Brown, James R. *The Story of Kings County California*. Hanford, California: A. H. Cawston Publishing, 1940.

Bruchey, Stuart (ed.). *Cotton and the Growth of the American Economy: 1790–1860*. New York: Harcourt, Brace & World, 1967.

Bryant, Jonathan M. *How Curious a Land: Conflict and Change in Greene County, Georgia 1850–1885*. Chapel Hill, North Carolina: University of North Carolina Press, 1996.

Burton, Art. *Black, Red and Deadly*. Austin, Texas: Eakin Press, 1991.

Bush, George S. *An American Harvest: The Story of Weil Brothers-Cotton*. Englewood Cliffs, New Jersey: Prentice-Hall Inc., 1982.

California Department of Water Resources. *California State Water Project Atlas*. Sacramento, California, 1999.

_____. *Sacramento-San Joaquin Delta Atlas*. Sacramento, California, 1995.

Carson, J. H. *Early Recollections of the Mines and a Description of the Great Tulare Valley*. Stockton, California, 1852.

Castro, Doris Shaw. *James H. Carson's California, 1847–1853*. New York: Vantage Press, 1997.

Clawson, Marion. *Uncle Sam's Acres*. New York: Dodd, Mead & Company, 1951.

Clemings, Russell. *Mirage: The False Promise of Desert Agriculture*. San Francisco: Sierra Club Books, 1996.

Cohn, David L. *The Life and Times of King Cotton*. New York: Oxford University Press, 1956.

Cooper, Erwin. *Aqueduct Empire: A Guide to Water in California, Its Turbulent History and Its Management Today*. Glendale, California: Arthur H. Clark Company, 1968.

Creel, George. *Rebel at Large: Recollections of Fifty Crowded Years*. New York: G. P. Putnam's Sons, 1947.

Cross, William T. and Cross, Dorothy E. *Newcomers and Nomads in California*. Stanford, California: Stanford University Press, 1937.

Culver, John C. and Hyde, Henry. *American Dreamer: A Life of Henry A. Wallace.* New York: W. W. Norton & Company, 2000.

Curtis, Alice Turner. *The Story of Cotton.* Philadelphia: Penn Publishing Company, 1916.

Daniel, Cletus. *Bitter Harvest: A History of California Farmworkers 1870–1941.* Berkeley, California: University of California Press, 1982.

Davis, Margaret Leslie. *Rivers in the Desert: William Mulholland and the Inventing of Los Angeles.* New York: HarperCollins, 1993.

De L. Welch, Marie. *This is Our Own.* New York: MacMillan Company, 1940.

deBuys, William and Myers, Joan. *Salt Dreams: Land & Water in Low-Down California.* Albuquerque, New Mexico: University of New Mexico Press, 1999.

De Roos, Robert. *The Thirsty Land: The Story of the Central Valley Project.* Stanford, California: Stanford University Press, 1948.

Dickson, Harris. *The Story of King Cotton.* New York: Funk & Wagnalls, 1937.

Dies, Edward Jerome. *Titans of the Soil: Great Builders of Agriculture.* Chapel Hill, North Carolina: University of North Carolina Press, 1949.

Dodge, Bertha S. *Cotton: The Plant That Would be King.* Austin, Texas: University of Texas Press, 1984.

Douglas, William O. *The Court Years, 1939–1975: The Autobiography of William O. Douglas.* New York: Random House, 1980.

Downey, Sheridan. *Onward America.* Sacramento, California: Larkin Printing Company, 1933.

_____. *They Would Rule the Valley.* San Francisco, 1947.

Dumke, Glenn S. *The Boom of the Eighties in Southern California.* San Marino, California: Huntington Library, 1944.

Dumond, Dwight Lowell (ed.). *Southern Editorials on Secession.* New York: Century Company, 1931.

Dunne, John Gregory. *Delano: The Story of the California Grape Strike.* New York: Farrar, Straus & Giroux, 1967.

Ellison, William Henry. *A Self-governing Dominion: California 1849–1860.* Berkeley, California: University of California Press, 1950.

Fellmeth, Robert C. (project dir.). *Politics of Land: Ralph Nader's Study Group Report on Land Use in California.* New York: Grossman Publishers, 1973.

Ferriss, Susan and Sandoval, Ricardo. *The Fight in the Fields: Cesar Chavez and the Farmworkers Movement.* New York: Harcourt, Brace and Company, 1997.

Findley, Paul. *The Federal Farm Fable.* New Rochelle, New York: Arlington House, 1968.

Finnerty, Margaret; Blanc, Tara; and McCann, Jessica. *Del Webb: A Man. A Company.* Phoenix, Arizona: Heritage Publishers, 1999.

Fitch, Noel Riley. *Appetite for Life: The Biography of Julia Child.* New York: Doubleday, 1997.

Fleming, Douglas K. *Thoughts of a Cotton Man: From Lamar Fleming's Collected Papers.* Seattle: AlphaGraphics, 1994.

Fleming, Lamar, Jr. *Growth of the Business of Anderson, Clayton & Company*
Houston: Texas Gulf Coast Historical Association, 1966.

Folsom, James M. *Heros and Martyrs of Georgia: Georgia's Record in the Revolution of 1861*. Macon, Georgia: Burkey, Boykin & Company, 1864.

Foote, Shelby, *The Civil War: A Narrative*. New York: Random House, 1958.

Fossedal, Gregory A. *Our Finest Hour: Will Clayton, the Marshall Plan and the Triumph of Democracy*. Stanford, California: Hoover Institution Press, 1993.

Frederick, Kenneth D. with Hanson, James C. *Water for Western Agriculture*. Washington, D.C.: Resources for the Future, 1982.

Freeman, Jane and Sanberg, Glenn. *Silver Anniversary Jubilee: A History of Sun City, Arizona*. Sun City, Arizona: Sun City Historical Society, 1984.

Fremont, John Charles. *Geographical Memoir*. San Francisco: The Book Club of California, 1964.

Garside, Alston Hill. *Cotton Goes to Market: A Graphic Description of a Great Industry*. New York: Frederick A. Stokes Company, 1935.

_____ (ed.). *Cotton Year Book of the New York Cotton Exchange, 1935*. New York, Van Rees Press, 1935.

Garwood, Ellen Clayton. *Will Clayton: A Short Biography*. Austin, Texas: University of Texas Press, 1958.

Gates, Paul W. *The Farmer's Age: Agriculture, 1815–1860*. New York: Harper Torchbooks, 1968.

_____. *The Jeffersonian Dream: Studies in the History of American Land Policy and Development*. Albuquerque, New Mexico: University of New Mexico Press, 1996.

_____. *Land and Law in California: Essays on Land Policies*. Ames, Iowa: Iowa State University Press, 1991.

Gentry, Byron. *The Way the Ball Bounces*. San Antonio, Texas: Naylor Company, 1962.

Gist, Brooks D. *Empire out of the Tules*. Tulare, California: Gist Publishing, 1976.

_____. *The Years Between*. Tulare, California: Gist Publishing, 1952.

Goldschmidt, Walter. *As You Sow*. New York: Harcourt, Brace and Company, 1947.

Goodall, Merrill R.; Sullivan, John D.; and De Young, Timothy. *California Water: A New Political Economy*. Montclair, New Jersey: Allanheld, Osmun and Company, 1978.

Goodstein, Judith R. *Millikan's School: A History of the California Institute of Technology*. New York: W. W. Norton & Company, 1991.

Goodwin, Doris Kearns. *No Ordinary Time: Franklin and Eleanor Roosevelt: The Home Front in World War II*. New York: Simon & Schuster, 1994.

Gottlieb, Robert and FitzSimmons, Margaret. *Thirst for Growth: Water Agencies as Hidden Government in California*. Tucson, Arizona: University of Arizona Press, 1991.

Gottlieb, Robert and Wolt, Irene. *Thinking Big: The Story of the Los Angeles*

Times, Its Publishers and Their Influence on Southern California. New York: G. P. Putnam's Sons, 1977.

Green, Constance McL. *Eli Whitney and the Birth of American Technology.* Boston: Little, Brown and Company, 1956.

Greer, T. Keister. *The Great Moonshine Conspiracy Trial of 1935.* Rocky Mount, Virginia: History House Press, 2002.

Gregory, James N. *American Exodus: The Dust Bowl Migration and Okie Culture in California.* New York: Oxford University Press, 1989.

Griswold, A. Whitney. *Farming and Democracy.* New Haven, Connecticut: Yale University Press, 1952.

Hake, S. Johnson; Kerby T. A.; and Hake, K. D. *Cotton Production Manual.* Berkeley, California: University of California, Division of Agriculture and Natural Resources, 1996.

Halberstam, David. *The Powers That Be.* New York: Alfred A. Knopf, 1979.

Hamilton, David E. *From New Day to New Deal: American Farm Policy from Hoover to Roosevelt, 1928–1933.* Chapel Hill, North Carolina: University of North Carolina Press, 1991.

Harding, S. T. *Water in California.* Palo Alto, California: N-P Publications, 1960.

Harris, Tom. *Death in the Marsh.* Washington, D.C.: Island Press. 1991.

Haslam, Gerald. *The Other California.* Reno, Nevada: University of Nevada Press, 1994.

Hobhouse, Henry. *Seeds of Change: Five Plants That Transformed Mankind.* New York: Harper & Row, 1986.

Holliday, J. S. *Rush for Riches: Gold Fever and the Making of California.* Berkeley, California: University of California Press, 1999.

Hundley, Norris, Jr. *The Great Thirst: Californians and Water, 1770s–1990s.* Berkeley, California: University of California Press, 1992.

Igler, Robert. *Industrial Cowboys: Miller & Lux and the Transformation of the Far West, 1850–1920.* Berkeley, California: University of California Press, 2001.

Irvine, Leigh H. *A History of the New California.* New York: Lewis Publishing Company, 1905.

Jacobson, Timothy Curtis and Smith, George David. *Cotton's Renaissance: A Study in Market Innovation.* Cambridge, England: Cambridge University Press, 2001.

Jamieson, Stuart. *Labor Unionism in American Agriculture.* New York: Arno Press, 1976.

Jelinek, Lawrence J. *Harvest Empire: A History of California Agriculture.* San Francisco: Boyd & Fraser Publishing Company, 1979.

Johnson, Hannibal B. *Black Wall Street.* Austin, Texas: Eakin Press, 1998.

Johnson, Stephen; Haslam, Gerald; and Dawson, Robert. *The Great Central Valley: California's Heartland.* Berkeley, California: University of California Press, 1993.

Kahrl, William L. (ed.). *The California Water Atlas*. Sacramento, California: State of California, 1979.

_____. *Water and Power*. Berkeley, California: University of California Press, 1983.

Kaupke, Charles L. *Forty Years on Kings River 1917–1956*. Fresno, California: Kings River Water Association, 1957.

Kelley, Robert. *Battling the Inland Sea: American Political Culture, Public Policy, & the Sacramento Valley, 1850–1986*. Berkeley, California: University of California Press, 1989.

Kincaid, Robert. *The Wilderness Road*. Harrogate, Tennessee: Lincoln Memorial University, 1996.

Klein, Joe. *Woody Guthrie: A Life*. New York: Delta, 1999.

King, William F. *The San Gabriel Valley: Chronicles of an Abundant Land*. Windsor Publications Inc., 1990.

Kirkendall, Richard S. *Social Scientists and Farm Politics in the Age of Roosevelt*. Columbia, Missouri: University of Missouri Press, 1966.

Kotkin, Joel and Grabowicz, Paul. *California Inc*. New York: Rawson, Wade Publishers, 1982.

Kramer, Mark. *Three Farms: Making Milk, Meat and Money from the American Soil*. Boston: Atlantic-Little, Brown, 1977.

Lange, Dorothea and Taylor, Paul Schuster. *An American Exodus: A Record of Human Erosion*. New York: Arno Press, 1975.

Latta, F. F. *Black Gold in the Joaquin*. Caldwell, Idaho: Caxton Printers, 1949.

_____. *Handbook of Yokuts Indians*. Oildale, California: Bear State Books, 1949.

_____. *Tailholt Tales*. Santa Cruz, California: Bear State Books, 1976.

Lawren, William. *The General and the Bomb: A Biography of General Leslie R. Groves, Director of the Manhattan Project*. New York: Dodd, Mead & Company, 1988.

Lemann, Nicholas. *The Promised Land: The Great Black Migration and How it Changed America*. New York: Alfred A. Knopf, 1991.

LeVeen, E. Phillip and King, Laura B. *Turning Off the Tap on Federal Water Subsidies*. San Francisco: Natural Resources Defense Council and California Rural Legal Assistance Foundation, 1985.

Litchfield, P. W. *Industrial Voyage: My Life as an Industrial Lieutenant*. Garden City, New York: Doubleday & Company, 1954.

Litwack, Leon F. *Trouble in Mind: Black Southerners in the Age of Jim Crow*. New York: Alfred A. Knopf, 1998.

Loftis, Anne. *Witnesses to the Struggle: Imaging the 1930s California Labor Movement*. Reno, Nevada: University of Nevada Press, 1998.

Maass, Arthur. *Muddy Waters: The Army Engineers and the Nation's Rivers*. New York: Da Capo Press, 1974.

Maass, Arthur and Anderson, Raymond L. *And the Desert Shall Rejoice: Conflict, Growth and Justice in Arid Environments*. Malabar, Florida: Robert E. Krieger Publishing Company, 1986.

Masumoto, David Mas. *Country Voices: The Oral History of a Japanese American Family Farm Community*. Del Rey, California: Inaka Publications, 1987.

Mayfield, Thomas Jefferson. *Indian Summer*. Berkeley, California: Heyday Books, 1993.

McDougal, Dennis. *Privileged Son: Otis Chandler and the Rise and Fall of the L.A. Times Dynasty*. Cambridge, Massachusetts: Perseus Publishing, 2001.

McFarland, J. Randall. *Water for a Thirsty Land*. Selma, California: Consolidated Irrigation District, 1996.

McPherson, James M. *Battle Cry of Freedom*. New York: Oxford University Press, 1988.

McWilliams, Carey. *The Education of Carey McWilliams*. New York: Simon & Schuster, 1979.

_____. *Factories in the Field*. Boston: Little, Brown and Company, 1939.

_____. *Ill Fares the Land*. Boston: Little, Brown and Company, 1942.

_____. *Southern California Country*. New York: Duell, Sloan & Pearce, 1946.

Mencken, H. L. *The American Scene: A Reader*. New York, Vintage Books, 1982.

Menefee, Eugene. *History of Tulare and Kings Counties*. Los Angeles: Historic Record Company, 1913.

Merlo, Catherine M. *Legacy of a Shared Vision: The History of Calcot Ltd.* Bakersfield, California: Sierra Printers Inc., 1995.

Michael, Jay and Walters, Dan with Weintraub, Dan. *The Third House: Lobbyists, Money, and Power in Sacramento*. Berkeley, California: Berkeley Public Policy Press, 2002.

Miller, May Merrill. *First the Blade*. New York: Alfred A. Knopf, 1938.

Millikan, Robert A. *The Autobiography of Robert A. Millikan*. New York: Prentice-Hall Inc., 1950.

Mitchell, Annie R. *King of the Tulares*. Visalia, California: Visalia Times Delta, 1941.

_____. *Land of the Tules*. Visalia, California: Visalia Times Delta, 1949.

_____. *Visalia: Her First Fifty Years*. Exeter, California: self-published, 1963.

_____. *The Way it Was*. Fresno, California: Valley Publishers, 1976.

Mitchell, Don. *The Lie of the Land: Migrant Workers and the California Landscape*. Minneapolis: University of Minnesota Press, 1996.

Mitchell, Greg. *The Campaign of the Century: Upton Sinclair's Race for Governor of California and the Birth of Media Politics*. New York: Random House, 1992.

Mitchell, Margaret. *Gone with the Wind*. New York: Warner Books, 1999.

Mitchell, Ruth Comfort. *Of Human Kindness*. New York: D. Appleton-Century Company, 1940.

Morgan, Dan. *Merchants of Grain*. New York: Viking Press, 1979.

_____. *Rising in the West: The True Story of an "Okie" Family from the Great Depression Through the Reagan Years*. New York: Alfred A. Knopf, 1992.

Morgan, Wallace M. *History of Kern County California*. Los Angeles: Historic Record Company, 1914.

Muir, John. *The Mountains of California*. New York: Dorset Press, 1988.

Mulholland, Catherine. *The Owensmouth Baby*. Northridge, California: Santa Susana Press, 1987.

Nader, Ralph and Smith, Wesley J. *No Contest: Corporate Lawyers and the Perversion of Justice in America*. New York: Random House, 1996.

Newmark, Harris. *Sixty Years in Southern California 1853–1913*. Boston: Houghton Mifflin Company, 1930.

Nordhoff, Charles. *California for Travellers and Settlers*. Berkeley, California: Ten Speed Press, 1973.

Norris, Frank. *The Octopus*. New York: Penguin Books, 1986.

Olmsted, Frederick Law. *The Cotton Kingdom. A Traveller's Observations on Cotton and Slavery in the American Slave* States. New York: Alfred A. Knopf, 1953.

Opie, John. *The Law of the Land: Two Hundred Years of American Farmland Policy*. Lincoln, Nebraska: University of Nebraska Press, 1994.

Packard, Walter E. *The Economic Implications of the Central Valley Project*. Los Angeles: Adcraft, 1942.

Partridge, Elizabeth (ed.). *Dorothea Lange: A Visual Life*. Washington: Smithsonian Institution Press, 1994.

Pineda, Manuel and Perry, E. Caswell. *Pasadena Area History*. Pasadena, California: Historical Publishing Company, 1972.

Pisani, Donald J. *From the Family Farm to Agribusiness*. Berkeley, California: University of California Press, 1984.

_____. *Water, Land, and Law in the West: The Limits of Public Policy, 1850–1920*. Lawrence, Kansas: University Press of Kansas, 1996.

Preston, William L. *Vanishing Landscapes: Land and Life in the Tulare Lake Basin*. Berkeley, California: University of California Press, 1981.

Raper, Arthur F. *Preface to Peasantry*. New York: Atheneum, 1968.

_____. *Tenants of the Almighty*. New York: MacMillan Company, 1943.

Reisler, Mark. *By the Sweat of Their Brow: Mexican Immigrant Labor in the United States, 1900–1940*. Westport, Connecticut: Greenwood Press, 1976.

Reisner, Marc: *Cadillac Desert: The American West and Its Disappearing Water*. New York: Penguin Books, 1993.

Rhodes, James Ford. *History of the Civil War 1861–1865*. New York: Frederick Ungar Publishing Company, 1961.

Rhodes, Richard. *Farm: A Year in the Life of an American Farmer*. Lincoln, Nebraska: Bison Books, 1997.

Rice, Thaddesu Brockett and Williams, Carolyn White. *History of Greene County Georgia: 1786–1886*. Washington, Georgia: Wilkes Publishing Company, 1973.

Richards, Henry I. *Cotton and the AAA*. Washington, D.C.: The Brookings Institution, 1936.

Rider, Fremont. *Rider's California: A Guide-Book for Travelers*. New York: MacMillan Company, 1925.

Rodengen, Jeffrey L. *The Legend of Goodyear: The First 100 Years*. Fort Lauderdale, Florida: Write Stuff Syndicate Inc., 1997.

Rolle, Andrew F. *California: A History*. Wheeling, Illinois: Harlan Davidson, 1998.

Rowley, William D. *Reclaiming the Arid West: The Career of Francis G. Newlands*. Bloomington, Indiana: Indiana University Press, 1996.

Russell, Albert R. *U.S. Cotton and the National Cotton Council, 1938–1987*. Memphis, Tennessee: National Cotton Council, 1987.

Sandburg, Carl. *The American Songbag*. New York: Harcourt Brace Jovanovich, 1990.

Schetter, Clyde E. *Litchfield Park: The Story of a Town*. Litchfield Park, Arizona: Litchfield Park Library Association, 1976.

Sisk, B.F. *The Memoir of Bernie Sisk*. Fresno, California: Panorama West, 1980.

Sitton, Tom and Deverell, William (eds.). *Metropolis in the Making: Los Angeles in the 1920s*. Berkeley, California: University of California Press, 2001.

Smith, C. Wayne and Cothren, J. Tom. *Cotton: Origin, History, Technology and Production*. New York: John Wiley & Sons, 1999.

Smith, Robert H. with Crowley, Michael K. *Dead Bank Walking: One Gutsy Bank's Struggle for Survival and the Merger That Changed Banking Forever*. Winchester, Virginia: Oakhill Press, 2000.

Smith, Wallace. *Garden of the Sun*. Fresno, California: Max Hardison—A-1 Printers, 1956.

Smythe, William E. *The Conquest of Arid America*. Seattle: University of Washington Press, 1970.

Snyder, Robert E. *Cotton Crisis*. Chapel Hill, North Carolina: University of North Carolina Press, 1984.

Solkoff, Joel. *The Politics of Food*. San Francisco: Sierra Club Books, 1985.

Southern Historical Publication Society. *The South in the Building of the Nation*. Richmond, Virginia: L. H. Jenkins, 1909.

Starr, Kevin. *Commerce and Civilization: Claremont McKenna College, The First Fifty Years, 1946–1996*. Claremont, California: Claremont McKenna College, 1998.

_____. *The Dream Endures: California Enters the 1940's*. New York: Oxford University Press, 1997.

_____. *Endangered Dreams: The Great Depression in California*. New York: Oxford University Press, 1996.

_____. *Material Dreams: Southern California Through the 1920s*. New York: Oxford University Press, 1996.

Steffens, Lincoln. *The Letters of Lincoln Steffens*. New York: Harcourt, Brace and Company, 1938.

Stegner, Wallace. *Beyond the Hundredth Meridian: John Wesley Powell and the Second Opening of the West*. New York: Penguin Books, 1992.

Stein, Walter J. *California and the Dust Bowl Migration*. Westport, Connecticut: Greenwood Press, 1973.

Steinbeck, John. *The Grapes of Wrath*. New York: Penguin Books, 1976.

_____. *In Dubious Battle*. New York: Penguin Books, 1992.

Stone, Irving. *Men to Match My Mountains*. New York: Doubleday, 1956.

Street, James H. *The New Revolution in the Cotton Economy*. Chapel Hill, North Carolina: University of North Carolina Press, 1957.

Taylor, Paul S. *Essays on Land, Water and the Law in California*. New York: Arno Press, 1979.

_____. *On the Ground in the Thirties*. Salt Lake City, Utah: Gibbs M. Smith Inc., 1983.

Thomson, John and Dutra, Edward A. *The Tule Breakers*. Stockton, California: University of the Pacific, 1983.

Treadwell, Edward F. *The Cattle King*. Santa Cruz, California: Western Tanager Press, 1981.

Turner, John. *White Gold Comes to California*. Fresno, California: Book Publishers Inc., 1981.

Twain, Mark. *The Autobiography of Mark Twain*. New York: Harper & Brothers, 1924.

Vandergriff, A. L. *Ginning Cotton: An Entrepreneur's Story*. Lubbock, Texas: Texas Tech University Press, 1997.

Vandiver, Frank E. *Black Jack: The Life and Times of John J. Pershing*. College Station, Texas: Texas A&M University Press, 1977.

Wahl, Richard W. *Markets for Federal Water: Subsidies, Property Rights and the Bureau of Reclamation*. Washington, D.C.: Resources for the Future, 1989.

Walters, Katherine Bowman. *Oconee River: Tales to Tell*. Spartanburg, South Carolina: The Reprint Company, 1995.

Warren, Robert Penn. *Segregation: The Inner Conflict in the South*. New York: Random House, 1956.

Watkins, T.H. *The Hungry Years: A Narrative History of the Great Depression in America*. New York: Henry Holt and Company, 1999.

_____. *Righteous Pilgrim: The Life and Times of Harold L. Ickes 1874–1952*. New York: Henry Holt and Company, 1990.

Weaver, John D. *Los Angeles: The Enormous Village 1781–1981*. Santa Barbara, California: Capra Press, 1980.

Weber, Devra. *Dark Sweat, White Gold: California Farm Workers, Cotton, and the New Deal*. Berkeley, California: University of California Press, 1996.

Williams, Gregory. *The Story of Hollywoodland*. Los Angeles: Papavasilopoulos Press, 1992.

Williams, Tennessee. *27 Wagons Full of Cotton and Other One-Act Plays*. Norfolk, Connecticut: New Directions, 1945.

Winter, Ella. *And Not to Yield*. New York: Harcourt, Brace & World, 1963.

Woodman, Harold D. *King Cotton and His Retainers: Financing and Marketing the Cotton Crop of the South, 1800–1925*. Columbia, South Carolina: University of South Carolina Press, 1990.

Workers of the Writers' Program. *The Central Valley Project*. Sacramento, California: California State Department of Education, 1942.

Worster, Donald. *A River Running West: The Life of John Wesley Powell.* New York: Oxford University Press, 2001.

———. *Rivers of Empire: Water, Aridity, and the Growth of the American West.* New York: Oxford University Press, 1992.

Zieger, Robert H. *American Workers, American Unions.* Baltimore: Johns Hopkins University Press, 1994.

Selected Scholarly Works

Ballard, Patricia Louise. "And Conflict Shall Prevail: Reclamation Law, Water Districts, and Politics in the Kings River Service Area of California—An Alternative Framework." Master's Thesis, UCLA, 1980.

Candee, Hamilton. "The Broken Promise of Reclamation Reform." *Hastings Law Journal,* Volume 40, no. 3, March 1989.

Chacon, Ramon. "Labor Unrest and Commercialized Agriculture: The Case of the 1933 San Joaquin Valley Cotton Strike." Undated typescript, Humboldt State University.

Constantine, John Hampton. "An Economic Analysis of California's One-Variety Cotton Law." Ph.D. Dissertation, University of California at Davis, 1993.

Constantine, John H.; Alston, Julian M.; and Smith, Vincent H. "Economic Impacts of the Calfironia One-Variety Cotton Law." *Journal of Political Economy,* Volume 102, no. 5, October 1994.

Findley, James Clifford. "The Economic Boom of the Twenties in Los Angeles." Ph.D. Dissertation, Claremont Graduate School, 1958.

Johns, Bryan Theodore. "Field Workers in California Cotton." Master's Thesis, UCLA, 1947.

Koppes, Clayton R. "Public Water, Private Land: Origins of the Acreage Limitation Controversy, 1933–1953." *Pacific Historical Review,* November 1978.

Musoke, Moses S. and Olmstead, Alan L. "A History of Cotton in California: A Comparative Perspective." University of California at Davis, Agricultural History Center, Working Paper Series no. 6., July 1980.

Pierson, Jay Dexter. "The Growth of a Western Town: A Case Study of Yuma, Arizona 1915–1950." Master's Thesis, Arizona State University, 1987.

Smith, Susan M. "Litchfield Park and Vicinity." Master's Thesis, University of Arizona, 1948.

Stemen, Mark. "Forty Acres and a Ditch: Cotton, Family Farming and the State in the Imperial Valley, 1902–1933." Typescript, 1995.

Wilson, Bryan J. "Westlands Water District and Its Federal Water: A Case Study of Water District Politics." *Stanford Environmental Law Journal,* 1987/1988.

Zimrick, Steven John. "The Changing Organization of Agriculture in the Southern San Joaquin Valley, California." Ph.D. Dissertation, Louisiana State University, 1976.

Permissions

Excerpts from various works were used courtesy of the following:

1   THE GRAPES OF WRATH by John Steinbeck, copyright 1939, renewed ©
     1967 by John Steinbeck. Used by permission of Viking Penguin, a division
     of Penguin Group (USA) Inc.

87   TULARE DUST Copyright 1972 (Renewed) Sony/ATV Songs LLC. All rights
     administered by Sony/ATV Music Publishing, 8 Music Square West,
     Nashville, TN 37203. All rights reserved. Used by permission.

155 BLOOD ON THE COTTON by Langston Hughes and Ella Winter. Copy-
     right 1934 by Langston Hughes. Copyright renewed 1961 by Langston
     Hughes. Reprinted by permission of Harold Ober Associates Inc.

175 GREENBACK DOLLAR New Words by Woody Guthrie © 1980 by Woody
     Guthrie Publications Inc. All rights reserved. Used by permission.

233 WINNIE-THE-POOH by A. A. Milne, illustrated by E. H. Shepard, copy-
     right 1926 by E. P. Dutton, renewed 1954 by A. A. Milne. Used by permission
     of Dutton Children's Books, a division of Penguin Young Readers Group, a
     member of Penguin Group (USA) Inc., 345 Hudson St., New York, NY 10014.
     WINNIE-THE-POOH © A. A. Milne. Copyright under the Berne Conven-
     tion. Published by Egmont Books Ltd. and used with permission.

409 GONE WITH THE WIND Copyright © 1936 by Macmillan Publishing Co., a
     division of Macmillan Inc. Copyright renewed © 1964 by Stephens
     Mitchell and Trust Co. of Georgia as executors of Margaret Mitchell
     Marsh. Reprinted by permission of William Morris Agency Inc. on behalf
     of the author.

Acknowledgments

NO BOOK, NO MATTER HOW SOLITARY THE PURSUIT, is debt free. Like any journey that takes six years, this one leaned on a considerable community of colleagues, scholars, editors, friends, spouses and children.

Because we're newspaper journalists with fulltime jobs, writing a book of this breadth required bosses who understood our passion. No boss did more to make this book happen than Paul Steiger, the managing editor of *The Wall Street Journal*, where Rick worked for fifteen years before joining Mark at the *Los Angeles Times* in 2002. When Rick needed time away from the paper, Paul came through with an extraordinary offer. His support can't be overstated. At the *Journal*, Larry Rout, Peter Gumbel and Jonathan Friedland were also incredibly accommodating while Rick engaged in some creative double-dipping in order to move the book along.

Working for a newspaper and writing a book might seem like divided loyalties. John Carroll and Dean Baquet, our bosses at the *Times*, understand that each pursuit informs and invigorates the other. They, as well as editors on the state desk where Mark works, showed much patience with the distractions of this book.

A network of historians—some leading academics, others of the backyard variety—helped with research and vetted the manuscript: Jonathan M. Bryant, David Bush, Doris Castro, Lewis Coleman, Bruce Gardner, David E. Hamilton, Bill Horst, Norris Hundley Jr., Suzanne Kauffman, William Mahar, Randy McFarland and Harold D. Woodman. Two of California's most eminent historians, Kevin Starr and Gary Kurutz of the State Library, recognized the import of this project and threw their backing behind it. As a result, our interview transcripts are to be housed in the library's special collections in Sacramento.

Digging back into history is never easy. But it would have been practically impossible in this case without the generosity of the *Corcoran Journal*'s managing editor, Jeanette Todd, who entrusted us with microfilm of the newspaper stretching back decades. Then there was the Sherrill family, which for years had preserved boxes and boxes of Fred Sherrill's papers—and happily shared the entire trove.

The only thing that proved more valuable was the company archive in Pasadena that Jim Boswell opened up for us. Rather than looking over our shoulders, he stayed away while we combed through drawers of material, including the lost voices of his mother and other family members that had been recorded years earlier by David Vowell. When it came time for Boswell to read our manuscript—something we allowed him to do for the sake of accuracy—he commended us for our thoroughness. But that hadn't changed the bottom line. "I wish it wasn't written," he said. He took particular issue with the title of the book, saying that it was "ludicrous" to call him the "King of California." "I'm going to have to quit every club I belong to," Boswell said. "I can never approach my friends again. They're going to ride my ass."

When it came to finding a narrative structure that would do justice to the sweep of the past without losing what we had witnessed firsthand in the present, Peter King, our colleague at the *Times*, offered a solution. Why not tell the present-day scenes as a year on the biggest farm in America? Now that was an idea! Other friends and colleagues served up needed doses of encouragement, gave concrete suggestions on the manuscript and provided comfortable places to stay while we were on the road doing research. Among them: Mitchel Benson, Matt Black, Wendy Bounds, Ellie Herman, Marty Kaplan, John Kirkpatrick, Kathryn Kranhold, Marc Lifsher, David McGowan, Matt Miller, Alan Murray, Sheila Muto, Bruce Orwall, Paul Shapiro and, especially, Rich Turner. Paul Wartzman, Rick's dad, was a tireless listener; he spent hours on the phone, hearing chapter after chapter read aloud, so that the rhythm of the words might be just right.

Of course, all of this could have been for naught if it wasn't for our agent, Kris Dahl. She took the bad news of one publisher getting cold feet two years into the project and magically turned it into good news: PublicAffairs, headed by Peter Osnos and among the last of the houses devoted to serious literary journalism, was eager to tell this story.

There is an impulse at this juncture to throw in all sorts of counterfeit praise. Passing along our deep gratitude to our editor, Robert Kimzey, requires none of that. Robert possesses every virtue that a writer looks for in an editor—a nimble mind, a keen eye, a wonderful touch, a deep reservoir of patience—and then to find out that he even did time in the San Joaquin Valley as a kid.

Finally, to moonlight as authors means constantly finding ways to rob Peter to pay Paul. No one had more stolen from them than our wives, Jacoby and Randye, and our children, Ashley, Joseph, Jacob and Emma and Nathaniel. Through it all, Coby and Randye played many roles: motivator, sounding board, editor. Our only hope is that when they've finished reading these pages, they'll conclude that their husbands weren't completely nuts.

Mark Arax and Rick Wartzman
June 2003

Index

PUBLICAFFAIRS is a publishing house founded in 1997. It is a tribute to the standards, values, and flair of three persons who have served as mentors to countless reporters, writers, editors, and book people of all kinds, including me.

I. F. STONE, proprietor of *I. F. Stone's Weekly,* combined a commitment to the First Amendment with entrepreneurial zeal and reporting skill and became one of the great independent journalists in American history. At the age of eighty, Izzy published *The Trial of Socrates,* which was a national bestseller. He wrote the book after he taught himself ancient Greek.

BENJAMIN C. BRADLEE was for nearly thirty years the charismatic editorial leader of *The Washington Post.* It was Ben who gave the *Post* the range and courage to pursue such historic issues as Watergate. He supported his reporters with a tenacity that made them fearless, and it is no accident that so many became authors of influential, best-selling books.

ROBERT L. BERNSTEIN, the chief executive of Random House for more than a quarter century, guided one of the nation's premier publishing houses. Bob was personally responsible for many books of political dissent and argument that challenged tyranny around the globe. He is also the founder and was the longtime chair of Human Rights Watch, one of the most respected human rights organizations in the world.

·   ·   ·

For fifty years, the banner of Public Affairs Press was carried by its owner, Morris B. Schnapper, who published Gandhi, Nasser, Toynbee, Truman, and about 1,500 other authors. In 1983 Schnapper was described by *The Washington Post* as "a redoubtable gadfly." His legacy will endure in the books to come.

Peter Osnos, *Publisher*